METHODS IN MOLECULAR

Series Editor
**John M. Walker
School of Life Sciences
University of Hertfordshire
Hatfield, Hertfordshire, AL10 9AB, UK**

For further volumes:
www.springer.com/series/7651

Cancer Cell Culture

Methods and Protocols

Second Edition

Edited by

Ian A. Cree

*Translational Oncology Research Centre,
Queen Alexandra Hospital, Portsmouth, UK*

Humana Press

Editor
Ian A. Cree, M.B. Ch.B., Ph.D., F.R.C.Path.
Translational Oncology Research Centre
Queen Alexandra Hospital
Portsmouth
UK
ian.cree@porthosp.nhs.uk

ISSN 1064-3745   e-ISSN 1940-6029
ISBN 978-1-61779-079-9   e-ISBN 978-1-61779-080-5
DOI 10.1007/978-1-61779-080-5
Springer New York Dordrecht Heidelberg London

Library of Congress Control Number: 2011926687

© Springer Science+Business Media, LLC 2011
All rights reserved. This work may not be translated or copied in whole or in part without the written permission of the publisher (Humana Press, c/o Springer Science+Business Media, LLC, 233 Spring Street, New York, NY 10013, USA), except for brief excerpts in connection with reviews or scholarly analysis. Use in connection with any form of information storage and retrieval, electronic adaptation, computer software, or by similar or dissimilar methodology now known or hereafter developed is forbidden.
The use in this publication of trade names, trademarks, service marks, and similar terms, even if they are not identified as such, is not to be taken as an expression of opinion as to whether or not they are subject to proprietary rights.
While the advice and information in this book are believed to be true and accurate at the date of going to press, neither the authors nor the editors nor the publisher can accept any legal responsibility for any errors or omissions that may be made. The publisher makes no warranty, express or implied, with respect to the material contained herein.

Printed on acid-free paper

Humana Press is part of Springer Science+Business Media (www.springer.com)

Preface

Cancer cell culture remains as important today as it was when the first edition of *Cancer Cell Culture: Methods and Protocols* was published 6 years ago. However, the emphasis of research using cell culture methods is shifting towards the use of primary cells rather than the generation of new cell lines, and the chapters in this edition reflect this.

As before, the basic concepts of cancer cell biology and culture are covered in the first few chapters, with an emphasis on safe working practice for both cells and laboratory researchers. These chapters contain the information critical to success – only by good practice and quality control will the results of cancer cell culture improve.

I am grateful to several authors from the last edition for updating some of their chapters to reflect improvements, particularly the rapid improvement in real-time PCR technology. In a change to the last edition, which is still available, I have included a series of chapters covering a series of different assay methods which allow the action of drugs and toxins on cells to be measured. Many of these can be used with primary cell preparations. There is a chapter reviewing high throughput screening of natural products, and a practical guide to the assessment of drug interactions. It is now possible to introduce genetic defects into normal cells with greater precision than ever before, and this is already delivering a completely new set of cell lines to researchers, including those in the pharmaceutical industry and academic laboratories.

In organising a book like this, one quickly realises the breadth and depth of research activity that depend on cell lines. Inevitably, some potential authors were too busy to contribute, and there were many I probably should have asked but did not as the list of chapters filled up all too quickly. There is no attempt here to claim to be comprehensive, but I hope that the new edition of this valuable book will prove useful to many laboratories. I am immensely grateful to those that found time to contribute and to Professor John Walker for his guidance during the difficult phases of this year-long process. My wife, Brigit had to put up with the amount of time I spent in my study, and did so very graciously – I am indebted to her again and I am very grateful for her support.

Portsmouth, UK   *Ian A. Cree*

Contents

Preface... v
Contributors... xi

1. Cancer Biology.. 1
   Ian A. Cree
2. Principles of Cancer Cell Culture...................................... 13
   Ian A. Cree
3. Storage of Cell Lines.. 27
   Katharine A. Parker
4. Characterization and Authentication of Cancer Cell Lines: An Overview.. 35
   Yvonne A. Reid
5. Online Verification of Human Cell Line Identity by STR DNA Typing...... 45
   Wilhelm G. Dirks and Hans G. Drexler
6. Cytogenetic Analysis of Cancer Cell Lines.............................. 57
   Roderick A.F. MacLeod, Maren Kaufmann, and Hans G. Drexler
7. Cell Culture Contamination... 79
   Glyn N. Stacey
8. Detecting Mycoplasma Contamination in Cell Cultures
   by Polymerase Chain Reaction... 93
   Cord C. Uphoff and Hans G. Drexler
9. Elimination of Mycoplasmas from Infected Cell Lines Using Antibiotics.. 105
   Cord C. Uphoff and Hans G. Drexler
10. Quality Assurance and Good Laboratory Practice........................ 115
    Louise A. Knight and Ian A. Cree
11. Generation of Lung Cancer Cell Line Variants by Drug Selection or Cloning.... 125
    Laura Breen, Joanne Keenan, and Martin Clynes
12. Isolation and Culture of Colon Cancer Cells and Cell Lines............ 135
    Sharon Glaysher and Ian A. Cree
13. Isolation and Culture of Melanoma and Naevus Cells and Cell Lines..... 141
    Julia K. Soo, Alastair D. MacKenzie Ross, and Dorothy C. Bennett
14. Isolation and Culture of Squamous Cell Carcinoma Lines................ 151
    Karin J. Purdie, Celine Pourreyron, and Andrew P. South
15. Isolation and Culture of Ovarian Cancer Cells and Cell Lines.......... 161
    *Christian M. Kurbacher, Cornelia Korn, Susanne Dexel, Ulrike Schween,
    Jutta A. Kurbacher, Ralf Reichelt, and Petra N. Arenz*
16. Establishment and Culture of Leukemia–Lymphoma Cell Lines............. 181
    Hans G. Drexler
17. Isolation of Inflammatory Cells from Human Tumours.................... 201
    Marta E. Polak

| 18 | Isolation of Endothelial Cells from Human Tumors 209
Elisabeth Naschberger, Vera S. Schellerer, Tilman T. Rau,
Roland S. Croner, and Michael Stürzl |
|---|---|
| 19 | Cellular Chemosensitivity Assays: An Overview 219
Venil N. Sumantran |
| 20 | Cell Sensitivity Assays: The MTT Assay 237
Johan van Meerloo, Gertjan J.L. Kaspers, and Jacqueline Cloos |
| 21 | Cell Sensitivity Assays: The ATP-based Tumor Chemosensitivity Assay 247
Sharon Glaysher and Ian A. Cree |
| 22 | Differential Staining Cytotoxicity Assay: A Review 259
Larry M. Weisenthal |
| 23 | Real-Time Cytotoxicity Assays 285
Donald Wlodkowic, Shannon Faley, Zbigniew Darzynkiewicz,
and Jonathan M. Cooper |
| 24 | Purification of Annexin V and Its Use in the Detection of Apoptotic Cells 293
Katy M. Coxon, James Duggan, M. Francesca Cordeiro, and Stephen E. Moss |
| 25 | Measurement of DNA Damage in Individual Cells Using the Single Cell Gel
Electrophoresis (Comet) Assay 309
Janet M. Hartley, Victoria J. Spanswick, and John A. Hartley |
| 26 | Molecular Breakpoint Analysis of Chromosome Translocations
in Cancer Cell Lines by Long Distance Inverse-PCR 321
Björn Schneider, Hans G. Drexler, and Roderick A.F. MacLeod |
| 27 | Cell Migration and Invasion Assays 333
Karwan A. Moutasim, Maria L. Nystrom, and Gareth J. Thomas |
| 28 | Angiogenesis Assays 345
V. Poulaki |
| 29 | Flow Cytometric DNA Analysis of Human Cancers and Cell Lines 359
Sarah A. Krueger and George D. Wilson |
| 30 | Expression Analysis of Homeobox Genes in Leukemia/
Lymphoma Cell Lines 371
Stefan Nagel and Hans G. Drexler |
| 31 | Measuring Gene Expression from Cell Cultures by Quantitative
Reverse-Transcriptase Polymerase Chain Reaction 381
Sharon Glaysher, Francis G. Gabriel, and Ian A. Cree |
| 32 | Proteomic Evaluation of Cancer Cells: Identification of Cell Surface Proteins ... 395
Samantha Larkin and Claire Aukim-Hastie |
| 33 | Development of Rituximab-Resistant B-NHL Clones: An In Vitro
Model for Studying Tumor Resistance to Monoclonal Antibody-Mediated
Immunotherapy 407
Ali R. Jazirehi and Benjamin Bonavida |
| 34 | Analysis of Drug Interactions 421
Irene V. Bijnsdorp, Elisa Giovannetti, and Godefridus J. Peters |
| 35 | Transfection and DNA-Mediated Gene Transfer 435
Davide Zecchin and Federica Di Nicolantonio |

36 Drug Design and Testing: Profiling of Antiproliferative Agents for Cancer Therapy Using a Cell-Based Methyl-[3H]-Thymidine Incorporation Assay 451
Matthew Griffiths and Hardy Sundaram

37 Feeder Layers: Co-culture with Nonneoplastic Cells 467
Celine Pourreyron, Karin J. Purdie, Stephen A. Watt, and Andrew P. South

38 Xenotransplantation of Breast Cancers 471
Massimiliano Cariati, Rebecca Marlow, and Gabriela Dontu

Appendix A *483*
Appendix B *487*
Index *493*

Contributors

PETRA N. ARENZ • *Medical Center Bonn-Friedensplatz, Friedensplatz, Bonn, Germany*

DOROTHY C. BENNETT • *Division of Basic Medical Sciences, Molecular and Metabolic Signalling Centre, St. George's, University of London, London, UK*

IRENE V. BIJNSDORP • *Department of Medical Oncology, VU University Medical Center, Amsterdam, The Netherlands*

LAURA BREEN • *National Institute for Cellular Biotechnology, Dublin City University, Dublin, Ireland*

BENJAMIN BONAVIDA • *Department of Microbiology, Immunology, and Molecular Genetics, Jonsson Comprehensive Cancer Center, David Geffen School of Medicine, University of California, Los Angeles, CA, USA*

MASSIMILIANO CARIATI • *Department of Research Oncology, School of Medicine, King's College London, London, UK*

JACQUELINE CLOOS • *Department of Paediatric Oncology, VU University Medical Center, Amsterdam, The Netherlands; Department of Haematology, VU University Medical Center, Amsterdam, The Netherlands*

JONATHAN M. COOPER • *The Bioelectronics Research Center, University of Glasgow, Glasgow, UK*

M. FRANCESCA CORDEIRO • *Division of Cell Biology, Institute of Ophthalmology, University College London, London, UK*

KATY M. COXON • *Division of Cell Biology, Institute of Ophthalmology, University College London, London, UK*

MARTIN CLYNES • *National Institute for Cellular Biotechnology, Dublin City University, Dublin, Ireland*

IAN A. CREE • *Translational Oncology Research Centre, Queen Alexandra Hospital, Portsmouth, UK*

ROLAND S. CRONER • *Division of Molecular and Experimental Surgery, University Medical Center Erlangen, Erlangen, Germany*

ZBIGNIEW DARZYNKIEWICZ • *Brander Cancer Research Institute, New York Medical College, Valhalla, NY, USA*

SUSANNE DEXEL • *L.a.n.c.e. Inc., Friedensplatz, Bonn, Germany*

FEDERICA DI NICOLANTONIO • *Laboratory of Molecular Genetics, Institute for Cancer Research and Treatment (IRCC), University of Torino Medical School, Candiolo, Turin, Italy*

WILHELM G. DIRKS • *DSMZ-Deutsche Sammlung von Mikroorganismen und Zellkulturen, Braunschweig, Germany*

GABRIELA DONTU • *Department of Research Oncology, School of Medicine, King's College London, London, UK*

HANS G. DREXLER • *DSMZ-Deutsche Sammlung von Mikroorganismen und Zellkulturen, Braunschweig, Germany*

JAMES DUGGAN • *Division of Cell Biology, Institute of Ophthalmology, University College London, London, UK*

SHANNON FALEY • *The Bioelectronics Research Center, University of Glasgow, Glasgow, UK*

FRANCIS G. GABRIEL • *Translational Oncology Research Centre, Queen Alexandra Hospital, Portsmouth, UK*

ELISA GIOVANNETTI • *Department of Medical Oncology, VU University Medical Center, Amsterdam, The Netherlands*

SHARON GLAYSHER • *Translational Oncology Research Centre, Queen Alexandra Hospital, Portsmouth, UK*

MATTHEW GRIFFITHS • *Cell Biology Department, Vertex Pharmaceuticals (Europe) Ltd, Abingdon, UK*

JANET M. HARTLEY • *Cancer Research UK Drug-DNA Interactions Research Group, UCL Cancer Institute, University College London, London, UK*

JOHN A. HARTLEY • *Cancer Research UK Drug-DNA Interactions Research Group, UCL Cancer Institute, University College London, London, UK*

CLAIRE AUKIM-HASTIE • *Division of Health and Social Care, Faculty of Health and Medical Sciences, University of Surrey, Guildford, UK*

ALI R. JAZIREHI • *Department of Surgery, Jonsson Comprehensive Cancer Center, David Geffen School of Medicine, University of California, Los Angeles, CA, USA*

GERTJAN J.L. KASPERS • *Department of Paediatric Oncology, VU University Medical Center, Amsterdam, The Netherlands; Department of Haematology, VU University Medical Center, Amsterdam, The Netherlands*

JOANNE KEENAN • *National Institute for Cellular Biotechnology, Dublin City University, Dublin, Ireland*

LOUISE KNIGHT • *Translational Oncology Research Centre, Queen Alexandra Hospital, Portsmouth, UK*

CORNELIA KORN • *Medical Center Bonn-Friedensplatz, Friedensplatz, Bonn, Germany*

SARAH A. KRUEGER • *Department of Radiation Oncology, William Beaumont Hospital, Royal Oak, MI, USA*

CHRISTIAN M. KURBACHER • *Medical Center Bonn-Friedensplatz, Friedensplatz, Bonn, Germany*

JUTTA A. KURBACHER • *Medical Center Bonn-Friedensplatz, Friedensplatz, Bonn, Germany*

SAMANTHA LARKIN • *Cancer Sciences Division, School of Medicine, University of Southampton, Southampton, UK*

ALASTAIR D. MACKENZIE ROSS • *Division of Basic Medical Sciences, Molecular and Metabolic Signalling Centre, St. George's University of London, London, UK*

RODERICK A.F. MACLEOD • *DSMZ-Deutsche Sammlung von Mikroorganismen und Zellkulturen, Braunschweig, Germany*

REBECCA MARLOW • *Department of Research Oncology, School of Medicine, King's College London, London, UK*

JOHAN VAN MEERLOO • *Department of Paediatric Oncology, VU University Medical Center, Amsterdam, The Netherlands; Department of Haematology, VU University Medical Center, Amsterdam, The Netherlands*

STEPHEN E. MOSS • *Division of Cell Biology, Institute of Ophthalmology, University College London, London, UK*

KARWAN A. MOUTASIM • *Cancer Sciences Division, University of Southampton, School of Medicine, Southampton, UK*

STEFAN NAGEL • *DSMZ-Deutsche Sammlung von Mikroorganismen und Zellkulturen, Braunschweig, Germany*

ELISABETH NASCHBERGER • *Division of Molecular and Experimental Surgery, University Medical Center Erlangen, Erlangen, Germany*

MARIA L. NYSTROM • *Cancer Sciences Division, University of Southampton, School of Medicine, Southampton, UK*

KATHARINE A. PARKER • *Translational Oncology Research Centre, Queen Alexandra Hospital, Portsmouth, UK*

GODEFRIDUS J. PETERS • *Department of Medical Oncology, VU University Medical Center, Amsterdam, The Netherlands*

MARTA E. POLAK • *Division of Infection, Inflammation and Immunity, University of Southampton School of Medicine, Southampton, UK*

V. POULAKI • *Department of Ophthalmology, Angiogenesis Laboratory, Massachusetts Eye and Ear Infirmary, Harvard Medical School, Boston, MA, USA*

CELINE POURREYRON • *Centre for Oncology and Molecular Medicine, University of Dundee, Ninewells Hospital & Medical School, Dundee, UK*

KARIN J. PURDIE • *Centre for Oncology and Molecular Medicine, University of Dundee, Ninewells Hospital & Medical School, Dundee, UK*

TILMAN T. RAU • *Division of Molecular and Experimental Surgery, University Medical Center Erlangen, Erlangen, Germany*

RALF REICHELT • *L.a.n.c.e. Inc., Friedensplatz, Bonn, Germany*

YVONNE A. REID • *ATCC, Manassas, VA, USA*

VERA S. SCHELLERER • *Division of Molecular and Experimental Surgery, University Medical Center Erlangen, Erlangen, Germany*

BJÖRN SCHNEIDER • *DSMZ-Deutsche Sammlung von Mikroorganismen und Zellkulturen, Braunschweig, Germany*

ULRIKE SCHWEEN • *L.a.n.c.e. Inc., Friedensplatz, Bonn, Germany*

JULIA K. SOO • *Division of Basic Medical Sciences, Molecular and Metabolic Signalling Centre, St. George's, University of London, London, UK*

ANDREW P. SOUTH • *Centre for Oncology and Molecular Medicine, University of Dundee, Ninewells Hospital & Medical School, Dundee, UK*

VICTORIA J. SPANSWICK • *Cancer Research UK Drug-DNA Interactions Research Group, UCL Cancer Institute, University College London, London, UK*

GLYN N. STACEY • *National Institute for Biological Standards and Control, South Mimms, Herts, UK*

MICHAEL STÜRZL • *Division of Molecular and Experimental Surgery, University Medical Center Erlangen, Erlangen, Germany*

VENIL N. SUMANTRAN • *Adjunct Professor, Department of Biotechnology, Indian Institute of Technology (IIT)-Chennai, 201, Bhupat & Jyothi Mehta School of Biosciences Chennai, Tamil Nadu, India*

HARDY SUNDARAM • *Cell Biology Department, Vertex Pharmaceuticals (Europe) Ltd, Abingdon, UK*

GARETH J. THOMAS • *Cancer Sciences Division, University of Southampton, School of Medicine, Southampton, UK*

CORD C. UPHOFF • *DSMZ-Deutsche Sammlung von Mikroorganismen und Zellkulturen, Braunschweig, Germany*

STEPHEN A. WATT • *Centre for Oncology and Molecular Medicine, University of Dundee, Ninewells Hospital & Medical School, Dundee, UK*

LARRY M. WEISENTHAL • *Weisenthal Cancer Group, Huntington Beach, CA, USA*

GEORGE D. WILSON • *Department of Radiation Oncology, William Beaumont Hospital, Royal Oak, MI, USA*

DONALD WLODKOWIC • *The BioMEMS Research Group, Department of Chemistry, University of Auckland, Auckland, New Zealand*

DAVIDE ZECCHIN • *Laboratory of Molecular Genetics, Institute for Cancer Research and Treatment (IRCC), University of Torino Medical School, Candiolo, Turin, Italy*

Chapter 1

Cancer Biology

Ian A. Cree

Abstract

The process of carcinogenesis involves a number of changes in cellular phenotype, which are largely based on acquired genetic changes in cells that are not terminally differentiated. The ability of cancer cells to grow and their failure to respond to the usual controls on such proliferation are obvious features, but they also evade cell death and most have no limits on their ability to replicate beyond the limits imposed by telomere length in normal cells. In addition, they are able to stimulate the formation of blood vessels to ensure a steady supply of oxygen and nutrients, and to invade normal tissues, sometimes subverting the normal processes within those tissues. Finally, it has become increasingly apparent that cancer cells undergo a process of selection which renders the immune system ineffective. Some of these characteristics are retained by cells in culture, and an understanding of the biological properties of cancer cells will assist in the design of experiments and the interpretation of their results.

Key words: Growth, Oncogene, Anti-oncogene, Apoptosis, Angiogenesis, Telomere, Invasion, Immunity

1. Introduction

Cancer cells are transformed cells – they have acquired a series of changes that permit them to form tumors, which behave in different ways according to the (mainly) genetic changes that underlie this transformation. It is now 10 years since Hanahan and Weinberg (1) published their seminal paper suggesting that there were six identifiable "Hallmarks" of cancer cells. These, they listed as follows:

- Self-sufficiency in growth signals
- Insensitivity to growth-inhibitory (antigrowth) signals
- Evasion of programmed cell death (apoptosis)
- Limitless replicative potential

- Sustained angiogenesis
- Tissue invasion.

In the intervening years, in many ways, they have been proved correct, though many would perhaps add one more – altered immunity, of which more later. This chapter is intended to provide a brief overview of cancer biology and way into the literature for those starting their cell culture – a number of key reviews are referenced and provide a more in-depth analysis than is possible here.

The traits Hanahan and Weinberg (1) identified were based on the understanding that cancers arose as a series of steps towards transformation, for which the initial model was probably retinoblastoma (2) (Fig. 1), but for which the complexity involved in most tumor types is typified by the pathogenesis of colorectal cancer (3, 4) (Fig. 2). While this initially seemed relatively simple, as more studies are performed, the idea that this is a strictly linear process seems to be breaking down, and it is likely that both haematogenous and solid tumors share a complex evolution as they develop over time within patients. Like evolution itself, which was first promulgated over by Darwin 150 years ago (5), natural selection seems to be the driving force behind the changes that result in the transformation of cells to form cancers.

Is there any limit on the number of changes involved? Several recent publications have tried to enumerate the genetic (6) and epigenetic (7) alterations involved. This is still a very considerable number (8), but given the 34,000 genes in the human genome, not perhaps as large as one might fear. The number of permutations arising from these alterations are, however, still very considerable, and that assumes that they are binary – on or off. In fact, there is now a body of evidence to suggest that many such changes

Fig. 1. Retinoblastoma results from two mutations, each of which results in the loss of one copy of the RB gene. (**a**) In sporadic retinoblastomas, which are not inherited and tend to affect a single eye, both copies are lost by mutation in the affected child. (**b**) Inherited retinoblastoma results from a single mutation occurring within a retinoblast which has already lost one copy of the gene through mutation in the gamete or germline of one parent.

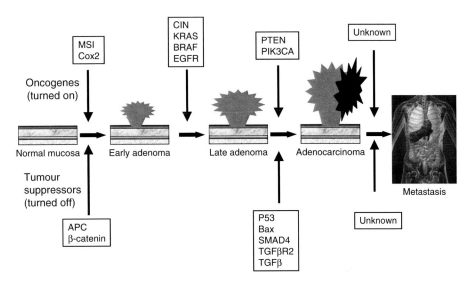

Fig. 2. The progression of colorectal carcinogenesis involves multiple genes, often in concert, to produce firstly a benign outgrowth from the mucosa, an adenoma, and finally an adenocarcinoma. The mechanisms involved in metastasis are less certain. *Abbreviations*: microsatellite instability (*MSI*), cyclooxygenase 2 (*Cox2*), chromosome instability (*CIN*); Genes: KRAS, BRAF, EGFR, PTEN, PIK3CA, APC, β-catenin, P53, Bax, SMAD4, TGFβR2, and TGFβ.

are anything but binary – gene dosage (i.e. the degree of gene amplification) makes a major difference to some tumors, and the degree of methylation of a gene is also very variable.

Like even simple multicellular organisms, many cancers are in fact very complex tissues, and the phenotypic behaviour of a cancer may not depend entirely on its genetic profile, but also on host factors which will vary according to the genotype of the host and the way in which that host phenotype is altered by the environment – including diet, smoking, and many other factors. It is therefore hardly surprising that those who have looked find immense heterogeneity between cancers (9), even those apparently of the same type arising within the same tissue – and indeed this can be seen within the same tissue in a single individual (e.g. prostate cancer (10)).

This chapter examines the hallmarks of cancer in relation to cell lines and primary cell cultures, highlighting the differences and similarities which should be taken into account when designing the experiments that require the methods contained within this book.

2. Self-sufficiency in Growth Signals

Normal cells do not grow unless told to. This even applies to stem cells, which have the potential to do what they please, but are under strict controls which break down in cancers. Indeed some authors consider that all cancers arise from stem cells (11).

The self-sufficiency in growth signals in cancer cells can be endogenous or exogenous. A good example of this of recent therapeutic relevance has become important in non-small cell lung cancer and colorectal cancer. In this tumor type, the epidermal growth factor receptor (EGFR) pathway is key to the behaviour of many (but not all) tumors (12, 13). Exogenous signalling includes autocrine production of amphiregulin and epiregulin, both ligands of EGFR, while endogenous signalling comes from acquired activating mutations of the EGFR gene. However, even the EGFR gene can be bypassed by an activating mutation of the KRAS gene, and in patients with activating mutations of EGFR treated with small molecule inhibitors, resistance occurs by the development of resistant mutations or activation of alternate pathways. Many tumor types have alterations of growth genes (oncogenes) acting downstream of such cell surface growth factor receptors (Fig. 3), some of which are transcription factors (e.g. c-myc) that activate many effector pathways within the cell.

In cell culture then, isolated neoplastic cells are usually able to manage quite well on their own, though they may depend on paracrine signals from other cells included in the tumor which are

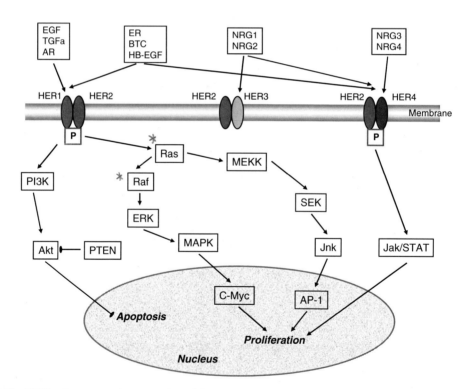

Fig. 3. The EGFR pathway comprises a number of ligands, and four cell membrane-located receptor tyrosine kinases which dimerise to transduce a signal via the PI3K, Ras–Raf, and STAT pathways. Whereas the PI3K pathway inhibits apoptosis, the Ras–Raf and STAT pathways lead to increased cell proliferation in most cell types. Ligands: EGF - Epidermal Growth Factor, AR - Amphiregulin, TGF - Transforming Growth Factor, ER - Epiregulin, NRG - Neuregulin, HB-EGF - Heparin-binding EGF, BTC - Betacellulin.

3. Insensitivity to Growth-Inhibitory (Antigrowth) Signals

always available to them. While this is less true of high-grade tumors, low-grade tumors may not grow unless these conditions are satisfied.

The potential consequences of cell growth for the organism necessitate the control of proliferation. The control mechanisms can be divided into those which turn off proliferation permanently and those which do so transiently, or until a signal is received by the cell, which reverses the switch. Loss of antigrowth control occurs by several mechanisms. The most obvious is by loss of the gene concerned – retinoblastoma is a case in point (14), and there is evidence that many anti-growth signals work via the RB protein, which inhibits the function of the E2F transcription factor, which is essential for proliferation (15). The anti-oncogenes p15 and p16 operate via this pathway, as does the cyclin kinase cdk4, and there is an important link to p53 (Fig. 4). Loss of these genes by mutation or sequestration, for example by viral proteins such as E6 and E7 of human papilloma virus (HPV) can lead to cancer.

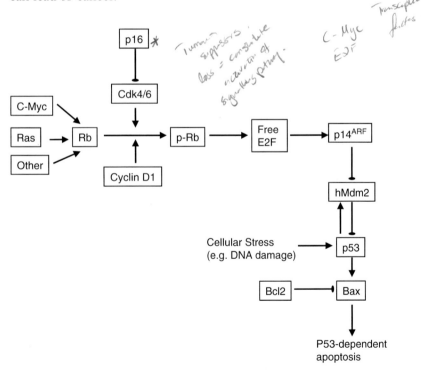

Fig. 4. The RB pathway is critical to many anti-growth signals. RB inhibits the function of the E2F transcription factor, which is essential for proliferation. The tumor suppressor genes p14, p16, and p53 all have functions within this pathway.

p53 is considered the key guardian of the genome (16), and is mutated in around 50% of cancer cases. In most other cancers, p53 function is compromised by dysregulation of associated signalling pathways. Normally, p53 expression is triggered by a variety of stimuli, including DNA damage, oxidative stress, depletion of heat shock proteins, and by activated oncogenes. This leads to growth arrest, usually via p21 or other regulatory molecules which mediate inhibition of cell cycle proteins. If the insult is sufficiently severe or not reparable, p53 activation leads to apoptosis via multiple intrinsic (e.g. bax, bid, puma, and noxa) and extrinsic mechanisms (e.g. Fas and Trail receptors).

Permanent growth arrest is mediated by differentiation, for example, by inactivation of the APC/beta-catenin pathway (Fig. 2) in colorectal cancer (17). This is one area in which there has been considerable progress over the last 10 years. Understanding of differentiation owes much to the human genome project which underpinned identification of many of the genes involved. With some exceptions, it seems that cancers do not inactivate these mechanisms by mutation or deletion as often as one might expect. This may reflect redundancy in the mechanisms involved, but it certainly seems that gene methylation and histone acetylation are important in silencing expression of differentiation genes in many cancers (7).

4. Evasion of Programmed Cell Death (Apoptosis)

Control of cell numbers within tissues is mediated largely by the process of apoptosis – known as programmed cell death or suicide. Cells which have fulfilled their function or are no longer required are removed from tissues without inflammation or other sequelae by this process. This is important in many situations throughout life – for example, the embryo loses the webs between toes and fingers, while in the adult showers of apoptoses occur in the endometrium during menstruation (18). Such processes are also very important in the immune system to avoid auto-immunity and select foreign antigen-reactive cells. Control of the apoptotic process is important, and is mediated by a balance of pro- and anti-apoptotic factors (19), as well as endogenous and exogenous triggers.

In cancer cells, these apoptotic controls are almost always affected, though it varies between tumor types and the pathways involved in their genesis. Exogenous triggers include the Fas–Fas Ligand system, and the TRAIL system, while the endogenous trigger tends to be mediated by p53, also the most commonly mutated anti-oncogene which has a particularly complex biology (16).

5. Limitless Replicative Potential and Senescence

Normally, cell replication is limited to 50–70 passages (20). However, cancer cells are immortalised and capable of exceeding this limit – which is also known as the "Hayflick phenomenon" or "replicative senescence", first recognised by Hayflick and Moorehead in the 1960s (21). The mechanism by which cancer cells do this involves replenishment of telomeric DNA by an enzyme, telomerase.

Telomeres are made up of multiple repeats of the TTAGGG sequence – usually over several thousand bases (22). As DNA polymerases are unable to complete replication of daughter DNA at the 5 end of linear DNA sequences, each cycle of replication leads to loss of DNA from the telomere, a process implicated in cellular senescence. In stem cells and germinative cells this is prevented by their expression of telomerase, which repairs the telomere in cells that express it – this is normal within stem cells, but not within somatic cells. Reactivation of telomerase is part of carcinogenesis, and probably a critical step in this process. Nearly all cancer cells show expression of telomerase – the few that do not have alternative mechanisms to prevent loss of telomeric DNA. However, telomerase is more likely to be expressed by the more malignant variants with higher replicative activity. The currently most favoured hypothesis is that expression of telomerase is triggered in dividing cells when telomere erosion triggers a DNA repair response via the ATM-p53 pathway, which should in turn trigger growth arrest (also known as stress-induced senescence). Such senescence is essentially a form of tumor suppression, and is an important barrier to carcinogenesis (20). Mutations in the p53 or RB pathway lead to reduced senescence – and inactivation of both seems to be required in human cells to prevent replicative senescence (20). This phenomenon is now of interest in many different cancers, including leukaemias (23) and solid tumors such as melanoma (24).

However, telomerase is far from an Achilles heel and drugs which block it do not necessarily kill cancers, which may still be able to replicate sufficiently to continue their growth for enough cycles to cause the death of the patient. Nevertheless such agents may prove useful and some are in development.

6. Sustained Angiogenesis

Tumors cannot attain a size of more than a millimetre or so in diameter without requiring a blood supply. There are several ways in which they can do so:

- Angiogenesis
- Vessel co-option
- Vascular mimicry.

The first is the most obvious – and the most commonly used by cancer cells, which commonly secrete pro-angiogenic cytokines such as vascular endothelial growth factor (VEGF) and basic fibroblast growth factor (bFGF) (25). These cytokines stimulate the production of new blood vessels from existing capillaries (by sprouting) and from circulating endothelial cell precursors. Co-culture experiments of cancer cells with human umbilical vein endothelial cells (HUVEC) are often used to study such interactions, though it should be noted that HUVEC may not be representative of endothelial cells in tissues (25).

Cancer cells may also invest vessels already present in tissues, growing in close apposition and co-opt them into the tumor. This is common and markers of new vessels have been used to show its extent within tumors. Growth around vessels is likely to be mediated by adhesion molecules (26), and the mechanisms are similar to those used in tissue invasion.

Vascular mimicry remains controversial, but the hydrodynamic forces within tumors caused in part by the presence of poorly formed "leaky" blood vessels lead to a flow of tissue fluid within the tumor that may play an important role in providing nutrients to cancer cells. In some cases, there is compelling evidence that slits and channels may form, lined by neoplastic cells with some characteristics of endothelial cells (27).

7. Tissue Invasion

The ability of cancer cells to invade adjacent tissue and spread to other tissues, usually via the lymphatic or bloodstream is the main reason underlying cancer mortality. Cancer cells have a number of characteristics which permit them to invade and metastasise. Firstly, they are usually able to survive and grow in an adherence-free cell culture system. Secondly, they show little evidence of polarisation, and thirdly they can alter shape easily to allow motility through tissues. The mechanisms are essentially similar to those known from the movement of immune cells through tissues, and recapitulate the migration of cells within the embryo.

Movement through tissues requires a loss of adhesion molecules such as ICAM-1, and the secretion of enzymes to break down the intercellular matrix and basement membrane collagen. Collagenases, now more commonly known as matrix metalloproteinases (MMPs), are of particular importance and are often secreted by cancer cells, thus easing their passage. Tissue inhibitors of metalloproteinases (TIMPs) within tissues are important inhibitors of this process, but to date, drugs which counteract MMPs have had little success in the clinic.

8. Altered Immunity

Although it was not one of Hanahan and Weinberg's initial list, there is now no doubt that the interaction between the immune system and cancer cells is actively managed by changes within cancer cells. The altered immune status of cancer cells is important for their survival. It has long been known that cancer cells express unique or embryologically derived antigens which can excite an immune response. A number of mechanisms of anti-cancer immunity have been described:

- Humoral immunity – complement-fixing anti-cancer antibodies
- Antibody-dependent cellular cytotoxicity (ADCC)
- NK cells
- Cytotoxic T lymphocytes.

Several hypotheses have been put forward to explain how cancer cells evade the immune system. The most recent and accepted of these combines earlier models (*see* ref. 28 for review) and is termed "immunoediting". This model implies that the immune system exerts a selective pressure on developing cancer cells that leads to them having some or all of the following characteristics:

- Loss of MHC class I or II (indeed this may change during metastasis)
- Secretion of cytokines and immune active enzymes (e.g. TGFβ, IDO)
- Active killing of lymphocytes by FasL expression.

These mechanisms have the effect of limiting the ingress of competent immune cells to the tumor and causing general systemic immunosuppression. Indeed, it has recently been established that dendritic cells within the tumor or within draining lymph nodes are maintained in an immature, immunosuppressive state, which may actually assist the tumor to evade the immune system (29).

9. Conclusion

Cancer cells are remarkable in many ways – they develop a series of changes during their development which allow them to survive and grow despite many checks and balances on this progression. Understanding of this biology is probably still far from complete, but is essential to successful cell and tissue culture of cancer cells.

References

1. Hanahan, D., and Weinberg, R. A. (2000) The hallmarks of cancer, *Cell 100*, 57–70.
2. Knudson, A. G., Jr. (1971) Mutation and cancer: statistical study of retinoblastoma, *Proc Natl Acad Sci USA 68*, 820–823.
3. Kinzler, K. W., and Vogelstein, B. (1996) Lessons from hereditary colorectal cancer, *Cell 87*, 159–170.
4. Soreide, K., Nedrebo, B. S., Knapp, J. C., Glomsaker, T. B., Soreide, J. A., and Korner, H. (2009) Evolving molecular classification by genomic and proteomic biomarkers in colorectal cancer: potential implications for the surgical oncologist, *Surg Oncol 18*, 31–50.
5. Darwin, C. (1859) The Origin of Species.
6. Wood, L. D., Parsons, D. W., Jones, S., Lin, J., Sjoblom, T., Leary, R. J., Shen, D., Boca, S. M., Barber, T., Ptak, J., Silliman, N., Szabo, S., Dezso, Z., Ustyankksy, V., Nikolskaya, T., Nikolsky, Y., Karchin, R., Wilson, P. A., Kaminker, J. S., Zhang, Z., Croshaw, R., Willis, J., Dawson, D., Shipitsin, M., Willson, J. K., Sukumar, S., Polyak, K., Park, B. H., Pethiyagoda, C. L., Pant, P. V., Ballinger, D. G., Sparks, A. B., Hartigan, J., Smith, D. R., Suh, E., Papadopoulos, N., Buckhaults, P., Markowitz, S. D., Parmigiani, G., Kinzler, K. W., Velculescu, V. E., and Vogelstein, B. (2007) The genomic landscapes of human breast and colorectal cancers, *Science 318*, 1108–1113.
7. Hoque, M. O., Kim, M. S., Ostrow, K. L., Liu, J., Wisman, G. B., Park, H. L., Poeta, M. L., Jeronimo, C., Henrique, R., Lendvai, A., Schuuring, E., Begum, S., Rosenbaum, E., Ongenaert, M., Yamashita, K., Califano, J., Westra, W., van der Zee, A. G., Van Criekinge, W., and Sidransky, D. (2008) Genome-wide promoter analysis uncovers portions of the cancer methylome, *Cancer Res 68*, 2661–2670.
8. Salk, J. J., Fox, E. J., and Loeb, L. A. (2009) Mutational Heterogeneity in Human Cancers: Origin and Consequences, *Annu Rev Pathol*.
9. Cree, I. A., Neale, M. H., Myatt, N. E., de Takats, P. G., Hall, P., Grant, J., Kurbacher, C. M., Reinhold, U., Neuber, K., MacKie, R. M., Chana, J., Weaver, P. C., Khoury, G. G., Sartori, C., and Andreotti, P. E. (1999) Heterogeneity of chemosensitivity of metastatic cutaneous melanoma, *Anticancer Drugs 10*, 437–444.
10. Mehra, R., Han, B., Tomlins, S. A., Wang, L., Menon, A., Wasco, M. J., Shen, R., Montie, J. E., Chinnaiyan, A. M., and Shah, R. B. (2007) Heterogeneity of TMPRSS2 gene rearrangements in multifocal prostate adenocarcinoma: molecular evidence for an independent group of diseases, *Cancer Res 67*, 7991–7995.
11. Rosen, J. M., and Jordan, C. T. (2009) The increasing complexity of the cancer stem cell paradigm, *Science 324*, 1670–1673.
12. Perez-Soler, R. (2009) Individualized therapy in non-small-cell lung cancer: future versus current clinical practice, *Oncogene 28 Suppl 1*, S38–45.
13. Heinemann, V., Stintzing, S., Kirchner, T., Boeck, S., and Jung, A. (2009) Clinical relevance of EGFR- and KRAS-status in colorectal cancer patients treated with monoclonal antibodies directed against the EGFR, *Cancer Treat Rev 35*, 262–271.
14. Knudson, A. G., Jr. (1978) Retinoblastoma: a prototypic hereditary neoplasm, *Semin Oncol 5*, 57–60.
15. Hallstrom, T. C., and Nevins, J. R. (2009) Balancing the decision of cell proliferation and cell fate, *Cell Cycle 8*, 532–535.
16. Vousden, K. H., and Lane, D. P. (2007) p53 in health and disease, *Nat Rev Mol Cell Biol 8*, 275–283.
17. Phelps, R. A., Broadbent, T. J., Stafforini, D. M., and Jones, D. A. (2009) New perspectives on APC control of cell fate and proliferation in colorectal cancer, *Cell Cycle 8*, 2549–2556.
18. Kerr, J. F., Wyllie, A. H., and Currie, A. R. (1972) Apoptosis: a basic biological phenomenon with wide-ranging implications in tissue kinetics, *Br J Cancer 26*, 239–257.
19. Fulda, S. (2009) Tumor resistance to apoptosis, *Int J Cancer 124*, 511–515.

20. Zuckerman, V., Wolyniec, K., Sionov, R. V., Haupt, S., and Haupt, Y. (2009) Tumour suppression by p53: the importance of apoptosis and cellular senescence, *J Pathol 219*, 3–15.
21. Hayflick, L., and Moorhead, P. S. (1961) The serial cultivation of human diploid cell strains, *Exp Cell Res 25*, 585–621.
22. De Boeck, G., Forsyth, R. G., Praet, M., and Hogendoorn, P. C. (2009) Telomere-associated proteins: cross-talk between telomere maintenance and telomere-lengthening mechanisms, *J Pathol 217*, 327–344.
23. Keller, G., Brassat, U., Braig, M., Heim, D., Wege, H., and Brummendorf, T. H. (2009) Telomeres and telomerase in chronic myeloid leukaemia: impact for pathogenesis, disease progression and targeted therapy, *Hematol Oncol 27*, 123–129.
24. Ha, L., Merlino, G., and Sviderskaya, E. V. (2008) Melanomagenesis: overcoming the barrier of melanocyte senescence, *Cell Cycle 7*, 1944–1948.
25. Jain, R. K., Duda, D. G., Willett, C. G., Sahani, D. V., Zhu, A. X., Loeffler, J. S., Batchelor, T. T., and Sorensen, A. G. (2009) Biomarkers of response and resistance to anti-angiogenic therapy, *Nat Rev Clin Oncol 6*, 327–338.
26. Ramjaun, A. R., and Hodivala-Dilke, K. (2009) The role of cell adhesion pathways in angiogenesis, *Int J Biochem Cell Biol 41*, 521–530.
27. Folberg, R., Hendrix, M. J., and Maniotis, A. J. (2000) Vasculogenic mimicry and tumor angiogenesis, *Am J Pathol 156*, 361–381.
28. Polak, M. E., Borthwick, N. J., Jager, M. J., and Cree, I. A. (2009) Melanoma vaccines-the problems of local immunosuppression, *Hum Immunol*.
29. Polak, M. E., Borthwick, N. J., Gabriel, F. G., Johnson, P., Higgins, B., Hurren, J., McCormick, D., Jager, M. J., and Cree, I. A. (2007) Mechanisms of local immunosuppression in cutaneous melanoma, *Br J Cancer 96*, 1879–1887.

Chapter 2

Principles of Cancer Cell Culture

Ian A. Cree

Abstract

The basics of cell culture are now relatively common, though it was not always so. The pioneers of cell culture would envy our simple access to manufactured plastics, media and equipment for such studies. The prerequisites for cell culture are a well lit and suitably ventilated laboratory with a laminar flow hood (Class II), CO_2 incubator, benchtop centrifuge, microscope, plasticware (flasks and plates) and a supply of media with or without serum supplements. Not only can all of this be ordered easily over the internet, but large numbers of well-characterised cell lines are available from libraries maintained to a very high standard allowing the researcher to commence experiments rapidly and economically. Attention to safety and disposal is important, and maintenance of equipment remains essential. This chapter should enable researchers with little prior knowledge to set up a suitable laboratory to do basic cell culture, but there is still no substitute for experience within an existing well-run laboratory.

Key words: Cell culture, Serum, Medium, Buffer, Adherence, Equipment, Plastics

1. Introduction

Cell culture is an important tool for biomedical scientists and is widely practiced within academic, hospital and industry laboratories. It permits everything from the discovery of new molecules and their function, to high throughput testing of potential drugs – and not just for cancer. Culture of normal cells is now relatively routine and it is possible to obtain non-neoplastic cells (e.g. inflammatory cells, endothelial cells) from tumours for study. Culture of neoplastic cells directly from human tumours is routine, though multiple methods exist. However, most experiments are done with cell lines, grown from tumours and often passaged many times in media containing serum and other additives to promote their growth.

2. History of Cancer Cell Culture

The history of cell culture, reviewed in the previous edition of this book by Langdon (1), goes back 125 years to a publication by William Roux who successfully cultured chick embryo tissue in saline for several days. The first human experiments were also performed in the nineteenth century when Ljunggren (1898) showed that human skin could survive in ascitic fluid. Ross Harrison (1907) is regarded as the father of tissue culture, as he introduced tissue from frog embryos into frog lymph clots and showed that not only did the tissue survive, but nerve fibres grew out from the cells. The lymph was quickly replaced by plasma, and then by more systematic studies to identify factors required for cell growth in culture. The first cancer cells were cultured by Losee and Ebeling (1914) and the first continuous rodent cell line was produced by William Earle in 1943 at the National Cancer Research Institute. In 1951, George Gay produced the first human continuous cell line from a cancer patient, Helen Lane, and HeLa cells are still used very extensively. The next two decades saw a huge expansion in the number of studies which defined the media we still use today. Serum free media started with Ham's fully defined medium in 1965, and in the 1970s, serum-free media were optimised by the addition of hormones and growth factors. There are now thousands of cell lines available, and for some, many stable variants have also been produced.

3. Definitions and Terms

Some common definitions are listed in Table 1 (1). "Cell culture" refers to the maintenance in vitro of disaggregated cells, while "organ culture" refers to a culture of a non-disaggregated tissue. The term "tissue culture" encompasses both terms. The initial culture is known as a "primary culture" and undergoes multiple "sub-cultures" or "passages" to produce a "cell line". Cell lines may be continuous (capable of unlimited growth) or finite, characterised by senescence after a limited number of population doublings. Continuous cell lines are "transformed" – this seems to equate to their acquisition of telomerase activity in most instances (see Chapter 1) and may occur during or at the time of generation of a cell line. There are now artificial methods by which cells can be "immortalised" (synonymous with transformation),

Table 1
Definitions of terms used in cancer cell culture, adapted from Langdon (1)

Term	Definition
Cell culture	Maintenance of dissociated cells in vitro
Tissue culture	Maintenance of tissue explants in vitro
Primary culture	The initial culture of cells dissociated directly from tumour or obtained from the blood or a malignant effusion
Cell line	Cells sub-cultured beyond the initial primary culture
Finite cell line	A cell line with a limited life span, one that undergoes senescence after a defined number of doublings
Continuous cell line	A culture which is apparently capable of an unlimited number of population doublings; often referred to as immortal cell culture
Clone	The cells derived from a single cell of origin
Immortalization	Enabling cells to extend their life in culture indefinitely
Lag phase of growth	Initial slow growth phase which occurs when cells are sub-cultured
Log phase of growth	Most rapid phase of exponential cell growth
Plateau phase of growth	Slowing of cell growth when cells become confluent
Population doubling time	Time taken for cell number to double
Cell bank	Repository of cancer cell lines and materials derived from them
Tissue bank	Repository of tissue samples from patients
Substrate	The matrix on which an adherent cell culture is grown
Passage	Subculture of cells from one container to another
Confluent	Situation where cells completely cover the substrate

for example, hTERT transfection, and these methods may be applicable to non-neoplastic cells.

Cell lines may exist as adherent cell cultures, grown on a "substrate" such as plastic or glass, or they may grow "in suspension" This tends to go with their cell of origin, so most lymphocyte cell lines will grow readily in suspension, while the reverse is true of carcinoma cell lines. The coating of plastic or glass with proteins or other molecules can promote adherence and change the phenotype of cells in culture, even after many passages. All grow in a cell culture "medium"– plural media – which can be made up from its constituents, but is more commonly bought from a commercial manufacturer.

4. Basic Requirements of Cell Culture

The requirements of cell culture were defined during the last century, and have not changed, despite our increasing knowledge of the molecular biology of the cell on which these requirements are based. Nutrients are provided, as well as oxygen and a means for the removal of carbon dioxide. Most cells require a substrate and this is now nearly always plastic, though growth of cells on glass coverslips is still a useful method.

4.1. Media

A number of common media are listed in Appendix 1, and most are based on a basal saline medium to which various components are added. To grow, cells require amino acids, vitamins, metal ions, trace elements, and an energy source, usually provided as glucose. Other additives include buffer (HEPES is particularly common) and phenol red to indicate the acidity of the culture. Eagle's basal medium (BME) was modified by adjusting the amino acid concentration to produce Eagle's minimum essential medium (MEM). This was further modified by Dulbecco who quadrupled the amino acid and vitamin concentration to produce Dulbecco's minimum essential medium (DMEM). Iscove's modification of DMEM is designed to support haematopoietic cells and contains even more amino acids and vitamins, as well as selenium, pyruvate and HEPES buffer.

Morgan et al. (2) used a medium that supported an even wider range of cells and is now known as Medium 199. Less complex versions were produced, including one by the Connaught Medical Research Laboratory (CMRL 1066) which was one of the first to examine serum-free growth of cell lines to examine the factors in serum required for the growth of certain cell types. Ham's nutrient mixtures F10 and F12 were developed for the growth of Chinese Hamster Ovary (CHO) cells with our without serum supplementation. However, a 50:50 mixture of Ham's F12 and DMEM is now widely used for serum-free cell culture as this combines the use of trace elements and vitamins in the F12 with the nutrients present in DMEM. Ham's group went on to design a series of media (MCDB) for a series of individual cell lines.

A basic medium described by McCoy et al. in 1959 (3) has also been widely used and was the basis for one developed by Moore et al. (4, 5) at Roswell Park Research Institute – RPMI1640 is perhaps the most widely used cell culture medium in use today.

5. Serum-Containing Media

Serum is often added to the media described above and in Appendix 1 to replace many requirements which are missing, even from the most sophisticated media. These include proteins, lipids and carbohydrates as well as growth factors and attachment factors. The concentration of components present is often variable and the practice of using newborn or fetal bovine sera has become common. Concentrations of 5–20% are normally used, depending on the cell type. There is batch to batch variation in the content of factors present and it is often best to stick with one supplier found to work well with a particular cell line. Serum can be divided into aliquots and frozen at –20°C following heat treatment for 30 min at 56°C (if necessary) to remove complement components.

6. Serum-Free Media

The inherent variability in the composition of sera can lead to variation in results obtained using media which contain sera. The uncertainty serum supplementation introduces to cell culture has lead many researchers to explore the use of serum-free media. Two different approaches have been taken – one, espoused by Sato and colleagues (6, 7), was to add specific supplements, while the other route, taken by Ham's group (8, 9), was to increase the concentration of various components of existing basal media. The addition of insulin, transferrin, selenite, hydrocortisone and cytokines has become common, with varying degrees of success. The addition of albumin to improve the protein content of media has proven useful, and components to promote adhesion signalling such as fibronectin are often useful. Other additives include prostaglandins, hormones and triiodothyronine (10–13). One useful tip for transferring cells from serum-containing media to serum-free media is to gradually reduce the serum content, as this allows cells to adapt gradually to changing conditions (14–16).

7. Obtaining Media

Media can be obtained commercially or made up from constituents. It is important to choose appropriately for the cell line and experimental design – shelf life varies from 9–12 months for made up media, to 2–3 years for powdered media that can be made up with sterile water. A good compromise is often to buy bottled 10× concentrated media with an extended shelf life (12–24 months) that is diluted with sterile water before use.

7.1. Substrate Specificity

Cell culture plates come in many different sizes and compositions. The majority in use allow adherence and are made of polystyrene, but in some circumstances it can be useful to use polypropylene which prevents most cells adhering to the plastic.

Plastics designed to promote adherence have a charged surface resulting from chemical treatment or irradiation. If cell adhesion is desired, it is often possible to improve this by coating plates with fibronectin or collagen. Several commercial preparations are available, but pre-coating plates with serum or conditioned medium can also be effective. Three-dimensional growth within collagen or more complex mixtures such as Matrigel can be used, and dual cell cultures either with direct contact or separation by a filter can allow the study of interactions under defined experimental conditions. The choice of well size or culture flask is dictated largely by the number of cells required for a particular assay and the time for which they will be in culture before passage. Cell culture has been performed successfully in 384-well and even in 1,536-well plates. At the other end of the scale are large systems with enormous surface areas capable of producing many millions of cells in a single passage.

7.2. Environment

Most cell culture is performed in incubators at 37°C with a tolerance of one degree from this. Above 40°C, cells will usually die rapidly. Culture media are usually designed with buffering capacity, but require an atmosphere of 5% CO_2 to ensure optimal performance and are kept at pH 7.2–7.4. Acidification of the media can be seen by changes in the colour of phenol red, if present. This is yellow at pH 6.5, orange at pH 7.0, red at pH 7.4 and purple at pH 7.8, providing an instant readout to the investigator of the pH of cultures. The use of 5% CO_2 in air means that cells are exposed to much higher partial pressures of oxygen than would be the case in tissues. Hypoxic chambers are available to control this, if necessary. Humidity is also critical, particularly in microplates, where drying artefact is to be avoided. Finally, sterility is all-important! These requirements dictate the necessity for a considerable outlay in terms of equipment to support cell culture – if any one of these facets is ignored, the result will usually be a collection of failed experiments.

8. Cell Lines

8.1. Cancer Cell Line Collections

There are thousands of cell lines available, and most can be accessed from culture collections. Table 2 lists a number of culture collections and companies that provide verified cell lines for research use. There are often constraints on the use of such lines imposed by the originating laboratory or the collection. In general these regulate to what extent cell lines can be used for purposes other than research in the institution purchasing the cell line. In addition to academic not-for-profit culture collections, a number of companies will now supply cell lines. These are sometimes

Table 2
Major cell line banks

Bank	Website	Location	Contents
American Type Culture Collection (ATCC)	www.atcc.org	USA	Contains 950 cancer cell lines, and expanding collection of hTERT immortalized cell lines from human tissue
European collection of animal cell culture (ECACC)	www.hpacultures.org.uk	UK	Contains 1,100 cell lines originating from over 45 different species
Deutsche Sammlung von Mikroorganismen und Zellkulturen GmbH (German Collection of Microorganisms and Cell Cultures)	www.dsmz.de	Germany	Contains 652 human and animal cell lines, and a separate collection of 586 human leukaemia–lymphoma cell lines
Banca Biologica e Cell Factory (Interlab cell line collection)	www.iclc.it	Italy	Lists 268 cell lines.
HyperCLDB	//bioinformatics.istge.it/hypercldb	Italy	Hyperlink database from the Interlab project, allowing rapid searches
Japanese Collection of Research Bioresources (JCRB)	cellbank.nibio.go.jp	Japan	Contains more than 1,000 cell lines.
Asterand	www.asterand.com	USA	Commercial supplier of Human Primary Cells/Cell lines, including breast, prostate and haematopoietic cell lines.
RIKEN gene bank	www.brc.riken.go.jp	Japan	Contains over 1,000 human and animal cell lines

more appropriate for commercial use or where the purchaser has little control over their use.

Cell culture collections will characterise cell lines new to them using techniques such as DNA fingerprinting and expression analysis to ensure that the identity of the cell line can be verified. They will then grow the cell line for a few passages to provide a Master Cell Bank, stored in liquid nitrogen, which will in turn provide material to Working Cell Banks from which researchers will be sent aliquots for their work. The samples stored in the Working Cell Bank will be authenticated and certified free of *Mycoplasma* and other common contaminating organisms.

More recently, many large academic hospitals have developed tissue banks, repositories of samples from patients, often but not always including frozen tissue samples stored in liquid nitrogen from which viable cells may be obtained.

8.2. General Growth Characteristics

When a new cell line is obtained, it is important for the laboratory to document its growth characteristics, including morphology, adhesion and serum requirements. Most laboratories passage new lines to produce their own master and working cell line banks. Storage of carefully labelled aliquots of early passage cells in liquid nitrogen is standard within most experienced cell culture laboratories, as with time and many passages, any cell line can lose or gain characteristics. Changes in genotype are common and direct comparison of results between laboratories is sometimes difficult as a result.

8.3. Develop or Use Existing Lines?

With such a wealth of cell lines available, it might be thought unnecessary to propagate new cell lines from primary cell cultures, but this is still valuable. As an example, it has recently become clear that some non-small cell lung cancers have activating mutations of the EGFR gene (17). However, only a fraction of these mutations are available within cell lines and established cultures from lung cancers with unusual mutations would be very valuable to researchers.

Although few laboratories now develop their own cell lines, many produce variants of existing lines, for instance by exposure of cell lines to drugs at increasing concentrations to produce variants with enhanced resistance. This allows the study of resistance mechanisms. Technical methods now exist to allow stable transfection of cells so that they overexpress genes of interest (18), and this is even provided as a service by some companies. Some variants, for instance cell lines expressing luciferase (19) or green fluorescent protein (GFP) (20) provide a simple means of assessing their viability, and such markers can be linked to genes of interest to provide simple markers of the level of gene expression under different conditions (19).

9. Primary Cell Culture

Several chapters within this book deal with primary cell cultures. Cell lines have many advantages, but over time show genetic and phenotypic drift as a result of their adaptation to cell culture. There is much to be said for their reproducibility and ease of use, but there is increasing realisation that overdependence upon cell lines has some drawbacks. Many of these can be overcome by the use of primary cell cultures.

Cells can be extracted from tumours in the same ways used to produce cell lines – the options are essentially mechanical or enzymatic dissociation of cells, followed by culture in media with the characteristics required by the experiment. Further selection of dissociated cells by selective culture media, antibody-labelled magnetic beads, or density centrifugation can be helpful to provide a more homogeneous population of cells for study.

10. Basic Laboratory Design and Equipment

Cell culture requires a clean and well-designed laboratory with sufficient equipment to allow safe, sterile working. The design should be as ergonomic as possible – with areas designated for the handling of new material and cultures free of contaminants. Completion of work on clean material in an area should precede work on potentially contaminated material, and there is something to be said for having separate incubators for potentially contaminated cultures and for clean cultures. Shared facilities are most at risk and need firm management to ensure that all users respect the rules which are there to protect them and their work. The guidance issued for Category 2 laboratories based on the Advisory Committee on Dangerous Pathogens dates from 1985, but makes useful points on lighting, heating, work surfaces, flooring, hand washing and air pressure (negative to corridors) that should be considered carefully.

The key equipment needs are listed and discussed below, but there are many small items that may not be immediately obvious to those starting up such facilities. Perhaps the most underrated requirement is for storage space. While most consumables can be ordered "just in time", the sudden lack of media, plasticware, or liquid nitrogen can render several months' work worthless within a few days. Stock control is a necessary concern for any cell culture laboratory and investment in this aspect of working practices will pay dividends later.

10.1. Laminar Flow Cabinets

Laminar flow cabinets are essential to allow cultures to be passaged safely and reliably. The Class II or Class 100 safety cabinet is ideal, as this provides operator and sample protection by drawing air into a grill at the front of the hood, and recirculating the air through a filter to provide sterile laminar airflow onto the specimen. Laminar flow cabinets contain high efficiency particulate air (HEPA) filters which need to be cleaned or replaced on a regular basis. While extraction to the outer atmosphere is preferred by most, cabinets with extraction to the room via a second or third HEPA filter work well. Cabinets extracting to the outer atmosphere may need a pump on the roof of the building to maintain a negative air pressure within any ducting required.

Laminar flow cabinets obviously require laminar air flow, yet this is often interrupted by equipment placed in the hood, rendering the cabinet a lot less effective. The aim should be to have as little in the cabinet as possible when working – a set of drawers on wheels beside the cabinet can prevent the need to have too much inside it. Environmental testing with agar plates exposed to the air within the hood can be useful, and if cells which may have a higher level of contamination (e.g. with known pathogens) are to be used, then it is wise to consider a higher level of containment and to seek advice from a microbiologist. Finally, all laminar flow cabinets require regular servicing – however costly and unnecessary this may seem.

10.2. Centrifuges

Cells do not require high speed centrifugation, and for short periods up to 40 min, refrigeration is also unnecessary. Most cell culture laboratories have multiple centrifuges of benchtop type with sealed buckets to prevent aerosol contamination. It is useful to purchase a type that has the ability to take multiple sizes of tubes, and some will even spin microplates. Tubes should not be overfilled and the centrifuge balanced carefully to avoid strain on the rotor as well as aerosol production. Think about where centrifuges are sited to avoid overheating, and to allow easy access without the need to juggle precious samples. Spills should not be ignored, but cleaned up immediately according to the manufacturer's instructions. A clear area of bench beside the centrifuge helps enormously, and we have always sited our centrifuges close to the laminar flow hoods to minimise the need to carry samples half-way round the laboratory. Again, servicing is essential to ensure that the centrifuge is performing within its expected limits – the engineer will check that rotors are not cracked, that there is no corrosion, and that electric motor brushes (if present) are replaced when necessary.

10.3. Incubators

Cell cultures require a carefully regulated environment in which to grow. This usually means that one of the larger purchases within the cell culture lab is likely to be at least two CO_2 incubators with temperature (animal cultures are usually performed at 37°C) and humidity control. Some more expensive versions have copper coating or HEPA filters to provide microbiological control, but placing copper filings in the water tray or adding a treatment fluid to the tray will also help. Why two? Well, they do break down, need regular (weekly) cleaning and frequent servicing!

10.4. Microscopes

It is usually necessary to look at cells within flasks or microplates, and this is virtually impossible with a standard upright microscope.

Even an inexpensive inverted microscope will prove a useful addition to the laboratory. It should have a series of lenses from ×2 to ×200 magnification and phase contrast as well as direct light facilities. Fluorescence can be very useful and should be considered, though it adds significantly to the cost.

10.5. Counting Cells

The ability to count cells is essential – in the past this was usually performed using a haemocytometer and microscope, but recently a number of simple instruments have come onto the market which provide greater accuracy and are of similar cost to an average upright microscope. Most use a CCD camera system and a manufactured disposable slide chamber or cassette which can be filled in the laminar flow hood. An alternative is a flow based cell counter, though these tend to be more expensive.

10.6. Storage and Other Facilities

The average cell culture laboratory requires several refrigerators, −20°C and −80 °C freezers, and liquid nitrogen storage. Inventory control is essential to keep these under control. The plasticware required will fill a large number of drawers, cupboards, all of which should be carefully labelled with their contents. Pipettes should be of good quality and regularly serviced – automated pipettes are often more accurate, even in the hands of an experienced operator, and should be considered. It is worthwhile having a strict policy to determine which pipette can be used in which hoods to further control contamination.

10.7. Work Surfaces and Flooring

Benches, walls and flooring should be easy to clean. If an older laboratory cannot be redeveloped, then disposable or easily wiped surfaces fixed to existing benches should be considered. Walls should be capable of withstanding a variety of chemicals including cleaning fluids, and floors should be resistant to cracking if liquid nitrogen is in use. Continuous flooring with a coved skirting board will reduce dust, and windows should be sealed unless required as fire exits. Benches abut walls and need similar consideration – many are now offered with a lip to prevent fluid spills from finding a way between wall and bench. Safe working heights for standing or sitting should be considered, and we recommend having at least two electrical sockets fitted with on/off indicator lights every metre or so. Some under-bench sockets and high wall mounted sockets may be required for fridges and freezers.

11. Quality Control

All cell culture laboratories should be run according to Good Laboratory Practice Guidelines. A number of organisations provide accreditation of clinical laboratories, and although most cell culture is probably done within research laboratories, similar principles should be applied. The major issues are infection and contamination such that the results obtained are inaccurate.

11.1. Infection and Microbial Contamination

Cell cultures may be produced within endogenous (mainly viral) infections, but most infections are acquired.

12. Bacterial or Fungal

Bacterial or fungal infections of cell cultures are usually obvious – the phenol red, if present, will turn yellow as the infection uses up available nutients and acidifies the medium, and under the inverted microscope the cells will be replaced by hyphae, yeast, or colonies of bacteria. However, it is common practice to add antibiotics to cultures and this can mask low level infection for a considerable time.

13. *Mycoplasma*

The most common and most missed infection in cell culture laboratories is probably *Mycoplasma*, so much so that we have devoted an entire chapter to it in this volume. Several species are involved and their effects are insidious. The indicators that there might be a problem include reduced growth rate, morphological changes, chromosomal aberration, and altered metabolism. There are several ways of testing for *Mycoplasma* – and many manufacturers provide kits. It is possible to treat *Mycoplasma* infection with antibiotics, but avoidance is the best policy. Most laboratories simply dispose of infected cultures and start again.

14. Viruses

Viral infection is also insidious – some cultures contain viruses in any case, either integrated into their genome, or as endogenous non-lethal infections. In many cases, these are not regarded as infections, and viral transformation of cells is a time-honoured method of producing continuous cell lines. Bovine serum may contain viruses – particularly bovine viral diarrhoea virus (BVDV), though manufacturers are aware of this and test for it.

15. Antibiotics

The commonest antibiotics used in cell culture are penicillin and streptomycin. In some circumstances, particularly primary cell culture methods, this may be necessary and the addition of antifungal agents may also be desirable. However, in many cases, these agents are not necessary and leaving them out has the advantage that they cannot interfere with experimental results, particularly if potential drugs are being tested. Equally overuse of antibiotics leads to the selection of resistant organisms in cell cultures, just as it does in the population at large.

15.1. Contamination of Cell Lines

Many cell lines look the same under the microscope and confusion of one vial with another or mislabelling is all too easy. Working with one cell line rather than several at a time will help, but may not be feasible. Good practice will help, and it is generally best to use authenticated cell lines obtained from a tissue culture collection.

16. Safety

16.1. Risk Assessment

The performance of risk assessment is a legal requirement in many countries. The purpose is to identify and mitigate risks to individuals. Cell lines can be classified as low risk (e.g. non-human/non-primate continuous cell lines and those with finite lifespans), medium risk (poorly characterised mammalian cell lines) and high risk (primary cell cultures, human cell lines, or those with experimental or endogenous infections). Category 2 containment is generally sufficient, but category 3 is required for cell lines containing HIV and some other viral pathogens.

16.2. Maintenance

The correct maintenance of the infrastructure and equipment within a laboratory is essential. This should include the equipment listed above and should be performed at regular intervals.

16.3. Disinfection and Waste Disposal

Disinfection of culture waste minimises the risk of harm to individuals and is necessary before equipment is serviced.

Hypochlorites are good general purpose disinfectants, active against bacteria and viruses, but can corrode metal surfaces – for example in centrifuges. They need to be made up daily before use, and commercial product instructions should be followed.

Alcohol is effective against all except non-enveloped viruses and can be used on most surfaces. Aldehydes such as formalin or glutaraldehyde are used for fumigation and are irritants. They can be used on metals. Phenolic disinfectants are not active against viruses, and are of limited use.

Waste disposal may be handled by a central contractor, but it is important to know how they do this – regular disposal is necessary. Tissue culture waste and smaller used plastics such as pipette tips should be decontaminated overnight in hypochlorite and then incinerated. Larger items can be bagged for incineration with care taken to ensure that any edges cannot penetrate the bag. Sharps boxes are useful for pipette tips and needles.

References

1. Langdon, S. P. (2004) Basic principles of cancer cell culture, *Methods Mol Med 88*, 3–15.
2. Morgan, J. F., Morton, H. J., and Parker, R. C. (1950) Nutrition of animal cells in tissue culture; initial studies on a synthetic medium, *Proc Soc Exp Biol Med 73*, 1–8.
3. McCoy, T. A., Maxwell, M., and Kruse, P. F., Jr. (1959) Amino acid requirements of the Novikoff hepatoma in vitro, *Proc Soc Exp Biol Med 100*, 115–118.
4. Moore, G. E., and Pickren, J. W. (1967) Study of a virus-containing hematopoietic cell line and a melanoma cell line derived from a patient with a leukemoid reaction, *Lab Invest 16*, 882–891.
5. Moore, G. E., and Glick, J. L. (1967) Perspective of human cell culture, *Surg Clin North Am 47*, 1315–1324.
6. Bottenstein, J. E., and Sato, G. H. (1979) Growth of a rat neuroblastoma cell line in serum-free supplemented medium, *Proc Natl Acad Sci USA 76*, 514–517.
7. Sato, J. D., Cao, H. T., Kayada, Y., Cabot, M. C., Sato, G. H., Okamoto, T., and Welsh, C. J. (1988) Effects of proximate cholesterol precursors and steroid hormones on mouse myeloma growth in serum-free medium, *In Vitro Cell Dev Biol 24*, 1223–1228.
8. Ham, R. G. (1963) An improved nutrient solution for diploid Chinese hamster and human cell lines, *Exp Cell Res 29*, 515–526.
9. Bettger, W. J., and Ham, R. G. (1982) The nutrient requirements of cultured mammalian cells, *Adv Nutr Res 4*, 249–286.
10. Barnes, D., and Sato, G. (1979) Growth of a human mammary tumour cell line in a serum-free medium, *Nature 281*, 388–389.
11. Prasad, K. N., and Kumar, S. (1975) Role of cyclic AMP in differentiation of human neuroblastoma cells in culture, *Cancer 36*, 1338–1343.
12. Allegra, J. C., and Lippman, M. E. (1978) Growth of a human breast cancer cell line in serum-free hormone-supplemented medium, *Cancer Res 38*, 3823–3829.
13. Murakami, H., and Masui, H. (1980) Hormonal control of human colon carcinoma cell growth in serum-free medium, *Proc Natl Acad Sci USA 77*, 3464–3468.
14. Evan, G. I., Wyllie, A. H., Gilbert, C. S., Littlewood, T. D., Land, H., Brooks, M., Waters, C. M., Penn, L. Z., and Hancock, D. C. (1992) Induction of apoptosis in fibroblasts by c-myc protein, *Cell 69*, 119–128.
15. Harrington, E. A., Bennett, M. R., Fanidi, A., and Evan, G. I. (1994) c-Myc-induced apoptosis in fibroblasts is inhibited by specific cytokines, *EMBO J 13*, 3286–3295.
16. Fernando, A., Glaysher, S., Conroy, M., Pekalski, M., Smith, J., Knight, L. A., Di Nicolantonio, F., and Cree, I. A. (2006) Effect of culture conditions on the chemosensitivity of ovarian cancer cell lines, *Anticancer Drugs 17*, 913–919.
17. Heist, R. S., and Christiani, D. (2009) EGFR-targeted therapies in lung cancer: predictors of response and toxicity, *Pharmacogenomics 10*, 59–68.
18. Arena, S., Isella, C., Martini, M., de Marco, A., Medico, E., and Bardelli, A. (2007) Knock-in of oncogenic Kras does not transform mouse somatic cells but triggers a transcriptional response that classifies human cancers, *Cancer Res 67*, 8468–8476.
19. Balaguer, P., Boussioux, A. M., Demirpence, E., and Nicolas, J. C. (2001) Reporter cell lines are useful tools for monitoring biological activity of nuclear receptor ligands, *Luminescence 16*, 153–158.
20. Crook, T. J., Hall, I. S., Solomon, L. Z., Birch, B. R., and Cooper, A. J. (2000) A model of superficial bladder cancer using fluorescent tumour cells in an organ-culture system, *BJU Int 86*, 886–893.

Chapter 3

Storage of Cell Lines

Katharine A. Parker

Abstract

The successful storage of cell lines depends upon many factors, including the condition of the cells to be frozen and the experience of the operator. Attempting to freeze down unhealthy, contaminated or poorly labelled cells can have huge implications for a research laboratory. This chapter outlines the importance of good record keeping, vigilant monitoring, aseptic technique, and high-quality reagents in the successful storage and downstream propagation of cell lines.

Key words: Cell lines, Liquid nitrogen, Recovery cell culture freezing medium, Aseptic, Dewar, Cryopreservation, Cryovial

1. Introduction

Success in freezing down cells for long-term storage largely relies upon the simplicity of the protocol in enabling the operator to work at speed. Modern reagents that have been developed without the inclusion of harsh chemicals mean that steps such as centrifugation to remove these chemicals from direct contact with cells can be excluded from the protocol altogether. The protocol for storage of cell lines requires much more than simply harvesting cells. To be able to store cells effectively and then recover a high enough proportion for quality research, storage conditions must be monitored on a permanent basis and excellent records kept.

To store cells you first need the correct laboratory equipment. At the very least, you need one liquid nitrogen storage dewar (but preferably two), a regular delivery of liquid nitrogen to keep the dewar topped up, personal protective equipment (cuffed liquid nitrogen gloves and a visor), internal thread cryovials, a database or excel spreadsheet, and an array of tissue culture reagents. When the facilities are in place, this then becomes a very rapid and reliable way of storing precious cell lines for future research.

Traditional methods for harvesting cells use proteolytic trypsin digestion to lift cells from the surface of tissue culture flasks. The drawback of this method is that it requires extra steps to inactivate the enzyme using foetal bovine serum (which adds to the cost) and then washing of the cell suspension with pre-warmed media. A more modern, and efficient, method is to use a recombinant enzyme substitute for trypsin, which is combined with mild EDTA chelation to remove magnesium and calcium ions, gently breaking the bonds holding the cells to the plastic substrate (1, 2). For this reason, wash media (HBSS) should not contain any magnesium or calcium ions.

The process outlined, in this chapter, involves harvesting cells from the tissue culture vessels in which they are grown, usually this is the plastic of polystyrene tissue culture flasks, using a recombinant enzyme-based dissociation reagent called TrypLE™ (Invitrogen). This is then followed by the preparation of cells for storage in recovery cell culture freezing media, a modern premixed solution containing glycerol that buffers cells gently through the freezing process and ensures maximum recovery. The process is completed with the thawing process. With the invention of enzyme-free reagents, this is a simple one-step process without the need for centrifugation. In summary, every step is optimised to avoid damaging the cells. Excessive pipetting and centrifugation of cells leads to irreparable damage and poor survival rates.

2. Materials

1. Sterile disposable polypropylene universal tubes (30 mL capacity, with white labels).
2. Sterile disposable serological pipettes (1.0–10.0 mL).
3. Barcoded cryovials (1.0 mL internal thread) plus cap inserts.
4. Indelible laboratory marker pen.
5. 1°C Freezing container (5100 Cryo "Mr Frosty"; Nalgene).
6. Sterile disposable plastic Pasteur pipettes, 3.0 mL and capillary stem.
7. Recovery cell culture freezing medium (oxygen sensitive should be spilt into 1.0 mL aliquots and stored at −20°C to reduce freeze-thawing and the amount of air coming into contact with the medium).

8. Recombinant-enzyme cell dissociation solution "TrypLE™" (Invitrogen) that should be divided into smaller aliquots to reduce pH alteration and stored at +4°C.
9. Hanks Balanced Salt Solution with phenol red and without magnesium and calcium (HBSS; 100 mL bottles can be stored at +4°C).
10. Sterile disposable Petri dishes.
11. Inverted microscope.
12. Two liquid nitrogen storage vessels (dewars) so that if one fails for some reason you always have a backup.

3. Methods

3.1. Harvesting Cells

1. Select cells for storage, cells should be a maximum of 80% confluent (see Note 1).
2. Using HBSS prewarmed to 37°C in an incubator, wash the cells (3.0 mL for a standard 25 cm^3 (T25) flask and 5.0 mL for a standard 75 cm^3 (T75) flask) to remove any debris (see Note 2).
3. Add TrypLE™ to the flask, 1.0 mL for a T25 flask and 2.0 mL for a T75 flask by holding the flask at approximately a 20° angle from the working surface, rocking gently to ensure the whole monolayer of cells is covered. Replace the lid and place into an incubator for 1 min (see Note 3).
4. After 1 min take out the flask and holding it flat, gently tap once on a clean, flat surface to dislodge any remaining cells. Check the cells have lifted sufficiently under an inverted laboratory microscope (see Note 4).
5. When the cells have lifted, add prewarmed media or HBSS to the flask and remove to a sterile universal tube for centrifugation. For maximum recovery of cells, the volume of media used should be spilt into two and two washes performed (see Note 5).
6. Spin the cells at $1,000 \times g$ for 5 min to form a good cell pellet (see Note 6).

3.2. Freezing Cells

1. Cells to be stored should be harvested immediately prior to use, using nonenzymatic reagents, to ensure maximum viability.
2. Prepare a freezing container (e.g. "Mr Frosty") for use by bringing to room temperature and ensuring it contains the correct level of isopropanol (see Note 7).

3. Clearly label pre-barcoded cryovials using an indelible laboratory marker with cell line designation (e.g., MCF-7 or SNB-19), passage number, date of storage, operator initials, concentration of cells, and size of flask frozen down (e.g. T25 or T75) (see Note 8).

4. Centrifuge harvested cells at $1,000 \times g$ for 5 min to form a good pellet.

5. Using a second sterile capillary stem pastette, carefully remove the supernatant and discard (see Note 9).

6. Pipette 1.0 mL recovery cell culture freezing medium (Invitrogen) on to the top of the cell pellet using a serological pipette. Then taking a new sterile pastette (again a capillary tube style is best) gently aspirate the pellet to mix, before carefully transferring to a cryovial.

7. Transfer the cryovials to the freezing container, screw the lid on firmly and place into a −80°C freezer immediately (see Note 10).

8. After 24 h, transfer the frozen cryovials to a liquid nitrogen dewar, ideally spitting cells between two dewars, making a note of the exact location.

3.3. Thawing Cells

1. When thawing cells for use, it is important to use aseptic techniques at all times and work quickly to ensure maximum recovery of viable cells (see Note 11).

2. Identify the cryovials to be thawed and ensure you update the dewar records to show which cells you will be removing (see Note 12).

3. Place the items required for thawing cells into the laminar flow cabinet, ready for use. Media does not need to be pre-warmed (see Note 13).

4. Clearly label tissue culture flasks and make sure you work only with one cryovial of cells at a time in the laminar flow cabinet – contamination of cells at this point is possible and worth avoiding by good technique (see Note 14).

5. Pipette the appropriate media into a clean, sterile flask, ready for the cryovial to be thawed. Replace the lid at all times when not in use.

6. Place the cryovial to be thawed into a sterile Petri dish and warm in a 37°C incubator, checking at regular intervals (see Note 15). Never spray cryovials with alcohol or biocleanse before placing them in a laminar flow hood, instead wipe them gently with a paper towel soaked in alcohol (making sure you do not wipe off the indelible marker).

7. When the cryovial contents are almost completely thawed, take them and place in the laminar flow hood.

8. Using a sterile pastette, aspirate the contents into the corresponding prelabelled flask, rock gently to mix, and place into the incubator. It is not necessary to centrifuge cells that have just been thawed from liquid nitrogen storage (see Note 16).

9. After 24 h, check the cells have adhered (if suspension, then wash cells as per normal protocol) and replace the media with fresh media (see Note 17).

4. Notes

1. When harvesting cells for storage it is important to ensure the growth phase is exponential, so that upon resurrection the cells are programmed to start dividing rapidly. Cells that have gone into full confluence for a long time do not always survive the freezing process; however, cells that are scant in number upon thawing can be rescued (see Note 16). If you are attempting to freeze cells that are fully confluent (which usually only happens by accident, if you check them each day), it is better to split them equally into two cryovials for storage so that they have room to grow when thawed.

2. It is preferable to warm HBSS and other tissue culture reagents in an incubator; however, they can be warmed in a water bath ensuring that they are placed inside sealed plastic bags first and then anchored to the side to avoid immersion. If there is any chance that water from the bath has come into contact with the reagent bottle, then the bottle must be discarded. When performing tissue culture always remember that the cost of losing a cell line to infection is far greater than the cost of a bottle of wash media.

3. For stubborn cells that are difficult to lift or have a double-layer effect of cells (e.g., LS174T colon cells), TrypLE™ can be used straight from the fridge.

4. If there are any cells that have not lifted, this is probably due to the cells not being washed adequately or being too confluent prior to harvesting. It is, however, completely safe to place the flask back into the incubator for a further minute, before checking under the microscope, and this can safely be repeated up to a maximum of 20 min.

5. The volume of wash or media depends on the user. The author recommends 5.0 mL for a T25 and 10.0 mL for a T75; however, the important thing is that you wash the flask out a second time to ensure all cells are removed.

6. Cells should be spun and processed as quickly as possible. Cooled centrifuges are not recommended as the temperature fluctuations between room and centrifuge are too great. If this is not possible, for example if you have a fire alarm and have to evacuate the building quickly, you can store the cells in the fridge briefly, but make a note in the dewar log.

7. You will need to keep a record of when the isopropanol is changed, ideally this should be after every three uses.

8. One of the most important pieces of information when freezing down cells is the size of the vessel from which they were harvested. If you know that everyone in your laboratory harvests cells at 70–80% confluent, then you can safely assume that a competent operator can thaw cells into the same size vessel. A popular method is to thaw all cells, no matter what number or origin, into a T25 to ensure survival, but if you are quite experienced this is not necessary. If, however, you are new to tissue culture, are thawing cells from another laboratory or a cell line bank, then always thaw cells into a T25 to be on the safe side. When freezing down cells, label all cryovials to be used before commencing tissue culture. Used an indelible laboratory marker and if there are a number of operators sharing a dewar, cryovials are available with different colour caps or coloured inserts that fit into a clear cap. Passage numbers are always sequential, they start at zero if you initiate the culture yourself from biopsy, and if you buy in cell lines it will tell you what passage they have been supplied at. Every time you split a flask of cells, the passage number increases by one. When you resurrect a cryovial of cells the passage number is the same as when they were frozen down. When making up batches of cells for storage, store cells at every passage to ensure you have plenty of backups.

9. If you accidentally disturb the pellet when removing the supernatant, you can recentrifuge the cells. Always use fresh pastettes, and if doing multiple pellets at the same time, pipette the recovery cell culture media on to the top of all pellets before starting to aspirate one by one.

10. The container will gradually freeze the cells over a 24-h period before transfer to a liquid nitrogen dewar, and by using the isopropanol as a temperature buffer, this will be slow enough that the cells do not suffer any damage. Cells should be routinely transferred to liquid nitrogen storage after 24-h, unless stored prior to a weekend.

11. The introduction of a pathogen to a cell line, through poor aseptic technique, can quickly mean whole batches of cells are infected. For this reason, it is important to always use separate tissue culture and wash media for each cell line and bank stocks of cells as regularly as possible.

12. In the interests of working quickly, it is important to know the location of the cells you require before attempting to open a liquid nitrogen dewar. Liquid nitrogen rapidly escapes the vessel once opened, raising the temperature inside, so to avoid compromising your cells and those of others, good organisation is the key. Plastic inserts that fit inside the cap of the cryovial can be written on and mean that they are easily visible, without having to remove them from the dewar racks. Any step taken to minimise the amount of time vials are removed from the dewar reduces the likelihood of damage to stored cells. It is important to note that if you have a cryovial with contents unknown, for whatever reason, it should not be used. There is no way of confirming the contents, no matter the experience of the operator, and the downstream implications of mixing up cell lines are huge.

13. Because the cells are coming directly from cold storage it is not essential that media is prewarmed.

14. An up to date log of cells stored in liquid nitrogen should be maintained, to ensure that cryovials can be located easily and batches of any one cell line do not run too low. The easiest way to do this, but perhaps not the most simple to set up, is to use barcoded cryovials. Using barcodes means an extra safety precaution in the event of smudged handwriting or operator error. Barcoded cryovials can either be used in conjunction with a barcode scanner and database, or simply on their own with an excel spreadsheet or log book. The barcodes are guaranteed never to come off, but it is important that you do not rely upon these alone because it is not unknown for the barcodes to be written down incorrectly. Allocate one box, or even one straw, per operator so that cells are easy to find. When you have used a cryovial of cells, make sure you update the records to reflect this. This ensures that the batches of cells do not run out.

15. If you have access to an autoclave, freshly sterilised small glass beakers are more cost effective than Petri dishes. Never do more than one cryovial at a time, to avoid getting cells mixed up. When using RCCFM you can put cells straight into a new tissue culture flask without centrifuging to remove the media in which they were frozen, this avoid extra damage and maximises cell recovery.

16. When cell numbers are low or there has been a problem with the storage process you can rescue a cell line by reconstituting in 1.0 mL of media and seeding in a "droplet" shape on the surface of a tissue culture flask. Without disturbing the droplet, place into an incubator for 24 h. Seeding cells in a small circle ensures that the cells remain in close contact with each other which is essential for cross exchange of growth signals

between neighbouring cells and avoids overdiluting any growth factors produced.

17. According to which cell line you are using, you may find that 48 h may be required for sufficient adherence of cells.

References

1. Prowse, A., Wolvetang, E., and Gray, P. (2009) A rapid, cost-effective method for counting human embryonic stem cell numbers as clumps, *Biotechniques* **47**, 599–606.
2. Ulyanova, T., Scott, L. M., Priestley, G. V., Jiang, Y., Nakamoto, B., Koni, P. A., and Papayannopoulou, T. (2005) VCAM-1 expression in adult hematopoietic and nonhematopoietic cells is controlled by tissue-inductive signals and reflects their developmental origin, *Blood* **106**, 86–94.

Chapter 4

Characterization and Authentication of Cancer Cell Lines: An Overview

Yvonne A. Reid

Abstract

Studies of the same cell lines by different laboratories are common in the literature and often show different results with the same methodology. Use of best cell culture practices is essential to ensure consistent and reproducible results. Assay outcomes are easily influenced by many factors including changes in functionality, morphology, doubling time of cells, passage numbers, microbial contamination, and misidentification of cells. Simple observation, monitoring, and documentation of cell morphology and behavior, including growth rates, provide early warning and should be standard practice. Changes may indicate microbial contamination, genotypic drift due to high passage number, or cross-contamination with another cell line. Rapid molecular methods allow the identification of microbial and cross-contamination. Increasingly, authentication of cell lines is a prerequisite for scientific publication to avoid erroneous results entering the literature.

Key words: Cell line, Characterization, Authentication

1. Introduction

Over the last few decades, there has been a progressive increase in the use of cell systems, particularly as both tools and models for basic research and industrial applications. New therapies in which cells and tissue engineering play a pivotal role are becoming more widely used. The use of cell systems in systems biology, the human genome project, the emerging fields of metabolomics and proteomics, and the discovery of new biomarkers for disease will lead to further significant developments in biomedical science (1). These cell systems have relied heavily on the use of cell lines. Very often the same cell line is studied, simultaneously, by several research laboratories throughout the world, with the expectation that each laboratory is using the same cell line and that the results are directly comparable. However, many studies have shown that this is often not the case. Thus, regardless of the cell system adopted, the use of best cell culture practices to ensure consistent and reproducible results are paramount. Reliable assay outcomes are easily influenced by many factors including changes in functionality, morphology, doubling time of cells, passage numbers, microbial contamination, and misidentification of cells. These inconsistencies in cell characteristics, authenticity, and quality have led to false results and have provided invalid comparisons of data among laboratories with similar research interests.

Observing, monitoring, and documenting cell morphology and behavior are important tools in tracking changes and monitoring the health of the cells in culture. For example, the mouse 3T3 cells grow as multipolar fibroblasts at low density, but become epithelial-like at confluency. Morphological changes were also observed as the medium, and substrate were changed when culturing these cells (2). As the population doubling levels (PDL) increase, cells become granular and highly vacuolated. A change in morphology is sometimes an early sign of deterioration of the culture and often indicates that the cells are differentiating (3), are contaminated with another cell line, or are undergoing crisis. It is highly recommended that morphological observations are made at the same cell density, growth stage, medium, and substrate.

2. The Growth Curve

Establishing a growth curve is another useful tool for determining the optimal growth condition of a cell line. Plotting the total number of viable cells (y-axis) vs. days in culture (x-axis) gives a growth curve that is characteristic for a given cell line. There are three stages of the growth curve (1) the lag phase: the period of adaptation to culturing conditions; (2) the exponential phase: the cells are most reproducible and viable and may be the best stage for removing samples for experimental procedures; and (3) the stationary phase: the growth rate is reduced or ceases due to nutrient depletion or due to contact inhibition of the cells. However, during this phase, there may be a relative increase in the synthesis of specialized proteins compared with structural proteins (4). In some cases, harvesting cells at this stage of the growth curve may be best for experimental use. The growth curve provides several pieces of information that is necessary to optimize the growth of the cell line (1) the cell concentration range at which the cells are subcultured is determined at the upper part of the exponential phase before entering into the stationary phase; (2) the inoculum at subculture is the cell concentration range at the lower end of the exponential phase; and (3) the population doubling time (PDT) is determined from the slope at the linear stage of the mid-exponential phase. Changes in the number of cells inoculated at subcultivation may result in changes in the slope of the exponential phase and consequently change in the PDT. Given that the growth conditions are constant, these phases are characteristic of each cell line and will result in consistent measurements for each phase with each subculture (5). Consistent recording and application of growth curve parameters for each cell line will give rise to consistent and reproducible results and better comparison of scientific data among laboratories.

3. Contamination

Microbial contamination (bacteria, fungi, *Mycoplasma*, and viruses) of cell culture systems continues to be problematic. In a majority of cases, microbial contamination is overt and is visible with the naked eye. However, low levels of contamination of the more fastidious microorganisms are often missed by many observers. By including a series of microbiological tests, most common microorganisms can be detected. For example, bacteriological media (i.e. blood agar, thioglycollate, trypticase soy, brain heart infusion, Sabouraud, and YM broths plus nutrient broths with 2% yeast extract) tests incubated at 37 and 26°C, under aerobic and anaerobic conditions, will detect the most fastidious organisms known to infect cell cultures and media (6).

Due to the insidious nature of *Mycoplasma* infection and its harmful impact on numerous cell functions, such contamination requires careful examination. The presence of some *Mycoplasma* species may produce cytopathic effects (7). Other species may metabolize and proliferate extensively without inducing noticeable changes in morphology. In either case, *Mycoplasma* infection can invalidate research findings by interfering with studies of metabolism, receptors, virus–host interactions, and cell divisions (6). There are numerous well-established methods for detecting *Mycoplasma* in cell culture, including direct cultivation, indirect flurochrome staining (Hoechst or DAPI), DNA hybridization, and PCR.

The three tests most commonly performed are the agar method (direct), Hoechst stain (indirect), and PCR. The standard agar culture test currently in use, described by Macy in 1979 (8), involves inoculating agar with samples from cell cultures, incubation under aerobic and anaerobic conditions, for examination after 4 weeks. Detection of *Mycoplasma* using conventional culture method remains the gold standard and is described by Code of Federal Regulations 21 CFR Ch.1 (4-4-01 Edition) §610.30 Subpart D – Mycoplasma; European Pharmacopeia (2.6.7 'Mycoplasma'); CBER, FDA Points to Consider in the

Characterization of Cell Lines Use to Produce Biologicals. The second method involves the flurochrome Hoechst 33258®. This method is robust in detecting more than 40 different *Mycoplasma* species, including *Mycoplasma hyorhinis*, but sensitive enough to detect few *Mycoplasma* per cell (9). The third method employs a PCR assay for the detection of *Mycoplasma* and *Acholeplasma*. The procedure involves nested amplification of primers designed specifically to the spacer regions of the 16S and 23S ribosomal RNA regions of *Mycoplasma* and *Acholeplasma*. This method is very sensitive and less time consuming than the agar and Hoechst stain methods (10).

The detection of endogenous and contaminating viruses in cell culture systems is most challenging since most infections are not obvious or have a distinctive cytopathic effect (CPE). However, absence of CPEs does not mean that the culture is virus free. In fact, persistence of latent infectious viruses in cell culture will remain undetected until appropriate test methods are implemented. Some of the more common viral detection methods are PCR, CPE, indirect immunofluorescent antibody (IFA), and enzyme immunoassay (EIA) (11).

4. Changes During Cell Culture

A fundamental misconception of cell culturists is that diploid or continuous cell lines will not change during propagation. However, many scientific publications have demonstrated the divergent effects of long-term culturing (or increase passaging) on cell line morphology, development, and gene expression (12–17). Cell lines placed in culture, long-term, are likely to undergo selective pressures that may lead to genotypic and phenotypic instabilities. These instabilities can give rise to an increased number of mutations in genomic DNA (gDNA) and mitochondrial DNA (mtDNA) and, subsequently, growth modulation (18, 19). The human Caco-2 cell line is frequently used as a model system for drug transport and toxicity studies. Several reports, including Hughes et al. (20–23), have demonstrated passage-related differences in growth rates and transepithelial electrical resistance (TEER). Caco-2 cells obtained from various Cell Banks and propagated at high passage (50 times) were found to have elevated paracellular permeability of the monolayer (consistent with decreasing TEER values); decrease in the transcellular permeability of the cells (consistent with a reduction in p-GP); and an increase in the proliferation rate of the cells when compared with cells at low passage. The effect of increased passage on cell culture characteristics is also demonstrated in other cell lines, including the prostate cell line, LNCaP (14), breast cell line, MCF-7 (24), and pluripotent ES cells (25).

Extensive cellular cross-contamination or misidentification of animal cell cultures is a long-standing problem in cancer and other biomedical research dating back to the 1950s (26). Although numerous publications have addressed this issue (27–31), the problem has persisted. The cost associated with invalidated research is estimated to be in the tens of millions of dollars (32). In most cases, the cells involved are not routinely tested for intra- or interspecies contamination, and the information pertaining to these cultures is assumed to be correct based upon the reputation of the source. The lack of routine testing was confirmed by a recent survey conducted by Buehring et al. (33) who announced that of the 483 respondents, 46% never tested for cell identity, 35% obtained cells lines from another laboratory, and 63% obtained at least one cell line from another laboratory rather than from a major repository. She also reported that over 220 publications from 1969 to 2004 in PubMed searches supported HeLa "contaminants" as models of ovarian cancer.

HeLa, a cell line derived from an invasive cervical carcinoma, has been shown to contaminate more than 90 cell lines (34, 35). More recently, Drexler et al. (36) have reported on receiving 500 leukemia cell lines and found 15% of the cell lines to be cross-contaminated; 59/395 were received directly from the original investigators and 23/155 were received from secondary investigators. The NCI-60 panel of 60 cancer cell lines, widely used for basic research and drug discovery, has some cross-contaminated cell lines. Based on SNP and karyotyping analyses, the multidrug-resistant MCF-7 (originally designated MCF-7/AdrR and later redesignated NCI/ADR-RES), a derivative of MCF-7 breast adenocarcinoma cell line, was found to be unrelated to MCF-7 but related to OVCAR-8 a human ovarian carcinoma cell line (37). SNP analyses at the Sanger Institute have demonstrated that the SNB19 and U251, glioblastoma cell lines, thought to be unrelated have been found to be related. Similar findings showed that the breast adenocarcinoma cell lines, MDA-MB-435, widely used in breast cancer research, have been shown by cDNA microarray analysis (38) and STR analysis (39) to be derived from the melanoma cell line, M14.

The most frequently used assay for determining the species of origin of animal cell lines are immunological tests, isoenzymology, karyotyping, and more recently cytochrome c subunit 1 (COI) (40, 41) for interspecies determination and STR (29) and SNP (42) analyses for intraspecies identification. Interspecies identification of animal cell lines by COI analysis relies on a multiplex PCR assay that rapidly identified 14 of the most common animal species in cell culture. The protocol involves amplification of a universal primer pair that targets a 650-bp COI consensus sequence. Each species is characterized by a specific amplicon size, which is used for identification. Procedure can be extended to identify unknown species by sequencing the amplicons and comparing them to sequences of voucher specimens (41).

Human cell lines are identified by either (short tandem repeat) STR or SNP (single-nucleotide polymorphism) analysis. STR analysis (DNA profiling) is the most commonly used method based on its simplicity, robustness, and informativeness. It involves the simultaneous amplification of several STR markers (about eight) plus amelogenin for gender determination. Using a multiplex PCR reaction, eight STR markers can discriminate one unique human cell line from another at a level of 1×10^{-9}. Each unique cell line has a distinct DNA pattern and the method can be used to detect contamination of human cells at a level as low as 10% (43).

Roland Nordone (44) has elevated the discussions pertaining to misidentification of cell line to a public policy issue. In 2008, he wrote a white paper on "Eradication of cross-contaminated Cell lines: a call for action." The "Call for Action" challenges the various stakeholders including the funding agencies, editors of journals, professional scientific societies, and scientists to take responsibility for their actions. The widespread use of misidentified cell lines has led to increase publications in refereed journals with erroneous results. As stakeholders, government and private funding agencies have a responsibility not to support grants and contracts that do not address authentication of cell lines; that the editors of journals request authors to authenticate their cell lines as prerequisite for publication; that professional societies support policies, conduct workshops, conferences, and training activities addressing the issues of cellular contamination and the adaptation of cell line authentication standards; that heads or laboratories and senior administrators ensure that staff members are well trained and are adhering to standard operating procedures which include authentication of cell lines (44) This responsibility extends to requiring scientists to issue published erratum when misidentification of cell lines are identified. Most important, the public should hold all the stakeholders accountable, especially when public funding is involved. The "Call for Action" has led to several events, including several journals (Cell Biochemistry and Biophysics, BioTechniques, In Vitro Cellular & Developmental Biology – Animal, AACR journals – Cancer Research, Cancer Epidemiology, Biomarkers & Prevention, Molecular Cancer Research, Clinical Cancer Research, and Cancer Prevention Research) are now requiring authentication of cell lines as a prerequisite for publication. In addition, an International Standard on the use of STR analysis for the identification of human cell lines is being drafted and sponsored by the ATCC® Standards Development Organization. Representative of the standard group are members of industry, academia, regulatory agencies, and cell banks.

It is highly recommended that, during the propagation of a cell line and certainly before publication, the functionality or a unique characteristic test is performed to confirm previous results. For example, the hTERT-immortalized cell line does not demonstrate increased polyploidy but remain diploid during propagation; that the mouse 3T3 fibroblast cells remain contact inhibited and can be induced to form adipocytes; that the mantle cell lymphoma cell lines continue to express the CD 20 marker and displays (11;14)(q13:q32) translocation.

5. Conclusion

As scientists embrace a systems biology approach to research, cellular cross-contamination, suboptimal growth conditions, microbial infections may all lead to errors in cellular characterization. Countless research projects may be producing erroneous results because researchers fail to recognize the importance of authentication and characterization testing of cell lines and, consequently, fail to incorporate these necessary practices in their research activities. Whether these changes are due to increased passages, suboptimal growth conditions, cellular cross-contamination or microbial contamination, it is important that a comprehensive approach be taken to reduce these changes. Ignoring any one of these factors will lead to inconsistent or irreproducible results; loss of cell lines; loss of time and money; misrepresentation of scientific information in the public domain; discordant or irreproducible results, private embarrassment, and public humiliation.

References

1. Hartung, T., Balls, M., Bardouille, C., Blanck, O., Coecke, S., Gstraunthaler, G., and Lewis, D. (2002) Good cell culture practice: ECVAM good cell culture practice task force report 1, *ATLA* **30**, 407–14.
2. Freshney, R.I. Culture of glioma of the brain.(1980) In: Thomas, D.G.T., and Graham, D.I. Eds. Brain tumours: Scientific basic, clinical investigation and current therapy. London Butterworths, pp. 21–50.
3. Bayreuther, K., Rodemann, H.P., Hommel, R., Dittmann, K., Albiez, M., and Francz, P.I. (1988) Human skin fibroblasts in vitro differentiate along a terminal cell lineage. *PNAS USA*, **85**, 5112–6.
4. Freshney, R.I. (2005) Analysis of Growth Curve, In: Culture of Animal Cells: A manual of basic techniques 5th Ed. New York: John Wiley & Sons Inc., pp. 351.
5. Freshney, R.I. (2000) Growth Curve. In: Masters JRW, ed. Animal Cell culture A Practical Approach, 3rd ed. Oxford University press, pp.12–13.
6. Hay, R.J. (1992) Continuous Cell lines - An International Workshop on Current Issues. *Develop Biol Standard*, **76**, 25–379.
7. Beary, S.J., and Walczak, E.M. (1983) Cytopathic Effect of Whole Cells and Purified Membranes of Mycoplasma hyopneumoniae. *Infect Immun*, **41**, 132–6.
8. Macy, M.L. (1979) Tests for mycoplasma contamination of cultured cells as applied at the ATCC. TCA Manual **5**, 1151–6.
9. Chen, T.R. (1977) In situ detection of mycoplasma contamination in cell culture by fluorescent Hoechst 33258 strain. *Exp Cell Res* **104**, 255–262.

10. Tang, J., Hu, M., Lee, S., and Roblin, R. (2000) A polymerase chain reaction based method for detecting *Mycoplasma/Acholeplasma* contaminants in cell culture. *J Microbiol Methods* **39**, 121–6.

11. Hay, R.J. (1998) Cell Line Banking and Authentication. *Dev Biol Stand* **93**, 15–19.

12. Briske-Anderson, M.J., Finley, J.W., and Newman, S.M. (1997) Influence of culture time and passage number on morphological and physiological development of Caco-2 cells. *Proc Soc Exp Biol Med.* 214; **3**, 248–57.

13. Chang-Liu, C-M., and Woloschak, G.E. (1997) Effect of passage number on cellular response to DNA-damaging agents: cell survival and gene expression. *Cancer Lett* **113**, 77–86.

14. Esquenet, M., Swinnen, J.V., Heyns, W., and Verhoeven, G. (1997) LNCaP prostatic adenocarcinoma cells derived from low and high passage numbers display divergent responses not only to androgens but also to retinoids. *J Steroid Biochem Mol Biol* **62**, 391–9.

15. Langeler, E.G., van Uffelen, C.J., Blankenstein, M.A., van Steenbrugge, G.J., and Mulder, E. (1993) Effects of culture conditions on androgen sensitivity of the human prostate cancer cell line LNCaP. *Prostate* **23**, 213–23.

16. Sambuy, Y., De Angelis, I., Ranaldi, G., Scarino, M.L., Stammati, A., and Zucco, F. (2005) The Caco-2 cell line as a model of the intestinal barrier: influence of cell and culture-related factors on Caco-2 cell functional characteristics. *Cell Biol Toxicol* **21**, 1–26.

17. Yu, H. (1997) Evidence for diminished functional expression of intestinal transporters in Caco-2 Cell monolayers at high passages. *Pharm Res* **14**, 757–62.

18. Matsumura, T. (1980) Multinucleation and polyploidization of aging human cells in culture. *Adv Exp Med Biol*, **129**, 31–38.

19. Dumont, P., Burton, M., Chen, Q.M., Gonos, E.S., Frippiat, C., Mazarati, J.B., Eliaers, F., Remacle, J., and Toussaint, O. (2000) Induction of replicative senescence biomarkers by sublethal oxidative stresses in normal human fibroblast. *Free Radical Biol Med*, **28**, 361–73.

20. Hughes, P., Marshall, D., Reid, Y., Parkes, H., and Gelber, C. (2007) The costs of using unauthenticated, over-passaged cell lines: how much more data do we need? *Biotechniques* **43**, 575–86.

21. Mossberg, P.I. (2005) Wise Marketing Consultancy. Study Report: Cell Lines and Their Use in Research.

22. Lu, S., Gough, A.W., Babrowski, W.F., and Stewart, B.H. (1996) Transport properties are not altered across Caco-2 cells with heightened TEER despite underlying physiological and ultrastructural changes. *J Pharm Sci*, **85**, 270–273.

23. Ranaldi, G., Consalvo, R., Sambuy, Y., and Scarino, M.I. (2003) Permeability characteristics of parental and clonal human intestinal Caco-2 cell lines differentiated in serum-supplemented and serum-free media. *Toxicol In Vitro* **17**, 761–767.

24. Wenger, S.L., Senft, J.R., Sargent, L.M., Bamezai, R., Bairwa, N., and Grant, S.G. (2004) Comparison of established cell lines at different passages by karyotype and comparative genomic hybridization. *Biosci Rep* **24**, 631–9.

25. Draper, J.S., Smith, K., Gokhale, H.D., Moore, E., Maltby, J., Johnson, L., Meisner, T., Zwakea, T.P., Thomson, J.A., and Andrews, P.W. (2004) Recurrent gain of chromosomes 17q and 12 in cultured human embryonic stem cells. *Nat Biotechnol* **22**, 53–4.

26. McCulloch, E.A., and Parker, R.C. (1957) In Begg, R.W. Ed Canadian cancer conference. Vol.2 New York: Academic Press.

27. Reid, Y.A., O'Neill, K., Chen, T.R. (1995) Cell cross-contamination of U-937. *J Leukoc Biol* **57**, 804.

28. Masters, J.R., Thomson, J.A., Daly-Burns, B., Reid, Y.A., Dirks, W.G., Packer, P., Toji, L.H., Ohno, T., Tanabe, H., Arlett, C.F., Kelland, L.R., Harrison, M., Virmani, A., Ward, T.H., Ayers, K.L., and Debenham, P.G. (2001) Short tandem repeat profiling provides an international reference standard for human cell lines. *PNAS* **98**, 8012–7.

29. Gilbert, D.A., Reid, Y.A., Gail, M.H., Pee, D., White, C., Hay, R.J., and O'Brien, S.J. (1990) Applications of DNA fingerprints for cell line individualization. *Am J Hum Genet* **47**, 499–514.

30. Durkin, A.S., Cedrone, E., Sykes, G., Boyles, D., Reid, Y.A. (2000) Utility of gender determination in cell line identity. *In Vitro Cell Dev Biol Anim* **36**, 344–7.

31. Markovic, O., Markovic, N. (1998) Cell cross-contamination in cell culture: the silent and neglected danger. *In Vitro Cell Dev Biol Anim* **34**, 1–8.

32. Gold, M.A. (1986) Conspiracy Of Cells, One Woman's Immortal Legacy and The Medical Scandal It Caused. Albany, NY: State Univ. of New York Press.

33. Buehring, G.C., Eby, E.A., and Eby, M.J. (2004) Cell Line cross-contamination: how aware are mammalian cell culturists of the problem and how to monitor it? *In Vitro Cell Dev Biol Anim* **40**, 211–5.

34. Nelson-Rees WA, Flandermeyer RR. HeLa cultures defined. Science, 191(4222):96–98, 1976.
35. Nelson-Rees, W.A., Daniels, D.W., and Flandermeyer, R.R. (1981) Cross-contamination of cells in culture. *Science* **212**, 446–52.
36. Drexler, H.G., Dirks, W.G., Matsuo, Y., and MacLeod, R.A.F. (2003) False leukemia-lymphoma cell lines: an update on over 500 cell lines. *Leukemia* **17**, 416–26.
37. Liscovitch, M., and Ravid, D. (2006) A case of misidentification of cancer cell lines: MCF-7/AdrR cells (re-designated NCI/ADR-RES) are derived from OVCAR-8 human ovarian carcinoma cells. *Cancer Lett* **245**, 350–2.
38. Ross, D.T., Scherf, U., Eisen, M.B., Perou, C.M., Rees, C., Spellman, P., Iyer, V., Jeffrey, S.S., Van de Rijn, M., Walthan, M., Pergamenschikov, A., Lee, J.C., Lashkari, D., Shalon, D., Myers, T.G., Weinstein, J.N., Botstein, D., and Brown, P.O. (2000) Systematic variation in gene expression patterns in human cancer cell lines. *Nat Genet* **24**, 227–35.
39. Rae, J.M., Creighton, C.J., Meck, J.M., Haddad, B.R., and Johnson, M.D. (2007) MDA-MB-435 is derived from M14 melanoma cells - a loss for breast cancer but a boon for melanoma research. *Breast Cancer Res Treat* **104**, 13–19.
40. Parodi, B., Aresu, O., Bini, D., Lorenzini, R., Schena, F., Visconti, P., Cesaro, M., Ferrera, D., Andreotti, V., and Ruzzon, T. (2002) Species identification and confirmation of human and animal cell lines: a PCR-based method. *BioTechniques* 32, 432–4, 436, and 438–440.
41. Cooper, J.K., Sykes, G., King, S., Cottrill, K., Inanova, N.V., Hanner, R., and Ikonomi, P. (2007) Species identification in cell culture: a two-pronged molecular approach. *In Vitro Cell Dev Biol Anim* **43**, 344–51.
42. Demichelis, F., Greulich, H., Macoska, J.A., Beroukhim, R., Sellers, W.R., Garraway, L., and Rubin, M.A. (2008) SNP panel identification assay (SPIA): a genetic-based assay for the identification. *Nucleic Acids Res* **36**, 2446–56.
43. Reid, Y.A., Gilbert, D.A., and O'Brien, S.J.(1990) The use of DNA hypervariable probes for human cell line identification. *American Type Culture Collection Newsletter* 10, 1–3.
44. Nardone, R.M. (2007) Eradication of cross-contaminated cell lines: a call for action. *Cell Biol Toxicol* **23**, 367–72.

Chapter 5

Online Verification of Human Cell Line Identity by STR DNA Typing

Wilhelm G. Dirks and Hans G. Drexler

Abstract

The main prerequisition for any research, development, or production programs involving cell lines is whether a cell line is authentic or not. Microsatellites in the human genome harboring short tandem repeat (STR) DNA markers allow the identification of individual cell lines at the DNA level. Polymerase chain reaction (PCR) amplification of eight highly polymorphic microsatellite STR loci and gender determination have been proven to be the best tools for screening the uniqueness of DNA profiles in an STR database. The main Biological Resource Centers (BRCs), ATCC, DSMZ, JCRB, and RIKEN, have generated large databases of STR cell line profiles for identity control. In cooperation with the Japanese BRCs, DSMZ has piloted the generation of the most comprehensive international reference database, which is linked to a simple search engine for interrogating STR cell line profiles. The tool of online verification of cell line identities is available on the respective homepages of JCRB and DSMZ (http://cellbank.nibio.go.jp/cellbank_e.html, http://www.dsmz.de/STRanalysis). The following sections describe a rapid, practical, inexpensive, and reliable method available to students, technicians, and scientists.

Key words: Authentication, Human cell lines, DNA STR typing, Quality control, Cross-contamination

1. Introduction

Given the large number of human diseases, cell lines as model systems have wide applications in the medical and pharmaceutical industry, whereby drug and chemical testing was first carried out exhaustively on in vitro systems to reduce the need for difficult and invasive animal experiments. The requirement for authentication of cell lines has a history almost as long as cell culture itself, presumably beginning when more than one cell line could be cultured continuously.

1.1. History of Cell Line Identification

The application of specific species markers including cell surface antigens and chromosomes showed that interspecies misidentification was a widespread problem (1, 2). Subsequently, it was shown that intraspecies contamination of human cell cultures was also a serious problem, which could be monitored by the innovation of isoenzymatic analysis (3). After extending this approach to multiple polymorphic isoenzymes, the persistence of specific marker chromosomes in long-term-passaged cell lines demonstrated the unique power of cytogenetics (4). Based on the detection of chromosomal markers, it was convincingly demonstrated that multiple cell lines under active investigation were actually derived from one source, namely the HeLa cell line (5). Furthermore, cross-contamination among established cell lines occurred at frequencies as high as 16–35% in the late 1970s (6). Recently, our department has demonstrated an incidence of 14% of false human cell lines (7–9) indicating intraspecies cross-contamination as a chronic problem and highlighting the badly neglected need for intensive quality controls for cell line authenticity.

1.2. DNA Typing Technologies

When compared with polymorphic isoenzymes or marker chromosomes, a much higher resolution in discrimination among human cell lines was achieved using restriction fragment length polymorphism (RFLP) of simple trinucleotide repetitive sequences (10), which lead subsequently to the concept of "DNA fingerprinting" (11). The principle of the method is based on the phenomenon that genomes of higher organisms harbor many variable number of tandem repeat (VNTR) regions, which show multiallelic variation among individuals (12). Sequence analysis demonstrated that the structural basis for polymorphism of these regions is the presence of tandem-repetitive, nearly identical DNA elements, which are inherited in a Mendelian way. Depending on the length of the repeats, VNTRs are classified into minisatellites consisting of 9 to >70 bp core sequences, and microsatellites which include all short tandem repeats (STRs) with core sizes from 1 to 6 bp. STR typing of tetrameric repeats is now the gold standard and an international reference technique for authentication of human cell lines (13). The ATCC, DSMZ, JCRB, and RIKEN cell line banks have each built their own STR databases based on the following specific set of STR loci: D5S818, D13S317, D7S820, D16S539, vWA, TH01, TPOX, CSF1PO, and Amelogenin (AMEL) for gender determination. Amel has meanwhile become the most suitable gene for gender determination of samples from human origin (14). Using specific primers in polymerase chain reaction (PCR) applications, the sequence of the X-chromosomal version (AMELX, Xp22.1–Xp22.3) yields a 209 bp amplicon, while the Y-chromosomal gene (AMELY, Yp11.2) yields a 215 bp DNA fragment, which may be easily separated by different electrophoresis techniques (15). Hence, samples from male sources will show two bands in a PAGE analysis (209 and 215 bp), while female-derived cell lines will show only one band (209 bp).

1.3. DNA Typing as Quality Control in Routine Cell Culture

Most facilities culturing cells use multiple cell lines simultaneously. Because of the complexity of experimental designs today and because of the fact that the broad use of cell lines in science and biotechnology continues to increase, the possibility of inadvertent mixture of cell lines during the course of day-to-day cell culture is always present. Based on the reputation of a laboratory, the information on an exchanged cell line within a scientific cooperation is normally thought to be correct. This is the main reason for the increasing problem of cross-contaminated cell cultures all over the world. However, cross-contaminations of cell cultures appear to be the "peccadillo" of scientists, since publication of alerting reports on misidentified or cross-contaminated cell lines are most often simply ignored by the scientific community.

The authentication of human cell lines by the scientific service of DSMZ demonstrates that cross-contamination remains a chronic problem among human continuous cell lines. The combination of easy and rapidly generated STR profiles and their authentication by an online screen at the reference database (http://www.dsmz.de/STRanalysis) constitute a major and novel advance in decreasing the use of false cell lines. Today a modern average laboratory applying molecular biology and cell culture techniques is equipped with PCR machines and capillary electrophoresis, which are essential for STR DNA profiling. Multiplex PCR targets multiple locations throughout the genome and is an ideal technique for DNA typing because the probability of identical alleles in two individuals decreases with an increase in the number of polymorphic loci examined. The advent of fluorescent labeling of PCR primers permits the multiplexing of STR loci which may have alleles that fall in the same size range by labeling the overlapping loci with different colored fluorescent dyes that can then be resolved spectrally. The technique consists of multiplexed STR PCR amplification of eight prominent and highly polymorphic minisatellite STR loci and one additional locus for gender determination by the detection of the Amelogenin gene. The combination of eight STRs increases the exclusion rate to a sufficient extent and allows the discrimination of one human cell line from another at the level of 10^8.

2. Materials

All solutions should be prepared in water that has a resistivity of 18.2 MΩ cm and total organic content of less than five parts per billion. This standard is referred to as "distilled water" in this text.

2.1. DNA Extraction

1. Phosphate-buffered saline (PBS): 140 mM NaCl, 27 mM KCl, 7.2 mM Na_2HPO_4, and 14.7 mM KH_2PO_4, pH 7.2, autoclave.
2. Absolute isopropanol and absolute ethanol.
3. TE 10/1: 10 mM Tris–HCl, 1 mM EDTA, pH 8.0 prewarmed to 50°C.
4. High Pure PCR Template Preparation Kit (Roche).
5. Water bath prewarmed to 72°C.
6. Standard tabletop microcentrifuge capable of $13,000 \times g$ centrifugal force.
7. Standard spectral photometer for the determination of DNA concentration.

2.2. Multiplex STR Typing

1. Thermal cycler (any supplier).
2. Taq DNA polymerase (any supplier).
3. The primers should be concentrated at 100 µM in TE (10/1) as stock solution and stored at −20°C, while working solutions should be aliquoted at 10 µM in small amounts (ca. 25–50 µL aliquots) and stored frozen at −20°C.
4. Standard tabletop microcentrifuge or centrifuge capable of 96-well plates.
5. Capillary electrophoresis unit (any supplier).

2.3. Commercial Kits for DNA Typing

Kits from commercial sources are now predominantly used by labs around the world for their ease of use and high discriminatory power. Since the major BRCs, ATCC, DSMZ, JCRB, and RIKEN, have built the STR databases based on a specific set of STR loci, customers can use the following kits for STR amplification deemed to be compatible with the use of databases and the online STR search engine. The application of commercial multiplex STR kits should be carried out strictly following the specific manuals of Promega Corporation and Applied Biosystems.

1. Promega Corporation, (Madison, WI, USA): PowerPlex®1.1, full compatibility over nine loci; Power Plex®1.2, full compatibility over nine loci; and PowerPlex® 16, full compatibility plus seven additional loci.
2. Applied Biosystems, (Foster City, CA, USA): AmpFlSTR® Identifiler™, full compatibility plus seven additional loci; AmpFlSTR® Profiler™, eight compatible plus two additional loci; and AmpFlSTR® COfiler™, six compatible loci plus two additional loci.

2.4. STR Primer Sequences and Genomic Location

Fluorescent labeling of the PCR primers permits the multiplexing of STR loci even when alleles fall into the same size range. Labeling the overlapping loci with different colored fluorescent dyes enables the analysis. Depending on the capillary electrophoresis available, group 1 of primers can be labeled by one specific dye, which should be different from group 2 primers or the respective dye for the size standard. Only one of a primer pair of an STR locus should be labeled, regardless if it is the forward or reverse primer.

STR primer group I (dye I)

D16S539, chromosome 16q22–q24; noncoding region:

5'-GGG GGT CTA AGA GCT TGT AAA AAG

5'-GTT TGT GTG TGC ATC TGT AAG CAT GTA TC.

D13S317, chromosome 13q22–q31; noncoding region:

5'-ACA GAA GTC TGG GAT GTG GAG GA

5'-GGC AGC CCA AAA AGA CAG A.

D5S818, chromosome 5q21–q31; noncoding region:

5'-GGT GAT TTT CCT CTT TGG TAT CC

5'-AGC CAC AGT TTA CAA CAT TTG TAT CT.

D7S820, chromosome 7q11.21–q22; noncoding region:

5'-ATG TTG GTC AGG CTG ACT ATG

5'-GAT TCC ACA TTT ATC CTC ATT GAC.

STR primer group II (dye II)

CSF1, chromosome 5q33.3–q34; 3'-UTR of c-fms proto-oncogene for CSF-1 receptor gene:

5'-AAC CTG AGT CTG CCA AGG ACT AGC

5'-TTC CAC ACA CCA CTG GCC ATC TTC.

TPOX, chromosome 2p23–2pter; intron 10 of human thyroid peroxidase gene:

5'-ACT GGC ACA GAA CAG GCA CTT AGG

5'-GGA GGA ACT GGG AAC CAC ACA GGT TA.

TH01, chromosome 11p15–p15.5; intron 1 of human tyrosine hydroxylase gene:

5'-ATT CAA AGG GTA TCT GGG CTC TGG

5'-GTG GGC TGA AAA GCT CCC GAT TAT.

vWA, chromosome 12p12-pter; 3'-UTR of van Willebrandt factor gene:

5'-CTA GTG GAT GAT AAG AAT AAT CAG TAT GTG

5'-GGA CAG ATG ATA AAT ACA TAG GAT GGA TGG.

AMEL, chromosome Xp22.1–p22.3, Yp 11.2 coding region of amelogenin:

5'-ACC TCA TCC TGG G CAC CCT GGT T

5'-AGG CTT GAG GCC AAC CAT CAG.

Alternate primer sequences for the amplification of above-mentioned STRs as well as the information on the PCR products and allele sizes are available at http://www.cstl.nist.gov/div831/strbase.

3. Methods

3.1. Prerequisitions for DNA Typing

General rules to avoid DNA carry over contamination should be strictly followed: DNA extraction should be carried out using equipment (pipettes, microcentrifuge, etc.) which is independent from the PCR set-up. Optimally, this laboratory is separated from those rooms where the PCR reaction is set up or the PCR products are analyzed. All reagents should be stored in small aliquots to provide a constant source of uncontaminated reagents. New aliquot batches should be tested and compared for quality prior to any use. Reamplifications should never be carried out. If possible, the place of setting up the reactions should be a PCR working station or a hood capable of irradiating used pipettes, tips, and tubes by UV light. Finally, it is highly recommended to wear gloves during the whole procedure. Furthermore, it is also fundamental to integrate the appropriate positive and negative controls (e.g., HeLa DNA and H_2O, respectively).

3.2. Preparation of High Molecular Weight DNA

The principle of this assay is that cells are lysed during a short incubation time with proteinase K in the presence of a chaotropic salt (guanidinium-hydrochloride), which immediately inactivates all nucleases. Nucleic acids bind selectively to glass fibers prepacked in the filter tube. Bound genomic DNA is purified in a series of rapid washing and spinning steps to remove inhibiting cellular components. Finally, low salt elution releases the DNA from the glass fiber cushion.

1. Cell culture suspensions (see Notes 1 and 2) containing $3–5 \times 10^6$ diploid cells are centrifuged in an Eppendorf tube at $1,000 \times g$ for 4 min in a 14-mL tube. The supernatant is removed with a disposable pipette and discarded. The remaining pellet is carefully resuspended in 5 mL PBS using a pipette. Repeat centrifugation.

2. After the washing step, the pellet is resuspended in 200 µL PBS by vortexing. Make sure that even tiny clumps of cells are carefully resuspended. Prewarm the water bath to 72°C.

3. For isolation of the genomic DNA, the commercially available DNA extraction kit from Roche is applied. 200 µL of well-mixed solution I (guanidinium-hydrochloride) is added to the sample solution and mixed by careful pipetting.

4. Add immediately 40 µL proteinase K, mix well using a vortex and incubate at 72°C for at least 10 min.

5. Add 100 µL of isopropanol to the sample, mix well and apply the whole mixture to a filter tube. Centrifuge for 1 min at 6,000×g.

6. Discard the flow through, add 500 µL of inhibitor removal buffer and centrifuge again for 1 min at 6,000×g.

7. Discard the flow through, add 500 µL of wash buffer and centrifuge again for 1 min at 6,000×g.

8. Repeat step 8.

9. Add a new tube and 200 µL of elution buffer preheated to 72°C and centrifuge for 1 min at 6,000×g. For maximum yield the elution step should be repeated using 100 µL elution buffer.

10. Using a standard spectral photometer, genomic DNA should be adjusted to 10 ng/µL per sample of a diploid cell line.

11. Genomic DNA (see Note 3) should be stored at 4°C temperature.

3.3. Hot Start Nonaplex Fluorescence PCR DNA Typing

The multiplex PCR described here contains eight different STR loci and gender determination. The amplification procedure and the parameters are optimized for an application in 0.2-mL reaction tubes in an i-Cycler thermal cycler (Bio-Rad). It is fundamental to integrate the appropriate positive and negative controls (e.g. HeLa DNA template and H_2O, respectively). Prepare a premaster mix calculated for 25 µL per reaction of each sample, plus one additional reaction every ten samples. For a single reaction, the components are as follows:

(a) 10 pmol of all primer pairs (1 µL of a 10 µM working solution containing all primer pairs) (see Notes 4–6)

(b) 2.5 µL 10× Hot start PCR buffer (any supplier)

(c) 1 µL dNTP (5 µM stock solution)

(d) 0.2 µL (1 U) hot start Taq polymerase (any supplier)

(e) 19.5 µL distilled water.

Genomic DNA should be added after the mastermix has been placed into all reaction tubes or to 96-well reaction plates. Finally, 1 µL of genomic DNA adjusted from 0.2 to 1 ng/µL is added to the reaction and the tubes/plates are centrifuged for 4 min at 600×g.

The STR PCR program is as follows:

Cycle 1	95°C for 3 min, one repeat
Cycle 2	94°C for 30 s 57°C for 30 s 72°C for 45 s, 30 repeats
Cycle 3	60°C for 15 min, one repeat

The application of commercial multiplex STR kits should be carried out strictly following the specific manuals of Promega Corporation and Applied Biosystems. If not using commercial kits, it is important to take care to use general amplification parameters. A "hot start" PCR should be always performed to activate the Taq DNA Polymerase and to prevent the formation of nonspecific amplification products. The number of cycles depends on the amount of DNA: 30 cycles are recommended for all samples.

3.4. DNA Fragment Detection and Allelic STR Lists

Since DNA possesses a constant mass-to-charge ratio, some form of separation matrix is needed to separate different sizes of DNA fragments by their molecular weight. In traditional gel electrophoresis, the requirement for a sieving matrix is met with polyacrylamide or agarose gels. The movement of larger DNA fragments is impeded relative to that of the smaller DNA fragments as the molecules migrate through the gel under the influence of an electric field. Polyacrylamide gels are no longer the only slab gel systems available for resolving STR alleles. A recent publication was able to demonstrate that even small agarose gels could have sufficient resolving power to type tetranucleotide repeats. Even dinucleotide repeats could be resolved with MetaPhor agarose and detected with SYBR Green staining (16).

Various automated fluorescence detection systems have been used for separation, detection, and typing of STR alleles. Full automation of the electrophoresis process with no need to pour the gel or manually pipette the samples onto the gel is one of the big advantages of capillary electrophoresis (CE). With the higher sensitivity of laser-induced fluorescence, sample preparation is no longer necessary. Samples are diluted in water or formamide and can be easily detected. Separation of STR alleles may be performed in a matter of minutes rather than hours. Using the CE developed by Beckman–Coulter, aliquots of 1 µL of the amplification products are combined with 0.25 µL of an internal size standard (Size standard kit 400, Beckman–Coulter) in a total volume of 30 µL of sample loading solution in a microplate. The samples are automatically loaded and analyzed using fragment analysis parameters, which have been established by the binning procedure after 6 months of DNA typing of cell lines at DSMZ (see Table 1). The fragment analysis software of CEQ 8000 enables the precise determination of detected alleles resulting in a genotype summary list. Allelic numbers of STR DNA fragments may be precisely determined by other technique or by allelic ladders in the ABI system. Once the sizes are known, respective allele numbers can be deduced from Table 1.

3.5. Online Evaluation of Suspicious STR Profiles

A simple search engine for interrogating STR cell line profiles has been made available on the homepage of DSMZ. Once the problem of false negatives due to discrepant representation of single

Table 1
Allele organization and sizes of amplified human STR loci

Allele	D5S818	D13S317	D7S820	D16S539	vWA	TH01	TPOX	CSF1PO	Amelogenin
3						169			
4						173			209 = X
5		164	212	266		177	220	287	215 = Y
6	114	168	216	270		181	224	291	
7	118	172	220	274		185	228	295	
8	122	176	224	278		189	232	299	
9	126	180	228	282		193	236	303	
9.3						*196*			
10	130	184	232	286	118	197	240	307	
11	134	188	236	290	122	201	244	311	
12	138	192	240	294	126	205	248	315	
13	142	196	244	298	130		252	319	
14	146	200	248	302	134		256	323	
15	150	204	252	306	138			327	
16	154				142			331	
17	158				146				
18					150				
19					154				
20					158				
21					162				
22					166				
23					170				

The table summarizes the nucleotide range and the number of known alleles of each STR loci (http://www.cstl.nist.gov/div831/strbase). Regular fragment sizes in base pairs of alleles are printed in bold, variant alleles are printed in italics

STR alleles – e.g., by losses of heterozygosity and bottlenecking selection – has been tackled and unambiguous search results are produced, human cell lines will need to be consistent with consensus STR reference data sets. STR profiles of all human cell lines distributed by DSMZ, JCRB, and RIKEN and one-third of the cell lines distributed by ATCC are now publicly accessible at http://www.dsmz.de/STRanalysis using an interactive database where match criteria have been arbitrarily set to 60%. Registered users simply login at the online STR analysis site on the DSMZ homepage and will be guided. Aided by simple prompts, users can input their own cell line STR data to retrieve best matches with authenticated cell lines listed on the database. Inevitably, reference profiles remain subject to revision until all commonly held cell lines have been STR typed across participating repositories. At present, about 2,342 such cell lines have been STR typed and are represented as reference sets on the database. Armed with this tool, online verification of cell line identity should prove a vital weapon to combat the havoc of cell line cross-contamination which has dogged cancer research since inception.

4. Notes

1. Prior to the isolation of genomic DNA, cell viability of samples should be analyzed using trypan blue exclusion assay. With the exception of factor-dependent hematopoietic cell lines, viability of a cultured cell line should not be below 85% at sampling.

2. To avoid protraction of PCR inhibitors from biological samples, it is of high importance to carry out a sophisticated resuspension the cell pellet prior to DNA isolation in PBS.

3. Quality control of isolated genomic DNA should be carried out by using 1% agarose gel electrophoresis. 300 ng of high molecular weight DNA should result in a single band, while DNA of apoptotic cells would show the specific DNA laddering.

4. If an individual constellation of genomic loci is of interest, the dye-labeled primer pairs should be tested in single reactions and the primer mixture adjusted for the generation of similar peak heights of the measured loci.

5. Working solutions of primers are stable for 1 month at 4°C temperature und should be kept in lightproof reaction tubes.

6. Degradation of primers will result in unspecific bands. Two bands of a diploid cell line should be detected within the sizes of each genomic locus as presented in Table 1.

References

1. Rothfels, F.H., Axelrad, A.A. and Simonovitch, L. (1959) The origin of altered cell lines from mouse, monkey, and man as indicated by chromosomes and transplantation studies. *Proc. Can. Cancer Res. Conf.*, **3**, 189–214.

2. Simpson, W.F. and Stuhlberg, C.S. (1963) Species identification of animal cell strains by immunofluorescence. *Nature*, **199**, 616–7.

3. Gartler, S.M. (1968) Apparent HeLa contamination of human heterodiploid cell lines. *Nature*, **217**, 750–751.

4. Miller, O.J., Miller, D.A. and Allerdice, P.W. (1971) Quinacrine fluorescent karyotypes of human diploid and heteroploid cell lines. *Cytogenetics*, **10**, 338–41.
5. Nelson-Rees, W.A. and Flandermeyer, R.R. (1976) HeLa cultures defined. *Science*, **191**, 96–8.
6. Nelson-Rees, W.A. (1978). The identification and monitoring of cell line specificity. *Prog Clin Biol Res*, **26**, 25–79.
7. MacLeod, R.A.F., Dirks, W.G., Matsuo, Y., Kaufmann, M., Milch, H., and Drexler, H. G. (1999) Widespread intraspecies cross-contamination of human tumor cell lines arising at source. *Int J Cancer*, **83**, 555–63.
8. Drexler, H.G., Dirks, W.G., and MacLeod, R.A.F. (1999) False human hematopoietic cell lines: cross-contaminations and misinterpretations. *Leukemia*, **13**, 1601–7.
9. Dirks, W.G., MacLeod, R.A., Jaeger, K., Milch, H., and Drexler, H.G. (1999) First searchable database for DNA profiles of human cell lines: sequential use of fingerprint techniques for authentication. *Cell Mol Biol*, **5**, 841–53.
10. Epplen, J.T., McCarrey, J.R., Sutou, S. and Ohno, S. (1982) Base sequence of a cloned snake W chromosome DNA fragment and identification of a male-specific putative mRNA in the mouse *Proc Natl Acad Sci USA*, **79**, 3798–802.
11. Jeffreys, A.J., Wilson, V. and Thein, S.L. (1985) Hypervariable minisatellite regions in human DNA. *Nature*, **314**, 67–73.
12. Nakamura, Y., Leppert, M., O'Connell, P., Wolff, R., Holm, T., Culver, M., Martin, C., Fujimoto, E., Hoff, M., Kumlin, E. and White, R. (1987) Variable number of tandem repeat (VNTR) markers for human gene mapping. *Science*, **235**, 1616–22.
13. Master, J. R., Thompson, J. A., Daly-Burns, B., Reid, Y. A., Dirks, W. G., Packer P., Toji, L. H., Ohno, T., Tanabe, H., Arlett, C. F., Kelland, L. R., Harrison, M., Virmani, A., Ward, T. H., Ayres, K. L., and Debenham, P. G. (2001) Short tandem repeat profiling provides an international reference standard for human cell lines. *Proc Natl Acad Sci USA*, **98**, 8012–7.
14. Nakahori, Y., Takenaka, O., and Nakagome, Y. (1991) A human X-Y homologous region encodes amelogenin. *Genomics*, **9**, 264–9.
15. Sullivan, K. M., Mannucci, A., Kimpton, C. P., and Gill, P. (1993) A rapid and quantitative DNA sex test: fluorescence-based PCR analysis of X-Y homologous gene amelogenin. *Biotechniques*, **15**, 636–41.
16. White HW, Kusukawa N. (1997) Agarose-based system for separation of short tandem repeat loci. *Biotechniques*, **22**, 976–80.

Chapter 6

Cytogenetic Analysis of Cancer Cell Lines

Roderick A.F. MacLeod, Maren Kaufmann, and Hans G. Drexler

Abstract

Cancer genes are often deregulated by genomic rearrangements. Accordingly, analysis of the participant chromosomes responsible now occupies a key role in characterizing and identifying cancer cell lines. Cytogenetics may also be used to study the nature and extent of chromosome breakage induced by radiation or chemicals ("clastogenesis"), to distinguish individual cells or clones within a tumor cell population and to monitor the stability of chromosome rearrangements. This chapter describes cytogenetic procedures for characterizing cancer cells in culture. Cell lines allow the use of a wider range of harvesting and hypotonic treatments to optimize metaphase chromosome preparations than that possible with primary cultures. This assists improved banding, fluorescence in situ hybridization (FISH), and Spectral Karyotyping (SKY) analysis for research, rendering cell lines ideal tools for oncogenomics, ideally in parallel with transcriptomic analysis of the same cells. The experience of the writers with more than 800 cell lines has shown that no single hypotonic harvesting protocol is adequate consistently to deliver satisfactory chromosome preparations. Thus, evidence-based protocols are described for hypotonic harvesting, rapid G-banding, and FISH and SKY analysis of cell cultures to allow troubleshooting and fine-tuning to suit the requirements of individual cell lines.

Key words: Cytogenetics, Chromosome, Hypotonic treatment, FISH, G-banding, SKY

1. Introduction

1.1. Background: The Utility of Cytogenetic Characterization

Numerous cancer cell lines have been established since HeLa in 1951. From human leukemia and lymphoma alone, probably the best documented group, more than 1,500 examples are known, and this does not include sister cell lines and subclones. The risk of cross-contamination and the difficulty of identifying cell lines are correspondingly high (1). Cytogenetic analysis has become a core element for identifying and characterizing cancer cell lines, mainly because it affords a rapid means of checking the species of origin, and for the unique key it provides to classifying cancer cells based on their distinct patterns of recurring chromosome rearrangements in both hematopoietic and, increasingly, in solid tumors as well (2). Recurrent chromosome changes provide a portal to mining underlying mutations at the DNA level in cancer, and cell lines are rich territory for characterizing them, and for testing new drugs to combat these mutations. This has led to the development of "intelligent" therapeutic approaches, which target specific cancer causing mutations such as the tyrosine kinase inhibitor imatinib (Gleevec), which inhibits the *BCR–ABL* fusion protein in chronic myeloid leukemia (CML). Imatinib, now used widely to treat CML and certain solid tumors with analogous mutations, including gastrointestinal stromal tumors, was initially developed using cell lines sourced from this institute (3).

Cancer changes reflect developmentally programmed patterns of gene expression and responsiveness within diverse cell lineages where dysregulation of certain genes facilitates evasion of existing antineoplastic controls, including those mediated by cell-cycle checkpoints or apoptosis. The tendency of cells to produce neoplastic mutations via chromosomal mechanisms, principally translocations, duplications, and deletions (http://atlasgeneticsoncology.org/, http://www.sanger.ac.uk/genetics/CGP/Census/), renders these changes microscopically visible, facilitating cancer diagnosis by chromosome analysis. Among the different types of alterations promoting cancer, those affecting chromosomal structure, notably recurrent chromosome translocations leading to gene fusion or upregulation upon which the neoplastic cell proliferation ultimately depends, combine the greatest informational content with the greatest stability. For example, cell lines established from CML patients with t(9;22)(q34;q11) causing fusion of *BCR* with *ABL* – the primary oncogenomic change in CML – retain this translocation in vitro (4). Although the usefulness of karyotype analysis for the characterization of cancer cell lines lies principally among those derived from tumors with stronger associations with specific chromosome rearrangements, e.g., hematopoietic (5), mesenchymal, and neuronal (6), an increasing variety of epithelial tumors have also been found to carry recurrent translocations (2).

Microscopic methods, such as cytogenetic analysis, require observations performed at the single-cell level, thus allowing detection of intercellular differences. Hence, a second virtue of cytogenetic data lies in the detection of distinct subclones and the monitoring of stability therein. With the exception of doublings in their modal chromosome number from $2n$ to $4n$ ("tetraploidization"), cell lines appear to be rather more stable than is commonly supposed (6–9). Indeed, the most intense phase of chromosomal rearrangement occurs in vivo, namely, physiological receptor gene rearrangement in T- and B-lineage lymphocytes (10, 11).

Cell lines also provide useful materials for studying clastogenesis. The role of chromosome breakage, notably in cancer, is currently a major research topic. Chromosome fragility occurs nonrandomly, clustered at so-called "fragile sites". However, to visualize these sites it is often necessary to culture cells in specific conditions conducive to their expression, procedures which are facilitated by use of cell lines. Our own data suggest a connection between sites of chromosome instability and loci bearing micro-RNA genes (9).

A further application of cytogenetic data is to identify and monitor cell line identities. At least 18% of new human tumor cell lines have been cross-contaminated by older, mainly "classic," cell lines, which tend to be those most widely circulated (12, 13). This problem, first publicized over 30 years ago (14) but neglected of late (15, 16), poses an insidious threat (17). While concordant DNA profiles of cell line and donor should be documented at the time of first publication, this ideal is still honored mainly in the breach. Thus, users wishing to check the identities of many, if not most, tumor cell lines in use are forced to relinquish DNA profiling for cytogenetics. Fortunately, karyotypes of many human tumor cell lines are published enabling suitably equipped users to check the identities of their own cultures by cytogenetic analysis. In addition, "quick and dirty" cytogenetic testing provides a rapid way to check the species of origin most commonly encountered among cell lines, such as mouse, rat, chinese or golden hamster, while experienced operators may be able to identify more exotic species.

1.2. Cytogenetic Methodology

In biology, few methods have proved more durable than cytogenetics. Over the last four decades, or so, tumor cytogenetics remained in the forefront thanks to a series of advances, both technical and informational. By revealing latent striations using a variety of staining procedures, "chromosome banding", it first became routine to distinguish and identify each of the 24 different human chromosomes (referred to as numbers 1–22, X, and Y) in the early 1970s. The original standard methods were principally Q(uinacrine)-banding (18), and G(iemsa)-banding (19). A further modification, trypsin G-banding (20), has gained the widest currency since its introduction in 1973 because of its speed and simplicity. Banding techniques were instrumental in the identification of the "Philadelphia chromosome" (Ph) marker and its origin via a reciprocal translocation, t(9;22)(q34;q11) (21), a mechanism not guessed when the Ph was first observed as an insignificant dot-like chromosome present in unbanded bone marrow chromosome preparations of CML patients more than a decade earlier (22). This observation ushered in the realization that cancer is caused by somatic gene alterations.

Simultaneous improvements in speed, sensitivity, and accuracy followed in the wake of computer-aided image analysis in the early 1990s, which enabled complex tumor karyotypes to be at last confronted, if not confounded. However, the resolution of tumor karyotypes was even then still often hampered by the presence of complex marker chromosomes yielding abnormal banding patterns.

The advent of fluorescence in situ hybridization (FISH) during the late 1980s and 1990s (23, 24) represented the next advance in cytogenetics. FISH exploits the stability and specificity of DNA–DNA hybrids formed after exposure of chromosomes to homologous DNA under renaturating conditions. FISH required the availability of nonisotopically labeled deoxynucleotides combined with a straightforward method for their efficient incorporation into DNA by nick translation. This need prompted the commercial development of chromosome library ("painting") probes specific for each of the 24 different human chromosomes. Pairwise combinations of painting probes when contrastingly labeled – usually red versus green with the remaining chromosomes counterstained in blue – may be used to resolve chromosome translocations. However, the components of complex "marker" karyotypes are by definition unknown, demanding time and resource-hungry approaches based on trial-and-error. This Gordian knot has now been cut following the availability of multicolor probe mixtures enabling each chromosome to be distinguished (reviewed in (25)). Analysis requires short pass chromatic visualization systems, either filter-based (e.g., from Cytovision or Metasystems), or spectrophotometric (namely, SKY from Applied Spectral Imaging).

Thanks to the technical and informational resources afforded the research community in the guise of "tilepath" bacterial artificial chromosome (BAC), or fosmid, clones which have been sequenced and positioned onto the human gene map, it is now possible to locate chromosome breakpoints in cancer cell lines with respect to potentially relevant genomic features, such as genes or regulatory regions. In this way, FISH serves to bridge the gap between classical cytogenetics and molecular biology.

In this chapter, we describe basic cytogenetic procedures which have been adapted in our laboratory for use with cell cultures. For those planning de novo cytogenetic analysis of tumor cell lines, it is convenient to split the task into the following steps: harvesting (see Subheadings 2.1. and 3.1), G-banding (see Subheadings 2.2, 2.3, and 3.2), and FISH (see Subheadings 2.3, 2.4, and 3.3).

2. Materials

Unless otherwise indicated, reagents may be stored up to 4 weeks at 4°C.

2.1. Harvesting

1. Cell culture(s) maintained in logarithmic phase.
2. *N*-Deacetyl-*N*-methylcolchicine (colcemid) 100× solution (Invitrogen): 4 µg/mL stock solution; store refrigerated for up to 1 year.

3. FUDR/uridine 100× stock solution. Mix one part 5-fluoro-2′-deoxyuridine (FUDR) (Sigma) (25 μg/mL) and three parts 1-β-d-Ribofuranosyluracil (uridine) (Sigma; 1 mg/mL); store refrigerated for up to 1 year.

4. Thymidine 100× stock solution: 1-(2-deoxy-β-d-ribofuranosyl)-5-methyluracil (thymidine) (Sigma). Dissolve 50 mg in 100 mL autoclaved TE buffer (10 mM Tris–HCl pH 7.5; 1 mM EDTA). Filter-sterilize through 0.22-μm filter.

5. Trypsin 0.5 g/L–EDTA 0.2 g/L (Invitrogen) for removal and dispersal of adherent cells; store at (−20°C) for up to 6 months.

6. Stock hypotonic solutions: KCl 5.59 g/L; or Na citrate 9.0 g/L. Working hypotonic solutions: mix KCl and Na citrate (e.g., 20:1, 10:1, 1:1, 1:10, 1:20, etc.) shortly before use, allowing time to reach desired temperature.

7. Fixative. Mix absolute methanol and glacial acetic acid at 3:1. Use fresh but can be stored up to 4 h at 4°C.

2.2. G-Banding Only

1. Slides (frosted ends for annotation). Wash mechanically overnight in warm ion-free detergent; rinse twice in deionized water; oven-dry, and leave overnight in ethanol (70%). Slides should then be polished using a lint-free cloth (or nonshredding tissue) and stored wrapped in aluminum foil at −20°C until use.

2. Phosphate-buffered saline (PBS): adjusted to pH 6.8 (Giemsa solution) or pH 7.2.

3. Trypsin stock solution (140×): dissolve 17.5 mg trypsin 1:250 (Difco) in PBS (pH 6.8). Store 500-μL aliquots at (−20°C) for up to 6 months.

4. Giemsa stain (1.09204.0500 Merck). Dissolve 5 mL in 100 mL PBS (pH 7.2) and filter before use.

5. Routine microscope with phase-contrast illuminator and the following objectives: ×10 (phase contrast), ×40 (phase contrast), and ×50 (brightfield dry) for slide evaluation and preliminary analysis.

2.3. G-Banding and FISH

1. Image analysis system for G-banding and FISH (see Note 1).

2. Laboratory oven for slide aging (G-banding) or slide drying (FISH).

3. Coplin jars, 100 mL (glass), for staining and washing.

4. 4× SSC: 35.1 g NaCl, 17.7 g Na citrate made up to 1 L. Adjust to pH 7.2.

5. 0.5× SSC, 2× SSC, and so forth: dilute from 4× SSC stock but monitor pH.

2.4. FISH Only

1. Ethanol: absolute, 90%, 70%. Can be used twice, thereafter discarded.
2. Pepsin stock solution: dissolve 250 mg pepsin (Sigma P7012) in 12.5 mL of deionized H_2O. Freeze 500-µL aliquots (–20°C) and store for up to 6 months.
3. Pepsin working solution: Dilute 500-µL stock solution in 100 mL deionized H_2O containing 1 mL of 1 N HCl; store at (–20°C) for up to 6 months.
4. Formaldehyde solution: 1% formaldehyde in PBS (pH 7.2) containing 50 mM $MgCl_2$.
5. Acetone, for use in mild pretreatment.
6. Hybridization buffer ("hybrisol"): Mix 5 mL deionized formamide (GenomeLab sample loading solution, Beckman–Coulter, Fullerton, CA), 1 ml 40% dextran sulfate, and 4 ml 250 mM Na_2HPO_4 in 5× SSC. (In our experience, commercial hybridization buffers are not entirely reliable because they may contain impure formamides capable of crippling fluorescent probe signals.). Store at room temperature (contains formamide).
7. Cold competitor DNA for prehybridization with probes containing repeat sequences: Cot-1 DNA, 1 µg/µL (Roche); store at (–20°C).
8. Nail varnish (clear).
9. Rubber cement.
10. Hybridization chamber: sealed container with an internal shelf to separate slides (above) from humidifier (e.g., water-impregnated towels (below)). Lidded stainless-steel instrument sterilization trays make admirable hybridization chambers, being readily sterilizable and both rustproof and heat resistant.
11. Hybridization bed: prewarmed freezer block kept in incubator at 37°C; use during application of probes to slides.
12. Wash solution: 4× SSC with 0.1% Tween-20, molecular biology grade (Sigma). Slides can be dipped into wash solution between any steps to prevent drying out.
13. Plastic coverslips for probe detection (Qbiogene).
14. Mounting medium: Dissolve 50 ng/mL 4′, 6-diamidino-2-phenylindole dihydrochloride (DAPI) in Vectashield antifade mounting medium (Alexis).
15. Coverslips: glass, grade 0.22 and 60 mm.
16. Chromosome painting probes: store at (–20°C) unless otherwise stated (see Note 2).
17. Research microscope with the following objectives with as large numerical apertures as budgets permit: ×10 (phase contrast for evaluating unstained preparations), ×50 Epiplan,

brightfield dry (for evaluating Giemsa-stained preparations), ×63 Planapochromat (oil), or equivalents from other manufacturers. We can specially recommend the ×100 Zeiss Apochromat (with 1.46 numerical aperture) oil objective; this is equally useful for both brightfield and FISH work. Ideally, a cytogenetics research microscope should be equipped with a motorized filter wheel and configured to an appropriate FISH imaging system (see Note 1).

3. Methods

3.1. Harvesting and Slide Preparation (See Note 3)

Mammalian cells in continuous culture typically divide every 1–3 days. The metaphase stage of mitosis, the only cell cycle stage when chromosomes are clearly visible, usually lasts less than 1 h, severely reducing the number of cells available for conventional cytogenetic analysis. Accordingly, the fraction of dividing cells must be enriched by exposure of growing cultures to colcemid or some other mitotic blocking agent for a few hours, or longer in the case of slow-growing cells. It is therefore important to ensure that cell cultures are in their logarithmic growth phase by feeding and, if necessary, diluting/seeding out. Neglect of this simple precaution is an all-too-common cause of failed harvests. It is difficult to overstate just how crucial initial harvesting and slide preparation is to subsequent success with both G-banding (see Subheading 3.2) and FISH (see Subheading 3.3). Cytogenetics textbooks often list harvesting protocols where hypotonic treatment is limited to the standard incubation (in 0.075 M KCl at ambient temperature for 7 min) with little discussion of possible options. In our experience, choice of hypotonic treatment is the main key to successful harvesting cancer cell lines for cytogenetic analysis (26). We find that hypotonic treatments that consistently yield good preparations with one cell line are often totally unsuitable for another of similar origin. It is therefore necessary to determine empirically which harvesting procedure is optimal for each cell line. This may be achieved by harvesting, in parallel, cell aliquots exposed to a range of hypotonic conditions (namely, with a variety of different buffers and incubation times and, if need be, incubation temperatures, etc. See Table 1 for an actual worked example).

Unlike hypotonic treatment, fixation permits standardization. Although some deterioration occurs, fixed cells can be stored for several years at −20°C until required. Immediately prior to slide-making, cell suspensions should be washed in fresh fixative. Slide-making is performed by dropping cell suspension onto ice-cold, precleaned slides held at a slight angle (about 1-in-10 to 1-in-20) on top of a prefrozen (−20°C) freezer cold block. Two drops aimed at the slide region immediately under

Table 1
Harvesting record sheet for RC-K8 (DSMZ ACC 561)

Date	Harvest no.	Colcemid time	Tube	KCl	Na Citrate	Other	Temp	Min	Q	Spr	Morph	Total	Untr	60°C GTG	−80°C FISH	−20°C Susp
	1a	3 h	10 mL	+	−	−	RT	7	B	AA	B	−	−	−	−	Reserve only
	1b			−	+	−		7	A	A	C	−	−	−	−	−
	1c			+	−	−	37°C	7	C	C	C	−	−	−	−	−
	1d			−	+	−		7	B	B	C	−	−	−	−	−

G-banding: *unsuitable*   Repeat: *yes*   Action: *discard; try KCl:Na Cit 20:1 and 1:1*

Date	Harvest no.	Colcemid time	Tube	KCl	Na Citrate	Other	Temp	Min	Q	Spr	Morph	Total	Untr	60°C GTG	−80°C FISH	−20°C Susp
	2a	3 h	10 ml	20	1	−	RT	7	A	AA	B	−	−	−	−	−
	2b			1	1	−		7	A	A	AB	16	1	8	7	Yes 2 ml
	2c			20	1	−		1	A	A	AB	−	−	−	−	−
	2d			1	1	−		1	A	AB	AB	−	−	−	−	−

G-banding: *yes, was OK.*   Repeat: *no*   Action: *mix tubes 2b and 2c discard rest*

Abbreviations: *Q* Quantity of metaphases is defined as follows "A" ≥1 metaphase per low power (~100×) microscope field; "B" ≥1 metaphase per ten low power fields; "C" ≤1 metaphase per row, *Spr* Spreading is defined as: "A", optimal with all or most chromosomes separately visible; "AA" (possibly usable for FISH), as "A", but mostly broken; "B" (usable), with most metaphases showing crossed-over chromosomes; and "C" (unusable), with no chromosomes separately visible, *Morph* Morphology: "A" (good), with parallel, solid, clearly separated chromatids; "B" (average); and "C" (poor) with amorphous or refractile chromatids when viewed under phase-contrast. Intermediate quantities and qualities are defined by "AB", "BC", etc. *Other abbreviations*: *GTG* G-banding, *Temp* Temperature, *Susp* Cell suspension, *Untr* Untreated. In the case of RC-K8, although the first harvest was discarded, it provided information to direct the choice of hypotonic buffers in second harvest towards a more satisfactory conclusion the frosted zone and at the lower middle, respectively, should result in figure-of-eight spreading patterns suitable for both G-banding and FISH. Once made, slides can be variously stored for a few years at −80°C, for short intervals at room temperature for FISH, or aged overnight at 60°C for G-banding.

Day 1

1. Add colcemid (final concentration 40 ng/mL) to growing cultures for 2–4 h.
2. As an alternative to colcemid treatment, incubate cells overnight with FUDR to improve chromosome morphology (see Note 4).

Day 2

3. Suspension cell cultures: aliquot cells (e.g., four times in 10-mL tubes), centrifuge (5 min at 400 g), and discard supernatant.
4. Adherent cell cultures: Shake vigorously to remove mitoses and retain supernatant in centrifuge tube (50 mL). Meanwhile, rinse remaining adherent cells with serum-free medium or PBS and discard wash. Add sufficient trypsin/EDTA to cover the cells and incubate briefly (5–15 min) with intermittent light agitation. When cells are ready (i.e., "rounded up"), shake vigorously and remove by rinsing with supernatant from the centrifuge tube. Then, centrifuge aliquots as with suspension cultures. (The serum present in the culture medium will act to inactivate residual trypsin activity.)
5. Resuspend cell pellets gently by manual agitation. Add 5–20 vol. from various working hypotonic solutions (20:1, 1:1, etc.). Incubate paired aliquots at (initially) room temperature for 1 min and 7 min, respectively (see Table 1 for example).
6. Centrifuge and discard supernatant. Resuspend cells gently and carefully add ice-cold fixative, at first dropwise, and then faster, until the tube is full.
7. Store refrigerated for 1–2 h.
8. Equilibrate to room temperature to minimize clumping, then centrifuge (5 min at 400 g). Repeat.
9. Store fixed cells overnight at 4°C.

Day 3

10. Next day, equilibrate to ambient temperature, then centrifuge (5 min at 400 g). Repeat twice.
11. Resuspend cells in sufficient fixative to yield a lightly opaque suspension. Typical cell concentrations range from 2 to 8 million cells per milliliter.

12. Remove four precleaned slides (one per harvest tube) from storage at (–20°C) and place on a plastic-covered freezer block held at a slight incline away from the operator by insertion of a pipette.
13. Locally humidify by breathing heavily over the slides.
14. Holding the pipette approx 30 cm above the slides, place two drops of cell suspension onto each slide, the first immediately below the frosted zone and the second about two-thirds along the slide. Do not flood.
15. Lift slides in pairs for speed. Breathe over them again to maximize spreading.
16. (Optional) To improve spreading, gently ignite residual fixative (with a camping stove or Bunsen burner). Do not allow slide to get hot, as this could spoil subsequent G-banding and FISH.
17. Label and air-dry. Stand slides vertically until dry.
18. Examine slides by phase-contrast microscopy and assess each hypotonic treatment individually (see Note 3).
19. Prepare slides from successful treatments, mixing cell suspensions if more than one is deemed adequate. Label.
20. Store unused cell suspensions at –20°C in sealed 2-mL microfuge tubes filled to the brim to exclude air. Under such conditions, suspensions remain stable for several years; we have performed FISH successfully using 5-year-old suspensions. Suspensions cryopreserved in this way must be thoroughly washed in fresh fixative prior to slide preparation. After sampling, suspensions should be refilled to the brim, marking the original level to control dilution.

3.2. Trypsin G-Banding (See Note 5)

Although several banding methods are in use, the standard procedure involves G-banding by trypsin pretreatment (20). G-Banding selectively depletes the chromatin of certain proteins to produce strong lateral bands after staining with Giemsa (Fig. 1a). Analysis of chromosomes harvested using the above-described technique should typically reveal some 300 bands, the absolute minimum required for detecting noncryptic rearrangements. However, with stretched or submaximally condensed (prometaphase) chromosome preparations up to 1,000 bands may be distinguished.

1. Fresh slides are unsuitable for immediate G-banding. Slides must be aged first. This is best achieved by baking overnight at 60°C in a dry oven. About six to eight slides containing an adequate supply of well-spread metaphases with good chromosome morphology should be prepared for each cell line.

2. First prepare three Coplin jars, one each for 500 µL trypsin in 70 mL PBS (pH 7.2), ice-cold PBS (pH 6.8) to stop enzymatic activity, and 5% Giemsa in PBS (pH 6.8).

3. A Coplin jar containing trypsin in PBS should be placed in a water bath at 37°C and equilibrated to 37°C before use.

4. A second Coplin jar containing PBS alone should be placed on ice nearby and allowed to equilibrate to about 4°C before use.

5. To estimate optimal trypsin incubation times, dip the first slide halfway into the trypsin for 10 s and, thereafter, the whole slide for another 10 s to test, in this case, for 10 s and 20 s trypsinization times, respectively.

6. Immediately, stop trypsin activity by immersion in ice-cold PBS for a few seconds.

7. Stain in Giemsa solution for 15 min.

8. Rinse briefly in deionized water and carefully blot-dry using paper towels.

9. Examine microscopically (see Note 6). Scan for likely metaphases at low power. Examine in more detail those selected at higher power using the Epiplan dry objective. From the chromosome banding quality, decide whether the suitable trypsin time lies within the 10–20 s range spanned by the test slide. If satisfactory, repeat steps 1–8. If unsatisfactory, repeat steps 1–8 using longer (e.g., 30–45 s) or shorter (e.g., 3–6 s) trypsin test times, as appropriate until the optimal incubation time becomes apparent.

3.3. FISH (See Notes 7 and 8)

Chromosome painting describes FISH using mixtures of DNA covering an entire single chromosome ("whole chromosome probe"). Painting probes may used singly or in color-contrasted mixtures – the latter maximizing the informational possibilities, e.g., by confirming a translocation suspected after G-banding. Hybridization with painting probes for chromosomes 3 and 7 is shown in Fig. 1b to illustrate a chromosome rearrangement in B-cell lymphoma. Whichever probe combination is adopted, it is usually necessary to counterstain the chromosomes. The standard counterstain is 6-diamidino-2-phenylindole dihydrochloride (DAPI), which yields deep blue color, more intense at the centromeres, in particular those of chromosomes 1, 9, and 16, and in the terminal long-arm region of the Y chromosome. In better preparations, DAPI generates negative G-bands that most image analysis programs can convert into G-bands. Painting probes can be produced by selective PCR amplification of human chromosomal material retained by monochromosomal human/rodent hybrid cell lines. By exploiting human-specific repeat sequences (e.g., Alu) as primer targets, it is possible to amplify human DNA selectively. Such probes inevitably include significant amounts of human repeat DNA hybridizing indiscriminately across the genome,

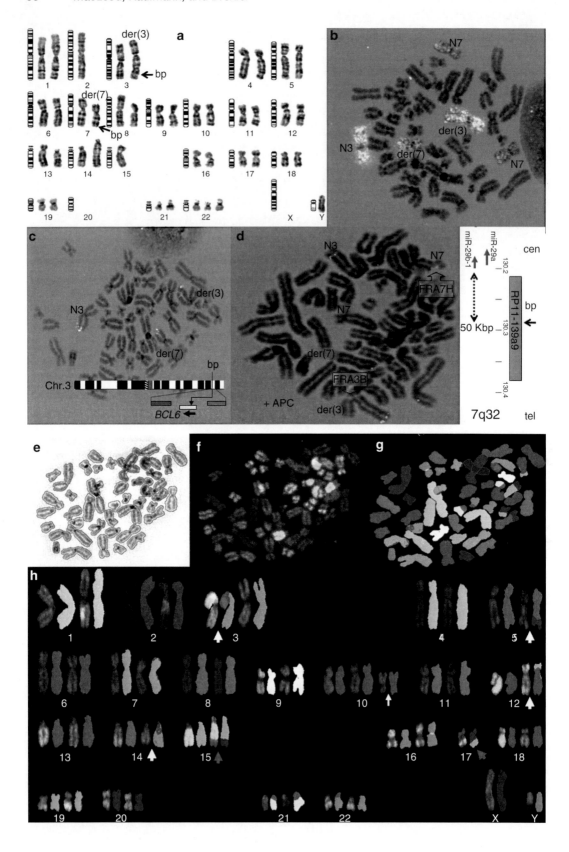

which must be suppressed. This is achieved by preincubating probe material together with unlabeled ("cold") human DNA enriched for repetitive sequences by a two-step denaturation–renaturation process. During renaturation, the most highly repetitive sequences (Cot-1 DNA) are the first to reanneal, allowing more complex, slower reannealing DNA to be digested away using single-strand-specific DNase-1. For this reason, most commercial painting probes include Cot-1 DNA.

Single-locus probes can be produced by labeling large-insert clones obtained from genomic resource centers, notably BACPAC Resources (http://bacpac.chori.org/) which carries complete

Fig. 1. Cytogenetic analysis of cancer cell lines. (**a**)–(**c**): Analysis of a cell line (RC-K8) established in 1984 established from a patient with diffuse large B-cell lymphoma (DLBCL). (**a**) G-banded karyotype: Metaphase cells were prepared from RC-K8 cells as described in Subheading 3.1 using hypotonic treatment specified in Table 1, and slide preparations aged overnight at 60°C for G-banding performed as described in Subheading 3.2. Metaphase cells were analyzed and the chromosomes arranged to form the karyogram using a Quips image analysis system (Applied Imaging) configured to an Axioplan photomicroscope using a x63 Planapochromat objective (Zeiss). The consensus karyotype was found to be: 47,XY,der(Y)t(Y;5)(qter;?p15),der(?1)t(1;13)(p32;q11),del(1)(q11),del(2)(p11p13), dup(2)(p11–12p14–15), t(3;7)(q27;q32), der(5)inv(5)(q15q32)add(5)(q32), inv(5)(p?13q?32), der(8)t(8;8)(p12;q11), del(9)(p24)x2, der(10)t(9;10)(?p24;p11), -13, t(11;14)(q23;q32), der(14)t(X;14)(p11;p11), -15, der(15)t(X;15)(p11.3;p11), add(17)(p11), der(20)t(13;20)(q11;q13), +2mar. In this case, G-banding revealed the presence of a candidate primary change known to be recurrent in DLBCL, a translocation involving a breakpoint at 3q27 present as der(3) (*arrowhead*). Such translocations target deregulation of a gene (*BCL6*) that plays a central role in B-cell development. A putative der(7) partner chromosome with a breakpoint at 7q32 was also identified (*arrow*). (**b**) The t(3;7)(q27;q32) rearrangement was confirmed by FISH using whole chromosome painting probes for chromosomes 3 and 7. Note presence of both der(3) and der(7) showing the rearrangement to be reciprocal, together with two normal copies of chromosomes 7. (**c**) Characterization of the t(3;7) by FISH was performed using three RP11 library BAC clones: 88p6 (centromeric of *BCL6*), 211g3 (*Spectrum gold* and which straddles *BCL6*), and 632m3 (*Spectrum green* lying telomeric of *BCL6*). The results clearly show that the t(3;7) has split the probe at *BCL6*. Recurrent 3q27 breakpoint translocations in DLBCL typically deregulate *BCL6* by promoter exchange. (**d**) In t(3;7), the partner sequence at 7q32 lies outwith any known gene coding sequence (9), and while BCL6 is also dysregulated, precisely how remains unclear as canonical mRNA fusion is absent (9). The 7q32 breakpoint lies within a common chromosomal fragile site, FRA7H, expressed in RC-K8 cells treated with aphidicolin (APC) and located about 50 kbp telomeric of the miR-29 cluster (*inset*). Interestingly, after APC treatment RC-K8 cells preferentially express FRA7H in normal homologues (e.g., see Fig. 1d), implying greater stability therein. Note additional FRA3B present in the der(3); FRA3B is generally regarded the most common fragile site in man. This image also shows the utility of cell lines for those wishing to investigate chromosome fragility and its relation to cancer. FISH images were captured using a cooled CCD camera configured to a Smart Capture imaging system (Applied Imaging). Color images were captured separately, and the contrast-enhanced images merged. (**e**)–(**h**) Images shows spectral karyotyping (SKY) of an acute myeloid leukemia (AML) cell line (AP-1060) established in 1998 from the bone marrow of a 45-year-old man with acute promyelocytic leukemia (AML FAB M3) at fourth relapse. (**e**) Pseudo-G-banded metaphase produced by reversal of DAPI fluorescent image. (**f**) The same metaphase visualized by SKY; note suboptimal contrast due to imperfect color reproduction. (**g**) SKY pseudocolored metaphase; note improved discrimination; (**h**) Full SKY karyotype: original images on the *left*, and pseudo-colored on the *right*. Rearrangements are indicated by *arrows*. Note exchange between chromosomes 15 and 17 (*red arrows*) due to formation of t(15;17)(q22;q11.2), which fuses two genes, *PML* and *RARA*, leading to formation of a hybrid *PML-RARA* gene, which yields hybrid protein characterizing this type of leukemia. The SKY system uses a complex probe mixture where different fluors are mixed in 24-chromosome specific combinations. Special software assists portrayal on-screen by pseudocoloring. All images were visualized using Zeiss photomicroscopes configured to the aforementioned image analysis systems. (For cell line details see refs. 32, 33) *Abbreviations*: *apc* aphidicolin, *bp* breakpoint, *der* derivative chromosome, *N* Normal homologue.

tilepath clones for humans and model organisms. Our current labeling protocol is given elsewhere (26). Alternately, labeled probes for common cancer gene loci are available commercially for a variety of neoplastic loci: FISH analyses using probes covering the BCL6 locus at chromosome 3 band q27 and a noncoding fragile site region (FRA7H) on chromosome 7 band q32 are depicted in Fig. 1c, d. These and similar probes are produced by harvesting DNA from mapped bacterial artificial chromosome (BAC) clones made available in the wake of the Human Genome Project and identified on genome browsers, such as that hosted by the University of California at Santa Cruz (http://genome.ucsc.edu/). BAC clones used to prepare such probes contain repeat sequences that require suppression by prehybridization with Cot-1 DNA. The posthybridization stringency wash, which can be performed at either low temperatures including formamide, which lowers the stability of the DNA double helix, or at higher temperatures using low SSC concentrations alone, is critical to success. Stringency washing allows the operator to control the balance of probe signal intensity against background. The stability of DNA–DNA hybrids on FISH slides allows repeated cycles of stringency washing. For those starting with untested FISH probes, it is feasible to start off using a less stringent wash, which, if yielding unacceptable background levels, can be repeated at higher stringencies (i.e., at lower salt concentrations), the highest stringencies imposed by washes performed in water alone.

The FISH protocol described below is applicable to a wide variety of probes and, therefore, useful for those intending to combine probes from different sources. Indirectly labeled probes (e.g., with digoxigenin or biotin) require additional detection steps that can be plugged into the following protocol. In our hands, the same protocol also works for complex probe mixes, such as those used for M-FISH or SKY. The accompanying SKY images (Fig. 1e–h) were produced using the same protocol, while in our hands the "official" SKY protocol supplied by the manufacturer took longer and yielded inferior results.

Day 1

1. Use either fresh (1–7 days old) or archival slides stored at (−80°C).
2. Whenever present, extraneous background signal can be reduced by preincubation in pepsin solution for 2 min at 37°C. (see Note 7.)
3. Slide dehydration. Pass slides sequentially through an alcohol series for 2 min in 70% (two times), 90% (two times), and 100% ethanol in Coplin jars.
4. Desiccate slides overnight at 42°C in a dry oven.

Day 2

5. Deproteinize in acetone for 10 min (to minimize background autofluorescence).
6. Slide denaturation. Place slides for 2 min at 72°C in 30 mL of 2× SSC plus 70 mL formamide. The temperature of this step is critical. Therefore, avoid denaturing too many slides simultaneously. If a high throughput is desired, slides should be prewarmed. Quench in prechilled (–20°C) 70% ethanol for 2 min.
7. Repeat step 3 (the alcohol series).
8. Varnish slide label (to prevent subsequent eradication).
9. Place slide on prewarmed block at 37°C.
10. Remove probe from the freezer noting the concentration of labeled DNA. Add excess Cot-1 DNA (20–50× probe).
11. Probe denaturation: Place desired volume of probe into microfuge tube (sterile) and incubate in a "floater" for 5 min at 72°C in a water bath. (Important: If recommended by manufacturer, omit probe denaturation.)
12. Probe prehybridization. Collect probe by brief centrifugation, then incubate for 15–60 min at 37°C in a second water bath.
13. Probe application. Using shortened micropipette tips (sterile), carefully drop 8–12 mL of probe (making up the volume with hybrisol, if necessary) onto each slide half. Thus, two hybridizations may be performed on each slide (separated by a drop of hybrisol, to inhibit mixing). Cover slides carefully with glass coverslips, tapping out any bubbles, and seal with rubber cement.
14. Hybridization. Place slides carefully in moistened and sealed hybridization chamber. Leave overnight (or up to 72 h) in incubator (preferably humidified) at 37°C.

Day 3

15. After hybridization, carefully remove rubber cement and coverslips in 2× SSC using tweezers.
16. Stringency washing. Wash slides for 5 min at 72°C in 0.5× SSC.
17. (Optional) For use with digoxigenin-labeled probes; briefly prewash in wash solution at ambient temperature and shake to remove excess liquid. Important: Do not allow slides to dry out until dehydration (step 18). To each slide, apply 40 mL antidigoxigenin antibody labeled with FITC (Qbiogene) and cover with plastic coverslip. Incubate for 15–30 min at 37°C in hybridization chamber. Wash for 5 min

(three times) in wash solution at room temperature in subdued light.

18. Dehydration (alcohol series): Dehydrate slides as described in step 3, but performed in subdued light.

19. Mounting and sealing. Using shortened micropipette tips, to ensure even bubble-free coverage, carefully place three 30-mL drops of DAPI/Vectashield mountant along the slide. Apply coverslip and tap out any large bubbles using the blunt end of a pencil or equivalent. Seal with nail varnish. Allow varnish to dry.

20. Visualization. Slides should be visualized at high power under oil immersion with a ×63 objective with a high numerical aperture (see Note 8).

21. Analysis and interpretation (see Notes 8 and 9).

4. Notes

1. *Image analysis systems.* These tools are now essential kit for analyzing and documenting both banded and FISH preps. For further information, refer to the Web site of Applied Spectral Imaging (http://www.spectral-imaging.com/), Genetix (http://www.genetix.com/en/home/index.html), Metasystems (www.metasystems.de/), or Zeiss (http://www.zeiss.com/micro), which supply a variety of such systems. Imaging systems confer significant benefits, including amplification of weak signals, merging of differently colored signals, contrast enhancement, background reduction, generation of G-bands from DAPI counterstain, and rapid documentation and printing. While SKY, supplied by Applied Spectral Imaging, uses spectrophotometric separation to distinguish fluorescence excitation and emission spectra, the remaining systems use filters. Having compared all three systems, for multicolor FISH we obtained the best results with SKY (see Fig. 1e–h), though other systems also have their adherents.

2. *FISH probes.* Resolving complex rearrangements discovered by G-banding demands recourse to chromosome painting. Unfortunately, commercial painting and satellite DNA probes all-too-often yield unsatisfactory results because, e.g., as source DNA libraries become exhausted, suppliers are tempted to amplify their stocks using PCR-based methods that favor repeat DNA sequences, which tend to cross-hybridize onto other chromosomes at the expense of informative chromosome-specific regions. Thus, it is necessary first to calibrate new DNA probes using normal chromosomes. This effort is

usually well invested. Some probes generate unnecessarily bright signals. Knowing this beforehand allows such probes to be "stretched" by dilution with hybrisol. All too often, probes arrive that yield inadequate or inappropriate signals. Timely troubleshooting not only facilitates refund or replacement but could also prevent the pursuit of false trails inspired by ambiguous probes that hybridize to more than one region.

3. *Slide-making*. Slides for analysis should fulfill three criteria: sufficient metaphases, adequate chromosome spreading, and good morphology (i.e., large but undistended chromatids lying in parallel). To document progress in harvesting procedures and aid evidence-based searches for their improvement, we use a standard data sheet that records progress toward these ideals. An actual example is shown in Table 1, which presents harvesting data for the human B-cell lymphoma cell line RC-K8, the subsequent G-banding and FISH analysis of which are presented in Fig. 1a–c. In this case, reasonable preparations were only obtained at the second attempt using the standard protocol (see Subheading 3.1, step 3). Although, at the second attempt, all four hypotonic combinations yielded adequate numbers of metaphases, only tubes -b and -c yielded satisfactory spreading and morphology and were mixed for subsequent slide preparation. A total of 16 slides were prepared: 7 for G-banding, 1 for Giemsa staining alone (to check for the presence of smaller chromosomal elements that G-banding sometimes render invisible, such as so-called double minute chromosomes, which may harbor oncogenes), and 7 for FISH. In addition, the remaining cell suspension in fixative was stored (–20°C) for future use.

Slides with sparse yields of metaphases are unsuitable for FISH where probe costs are often critical. For slowly dividing cell lines (doubling times >48 h), colcemid times can be increased first to 6 h, then to 17 h (overnight), simultaneously reducing colcemid concentrations by a half to minimize toxicity. However, paucity of metaphases is usually the result of depletion by overly harsh hypotonic treatments. Paradoxically, We find that reducing hypotonic exposures to 1 min and, if necessary, performing this step in microfuge tubes to facilitate speedy centrifugation to reduce total hypotonic times may improve spreading and yield by enabling survival of fragile cells, which might otherwise be lost. Tight metaphases with an excess of overlapping chromosomes might be useful for FISH but are unsuitable for G-banding. In such cases, spreading can sometimes be improved by harsher hypotonic treatment, whether by increasing the proportion of KCl to 100% or by increasing the hypotonic time up to 15 min, or by performing the latter at 37°C instead of ambient temperature.

Gentle flaming often assists spreading and, contrary to received wisdom, has little or no deleterious effect on G-banding or FISH. In our hands, "dropping from a height" brings scant improvement in spreading, although "offensively" heavy breathing, performed both immediately before and after dropping, is beneficial, by increasing local humidity levels. Excessive spreading, on the other hand, is often cured by reducing the proportion of KCl, or by reducing hypotonic treatment times, or by retaining more of the original medium from the first centrifugation (Subheading 3.1, steps 3 and 4).

4. *Harvesting with FUDR.* As a general rule, the best morphologies are produced by hypotonics containing 50% or less Na citrate. Excessive amounts of the latter tend to yield fuzzy irregular morphologies that produce disappointing results with G-banding and FISH alike. Some types of cell, and derived cell lines alike, consistently yield short stubby chromosomes that appear refractory to all attempts at improvement. In such cases, it may be helpful to try FUDR pretreatment. Accordingly, treat cultures overnight with FUDR/uridine. The next morning, resuspend in fresh medium with added thymidine to reverse the blockade and harvest 7–9 h later.

5. *G-Banding.* As a general rule, "good" chromosomes yield good G-banding. Exceptions include chromosomes that are too "young" (puffed up or faint banding) or "over the top" (poor contrast or dark banding). Artificial aging by baking overnight at 60°C not only speeds up results but also tames variations in trypsin times due to variations in temperature or humidity. For those requiring a same-day result, aging times could be shortened to 60–90 min by increasing the hot plate/oven temperature to 90°C. Trypsin G-banding is a robust technique, and problems unconnected with poor chromosome morphology are rare. Those used to working with one species should note, however, that chromosomes of other species could exhibit higher/lower sensitivities to trypsin. Losses in tryptic activity occur after about 6 months among aliquots stored at ($-80°C$), which should then be discarded in favor of fresh stocks.

6. *Karyotyping.* G-Banding lies at the center of cytogenetic analysis. The ability to recognize each of the 24 normal human chromosome homologs necessarily precedes analysis of rearrangements. Because the majority of human cancer cell lines carry chromosome rearrangements, the choice of cell lines for learning purposes is critical. Learning should be performed using either primary cultures of normal unaffected individuals (e.g., lymphocyte cultures) or B-lymphoblastoid cell lines known to have retained their diploid character. Those intent on acquiring the ability to perform karyotyping are strongly advised to spend some time in a laboratory where such skills

are practiced daily (e.g., a routine diagnostic laboratory). The karyotype depicted in Fig. 1a shows several rearrangements other than the t(3;7)(q27;q21) picked out for illustration, including structural rearrangements affecting chromosomes 5 (both homologs), 8 (both homologs), 13, 14, 15, 17, 21, while chromosomes X, 2, 15, 20, show losses. Loss of chromosome X shown in the cell illustrated was attributable to preparative artifact, from its presence in other cells.

7. *FISH signals and noise.* FISH experiments are sometimes plagued by high background signals, or "noise." Large insert clones, such as BACs, may cover 200 kbp and often include DNA sequences which cross-hybridize. Commercial probes are usually, but by no means always, relatively free of this problem. Increasing the wash stringency (Subheading 3.3, step 16) by reducing the SSC concentration to 0.1× might help. Alternately, adding Cot-1 DNA to the hybridization mix might help to reduce hybridization noise. Among noncommercial probes, excessive noise can often be cured by reducing the probe concentration. Normal DNA concentrations for single-locus probes should range from 2–6 ng/μL to 10–20 ng/μL for painting probes. Assuming that it is not the result of "dirty" slides, nonspecific noise may be caused by either autofluorescence or protein–protein binding after antibody staining, which can often be reduced by additional slide pretreatment in pepsin solution (Subheading 3.3, step 2). Incubate slides for 2 min in acidified pepsin solution at 37°C. Rinse in PBS (pH 7.2) for 3 min at room temperature. Postfix slides, held flat, in 1% formaldehyde solution for 10 min at room temperature using plastic coverslips. Rinse in PBS (pH 7.2) for 3 min at room temperature. Continue with step 3 of Subheading 3.3. Weak FISH signal intensity might arise because the probe itself is inherently weak, the wash too stringent, or the chromosomes insufficiently denatured. To test for these alternatives, repeat the stringency wash (Subheading 3.3, step 16) but with either 2× or 1× SSC in the wash buffer. In parallel, repeat the slide denaturation (Subheading 3.3, step 6) increasing the denaturation time to 4 min. When neither modification brings any improvement and the probe is new and untested or old and infrequently used, it is likely that the probe is inherently weak. (Even large-insert clones sometimes deliver puzzlingly weak signals, usually attributed to the inaccessibility of their chromosomal targets.) For those equipped with advanced imaging systems incorporating a camera of high sensitivity, it is often possible to capture images from probe signals invisible to the naked eye. In the case of new commercial probes, the supplier should be contacted. Probes with larger targets often cross-hybridize to similar DNA sequences present on other chromosomes. It

is important first to identify patterns of cross-hybridization by FISH onto normal chromosomes to avoid misinterpreting the latter as rearrangements. Some resource centers, notably BAC/PAC Resources (http://bacpac.chori.org/), helpfully list cross-hybridization patterns for some clones.

8. *FISH analysis.* The first aim of FISH is to characterize those rearrangements of interest present that resist analysis by G-banding. This inevitably requires both intuition and luck. Clearly, the need for the latter is reduced where G-banding is optimized. In the case of the t(3;7), as illustrated here (Fig. 1a–d), while the der(3) partner serves as index chromosome due to its carrying a rearrangement at 3q27, which is recurrent in DLBCL, the identity of its partner is a matter of conjecture due to the promiscuous nature of 3q27 rearrangements and the significant number of additional rearrangements present. Indeed, in the original report the rearrangement in RC-K8 was described as t(3;4), which the accompanying photographic documentation fortunately shows to be the same rearrangement (27, 28). The apparent discrepancy is simply due to the original (reference) cytogenetic investigation predating the advent of FISH. Indeed, though chromosome painting probes (such as those shown in Fig. 1b) became widely available in the mid-1990s, the advent of mapped BAC clones enabling do-it-yourself construction of panels to detect specific oncogene alterations (such as that for BCL6 shown in Fig. 1c, d) came nearly a decade later in the wake of the Human Genome Project data and resources.

The most difficult rearrangements to resolve are unbalanced ones involving multiple chromosomes. Sometimes, originally reciprocal translocations appear unbalanced because of loss or additional rearrangement of one partner. In such cases, the identity of the "missing partner" might be often guessed at from among those chromosomes where one or more homologues appear to be missing. Having identified the chromosomal constituents of cryptic rearrangements, the next task is to reconcile FISH with G-banding data enabling breakpoint identification. In cases where chromosome segments are short or their banding patterns nondescript, this aim might be frustrated. The International System for Chromosome Nomenclature (ISCN) enables almost all rearrangements to be described with minimal ambiguity in most cases (29). This system was updated in 1991 for cancer cells (30), and in 1995, 2005, and 2009 for FISH (31). The most efficient way to detect and analyze multiple chromosome rearrangements is to combine G-banding with multiplex FISH, of which SKY is the form most widely cited (Fig. 1e–h).

9. *Use of cytogenetic data.* Having successfully completed cytogenetic analysis of a tumor cell line to the point of ISCN karyotyping,

what to do with the data? The first question to be addressed is identity: Has the cell line in question been karyotyped previously and, if so, does the observed karyotype correspond with that previously reported? In our experience, complete correspondence between cell line karyotypes is rare, even where their identity has been confirmed by DNA fingerprinting. First, among very complex karyotypes, complete resolution might be unnecessary and is, indeed, rarely achieved. This leaves significant scope for uncertainty and differences in interpretation. Wherever possible, refer to the original journal or reprint, as photocopies seldom permit reproduction of intermediate tones, which are the "devil in the detail" of G-banding. Second, a minority of cell lines might evolve karyotypically during culture in vitro. This instability could effect numerical or structural changes. Such a cell line is CCRF-CEM, derived from a patient with T-cell leukemia, and which has spawned a multitude of subclones – all cytogenetically distinct (12) and, sometimes following cross-contamination events, masquerading under aliases. Those wishing to compare their karyotypes with those derived at the DSMZ can refer to either "catalogues raisonnés" of human leukemia lymphoma cell lines (32, 33) or the DSMZ Web site, which features an interactive searchable database of all types of (mainly) human cancer cell lines (http://www.dsmz.de/).

Acknowledgment

We wish to thank Maren Kaufmann for her expert technical input and suggestions, many of which are silently incorporated into the text.

References

1. MacLeod, R.A.F., Nagel, S., Scherr, M., Schneider, B., Dirks, W. G., Uphoff, C. C., et al. (2008) Human leukemia and lymphoma cell lines as models and resources. *Current Medicinal Chemistry* **15**, 339–359.

2. Mitelman, F., Johansson, B., and Mertens, F. (2007) The impact of translocations and gene fusions on cancer causation. *Nat Rev Cancer* **7**, 233–45.

3. Deininger, M., Buchdunger, E., Druker, B. J. (2005) The development of imatinib as a therapeutic agent for chronic myeloid leukemia. *Blood* **105**, 2640–53.

4. Drexler, H. G, MacLeod, R. A. F., and Uphoff, C. C. (1999) Leukemia cell lines: in vitro models for the study of Philadelphia chromosome-positive leukemia. *Leukemia Res* **23**, 207–15.

5. Drexler, H. G., MacLeod, R. A. F., Borkhardt, A., and Janssen, J. W. G. (1995) Recurrent chromosomal translocations and fusion genes in leukemia–lymphoma cell lines. *Leukemia* **9**, 480–500.

6. Marini, P., MacLeod, R. A. F., Treuner, C., Bruchelt, G., Böhm, W., Wolburg, H., et al. (1999) SiMa, a new neuroblastoma cell line combining poor prognostic cytogenetic markers with high adrenergic differentiation. *Cancer Genet Cytogenet* **112**, 161–4.

7. Drexler, H. G., Matsuo, Y., and MacLeod, R. A.F. (2000) Continuous hematopoietic cell lines as model systems for leukemia–lymphoma research. *Leukemia Res* **24**, 881–911.

8. Tosi, S., Giudici, G., Rambaldi, A., Scherer, S.W., Bray-Ward, P., Dirscherl, L., et al. (1999) Characterization of the human myeloid

8. leukemia-derived cell line GF-D8 by multiplex fluorescence in situ hybridization, subtelomeric probes, and comparative genomic hybridization. *Genes Chromosomes Cancer* **24**, 213–21.

9. Schneider, B., Nagel, S., Kaufmann, M., Winkelmann, S., Drexler, H. G., and MacLeod, R. A. F. (2008) t(3;7)(q27;q32) fuses BCL6 to a non-coding region at FRA7H near miR-29. *Leukemia* **22**, 1262–6.

10. Vanasse, G. J., Concannon, P., and Willerford, D. M. (1999) Regulated genomic instability and neoplasia in the lymphoid lineage. *Blood* **94**, 3997–4010.

11. Küppers, R. and Dalla-Favera, R. (2001) Mechanisms of chromosomal translocations in B cell lymphomas. *Oncogene* **20**, 5580–94.

12. MacLeod, R. A. F., Dirks, W. G., Matsuo, Y., Kaufmann, M., Milch, H., and Drexler, H. G. (1999) Widespread intraspecies cross-contamination of human tumor cell lines arising at source. *Int J Cancer* **83**, 555–63.

13. Drexler, H. G., Dirks, W. G., and MacLeod, R. A.F. (1999) False human hematopoietic cell lines: cross-contaminations and misinterpretations. *Leukemia* **13**, 1601–7.

14. Nelson-Rees, W. A., Daniels, D. W., and Flandermeyer, R. R. (1981) Cross-contamination of cells in culture. *Science* **212**, 446–52.

15. Markovic, O. and Markovic, N. (1998) Cell cross-contamination in cell cultures: the silent and neglected danger. *In Vitro Cell Dev Biol Anim* **34**, 1–8.

16. MacLeod, R. A. F., Dirks, W. G., and Drexler, H. G. (2002) Persistent use of misidentified cell lines and its prevention. *Genes Chromosomes Cancer* **33**, 103–5.

17. Stacey, G.N., Masters, J. R., Hay, R.J., Drexler, H. G., MacLeod, R. A. F., and Frechney, R. I. (2000) Cell contamination leads to inaccurate data: We must take action now. *Nature* **403**, 356.

18. Caspersson, T., Zech, L., and Johansson, C. (1970) Differential binding of alkylating fluorochromes in human chromosomes. *Exp Cell Res* **60**, 315–19.

19. Sumner, A. T., Evans, H. J., and Buckland, R. A. (1971) New technique for distinguishing between human chromosomes. *Nature New Biol* **232**, 31–2.

20. Seabright, M. (1973) Improvement of trypsin method for banding chromosomes. *Lancet* **1**, 1249–50.

21. Rowley, J. D. (1973) A new consistent chromosomal abnormality in chronic myelogenous leukaemia identified by quinacrine fluorescence and Giemsa staining. *Nature* **243**, 290–3.

22. Nowell, P. C. and Hungerford, D. A. (1960) A minute chromosome in human granulocytic leukemia. *Science* **132**, 1497.

23. Cremer, T., Lichter, P., Borden, J., Ward, D. C., and Manuelidis, L. (1988) Detection of chromosome aberrations in metaphase and interphase tumor cells by in situ hybridization using chromosome-specific library probes. *Hum Genet* **80**, 235–46.

24. Lichter, P., Cremer, T., Borden, J., Manuelidis, L., and Ward, D. C. (1988) Delineation of individual human chromosomes in metaphase and interphase cells by in situ suppression hybridization using recombinant DNA libraries. *Hum Genet* **80**, 224–234.

25. Lichter, P. (1997) Multicolor FISHing: what's the catch? *Trends Genet* **12**, 475–79.

26. MacLeod, R. A. F., Kaufmann, M., and Drexler, H.G. (2007) Cytogenetic harvesting of commonly used tumor cell lines. *Nature Protocols* **2**, 372–82.

27. Kubonishi, I., Niiya, K., and Miyoshi, I. (1985) Establishment of a new human lymphoma line that secretes plasminogen activator. *Jpn J Cancer Res* **76**, 12–5.

28. Kubonishi, I., Niiya, K., Yamashita, M., Yano, S., Abe, T., Ohtsuki, Y., and Miyoshi, I. (1986) Characterization of a new human lymphoma cell line (RC-K8) with t(11;14) chromosome abnormality. *Cancer* **58**, 1453–60.

29. ISCN 1985: An International System for Human Cytogenetic Nomenclature: report of the Standing Committee on Human Cytogenetic Nomenclature. *Karger, Basel*.

30. ISCN 1991: Guidelines for Cancer Cytogenetics: Supplement to an International System for Human Cytogenetic Nomenclature. Karger, Basel.

31. ISCN 2009: An International System for Human Cytogenetic Nomenclature: Recommendations of the International Standing Committee on Human Cytogenetic Nomenclature. Published in collaboration with "Cytogenetic and Genome Research", Karger, Basel, in press, 2009.

32. Drexler, H. G. (2001) The Leukemia-Lymphoma Cell Line FactsBook, Academic Press, San Diego.

33. Drexler, H. G. (2009) Guide to Leukemia-Lymphoma Cell Lines, eBook (available from the author), DSMZ, Braunschweig, Germany.

Chapter 7

Cell Culture Contamination

Glyn N. Stacey

Abstract

Microbial contamination is a major issue in cell culture, but there are a range of procedures which can be adopted to prevent or eliminate contamination. Contamination may arise from the operator and the laboratory environment, from other cells used in the laboratory, and from reagents. Some infections may present a risk to laboratory workers: containment and aseptic technique are the key defence against such risks. Remedial management of suspected infection may simply mean discarding a single potentially infected culture. However, if a more widespread problem is identified, then all contaminated cultures and associated unused media that have been opened during this period should be discarded, equipment should be inspected and cleaned, cell culture operations reviewed, and isolation from other laboratories instituted until the problem is solved. Attention to training of staff, laboratory layout, appropriate use of quarantine for new cultures or cell lines, cleaning and maintenance, and quality control are important factors in preventing contamination in cell culture laboratories.

Key words: Cell culture, Contamination, *Mycoplasma*, Virus, Infection, Aseptic technique, Quality control, Quarantine

1. Introduction

Various kinds of microbial contamination can cause significant changes and often catastrophic loss of cell cultures for which a range of procedures can be adopted to prevent or eliminate contamination and thus will avoid significant waste of time and resources. However, one of the most powerful contributions to avoiding contamination is the skill of lab workers in aseptic technique, and their awareness of both the state of the cells they are using and any changes in the lab environment that may increase risk of contamination. Useful general guidance on good cell culture practice has been produced by Coecke et al. (1) and other international consensus guidance has also been produced for specialist areas such as human embryonic stem cells (2) and cell lines intended for the manufacture of medicines (3–6). In this chapter, the types, sources, and impact of microbial contamination will be considered along with potential ways of dealing proportionately with individual contamination events and more challenging situations that researchers may encounter.

2. Sources and Forms of Contamination

The human body operates a range of physical, biochemical, and immunological defences to exclude microorganisms from causing damaging and life-threatening infections allowing beneficial colonisation in areas such as the mucosal epithelia, e.g. mouth, gut, and genital tract. The monoseptic (i.e. single viable organism or cell) environment of in vitro cell culture is very different to the tissues of the human body. Its maintenance is dependent on the provision of a variety of physical containment/protection processes and chemical (i.e. antibiotic) controls to prevent or inhibit contamination. This state requires constant maintenance that must control infectious challenges from a variety of known and unknown sources.

2.1. Operator and Environment

As a source of contamination, the human operator is potentially the greatest hazard in the laboratory. Sloughing of skin scales from laboratory workers constantly seeds the environment with microorganisms and with the high levels of commensal microbes (i.e. normal inhabitants) in parts of the human mouth even just talking will generate potential contaminating aerosols. Furthermore, an individual laboratory worker with an infection will represent a much higher risk source of contamination.

Materials and staff coming into the laboratory and points of access to the environment (drains, ventilation grills, and unsealed ceiling voids) will inevitably provide a constant supply of fresh contamination which will accumulate on laboratory surfaces and multiply in damp areas (e.g. fridges, sinks, waterbaths, and humidified incubators).

Clearly procedural and physical control of materials coming into the laboratory, laboratory cleaning and aseptic technique will all be important activities that combined, will help to keep cell cultures free of contamination.

3. Other Cell Cultures

Primary cells, isolated directly from animal tissues, or their early passages are at a particularly high risk of contamination as the dissection process may promote contamination from the natural commensal flora in certain tissues and any subclinical infection.

The latter condition will be influenced by the species and tissues of the donor animal which will elevate the risk of contamination due to certain organisms. The risks from different species/tissues have been reviewed by Frommer et al. (7) and Stacey et al. (8), but due to changes in natural incidence of disease and new emerging pathogens, it is important to be aware of developing microbiological knowledge when evaluating risk of contamination. Where tissues and primary cells are handled the risk can be directly related to the number of samples handled from different individuals, the risk associated with any particular cohort of donors and also the geographical origin of the cells or tissue (9).

The likelihood of tissues/cells being contaminated with serious human pathogens is a significant issue for laboratory workers although fortunately a rare occurrence. Infectious organisms that could cause a serious infection in laboratory workers may be present in blood or brain of human or primate origin and include haemorrhagic fever viruses, T-lymphotrophic viruses, and hepatitis viruses (7). However, blood and tissues from a broad range of other mammalian species may also carry zoonotic infections that can lead to infection and even death in humans. Such species even include goats which can be a source of infection with Coxiella burnetii causing Q-fever in humans (7). Colonies of experimental rodents including guineapigs and rats may carry lymphocytic choriomeningitis virus, hantaan virus, or reovirus 3 without necessarily showing overt symptoms (10–12), and cell lines and antibody preparations processed from infected colonies may also become contaminated (13). Surprisingly, even cell lines from insects can be persistently infected with human pathogens such as yellow fever virus (14, 15). Frequent contaminants in bovine cell lines are noncytopathic strains of bovine viral diarrhoea virus, which also bind effectively to cells of non-bovine origin including primate cells but without causing infection (16). Another bovine virus known to have the potential to infect cells from a broad range of species is bovine polyoma virus and, unlike BVDV, it also has the potential to transform bovine and non-bovine cells (17). Viral contamination of cell lines may also arise with organisms which could not have been predicted based on the species and tissue of origin (18). For general cell culture work, it is wise to treat all cell cultures as potentially infectious and manage them accordingly using good cell culture practice and appropriate containment and waste disposal as outlined in Subheading 7.

4. Reagents

Any reagent is a potential carrier of contamination and when purchasing reagents and media for cell culture it is important to determine whether they are sterile or will require sterilisation. Certain reagents that are not sterilisable such as serum and growth factors may represent a high risk of contamination with virus, depending on the method of production (animals vs. recombinant DNA technology) and the source of the original materials. The risks can be evaluated and subsequently mitigated by gathering information on a number of factors including species and tissues of origin, geographical origin, elements of reagent processing likely to reduce contaminants, reliability of any sterilisation steps, and any testing for contaminants performed on the final reagent (18). Particular care should be taken with labile and relatively unprocessed animal-derived materials such as bovine serum or porcine trypsin, which are known to carry animal viruses that can infect some cell cultures (19, 20).

5. Consequences of Contamination

Invariably bacterial or fungal contamination will result in total loss of the affected culture and potentially increase the environmental contamination if not dealt with carefully. Recovery of cell lines from such overwhelming contaminations is undersirable as the contamination may survive in spite of antibiotic treatment only to reemerge later and in some cases with increased antibiotic resistance. Eradication of fungi is especially difficult as effective antifungal agents are often cytotoxic to the cell line and fungal spores are much more likely to persist in cultures and the environment only to cause intractable problems later.

Mycoplasma are very small Mollicutes (not bacteria) which are fastidious and generally require a source of cholesterol to survive. They can establish parasitic infections of cell cultures where they typically grow adhering to the cell membrane, although some species such as *M. penetrans* grow intracellularly. They are a widespread cause of cell culture contamination, originally arising from animal tissues, reagents of animal origin and even human operators. *Mycoplasma* infection can spread rapidly in the laboratory as contamination may not be obvious to the operator (i.e. no medium turbidity), is resistant to commonly used antibiotics and can survive in the laboratory environment in aerosols and splashes of culture medium. *Mycoplasma* has been shown to cause numerous types of damaging and permanent changes in cell lines including

- Enhanced secretion of cytokines by human lung fibroblasts (21),
- Changes in cell line karyology (22),
- Interference with identification and isolation procedures for retroviruses (23).

For reviews of the damaging effects that can be caused by *Mycoplasma* infection *see* refs. 24–26.

It is theoretically possible to eliminate *Mycoplasma* from cell lines using certain antibiotics (e.g. Ciprofloxacin™ and Myocoplasma Removal Agent™), but success rates for complete eradication are poor and the effective antibiotics are often genotoxic to mammalian cells (27, 28). Detection of *Mycoplasma* can be achieved using a number of methods and proprietary kits, and for a comparison of the different methods *see* ref. 29.

Persistent viral infections of cell lines may represent an infection risk to laboratory workers and this risk can be difficult to quantify since contaminants have been identified that would not be expected based on the species of the cell line (8). Such uncertain risks are difficult to address by comprehensive and expensive viral screening: containment and aseptic technique are the key defence against such risks. Cell culture infections with viruses which are not necessarily pathogenic for humans are also significant since they will clearly have some affect on the cell biology which could influence the validity of any data generated with that culture and could affect their performance and acceptability for manufacturing purposes.

6. Responses to Contamination Incidents

6.1. Remedial Management of Contamination Events

When an overwhelming bacterial or fungal contamination or a dramatic and unexpected cytopathic affect or other suspicious morphological change arises in cell cultures, the immediate response should be to discard the affected cultures. The event should be notified to all other users of the lab who might have utilised the same cells, reagents/media, or class II cabinet, so that they can be alert to any further contamination and if necessary be prepared to discard the reagents and cells they have been using. If the incident is isolated to one event, only one or a very small number of flasks were affected from a batch of cultures and no other lab users have reported problems, then it may be safe to assume that the contamination event arose due to chance introduction of organisms solely into the affected flasks and would be unlikely to recur. In this case, it should be safe to reinitiate fresh cultures and possibly even continue with the unaffected culture flasks. If an entire batch of cultures involving a number of flasks was affected, it will also be wise to discard any remaining bottles of media and reagents last used with that culture. In the majority of cases, such incidents are one-off chance events most likely due to contamination entering culture flask during passaging or other manipulation of cells in open processes (i.e. were cell suspensions are open to the lab air in the class II cabinet).

6.2. Suspected Ongoing or General Source of Contamination

Where contamination appears to be more widespread in the laboratory or recurs on a number of occasions, then this needs to be communicated to all lab staff rapidly and a coordinated plan put in place to prevent further spread of the contamination as soon as possible. Prompt action will hopefully prevent panic reactions which could worsen the situation, such as staff immediately moving apparently "contaminant-free" cells and media to another laboratory which may simply spread the problem. Assessment of such situations and their solution may well differ depending on the type of laboratory activity and local procedures and rules. The following suggestions give a sample of the kind of process to consider in dealing with significant contamination:

(a) All contaminated cultures and associated unused media which have been opened during this period should be discarded.

(b) Cabinets, incubators, waterbaths, and fridges should be inspected carefully to identify any obvious causes of contamination, such as fungal growth on damp surfaces, contaminated humidifying water trays in incubators, heavily contaminated waterbaths, and spill catchment trays under the work surfaces inside class II cabinets. On discovering very grossly contaminated areas, the situation should be cleaned up with care to avoid spreading the source of contamination more widely.

(c) The cell culture operations and contaminants in the lab should be reviewed immediately with regard to:

- Exactly which cultures were affected
- Whether the contamination events appear to be generally the same in the way that they affect the cells or morphology of colonies or cells or a mix of different organisms. In the former case, the contamination may potentially be from a common point source (such as a grossly contaminated reagent or cell line) and the latter presentation could indicate a breach in aseptic technique or containment (e.g. batch of damaged T flasks and failing class II cabinet). However, neither observation is specifically diagnostic for these causes of contamination.
- Whether there is any potential association with a particular culture, medium, reagent, location, equipment, staff-member, etc.
- Any features of the laboratory consumable supplies (i.e. those materials and lab hardware which come in direct contact with the cells) that have recently changed. It may also be useful to consider any recent significant changes or breakdowns in lab procedures or equipment that may have increased the risk of contamination.

(d) It is important to prevent carryover of any remaining residual contamination into new work and where the contamination is particularly serious and widespread, it would be wise to stop all work, and prior to reinitiating any new cell cultures, perform a careful laboratory clean which may be followed by fumigation if this is feasible to perform safely in the lab in question.

6.3. Recovery and Monitoring

The actual source of contamination often may not be resolved and therefore, the process of recovery from a major and/or widespread contamination incident will clearly demand careful monitoring to ensure that the original problem does not reappear. Staff should be encouraged to report any further incidents or suspicious observations so that any recurrence can be captured and controlled rapidly. If a particular type of microorganism was implicated, it may be possible to implement specific testing that will flag up its reappearance and differentiate it more quickly from other general contamination. Such procedures could be discussed with a local diagnostic microbiology laboratory.

7. Avoidance and Prevention

7.1. Training

The primary protection against contamination of cell cultures is a well trained and alert staff. Spotting contamination at the early stage with appropriate disposal of affected cultures will help to contain the spread of contamination. Where contamination is not directly evident or at very low levels staff training in aseptic technique will be the primary means to control its spread. Training in aseptic technique and identifying microbial contamination and appropriate disinfection and disposal of all routine waste will accordingly be vital and are an essential part of induction for new and inexperienced staff (1). A further important element in cell culture training is to impress on staff that routine cell culture can and should, in most cases, be performed without the addition of antibiotics. Good aseptic technique should eliminate the need for antibiotics except in a limited number of situations such as primary cell cultures where the risk of contamination of the cells is high, virus diagnostic testing where test cells are inoculated with heavily contaminated specimens and genetic modification of cells which require antibiotic selection.

7.2. Laboratory Layout

The laboratory layout can be developed to reduce the risk of contamination by placing cell culture cabinets out of main thoroughfares, placing waste disposal collection away from clean work areas and sterile media storage and obviously segregating any work with (or storage of) microorganisms from preparation of clean cell cultures. It is also valuable to segregate routine cell culture work from new cell lines and primary cell cultures which will be a significant source of contamination from persistent noncytopathic organisms or from the animals used to provide primary cells (see Subheading 7.4). Positioning of class II cabinets should be carefully planned to avoid interference by lab furniture/equipment, walls, benching, doors, and other cabinets and guidance on this is available in BSEN 12469 (8, 30).

Ideally lab air should be filtered as it enters the room and for critical work HEPA filter systems are required to provide very low levels of particulates which may carry contamination. For clean room operation, there are international standards to establish appropriate clean lab air as given in ISO 14644-8 (31).

7.3. Cleaning and Maintenance

Laboratory cleaning is at the frontline of preventive maintenance where contamination is concerned. A formal routine cleaning regime for the general laboratory will help to control the levels of environmental contamination and thereby help to reduce the risks of day-to-day contamination. In some countries, a documented disinfection and cleaning policy is required by law for all laboratories working with infectious microorganisms. While cell cultures are not innately infectious, they may carry undetected and potentially hazardous contamination which can be managed with appropriate containment and disinfection of waste and spills, and such procedures should be documented and available to all staff. Certain cell culture equipment could result in the spread of contamination from infected cultures or media should it fail (e.g. class II cabinets, centrifuges, and autoclaves) and obviously should receive appropriate maintenance to ensure containment of contamination (30, 32). Staff should also be aware of appropriate routine checks that will alert them to failures such as low flow rate in a cabinet, breakage in a centrifuge or inadequate sterilisation indicated by "indicator strips."

7.4. Quarantine

New cell lines and primary cell cultures or tissues represent one of the most common sources of contamination in the cell culture lab. Provision of a separate quarantine laboratory is ideal for this, however, an isolated class II cabinet can provide an alternative solution where space is at a premium although careful control of all processes outside of the cabinet will be crucial, including provision of separate or contained cold storage of in-use media and a separate incubator and centrifuge. In extreme situations where there is no spare class II cabinet for quarantine work, it may be possible to operate chronological quarantine by carrying out sterile work (e.g. media preparation) as the first process of the day, followed by culture of "screened" cell lines and finally quarantine cells followed by thorough disinfection of the cabinet. However, it should be emphasised that such an arrangement is not recommended as it presents the highest risk of contamination of the above quarantine solutions and would require very stringent control of quarantine cultures and any in-use media outside of the cabinet. In the manufacture of products from cell cultures (e.g. vaccines, biotherapeutics, and cell therapy), stringent controls will need to be in place to prevent contamination from new cells and often laboratories are dedicated to a single cell line to avoid risks associated with other work performed in the same environment.

The transfer of a culture out of quarantine status should only be permitted once the cells in question have tested negative in microbiological screening tests. The specification of these tests will vary depending on the nature of the laboratory work and the intended use of the cells. As a baseline, screening should include a test for *Mycoplasma* and sterility (i.e., bacteria and fungi) and microscopical observation of the cells for microorganisms or cytopathic effects that might be due to infection. In some circumstances, virological screening may be performed but comprehensive testing for abroad range of microbes will be highly expensive and generally not justifiable unless the application of the cells requires this (e.g. use in cell therapy and use for manufacture of medicines) (3–6). Accordingly, any claim that cells have been "screened" should be qualified and a procedure implemented in the laboratory to ensure that each new cell culture is subjected to an appropriate screening regime that meets local requirements.

As a consequence of the difficulty in predicting contaminating organisms and the significant number of predictable viral contaminants, it is not practical to test for and exclude all potential viral contaminants. These risks, therefore, should be managed through a combination of risk assessment of sources of cells and reagents and appropriate containment and waste disposal.

Quarantine may also need to be considered for storage of cryopreserved cell cultures. Storage of cells preserved during the quarantine phase should ideally be separated until screening tests enable them to be transferred to "screened" cell storage. Stored cells known to be persistently infected with microorganisms could represent a hazard to other stored cells and possibly laboratory workers accessing the storage vessel. It is known that the "ice"sludge, which can accumulate in liquid nitrogen storage vessels, is prone to gross contamination with bacteria and fungi (Fig. 1) (33). Transfer of pathogenic virus from accidentally damaged cryostorage containers to other samples has been tragically demonstrated in storage of patient tissues (34, 35) and the risks of such events considered and appropriate separate storage, secondary containment, or other measures adopted to manage the identified risk. Such contamination can be additionally controlled by careful cleaning of cryovials on thawing.

7.5. Routine Quality Control

No screening test is perfect and cultures which may have passed satisfactorily out of quarantine status may have had contamination that was originally below the level of detection or contamination may have occurred subsequently via the lab environment, contaminated

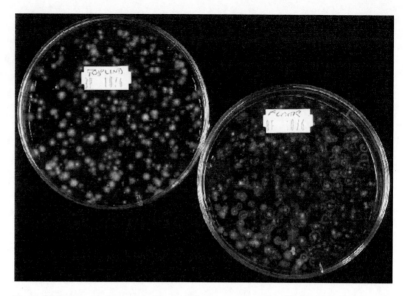

Fig. 1. Blood agar culture plate incubated at room temperature following exposure to vapour phase of liquid nitrogen in a storage vessel located in an unclean environment. Numerous colonies of bacteria and fungi can be seen emphasising the need to provide a clean environment for liquid nitrogen storage areas and thorough disinfection of the external surfaces of cryovials before transferring thawed cells into culture.

reagents, or other cultures which for some reason bypassed the normal screening process. Establishment of a cell banking regime that will provide a stock of low passage cultures that have been subjected to repeated quality control tests is a fundamental aspect of good cell culture practice. Lab staff should avoid long-term passaging of cultures which could propagate contamination, and is also undesirable both from the perspectives of genetic and phenotypic instability and increases the risk of cell line cross-contamination (1, 36). Periodic replacement of cultures with newly thawed from a cell bank (e.g. every 12 weeks) will significantly reduce these hazards. In addition, periodic testing of cell lines in use in the lab will give confidence that occult contamination, notably *Mycoplasma*, do not go unrecognised and where they do arise, are dealt with quickly so that the impact on any resulting work is minimised. However, even where sterility and *Mycoplasma* tests are performed some organisms will not necessarily be detected such as Mycobacteria and other fastidious bacteria and fungi (37, 38). Routine test methods have been reported elsewhere (39, 40) and for laboratories where there are more demanding regulatory issues, such as GLP safety testing of cells and products, pharmacopoeial methods have been established (41–44).

The products of cell culture may be used in a wide range of clinical products and therapeutic applications: from use of purified recombinant vaccines, to the implantation of cultured cells (e.g., stem cells) into immunocompromised or terminally ill patients. Contamination issues are clearly a crucial consideration in the cell cultures used for such products and cell banks used for production purposes are subjected to extensive characterisation and safety testing (3–5). The contamination hazards in cell culture processes for the manufacture of biological medicines from animal cell culture have been reviewed by Stacey et al. (18).

References

1. Coecke, S., Balls, M., Bowe, G., Davis, J., Gstraunthaler, G., Hartung, T., Hay, R., Merten, O-W., Price, A., Shechtman, L., Stacey, G., and Stokes, W. (2005) Guidance on Good cell culture practice: a report of the second ECVAM task force on good cell culture practice. *ATLA* **33**, 261–287.

2. International Stem Cell Initiative, Adewumi, O., Aflatoonian, B., Ahrlund-Richter, L., Amit, M., Andrews, P.W., Beighton, G., Bello, P.A., Benvenisty, N., Berry, L.S., Bevan, S., Blum, B., Brooking, J., Chen, K.G., Choo, A.B., Churchill, G.A., Corbel, M., Damjanov, I., Draper, J.S., Dvorak, P., Emanuelsson, K., Fleck, R.A., Ford, A., Gertow, K., Gertsenstein, M., Gokhale, P.J., Hamilton, R.S., Hampl, A., Healy, L.E., Hovatta, O., Hyllner, J., Imreh, M.P., Itskovitz-Eldor, J., Jackson, J., Johnson, J.L., Jones, M., Kee, K., King, B.L., Knowles, B.B., Lako, M., Lebrin, F., Mallon, B.S., Manning, D., Mayshar, Y., McKay, R.D., Michalska, A.E., Mikkola, M., Mileikovsky, M., Minger, S.L., Moore, H.D., Mummery, C.L., Nagy, A., Nakatsuji, N., O'Brien, C.M., Oh, S.K., Olsson, C., Otonkoski, T., Park, K.Y., Passier, R., Patel, H., Patel, M., Pedersen, R., Pera, M.F., Piekarczyk, M.S., Pera, R.A., Reubinoff, B.E., Robins, A.J., Rossant, J., Rugg-Gunn, P., Schulz, T.C., Semb, H., Sherrer, E.S., Siemen, H., Stacey, G.N., Stojkovic, M., Suemori, H., Szatkiewicz, J., Turetsky, T., Tuuri, T., van den Brink, S., Vintersten, K., Vuoristo, S., Ward, D., Weaver, T.A., Young, L.A. and Zhang, W. (2007) Characterization of human embryonic stem cell lines by the International Stem Cell Initiative. *Nat. Biotechnol.* **25**, 803–16.

3. Center for Biologics Evaluation and Research (1993) *Points to Consider in the Characterization of Cell Lines Used to Produce Biologicals*, Food and Drug Administration, Bethesda, MD, available at: http://www.fda.gov/cber/.

4. International Conference on Harmonisation (1997) ICH Topic Q 5 D Quality of Biotechnological Products: Derivation and Characterisation of Cell Substrates Used for production of Biotechnological/Biological Products (CPMP/ICH/294/95), ICH Technical Coordination, European Medicines Evaluation Agency, London.

5. World Health Organisation (1998) Requirements for the Use of Animal Cells as *in vitro* Substrates for the Production of Biologicals (Requirements for Biological Substances No. 50). *WHO Technical Report Series* No. 878. Geneva: World Health Organization, Geneva.

6. Knezevic, I., Stacey, G.N., Petricciani, J.P. and Sheets, R. (2010) Evaluation of Cell Substrates for the production of biological: revision of WHO recommendations. *Biologicals* **38**, 162–9.

7. Frommer, W., Archer, L., Boon. L., Brunius, G., Collins, C.H., Crooy, P., Doblhoff-Dier, O., Donikian, R., Economidis, J., and Frontali, C., et al. (1993) Safe Biotechnology recommendations for safe work with animal and human cell cultures concerning potential human pathogens. *Appl. Micr. Biotechnol.* **39**, 141–147.

8. Stacey, G.N., Tyrrel, D.A. and Doyle, A. (1998) Source Materials. In: Safety in Cell and Tissues Culture, Stacey, G.N., Hambleton, P.J. and Doyle, A. Kluwer Academic Publishers, Dordrecht, Netherlands, pp 1–25.

9. Cobo, F., Stacey, G.N., Hunt, C., Cabrera, C., Nieto, A., Montes, R., Cortes, J.L., Catalina, P., Barnie, A. and Concha, A. (2005) Microbiological control in stem cell banks: approaches to standardisation. *Appl. Microbiol. Biotechnol.* **68**, 456–66.

10. Kraft, V. and Meyer, B. (1990) Seromonitoring in small laboratory animal colonies : a five year survey 1984-1988. *Z. Versuchstierkol.* **33**, 29–35.

11. Lloyd, G. and Jones, N. (1984) Infection of laboratory workers with hantavirus acquired from immunocytomas propagated in laboratory rats. *J. Infect.* 12, 117–125.

12. Mahy, B.W., Dykewicz, C., Fisher-Hoch, S. Ostroff, S., Tipple, M. and Sanchez, A. (1991) Virus zoonoses and their potential for contamination of cell cultures. *Dev. Biol. Stand.* 75, 183–189.

13. Nicklas, W., Kraft, V. and Meyer, B. (1993) Contamination of transplantable tumour cell lines and monoclonal antibodies with rodent viruses. Lab. Anim. Sci., **43**, 296–300.

14. Stacey, G.N. and Possee, R. (1996) Safety aspects of insect cell cultures. *In:* Current

Applications of Cell Culture Engineering V2, Insect Cell Cultures-fundamental and applied aspects. Eds Vlak JM, de Gooijer CD, Tramper J and Mitenburger HG. Kluwer Acad Publishers, Dordrecht, Netherlands, pp 299–303.

15. Vaughn, J.L. (1991) Insect Cells: Adventitious Agents. *Dev. Biol. Stand.* **76**, 319-324.

16. Xue, W. and Minocha, H.C. (1996) Identification of bovine diarrhoea virus receptor in different cell types. Vet. Microbiol. 49, 67–79.

17. Schuurman, R., van Strien, A., van Steenis, B., van der Noorda, J. and Sol, C. (1992) Bovine polyomavirus, a cell-transforming virus with figenic potential. *J. Gen. Virol.* **73**, 2871–2878.

18. Stacey, G.N. (2007) Risk assessment of cell culture processes. In: Medicines from Animal Cells, Eds. Stacey, G.N. and Davis, J.M., J Wiley & sons, Chichester.

19. Erickson, G.A., Bolin, S.R. and Landgraf, J.D. (1991) Viral contamination of foetal bovin serum used for tissue culture: risks and concerns. *Dev. Biol. Stand.* **75**, 173–175.

20. Hallauer, C., Kronauer, G., and Siegl, G. (1971) Parvovirus contaminants of permanent human cell lines virus isolation from 1960–1970, *Arch. Gesamte Virusforsch.* **35**, 80–90.

21. Fabisiak, J.P., Weiss, R.D., Powell, G.A. and Dauber, J.H. (1993) Enhanced secretion of immune-modulating cytokines by human lung fibroblasts during in vitro infection with Mycoplasma fermentans. *Am. J. Respir. Cell Mol. Biol.* **8**, 358–364.

22. Polianskaia, G.G. and Efremova, T.N. (1993) The effect of Mycoplasma contamination and decontamination with ciprofloxacin on the karyotypic structure of the Chinese hamster V-79 lung cell line. *Tsitologia* **35**, 71–78.

23. Lipp, M., Koch, E., Brandner, G. and Bredt, W. (1979) Simulation and prevention of retrovirus--specific reactions by mycoplasmas. *Med. Microbiol. Immunol.* **167**, 127–36.

24. Del Guidice RA and Gardella RS. (1984) Mycoplasma infection of cell culture: effects, incidence and detection, In: In vitro Monograph 5, Uses and standardisation of vertebrate cell cultures, Tissue Culture Association, Gaithersberg MD, USA, 1984, pp104 –115.

25. McGarrity, G.I., Kotani, H. and Butler, G.H. (1993) Mycoplasma and tissue culture cells. In: Mycoplasmas: Molecular Biology and Pathogenesis, eds. Maniloff, J., ASM Press, pp 445–454.

26. Rottem, S. and Naot, Y. (1998) Subversion and exploitation of host cells by mycoplasma. *Trends Microbiol.* **6**, 436–440.

27. Shimada, H. and Itoh, S. (1996) Effects of new quinolone antibacterial agents on mammalian chromosomes. *J. Toxicol. Envir. Health.* **47**, 115–123.

28. Curry, P.T., Kropko, M.L., Garvin, J.R., Fiedler, R.D. and Thiess, J.C. (1996) In vitro induction of micronulclei and chromosome aberrations by quinolones: possible mechanisms. *Mut. Res.* **352**, 143–150.

29. Stacey, G. and Auerbach, J. (2007) Quality control procedures for stem cell lines, In: Culture of Human Stem Cells. Eds Freshney, I.R., Stacey, G.N. and Auerbach, J.M. J Wiley & Sons, Hoboken, New Jersey, pp 1–22.

30. BSEN 12469. (2000) Biotechnology. Performance criteria for microbiological safety cabinets, British Standards Institute, London, UK.

31. ISO 14644-8:2006 (2006) Cleanrooms and associated controlled environments -- Part 8: Classification of airborne molecular contamination. International Organization for Standardization (ISO), Geneva, Switzerland (http://www.iso.org/)

32. BS 2646-4 (1991) Autoclaves for sterilization in laboratories. Guide to maintenance. British Standards Institute, London, UK.

33. Fountain D., Ralston M., Higgins N., Gorlin J.B., Uhl. L., Wheeler C., Antin J.H., Churchill W.H. and Benjamin R.J. (1997) Liquid nitrogen freezers: A potential source of microbial contamination of hematopoietic stem cell components, *Transfusion* **37**, 585–591.

34. Hawkins AE, Zuckerman MA, Briggs M, Gilson RJ, Goldstone AH, Brink NS, Tedder RS. (1996) Hepatitis B nucleotide sequence analysis: linking an outbreak of acute hepatitis B to contamination of a cryopreservation tank. *J. Virol. Methods* **60**, 81–88.

35. Tedder, R.S., Zuckerman, M.A., Goldstone, A.H. et al. (1995) Hepatitis B transmission from a contaminated cryopreservation tank, *Lancet*, 346, 137–140.

36. Stacey, G.N. and Masters, J.R. (2008) Cryopreservation and banking of mammalian cell lines. *Nat. Protocols* **3**, 1981–1989.

37. Lelong-Rebel, I.H., Piemont, Y., Fabre, M. and Rebel, G. (2009) Mycobacterium avium-intracellulare contamination of mammalian cell cultures. *In Vitro Cell Dev. Biol. Anim.* **45**, 75–90.

38. Mowles, J.M., Doyle, A., Kearns, M.J. and Cerisier, Y. (1989) Isolation of an unusual fastidious contaminant bacterium from an adherent cell line using a novel technique. Advances in animal cell biology and technology for bioprocesses, Eds., Spier, R.E.,

Griffiths, J.B., Stephenne, J. and Crooy, P.J., Butterworths & Co., Tiptree, pp. 111–114.
39. Stacey, G.N. (2007) Quality Control of Human Stem Cell Lines. In: Human Embryonic Stem Cells (Human Cell Culture - Volume 6) Eds J. Masters, B. and Palsson, J. Thomson. Springer (Tokyo and New York), pp. 1–22.
40. Young, L, Sung, J., Stacey, G., and Masters JR. (2010) Detection of Mycoplasma in cell cultures. Nat. Protocols, in press.
41. US Food and Drugs Administration. (2005) Title 21, Code of Federal Regulations, Volume 7, revised April 2005, CFR610.12 (Sterility), FDA, Department of Health and Human Services.
42. US Food and Drugs Administration. (2005) Title 21, Code of Federal Regulations, Volume 7, revised April 2005, CFR610.30 (Test for Mycoplama), FDA, Department of Health and Human Services.
43. European Pharmacopeia. (2007). European Pharmacopeia section 2.6.1 (Sterility) (6th Edition), Supplement 8, Maisonneuve SA, Sainte Ruffine, pp. 5795-5797 (www.pheur.org).
44. European Pharmacopeia. (2007). European Pharmacopeia section 2.6.7 (Mycoplasma) (6th Edition), Supplement 6.1, Maisonneuve SA, Sainte Ruffine, pp. 3317–3322.

Chapter 8

Detecting Mycoplasma Contamination in Cell Cultures by Polymerase Chain Reaction

Cord C. Uphoff and Hans G. Drexler

Abstract

The detection of mycoplasmas in human and animal cell cultures is mandatory for every cell culture laboratory, because these bacteria are common contaminants, persist unrecognized in cell cultures for many years, and affect research results as well as the purity of cell culture products. The reliability of the mycoplasma detection depends on the sensitivity and specificity of the method and should also be convenient to be included in the basic routine of cell culture quality assessment. Polymerase chain reaction (PCR) detection is one of the acknowledged methodologies to detect mycoplasmas in cell cultures and cell culture products. Although the PCR offers a fast and simple technique to detect mycoplasmas, the method is also susceptible to errors and can produce false positive as well as false-negative results. Thus, the establishment and the routine application of the PCR assay require optimization and the inclusion of the appropriate control reactions. The presented protocol describes sample preparation, DNA extraction, PCR run, the analysis of the PCR products, and speciation of the contaminant. It also provides detailed information on how to avoid artifacts produced by the method. Established properly, PCR is a reliable, fast, and sensitive method and should be applied regularly to monitor the contamination status of cell cultures.

Key words: Cell lines, Contamination, Detection, Mycoplasma, PCR, Speciation

1. Introduction

1.1. Mycoplasma Contamination of Cell Lines

Contamination of cell cultures with bacteria, fungi, and yeasts represents a major problem in cell culture. Whereas these microorganisms are easily detected during routine cell culture by the turbidity of the culture and by observation under the inverted microscope, one class of bacteria regularly evades detection. These bacteria belong to the class of Mollicutes, commonly known as mycoplasmas. Mycoplasmas may persist undetected in cell cultures for a long time without visibly affecting the culture. Nevertheless, mycoplasmas can cause extensive alterations in the cell cultures (1).

The frequency of contamination is about 10–30% with pronounced variations in the reported series with regards to the laboratory, type of cell culture examined, and the origin of the culture, just to name a few examples. The infecting mycoplasmas are limited to a few species of the genera *Mycoplasma* and *Acholeplasma* with human, swine, and bovine as predominant natural hosts. The ubiquitous use of cell cultures and their transfer from one laboratory to another has lead to widespread dissemination of such infections. Once introduced into a cell culture laboratory, mycoplasmas are usually passed on to uninfected cultures by lapses in sterile cell culture technique. Hence, adequate detection methods need to be established and frequently employed in every laboratory applying cell cultures. Every incoming cell culture should be kept in quarantine until mycoplasma detection assays are completed and the infection status is determined. Positive cultures should either be discarded and replaced by clean cultures or cured with specific antibiotics (see Chapter 9). Only clean cultures should be used for research experiments and for the production of biologically active pharmaceuticals. Additionally, stringent rules for the prevention of further mycoplasma contamination of cell cultures should be followed (2).

1.2. Mycoplasma Detection

Polymerase chain reaction (PCR) provides a sensitive and specific option for the direct detection of mycoplasmas in cell cultures. PCR is useful for the routine screening of cell lines newly introduced into the laboratory, for initial analysis of primary cell cultures, and for the periodical monitoring of growing cell cultures. The advantages of the PCR assay are sensitivity, specificity, speed, cost efficiency, and the potential to screen a large number of samples. Furthermore, an objective result is obtained that is much easier to interpret than most other conventional assays as long as the appropriate control reactions are included in the PCR assay.

Some investigators have described the use of primer sequences that amplify highly conserved regions of the eubacterial 16S rRNA to include all mycoplasma species. But this approach leads to decreased specificity as other bacteria are also detected by the assay. Here, we describe the application of a mixture of oligonucleotides as primers for the amplification of sequences derived only from contaminating mycoplasmas (3). The advantage of this specific approach is that common airborne bacterial contaminations or contaminations that may be present in solutions – those used for washing steps, DNA extraction, and PCR reaction – or may be introduced by other materials will not be detected and will therefore not lead to false-positive results. Nevertheless, major emphasis should be placed on the preparation of the template DNA, the amplification of positive and negative control reactions, and the observance of general rules for the preparation of PCR reactions.

Various Taq DNA polymerase inhibitors can be present in cell cultures and the amount present increases with the age of the culture. To exclude these inhibitors, it is important that the DNA of the samples is extracted and purified. This can be achieved by conventional phenol–chloroform extraction and ethanol precipitation, column extraction, or by DNA binding to matrix. To control the integrity of the PCR reactions and the preceding template preparation steps, it is also essential to perform the appropriate control reactions; including internal, positive, and negative control reactions. The internal control represents DNA containing the same primer sequences but with an additional stretch of interspersed nucleotides, resulting in a gel band of different size than the expected amplicon of the contaminant. This internal control should be added to the PCR reaction in a limiting dilution to recognize any inhibitory components and to demonstrate the sensitivity of the PCR reaction for every sample. The following protocols detail an established PCR method for monitoring potential mycoplasma contamination in any laboratory (4).

2. Materials

2.1. Sample Collection and DNA Extraction Buffers

1. Sterile phosphate-buffered saline (PBS): 140 mM NaCl, 2.7 mM KCl, 7.2 mM Na_2HPO_4, 1.47 mM KH_2PO_4, pH 7.2; autoclaved.
2. Wizard DNA Clean-Up System (Promega, Madison, WI, USA).
3. 80% Isopropanol (v/v).
4. Double-distilled water prewarmed to 80°C.
5. Disposable 2-mL syringes.

2.2. PCR Reaction

1. GeneAmp 9700 thermal cycler (Applied Biosystems, Weiterstadt, Germany).
2. Platinum Taq DNA polymerase (Invitrogen, Karlsruhe, Germany) with 10× reaction buffer and 50 mM $MgCl_2$.
3. 6× Loading buffer: 0.09% bromophenol blue (w/v), 0.09% xylene cyanol FF (w/v), 60% glycerol (v/v), and 60 mM EDTA.
4. Deoxynucleotide triphosphate mixture (dNTP mix): mixture contains 5 mM each of deoxyadenosine triphosphate (dATP), deoxycytidine triphosphate (dCTP), deoxyguanosine triphosphate (dGTP), and deoxythymidine triphosphate (dTTP) (Invitrogen) in H_2O, placed in 50 µL aliquots and stored at –20°C.

5. Primers (any supplier) (see Note 1):

5′ Primers	3′ Primers
cgc ctg agt agt acg ttc gc w	gcg gtg tgt aca aga ccc ga r
cgc ctg agt agt acg tac gc	gcg gtg tgt aca aaa ccc ga
tgc ctg agt agt aca ttc gc r	gcg gtg tgt aca aac ccc ga
tgc ctg agt agt aca ttc gc	
cgc ctg agt agt atg ctc gc r	
cac ctg agt agt atg ctc gc	
cgc ctg ggt agt aca ttc gc	

Stock solutions: 100 µM in dH$_2$O, stored frozen at –20°C.

Working solutions: mix of forward primers at 5 µM each and mix of reverse primers at 5 µM each, placed in 25–50 µL aliquots, and stored frozen at –20°C.

6. Internal control DNA: Internal control DNA may be obtained from the DSMZ (German Collection of Microorganisms and Cell Cultures, Braunschweig, Germany; http://www.cell-lines.de). A limiting dilution should be determined experimentally by performing a PCR with a dilution series of the internal control DNA.

7. Positive control DNA: any mycoplasma-positive sample prepared as described below or may be obtained from the DSMZ.

8. 50× TAE buffer: 2 M Tris base, 5.71% glacial acetic acetate (v/v), 100 mM EDTA, pH 8.0.

9. Agarose (for DNA electrophoresis) (Serva, Heidelberg, Germany).

3. Methods

3.1. Sample Collection and Preparation of DNA

1. Prior to testing, the cells should have been in continuous culture for at least 2 weeks after thawing, grown without antibiotics, and the medium should not have been exchanged for at least 2–3 days after splitting to ensure that the titer of the mycoplasmas is above the detection level of the assay. For adherently growing cells, take directly 1 mL of the medium for the analysis. For suspension cell lines, stand culture flasks vertically allowing cells to settle for about 30 min prior to the removal of 1 mL of the supernatant. These procedures should yield optimum num-

bers of cells to detect mycoplasmas adhering to the eukaryotic cells to facilitate the detection of adherent mycoplasmas. Several mycoplasma species were also described to live intracellularly. Thus, it is also not necessary to centrifuge the sample to eliminate the eukaryotic cells. For collection of a series of samples over a period of time, the crude cell culture supernatants can be stored at 4°C for several days or frozen at –20°C for several weeks.

2. Centrifuge the cell culture suspension in an Eppendorf tube at $13,000 \times g$ for 5 min. Remove the supernatant with a disposable pipette and discard. Resuspend the remaining pellet in 1 mL PBS by vortexing.

3. Centrifuge the suspension again and wash one more time with PBS as described in step 2.

4. After the washing step, resuspend the pellet in 100 µL PBS by vortexing, and heat to 95°C for 15 min.

5. To isolate the DNA present in the solution, the commercially available Wizard DNA extraction kit from Promega is applied (see Notes 2 and 3). Add 1 mL of mixed DNA clean-up resin to the sample solution and mix by pipeting. Keep the solution at room temperature for a few minutes.

6. Fit the minicolumn to a syringe barrel after removing the plunger, transfer the solution into the syringe barrel, and insert into a 1.5 mL Eppendorf tube. Using the plunger, press the resin onto the minicolumn.

7. Detach the syringe from the minicolumn and remove the plunger. Discard the collected supernatant. Reattach the syringe and the minicolumn, insert into the empty tube, and fill the syringe with 2 mL 80% isopropanol for washing the bound DNA. Press the isopropanol through the minicolumn directly into a waste container.

8. Remove the syringe and centrifuge the minicolumn with the tube for 2 min at $13,000 \times g$ to remove the isopropanol.

9. Transfer the minicolumn to a new 1.5 mL tube and add 50 µL of preheated water onto the minicolumn. Centrifuge for 20 s at $13,000 \times g$ to collect the eluate quantitatively. The DNA is stored at –20°C until the PCR reaction is performed.

3.2. Polymerase Chain Reaction

The amplification procedure and the parameters described here are optimized for the application in 0.2 mL reaction tubes in an Applied Biosystems GeneAmp PCR System 9700 thermal cycler. An adjustment to any other equipment might be necessary (see Note 4). Amplified positive samples contain large amounts of target DNA that can contaminate subsequent PCR reaction.

Thus, established rules to avoid DNA carry over should be strictly followed:

(a) Separate the DNA extraction area from the PCR set-up area and the gel run locations.
(b) Store all reagents in small aliquots to provide a constant source of uncontaminated reagents.
(c) Avoid reamplifications.
(d) Reserve pipettes, tips, and tubes for their use in PCR only and irradiate the pipettes frequently with UV light.
(e) Follow strictly the succession of the PCR setup in the protocol.
(f) Wear gloves during the sample preparation and PCR setup.

It is also fundamental to integrate the appropriate control reactions, such as internal, positive and negative controls, and the water control reaction.

1. Prepare two master mixtures (calculated for 25 µL per sample, plus three additional reactions for positive and negative controls and surplus for pipetting variation) containing 1× PCR buffer (use the Taq DNA polymerase buffer recommended by the manufacturer) with 2.0 mM $MgCl_2$, 0.2 mM of each dNTP, 0.1 µM of each oligonucleotide (see Note 1), 1 U hot start Taq DNA polymerase, fill one master mix up to 21 µL per reaction and one master mix to 20 µL with dH_2O. Add 1 µL per reaction of a limiting dilution of the internal control solution to the 20 µL master mix (see Notes 5 and 6). Mix both master mixtures well by pipetting or vortex.

2. Transfer 21 µL of the master mixtures to labeled 0.2 mL PCR reaction tubes.

3. Store all reagents used for the preparation of the master mix and take out the samples of DNA to be tested and the positive control DNA. Do not handle the reagents and samples simultaneously. Add 4 µL of each DNA preparation to a reaction tube containing the reaction solutions without and with the internal control DNA.

4. Transfer the reaction tubes to the thermal cycler and start the PCR with an initial Taq DNA polymerase activation step at 96°C for 2 min.

5. After this activation step, perform 35 thermal cycles with the following parameters:

    Cycle step 1: 4 s at 94°C
    Cycle step 2: 8 s at 65°C
    Cycle step 3: 16 s at 72°C plus 1 s of extension time during each cycle.

6. The reaction is finished by a final amplification step at 72°C for 7 min and the samples are then cooled down to room temperature.

7. Prepare a 1.3% agarose–TAE gel containing 0.3 µg of ethidium bromide per mL. Submerge the gel in 1× TAE and add 12 µL of the amplification product (10 µL reaction mix plus 2 µL of 6× loading buffer) to each well and run the gel at 10 V/cm.

8. Visualize the specific products on a UV screen and document the result.

3.3. Interpretation of the Results

Figure 1 shows a representative ethidium bromide-stained gel with samples that produced the results as described in the following.

Ideally, all samples containing the internal control DNA show a band at 986 bp. This band might be more or less bright, but the band has to be visible, if no other bands are amplified (see Note 7). The absence of this band indicates that the reaction may be contaminated with Taq DNA polymerase inhibitors from the sample preparation. In this case, it is usually sufficient to repeat the PCR run with the same DNA solution as before. It is not necessary to collect a new sample. If the second run also shows no band for the internal control, the whole procedure has to be repeated with new cell culture supernatant.

Mycoplasma-positive samples show an additional band or only a single band at 502–520 bp, depending on the mycoplasma species. In the case of *Acholeplasma laidlawii* contamination and

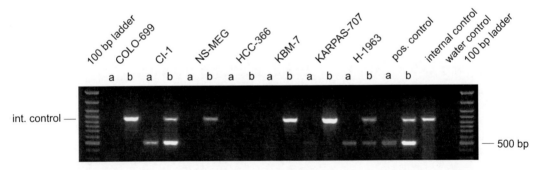

Fig. 1. PCR analysis of mycoplasma status in cell lines: a typical ethidium bromide-stained gel containing the reaction products following PCR amplification with the primer mix listed in Subheading 2.2. Products of about 510 bp were obtained; the differences in length reflect the sequence variation between different mycoplasma species. Shown are various examples of mycoplasma-negative and -positive cell lines. Two paired PCR reactions were performed: one PCR reaction contained an aliquot of the sample only (lanes a) and the second reaction contained the sample under study plus the control DNA as internal standard (lanes b). Cell lines COLO-699, NS-MEG, and KBM-7 are mycoplasma-negative; cell lines Cl-1 and H-1963 are clearly mycoplasma-positive. The analysis of cell line HCC-366 cannot be evaluated as the internal control was not amplified and is not visible. Additionally, the one single positive band did not appear. In this case, the analysis needs to be repeated. KARPAS-707 post-ciprofloxacin treatment shows a weak but distinctive band in the reaction without internal control and also needs to be repeated after a longer antibiotic-free period posttreatment, because the apparently low concentration might result from residual DNA of dead mycoplasmas.

applying the DSMZ internal control DNA, a third band might be visible between the internal control band and the mycoplasma-specific band. This is formed by hybridization of the complementary sequences of the single-stranded long internal control DNA and the single-stranded short wild-type Acholeplasma DNA form.

Contaminations of reagents with mycoplasma-specific DNA or PCR product are indicated by a band in the water control and/or in the negative control sample which are absolutely mandatory to be included. Very weak mycoplasma-specific bands can sometimes occur after the treatment of infected cell cultures with anti-mycoplasma reagents for the elimination of mycoplasma or when other antibiotics such as penicillin/streptomycin are applied routinely. In these cases, the positive reaction might either be due to residual DNA in the culture medium derived from dead mycoplasma cells or from viable mycoplasma cells which are present at a very low titer. Therefore, special caution should be taken when cell cultures are tested that were treated with antibiotics. Prior to PCR testing, cell cultures should be cultured for at least 2–3 weeks without antibiotics, or retested at frequent intervals.

3.4. Further Considerations

Although the method described is sufficient to detect mycoplasma contaminations, it might be of advantage to know the infecting mycoplasma species to determine the source of contamination. This PCR allows the identification of the mycoplasma species most commonly infecting cell cultures. In the case of a contamination detected by PCR, the PCR has to be repeated in a 50 µL volume and without internal control to amplify the mycoplasma-specific PCR fragment. This DNA fragment is then digested in parallel reactions with the restriction endonucleases *Asp*I, *Hpa*II,

Table 1
Restriction fragment patterns of the PCR amplicons digested with various restriction enzymes

	*Asp*I	*Hpa*II	*Xba*I	*Hae*III	*Sfu*I
A. laidlawii	–	–	–	–	436/81
M. arginini	–	–	266/253	–	–
M. bovis	303/213	–	–	–	–
M. fermentans	–	357/111/48	–	356/160	–
M. hominis	–	–	263/253	336/180	–
M. hyorhinis	303/213	–	263/253	–	–
M. orale	–	288/230	266/252	–	–

The numbers represent the sizes of the restriction fragments of the PCR product which was digested with the respective restriction enzyme; —, not digestable and *Xba*I (8 µL PCR reaction mix plus 1 µL of the appropriate 10× restriction enzyme buffer plus 1 µL of the restriction enzyme). Depending on the restriction pattern (see Table 1), the species can be determined directly. Otherwise, two more digestions with the restriction enzymes *Hae*III and *Sfu*I have to follow for the exact identification of the species. Of course, this analysis allows only the determination of those mycoplasma species that most often (>98%) occur in cell cultures and is not suitable for the global identification of all types of mycoplasma species. Cell culture infections are commonly restricted to the seven mycoplasma species listed in Table 1 (1, 5).

4. Notes

1. The primers can be designed using the standard IUPAC nucleotide code to incorporate two different nucleotides to form a mixture of two primers (Subheading 2.2). Do not use more than one accessory symbol per sequence. When all oligonucleotides are mixed and prepared in aliquots for use in the PCR reaction, remember that the molarities of the sequences containing accessory symbols are reduced by 50%. The primer solutions should be placed into small portions (25 µL aliquots) and stored frozen at –20°C to avoid multiple freeze-thawing cycles and to minimize contamination risks.

2. Any other DNA extraction and purification method that eliminates PCR inhibitors should work as well. We tested normal phenol/chloroform extraction and ethanol precipitation, High Pure PCR Template Preparation Kit from Roche (Mannheim, Germany), Invisorb Spin DNA MicroKit III from Invitek (Berlin, Germany), and Epicentre MasterPure Complete DNA & RNA Purification Kit (Biozym, Hessisch Oldendorf, Germany). Following these methods, the amplification of the mycoplasma sequences was similar to the one after the Wizard preparation when the DNA was dissolved or eluted in 50 µL of water. Applying phenol/chloroform extraction, both reagents need to be removed completely by ethanol precipitation, because they inhibit the Taq DNA polymerase even at low quantities.

3. Use of crude samples directly in the PCR reaction after heating to 95°C as described in some publications is not recommended. Even different dilutions of the unextracted samples should not be used, because endogenous substances might act as inhibitors of the Taq DNA polymerase at very low concentrations and cause false-negative results. Even after DNA extraction, the PCR reaction can be inhibited by the unidentified

substances. For this reason, internal control reactions need to be incorporated in the test series.

4. Although the described PCR method is very robust, the use of thermal cyclers other than the GeneAmp 9700 might require some modifications in the amplification parameters, for example, duration of the cycling steps, which are short in comparison to other applications. Also the concentrations of Mg^{2+}, primer, or dNTPs might need to be altered. The same is true if another Taq DNA polymerase is used, either polymerases from different suppliers or different kinds of Taq DNA polymerase.

5. The limiting dilution of the internal control DNA can be used maximally for 2 or 3 months when stored at 4°C. After this time, the amplification of the internal control DNA might fail even when no inhibitors are present in the reaction, because the DNA concentration might be reduced owing to degradation or attachment to the plastic tube.

6. Applying the internal control DNA, the described PCR method is competitive only for the group of mycoplasma species that carries primer sequences identical to the one from which the internal control DNA was prepared. The other primer sequences are not used up in the PCR reaction as a result of mismatches. Usually one reaction per sample is sufficient to detect mycoplasma in "naturally" infected cell cultures. To avoid the possibility of performing a competitive reaction and of decreasing the sensitivity of the PCR reaction, for example, after anti-mycoplasma treatment or for the testing of cell culture reagents, two separate reactions can be performed: First, run a reaction without internal control DNA to make all reagents available for the amplification of the specific product. Second, an additional reaction including the internal control DNA to demonstrate the integrity of the PCR reaction (Fig. 1).

7. During the analysis of heavily infected cultures, sometimes the internal control is not visible in the gel, owing to overwhelming amounts of target DNA in the sample compared with the trace amounts of the internal control DNA. But in these cases, the correct run of the PCR reaction is demonstrated by the appearance of the band that is indicative of mycoplasma positivity.

References

1. Drexler, H. G. and Uphoff, C. C. (2002) Mycoplasma contamination of cell cultures: Incidence, sources, effects, detection, elimination, prevention *Cytotechnol* **39**, 23–38.

2. Uphoff, C. C. and Drexler, H. G. (2001) Prevention of mycoplasma contamination in leukemia-lymphoma cell lines *Human Cell* **14**, 244–7.

3. Hopert, A., Uphoff, C. C., Wirth, M., Hauser, H., and Drexler, H. G. (1993) Specificity and sensitivity of polymerase chain reaction (PCR) in comparison with other methods for the detection of mycoplasma contamination in cell lines *J Immunol Methods* **164**, 91–100.
4. Uphoff, C. C. and Drexler, H. G. (2002) Detection of mycoplasma in leukemia-lymphoma cell lines using polymerase chain reaction *Leukemia* **16**, 289–93.
5. Uphoff, C. C. and Drexler, H. G. (2010) Contamination of cell cultures, mycoplasma. In: Encyclopedia of Industrial Biotechnology: Bioprocess, Bioseparation, and Cell Technology (Flickinger M. C., ed), Wiley-Blackwell, New York, pp. 3611–3630.

Chapter 9

Elimination of Mycoplasmas from Infected Cell Lines Using Antibiotics

Cord C. Uphoff and Hans G. Drexler

Abstract

Mycoplasma contamination may have multiple effects on cultured cell lines and this can also have a significant influence on the results of scientific studies, and certainly on the quality of cell culture products used in medicine and pharmacology. The elimination of mycoplasma contamination from cell cultures is an inevitable need when contaminated cell lines cannot be replaced by mycoplasma-free cell lines. The eradication of mycoplasmas with specific antibiotics has been shown to be one of the most efficient methods to permanently cleanse the cell culture. The effective antibiotics belong to the classes of fluoroquinolones, macrolides, and tetracyclines. The protocols describe the single use of diverse fluoroqinolones as well as combination therapies with antibiotics from different antibiotic classes or of antibiotics and antimicrobial peptides. Special emphasis has been placed to assure eradication of mycoplasma without the development of resistant mycoplasma strains and to treat heavily infected and damaged cultures as well as cultures which are sensitive to specific antibiotics. Experience has shown that no detectable alterations are found to affect eukaryotic cells during the course of the treatment. Antibiotic treatment of mycoplasma infected cultures appears to be the most practical, robust, and efficient approach.

Key words: Antibiotics, Cell culture, Cell lines, Elimination, *Mycoplasma*, Resistance, Treatment

1. Introduction

1.1. Mycoplasma Elimination: A Necessity

Although many mycoplasma contaminations of cell cultures remain undetected over years and thus seem to have no apparent influence on the growth or other characteristics of the cells, this is a considerable misinterpretation of the symbiotic relationship of eukaryotic cells and mycoplasmas. Mycoplasmas indeed have a multitude of different effects on various eukaryotic cells. However, these effects are not constantly seen in all infected cultures, but depend on the mycoplasma species, the type of cell culture, and the cell culture conditions. Nearly every parameter measured in cell culture or experimental investigations may be affected by mycoplasmas (1).

The spectrum of effects ranges from more or less marked growth inhibition, altered levels of protein, RNA, and DNA synthesis, induction of chromosomal aberrations, alteration of the membrane composition including surface antigen and receptor expression combined with alterations of signal transduction. The specificity of the infections is indicated by the adverse effects of induction or inhibition of activation, proliferation, and differentiation of hematopoietic cells, the induction or suppression of cytokine and growth factor expression, enhanced immunoglobulin secretion by B-cells, increase or decrease of virus propagation, and a variety of other parameters. Many of the effects mentioned above are attributable to the deprivation of essential culture medium components and the degradation of specific nucleic acids and their precursors, amino acids, lipids, and so forth. Moreover, the metabolites can be harmful for eukaryotic cells because of the production of acids from the fermentation (by fermentative mycoplasmas) or ammonia (by the arginine-hydrolyzing mycoplasmas) and is highly toxic to the eukaryotic cells (2).

Thus, mycoplasma contaminations cannot be regarded as harmless, but may have detrimental consequences. Moreover, contaminated cell cultures represent a source of infection for other cell cultures. Consequently, cell cultures harboring mycoplasmas should either be autoclaved and discarded immediately or taken into quarantine and cured of the infection (3).

1.2. Antibiotic Treatment of Mycoplasma-Infected Cell Cultures

Taking into account that 10–30% of all cell lines maintained in laboratories are infected with mycoplasmas, and many of these cannot be replaced by uncontaminated cell cultures, there is a clear need for effective and feasible methods for the elimination of mycoplasmas. A number of methods have been described to cure cell lines from the contaminants. These include physical, chemical, immunological, and chemotherapeutic procedures (2). Many of the elimination methods are of limited value for use in cell culture, because they are not always effective, and the mycoplasmas reappear after a while, are too laborious or require special equipment or facilities, or show detrimental effects on the eukaryotic cells (4). The application of chemotherapeutic reagents, such as antibiotics, currently seems to be the method of choice, because it is practicable for almost any cell culture laboratory.

Concerning morphology, biochemistry, and genetics, mycoplasmas exhibit special features which make them unique among the eubacteria. This is the reason why many antibiotic agents commonly used in cell culture, mainly penicillin and streptomycin, have no effect on the viability of mycoplasmas, although they may sometimes suppress their growth (see Note 1). Thus specific antimicrobial agents have to be applied.

Over recent years, three unique groups of antibiotics have been shown to be highly effective against mycoplasmas, both in human/veterinary medicine and in cell culture: fluoroquinolones, macrolides, and tetracyclines. Fluoroquinolones inhibit the bacterial DNA gyrase, essential for the replication of the DNA. Tetracyclines and macrolides inhibit protein synthesis by binding to the 30S and 50S ribosomal subunits, respectively. It is important to have a number of antibiotics available which act differently against mycoplasmas, because the development of resistant clones and the possible cross-reactivity with weakened eukaryotic cells can be overcome by the application of an antibiotic from an unrelated group of reagents. In addition, antimicrobial peptides such as alamethicin, gramicidin S, or surfactin have been shown to kill mycoplasmas in cell culture, although they have to be combined with other antibiotics to eliminate mycoplasmas permanently from cell cultures (5).

It should be mentioned that the antibiotics are not intended to be used as a general supplement in cell culture to avoid mycoplasma infections, but only for the short time period during anti-mycoplasma treatment. Here, we describe the use of several antibiotics and antimicrobial peptides for the treatment of mycoplasma-contaminated cells, the rescue of heavily infected cultures, salvage treatment of resistant cultures, and some pitfalls during and after treatment.

2. Materials

As many antibiotics are light sensitive, protect the stock and working solutions from light as well as the cell cultures containing the antibiotics as much as possible.

- Baytril (Bayer, Leverkusen, Germany) contains 100 mg/mL of enrofloxacin and is diluted 1:100 with RPMI 1640 medium immediately prior to the treatment. The dilution should be prepared freshly for every anti-mycoplasma treatment. This solution is used as 1:40 final dilution in cell culture (at 25 µg/mL).
- Ciprobay 100 (Bayer) contains 2 mg/mL ciprofloxacin and can be used 1:200 in cell culture (at 10 µg/mL). One milliliter aliquots should be taken in a sterile manner from the bottle and stored at 4°C.
- MRA (Mycoplasma Removal Agent, ICN, Eschwege, Germany) contains 50 µg/mL of a 4-oxo-quinolone-3-carboxylic acid derivative (specific type of antibiotic not disclosed) and is used in the treatment of cell cultures in 1:100 dilutions (at 0.5 µg/mL).
- Plasmocin (InvivoGen, San Diego, CA) contains two antibiotics (specific type of antibiotic not disclosed), of which one

is active against protein synthesis of bacteria and one inhibits the DNA replication (gyrase inhibitor). The antibiotics are applied concurrently and the mixture is a ready-to-use solution and applied 1:1,000 in the cell culture (at 25 μg/mL final concentration).

- BM-Cyclin (Roche, Mannheim, Germany) contains the macrolide tiamulin (BM-Cyclin 1) and the tetracycline minocycline (BM-Cyclin 2) both in lyophilized states. Dissolve the antibiotics in 10 mL sterile distilled H_2O, aliquot in 1 mL fractions and store at −20°C. Repeated freezing and thawing of the solutions is not critical for the activity of the antibiotics. The dissolved solutions can be used as 250-fold dilutions in cell culture (at 10 and 5 μg/mL, respectively).

- MycoZap (Lonza, Verviers, Belgium) is a combination of an antimicrobial peptide (MycoZap reagent 1) and an antibiotic (MycoZap reagent 2) (specific types of reagents not disclosed), which are employed consecutively. The solutions are ready to use.

3. Methods

3.1. General Considerations

1. Prior to treatment, aliquots of the contaminated cell line should be stored frozen, but separated from noninfected cultures, either at −80°C for short time (maximally 2–3 months) or preferably in liquid nitrogen in separate tanks (see Note 2). The ampoules have to be marked properly as mycoplasma-positive to prevent a mix-up of ampoules containing cured or infected cells. After successful cure, these mycoplasma-positive ampoules will be removed and the cells destroyed by autoclaving.

2. The mycoplasma infection often impedes the growth and negatively affects the general appearance of eukaryotic cells; in other words, the cells are objectively and subjectively not in the best of health. These impaired conditions are further aggravated by the exposure of the cells to the antibiotics, as the agents appear to affect those cells more than cells in a very good condition. Sometimes the desired anti-mycoplasma effect is offset by the negative effect of the antibiotics on the eukaryotic cells. Therefore, a few general rules should be followed to improve the culture conditions and to reduce the stress of infection and treatment on the eukaryotic cells:

    - Increase the FBS concentration to 20%, even if the cells grow well at lower concentrations.
    - Culture the cells at a medium cell density and keep this density almost constant over the period of treatment and a few weeks posttreatment.

- Observe the culture daily under the inverted microscope to recognize quickly any alteration in general appearance, growth, morphology, decrease in cell viability, detachment of cells, formation of granules, vacuoles, and so forth.
- In the case of deterioration of the cell culture, interrupt the treatment for a few days and let the cells recover (but this should only be a last resort).
- Detach frequently even slowly growing adherent cells to facilitate the exposure of all mycoplasmas to the antibiotic; the contaminants should not have the opportunity to survive in sanctuaries such as cell membrane pockets. It is similarly helpful to break up clumps of suspension cells by vigorous pipetting or using other reagents, for example, trypsin.
- Store the antibiotics at the recommended concentrations, temperatures, and usually in the dark, and do not use them after the expiration date.
- Prepare the working solutions freshly for every treatment and add the solution directly to the cell culture and not to the stored medium.
- Keep the concentration of the antibiotic constant during the treatment period; degradation of the antibiotic can be avoided by frequent exchange of the medium.

3. We recommend applying two types of treatment in parallel, a single antibiotic treatment and one of the combination therapies.

A schematic overview of the procedure is given in Fig. 1; an exemplary representation of the treatment with BM-Cyclin is shown in Fig. 2.

3.2. Antibiotic Treatment

3.2.1. Treatment with Fluoroquinolones and Plasmocin

1. Bring the cells into solution (detach adherent cells, break up clumps by pipetting or using other methods) (see Note 3); determine the cell density and viability by trypan blue exclusion staining. Seed out the cells at a medium density (see Note 4) in a 25 cm^2 flask or one well of a 6- or 24-well culture plate with the appropriate fresh and rich culture medium (10 mL for the flask and 2 and 1 mL for the wells, respectively). Add one of the following antibiotics to the cell culture and incubate for 2 days:
   - 25 µL of a 1 mg/mL solution enrofloxacin (Baytril) per mL medium
   - 10 µL of a 50 µg/mL solution MRA per mL medium
   - 5 µL of a 2 mg/mL solution ciprofloxacin (Ciprobay) per mL medium
   - 1 µL of a 25 µg/mL solution Plasmocin per mL medium
2. Remove all cell culture medium in flasks or wells containing adherent cells or after centrifugation of suspension cells. If

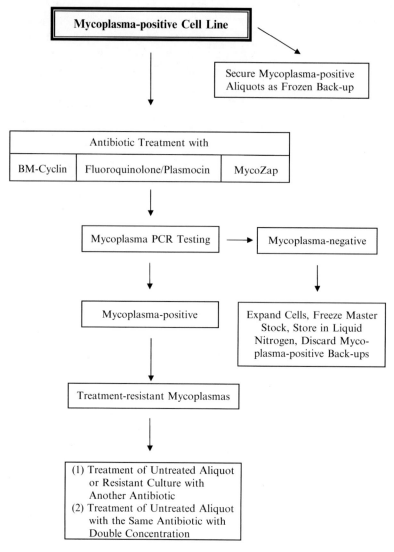

Fig. 1. Recommended scheme for mycoplasma eradication. An arsenal of different antibiotics can be used to treat mycoplasma-contaminated cell lines with a high rate of expected success. We recommend (1) cryopreservation of original mycoplasma-positive cells as back-ups and (2) splitting of the growing cells into different aliquots. These aliquots should be exposed singly to the various antibiotics. It is important to point out that quinolone-resistant cultures could still be cleansed by BM-Cyclin. Posttreatment mycoplasma analysis and routine monitoring with a sensitive and reliable method (e.g. by PCR) are of utmost importance.

applicable, dilute the cell cultures to the medium cell density. Add fresh medium and the same concentration of the respective antibiotic as used in step 1. Incubate for another 2 days (see Note 5).

3. Applying enrofloxacin or MRA, repeat step 2 another two times (altogether 8 days treatment). Employing ciprofloxacin or Plasmocin, repeat step 2 five times (altogether 14 days treatment). Proceed with protocol in Subheading 3.3.

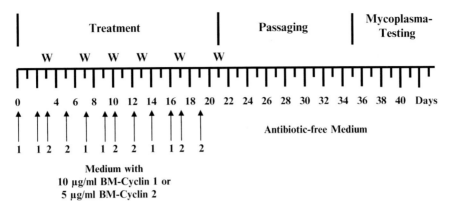

Fig. 2. Treatment protocol for BM-Cyclin. Antibiotics are given on the days indicated by *arrows*. Cells are washed (indicated by *w*) with PBS prior to the cyclical change of antibiotics to avoid the formation of resistant mycoplasmas due to low concentrations of the antibiotics. At the end of the decontamination period, cells are washed with PBS and suspended in antibiotic-free medium. After a minimum of 2 weeks posttreatment, the mycoplasma status of the cells is examined with sensitive and robust methods (e.g. by PCR).

3.2.2. Treatment with BM-Cyclin

1. Bring the cells into solution (detach adherent cells, break up clumps by pipetting or using other methods) (see Note 3); determine the cell density and viability by trypan blue exclusion staining. Seed out the cells at a medium density (see Note 4) in a 25 cm² flask or one well of a 6- or 24-well culture plate with the appropriate fresh and rich culture medium (10 mL for the flask and 2 and 1 mL for the wells, respectively). Add 4 µL of a 2.5 mg/mL solution BM-Cyclin 1 (tiamulin) per mL medium. Incubate the cell culture for 2 days.

2. Remove all cell culture medium in flasks or wells containing adherent cells or after centrifugation of suspension cells. If applicable, dilute the cell cultures to the medium cell density. Add fresh medium and the same concentration of BM-Cyclin 1 as used in step 1. Incubate for another day. This procedure will keep the concentration of the antibiotic approximately constant over the 3 day cycle applying tiamulin (see Note 5).

3. Remove the medium and wash the cells once with PBS to remove the residual antibiotic agent completely from the cells and loosely attached mycoplasmas (see Note 6). Seed out the cells at the appropriate density and add 4 µL of the 1.25 mg/mL solution BM-Cyclin 2 per mL medium. Incubate the culture for 2 days.

4. Remove the culture medium and substitute with fresh medium. Add the same concentration of BM-Cyclin 2 as used in step 3. Incubate the cell culture for 2 days to complete the 4 day cycle of minocycline treatment.

5. Repeat steps 1–4 twice (three cycles of BM-Cyclin 1 and BM-Cyclin 2 altogether). Proceed with protocol in Subheading 3.3.

3.2.3. Treatment with MycoZap

1. The activity of the antimicrobial peptide is influenced by the concentration of the fetal bovine serum (FBS). Thus, the FBS concentration of the cell culture medium should not exceed 5% during the treatment with MycoZap reagent 1. Add 500 μL MycoZap reagent 1 to 4.5 mL cell culture medium supplemented with maximally 5% FBS.

2. Bring the cells into solution (detach adherent cells, break up clumps by pipetting or using other methods) (see Note 3); determine the cell density and viability by trypan blue exclusion staining. Seed out 5×10^5 cells in 5 mL cell culture medium supplemented with maximally 5% FBS in a 25 cm² flask. Add 5 mL medium containing MycoZap reagent 1 prepared in step 1 and incubate the cells until the culture reaches the medium density, for at least 2 days.

3. Remove all cell culture medium in the flask containing adherent cells or after centrifugation of suspension cells. If applicable, dilute the cell cultures to the medium cell density. Add 9.5 mL fresh medium (containing the normal FBS concentration) and 0.5 mL of MycoZap reagent 2. Incubate for 2 days.

4. Repeat step 3 another two times (altogether 6 days of treatment) (see Note 5). Proceed with protocol in Subheading 3.3.

3.3. Culture and Testing Posttreatment

1. When the treatment is completed, remove the supernatant and seed out the cells as described in Subheading 3.2. Keep the cells at the higher cell density and use the enriched medium, but do not add any antibiotic compounds, also no penicillin/streptomycin or similar. The cells should then be cultured for at least another 2 weeks. Even if initially the cells appear to be in good health after the treatment, we found that the cells might go into a crisis after the treatment, especially following treatment with BM-Cyclin. Thus, the cell status should be frequently examined under the inverted microscope.

2. After passaging, test the cultures for mycoplasma contamination. If the cells are clean, freeze and store aliquots in liquid nitrogen. The cells in active culture have to be retested periodically to ensure continued freedom from mycoplasma contamination.

3. After complete decontamination, expand the cells and freeze master stocks of the mycoplasma-free cell line and store them in liquid nitrogen to provide a continuous supply of clean cells. Discard the ampoules with mycoplasma-infected cells.

3.4. Further Considerations

In our experience, it is of advantage to employ two types of treatments (one of the fluoroquinolones and BM-Cyclin or Plasmocin or MycoZap) in parallel, as usually at least one of the treatments is successful. In the rare event of resistance, cells of the untreated frozen back-up aliquots can be thawed and treated again with another antibiotic (Fig. 1). As MRA, ciprofloxacin, and enrofloxacin all belong to the group of fluoroquinolones, it is likely that the use of an alternative compound from the same group will produce the same end result (cure, resistance, or culture death). In the case of loss of the culture during or after the treatment, aliquots can be treated with MycoZap or MRA, as these treatments are usually well tolerated by the eukaryotic cells and show almost no effect on growth parameters during the treatment. These treatments are also recommended when the cells are already in a very bad condition prior to treatment, and the number of available cells would suffice only for one single treatment. Sometimes, the cells recover rapidly after starting the treatment due to the immediate reduction of mycoplasma load and the ensuing release of the cells from the mycoplasmal stranglehold.

If the cytotoxic effects of the antibiotics appear to be too strong, so that one might risk the loss of the cell culture (occasionally observed when BM-Cyclin is employed), the procedure can be interrupted for a short time, i.e., 1 or 2 days, to allow the cells sufficient time for recovery. This should be done when the first symptoms of serious cell culture deterioration begin to appear. If the cells are already beyond a certain degree of damage, it is difficult to reverse the progression of apoptosis.

4. Notes

1. The general use of antibiotics in cell culture is not recommended except under special circumstances and then only for short durations. Use of antibiotics may lead to lapses in aseptic technique, to the selection of drug-resistant organisms, and to the delayed detection of low-level infection with either mycoplasmas or other bacteria.

2. Storage in liquid nitrogen could be a potential contamination source of cell cultures with mycoplasmas, as mycoplasmas have been shown to survive in liquid nitrogen even without cryopreservation. Storage of the ampoules in the gaseous phase of the nitrogen is recommended to prevent contamination. Moreover, contaminated cell cultures and those of unknown status should be stored separate from noninfected cells, preferably in separate tanks to avoid inadvertent mix-up with noninfected cultures even after years of storage.

3. It is important to break up the clumps and clusters and to detach the cells from the surface of the culture vessels. Although the antibiotics are in solution and should be accessible to all parts of the cells, the membranes can be barriers

which cannot be passed by the antibiotics. Mycoplasmas trapped within clumps of eukaryotic cells or even in cavities formed by the cell membrane of a single cell can be protected from the antibiotic. This is also the reason for the advice to keep the concentration of the antibiotic constantly high by regularly exchanging the medium.

4. Depending on the growth rate of the cell line, which might be severely altered by the antibiotic, the cell density should be adjusted. If no data are available at all for a given cell culture, or if the cell culture is in a very bad condition, the cell density, growth rate, and viability should be recorded frequently to improve the condition of the culture.

5. Some cell lines are sensitive to a complete exchange of the medium. If the medium can only be exchanged partially, we found that 50% of the antibiotic concentration should be added to the remaining conditioned medium that already contains the antibiotic, whereas 100% of the antibiotic concentration is added for the fresh medium.

6. The concentration of the antibiotics should be kept at a constant level throughout the treatment period. Low antibiotic concentrations attributable to degradation in culture or dilution by passaging the cell culture may lead to the development of resistant mycoplasma strains. Thus, exchange of the complete medium is recommended.

References

1. Barile, M. F. and Rottem, S. (1993) Mycoplasmas in cell culture. In: Rapid Diagnosis of Mycoplasmas (Kahane I, Adoni A, eds), Plenum Press, New York, 155–93.
2. Drexler, H. G. and Uphoff, C. C. (2002) Mycoplasma contamination of cell cultures: Incidence, sources, effects, detection, elimination, prevention *Cytotechnol* **39**, 23–38.
3. Uphoff, C. C. and Drexler, H.G. (2001) Prevention of mycoplasma contamination in leukemia-lymphoma cell lines. *Human Cell* **14**, 244–7.
4. Uphoff, C. C., Meyer, C., and Drexler, H. G. (2002) Elimination of mycoplasma from leukemia-lymphoma cell lines using antibiotics. *Leukemia* **16**, 284–8.
5. Fehri, L. F., Wróblewski, H., and Blanchard, A. (2007) Activities of antimicrobial peptides and synergy with enrofloxacin against Mycoplasma pulmonis. *Antimicrob Agents Chemother* **51**, 468–74.

Chapter 10

Quality Assurance and Good Laboratory Practice

Louise A. Knight and Ian A. Cree

Abstract

Poor cell culture practice leads to poor science due largely to issues of cross-contamination between cell lines and of microbial contamination, but can be avoided by careful quality control and good laboratory practice. This chapter provides a brief and practical outline of the steps needed to mitigate the risks associated with poor cell culture practice. Good Laboratory Practice (GLP) is a set of principles that provides a framework within which laboratory studies are planned, performed, monitored, recorded, reported, and archived to ensure the reliability of data generated within a compliant laboratory. A key feature of this is the generation of quality-control methods and data management within the cell culture laboratory.

Key words: GLP, GCCP, Identity, Accuracy, Audit

1. Introduction

When used and handled correctly (see Note 1), cell lines are a useful tool in research, enabling scientists to investigate a variety of hypotheses; however, as the use of cell lines has grown, so have the practices of poor cell culture. The problems, highlighted in a number of the preceding chapters, include cross-contamination with another cell line, mycoplasma contamination, and over subculture leading to an alteration in cell characteristics. On the basis of submissions to major cell repositories during the last decade, between 18 and 36% of cell lines may be contaminated or misidentified (1), and although estimates vary, as many as 20% of scientific publications may report results based on contaminated or misidentified cell lines (2) (see Notes 2 and 3).

2. Good Practice

2.1. Good Laboratory Practice

Good Laboratory Practice (GLP) is a set of principles (3), (Table 1) that provides a framework within which laboratory studies are planned, performed, monitored, recorded, reported, and archived (see Note 4), thereby ensuring that data produced are reliable for the purposes of risk assessment. In the context of human pharmaceuticals, GLP is only applicable to preclinical toxicology studies that are used by regulatory receiving authorities to decide whether to grant a clinical trial authorisation or a marketing authorisation. GLP is also designed so that study events can be reconstructed if issues are identified at any point in the drug development process.

GLP studies can be performed in vivo and in vitro; if conducted in vitro using animal or human cell lines, the GLP principles would require that the study director had an appropriate amount of information on the cell line to assure that it is fit for purpose. This may include its identity, source, passage number, and data to ensure it is mycoplasma free. It is important to highlight, however, that even though this information is required, it does not have to be generated in a GLP facility, and as such there are no specific regulatory GLP requirements for the use of cell lines in GLP studies (MHRA – personal communication).

2.2. Good Cell Culture Practice

The first European Centre for the Validation of Alternative Methods (ECVAM) task force on Good Cell Culture Practice (GCCP) was established following proposals made at a workshop on the standardisation of cell culture procedures held during the 3rd World congress on alternatives and animal use in the life sciences. The proposal was that guidelines should be developed to define minimum standards in cell and tissue culture practices; the principles (Table 2) are analogous to the Organisation for Economic Co-operation and Development (OECD) principles of GLP, which cannot normally be fully implemented in basic research, including in vitro studies (4).

3. Methods

3.1. Auditing Cell Culture Practices in a Laboratory

Audit is defined in ISO 9001:2000 as "a systematic, independent and documented process for obtaining evidence and evaluating objectively the extent to which audit criteria are fulfilled" (5).

Table 1
A summary of the GLP principles and associated subsections

Part number	GLP principle	Subsection
I	Test facility organisation and personnel	Facility management's responsibility Study director's responsibilities Principal investigator's responsibilities Study personnel's responsibilities
II	Quality assurance programme	General Responsibilities of the quality assurance personnel
III	Facilities	General Test system facilities Facilities for the handling test, reference, and control items Archive facilities Waste disposal
IV	Apparatus, materials and reagents	–
V	Test systems	Physical/chemical Biological
VI	Test and reference items	Receipt, handling, sampling, and storage Characterisation
VII	Standard operating procedures	–
VIII	Performance of the regulatory study	Study plan Content of the study plan Conduct of the study Conduct of the regulatory study
IX	Reporting study results	General Content of the final report
X	Storage and retention of records and materials	–

A planned programme of auditing can help to test the effectiveness of a system, allowing weaknesses to be rectified and strengths to be built upon.

Table 3 (6) outlines a template for auditing cell culture practices, which can be adapted to the needs of individual laboratories. The GCCP principle that the section of the audit is particularly relevant is indicated in the table; however, this does not imply that the other principles should not be considered as there is a

Table 2
A summary of the GCCP principles

Number	GCCP principle
1.	Establishment and maintenance of a sufficient understanding of the in vitro system and of the relevant factors that could affect it
2.	Assurance of the quality of all materials and methods and of their use and application, to maintain the integrity, validity, and reproducibility of any work conducted
3.	Documentation of the information necessary to track the materials and methods used, to permit the repetition of the work, and to enable the target audience to understand and evaluate the work
4.	Establishment and maintenance of adequate measures to protect individuals and the environment from any potential hazards
5.	Compliance with relevant laws and regulations and with ethical principles
6.	Provision of relevant and adequate education and training for all personnel, to promote high-quality work and safety

considerable overlap between them. An example of an answer is also given in Subheadings 2.1 and 2.2.

4. Notes

1. http://lgcstandards-atcc.org

   In the technical support section, there are a number of online resources including useful technical bulletins and cell-biology resources.

2. For a summary of misidentified cell lines compiled by Dr. R. Ian Freshney and Dr. Amanda Capes-Davis, published on 20th June 2009 see the following Web site:

   http://www.hpacultures.org.uk/services/celllineidentityverification/misidentifiedcelllines.jsp

3. http://www.hpacultures.org.uk

   Here, you can find details on the European collection of cell cultures (ECACC), a Health Protection Agency (HPA) culture collection. There are also links to an extensive list of misidentified cell lines.

4. http://www.mhra.gov.uk

   In the "How we regulate" section, there is a "medicines" category that links to "inspection and standards." Information on GLP is provided here including The Good laboratory Practice Regulations – SI 1999/3106.

Table 3
Suggested template for an audit on cell line stores and cell culture practices in a laboratory

Area of audit	Findings	SOP number checked and/or other documentation	Compliant?	Action if not compliant	GCCP principle
1. Documentation					
a. Is the original cell line documentation available for the cell line?	E.g. The delivery note was filed, and each cell line received was indicated with a tick	Delivery note	Yes	N/A	3
b. Is there sufficient information to allow unequivocal identification of the cell line? *E.g. A record of the cell line identification code, number of vials received, lot number (if applicable)*					
c. Is the documentation signed to confirm that items ordered match items received?	It was signed by the receiver and there were no notes to indicate any problems				
d. Is the documentation date/time stamped?	The paperwork was date/time stamped				
e. Is there a Material transfer agreement (MTA) available?					
2. Cell line transportation					
a. Are there written procedures for cell line transport? *E.g. Guidelines for couriers, porters, packaging, labelling, and dispatch procedures*	E.g. No SOP was available for cell line transport	Details of what to do when cell lines that were transported were recorded in a laboratory book	No	Instructions in lab book should be transferred to an authorised SOP	3 4 5
b. Are these procedures readily available where required?					
c. Does the laboratory/transport staff know what to do if there is a problem with a specimen?					
d. Do sample transport arrangements meet current legislation and guidance?					

(continued)

Table 3 (continued)

Area of audit	Findings	SOP number checked and/or other documentation	Compliant?	Action if not compliant	GCCP principle
e. Is there evidence of procedures meeting all Health and Safety requirements? *E.g. Procedures in the event of spillage, incident reporting procedures*					
f. If the laboratory does not directly manage sample transport has there been an appropriate audit to ensure that all health and safety procedures are in place?					
3. Cell line reception					
a. Are there written procedures for sample receipt covering the following components: checking delivery note and cell line received; recording date/time of receipt; contacting supplier to confirm receipt when applicable; ensuring staff safety?					2
b. Is there a written procedure for rejection of incorrect cell line deliveries covering the following components: criteria of rejection; recording rejected cell lines; notification to supplier?					3
c. Has a record been made of whether the cell line was resurrected immediately or stored?					
d. Does the laboratory have a separate area for quarantine of new cell lines? If not what steps are taken to minimise the risk of contamination? *E.g. A separate safety cabinet, incubator, etc.*					

4. **Examination procedures**
   a. Is there evidence that new cell lines are characterized to confirm identity and that they are free from infection? 1
   b. Is there evidence that cell lines are regularly assessed to ensure that they are free from mycoplasma contamination? 2
   c. Are there documented standard operating procedures readily available and are they detailed enough to allow a consistency?
      List those checked
   d. Were procedures followed in accordance with documented method?
      List procedures checked
   e. Are records available for each stage of the procedure such that there is a complete audit trail?
      List all records examined

5. **Quality control**
   a. Are there written procedures for regular quality control tests/checks of the cell lines and associated materials? 2
      E.g. Authenticity, sterility, performance of controls 3
      List check procedures
   b. Are QC failures reported?

6. **Equipment**
   a. Where equipment was observed being used, is there evidence of routine maintenance and daily checks being performed as appropriate? 2
      List equipment checked 3
   b. Calibration and servicing where required
      List evidence e.g. calibration certificate etc.
   c. If equipment is not in use, is it labelled as such and a record of breakdown (if applicable) recorded?

(continued)

Table 3
(continued)

Area of audit	Findings	SOP number checked and/or other documentation	Compliant?	Action if not compliant	GCCP principle
7. Reagents a. Were all reagents properly labelled? b. Were all reagents dated and suitable for use? c. Were they used and stored correctly? d. If reagents were stored at ambient temperature, was this monitored? e. Were reagents rotated or changed regularly and was this recorded?					3
8. Disposal procedures a. Are procedures established for safe disposal of these samples? b. Were all waste reagents disposed of safely and according to regulations? *Are procedures in accordance with current legislation and guidelines?*					4 5
9. Reporting data a. Is there a written procedure for reporting results (whether it is for an internal report or for submission to a scientific journal) listing what details should be included? *E.g. Culture type, cell/tissue type, species, origin, description, product no., culture medium, serum, antibiotics, other additives, freq. of media change, flask type, culture plate type, subculture freq., subculture split ratio, detachment solution, usable passage range, passage no. at receipt, passage no. at use, maintenance conditions, storage conditions* b. Is a copy of the report able to be generated/viewed?					3

10. Stored cell lines 3

a. Are there documented procedures available for storage of cell lines including what information should be recorded? 5
   E.g. Culture ID; passage no; date frozen; freezing media, method of freezing, no. of vials frozen, no. remaining; location; viability; QC results
b. Are records/logs held of all cell lines stored?
c. Have the relevant specimens been retained in accordance with this procedure?
d. Is material readily accessible?
e. Are facilities for storage suitable for the purpose and in accordance with current legislation and guidelines?

11. Work environment

a. Were all areas kept clean and tidy?
b. Is there enough space available in all areas?

12. Staff training/competency 6

a. For those staff observed performing the tests, is there evidence of appropriate qualifications, staff training and education? 5
b. Is their competence level recorded for the tasks they were doing?
   List records examined
c. Is there evidence of adequate supervision, where appropriate?

The GCCP principle column is for information only and need not be included in a live audit

References

1. Hughes, P., Marshall, D., Reid, Y., Parkes, H., and Gelber, C. (2007) The costs of using unauthenticated, over-passaged cell lines: how much more data do we need? *BioTechniques* **43**, 575–586.
2. Nardone, R.M., Open Letter on Misidentified Cell Lines: http://www.hpacultures.org.uk/services/celllineidentityverification/NardoneOpenletter.jsp
3. The Good Laboratory Practice Regulations 1999, Statutory Instrument No. 3106.
4. Coecke, S., Balls, M., Bowe, G., Davis, J., Gstraunthaler, G., Hartung, T., et al. (2005) Guidance on Good Cell Culture Practice - A Report of the Second ECVAM Task Force on Good Cell Culture Practice. *ATLA* **33**, 261–287.
5. An approach to audit in the medical laboratory (2004) Clinical Pathology Accreditation (UK) Ltd: website:http://cpa-uk.co.uk/ (Retrieved January 13, 2010).
6. Vertical Assessment Form (2009), from Clinical Pathology Accreditation (UK) Ltd: website: http://cpa-uk.co.uk/ (Retrieved January 13, 2010).

Chapter 11

Generation of Lung Cancer Cell Line Variants by Drug Selection or Cloning

Laura Breen, Joanne Keenan, and Martin Clynes

Abstract

Clonal variants or subpopulations have been isolated from every major histological type of cancer, and cellular heterogeneity in lung cancer is a common occurrence. These subpopulations may exhibit differences in drug resistance and invasive potential. One therefore needs to consider the subpopulations as well as the tumour to overcome the barriers of drug resistance and metastasis for successful treatment. Isogenic variants of cancer cell lines can be very valuable in providing controlled human experimental systems to study clinically relevant parameters such as drug resistance and invasiveness. These variants can be established by selection based on a characteristic of the subpopulation or by isolating clonal subpopulations from a heterogeneous population. Drug-resistant variants can be generated by pulse selection, which usually generates low-level resistance, which may as well be clinically relevant, or by continuous exposure, which can be used to obtain high-level resistant variants. Clonal subpopulations may also be isolated based on morphological differences using simple cell-culture-based techniques.

Key words: Lung cancer, Isogenic, Variant, Clone, Resistance

1. Introduction

Cell line models of drug resistance provide a tool for investigating the mechanisms through which cells become resistant to chemotherapy and thereby identify mechanisms for overcoming the resistance. Many groups have developed their own models of drug resistance in vitro by treating cells either continuously or using weekly treatments or "pulses" to closely mirror the clinical setting where cancer patients usually receive chemotherapy once a week (1, 2).

Continuous selection involves treating a cell line with increasing concentrations of the agent; cells that survive in culture with the particular agent are building a resistance to it. A lung cancer cell line, DLKP-A, that is 300-fold resistant to the anticancer agent Adriamycin has been established using this method in our laboratory (3). With pulse selection, cells are treated once a week over a 6- to 10-week period and are allowed to recover from each drug treatment. We have successfully established a panel of lung cancer cell lines resistant to taxol and taxotere (4). These resistant cell lines show much more modest resistance than continuously selected cell lines, generally less than tenfold compared to the parent cell lines. Some of these cell line models were used in a proteomic study to identify proteins involved in resistance to taxanes (5).

There are other methods to induce drug resistance in lung cancer cell lines, such as cDNA transfection to increase expression of drug resistance genes, but here we focus on drug selection.

1.1. Generating Clonal Subpopulations of Human Cancer Lung Cell Lines

Early cloning techniques, i.e. generating a cell population that originates from a single cell, included growth in agar and agarose (6). However, these methodologies relied on the cells being able to grow in agar and their colony forming efficiency (CFE). Advanced methods of isolating subpopulations include use of FACS where certain prior knowledge of the subpopulation is needed. Current methods that can be easily carried out in most cell culture laboratories include using cloning rings, limiting dilution, or basing selection on a desired characteristic, e.g. drug resistance or invasion. The method of choice will depend on the CFE of the cells. For cells with a low CFE, cloning rings may be suitable, while for cells with a high CFE, limiting dilution may work very well.

Using limiting dilution, three clonal variants of a poorly differentiated lung cancer cell line displaying different morphological characteristics have been isolated in this laboratory (7). This method has also been successfully used to isolate p53-transfected lung cancer cells (8). In our laboratory, variants of a drug-resistant lung cancer cell line were isolated using cloning rings (9). Four subclones of differing invasive potentials have been isolated from the parental human lung adenocarcinoma based on their ability to migrate through matrigel (10).

2. Materials

1. Cell culture medium and trypsin/EDTA solution [0.25% trypsin (Gibco); 0.01% EDTA (Sigma)].
2. Cell culture flasks/plates (Costar).
3. Phosphate-buffered saline (Oxoid).
4. *p*-Nitrophenol phosphate (Sigma).

5. Sodium acetate buffer [0.1 M sodium acetate, Sigma; 0.1% Triton X-100 (BDH); pH 5.5].
6. NaOH 1 M (Sigma).
7. Cloning rings (Stainless steel, glass, or Teflon) with a smooth flat base (Costar).
8. Silicone grease (Costar).
9. Transwell inserts, 0.8-μm pores (BD biosciences).
10. ECM gel (Sigma).

3. Methods

3.1. Determination of IC50

1. Cells in the exponential phase of growth are harvested by trypsinisation, and cell suspensions containing 1×10^4 cells/mL are prepared in cell culture medium.
2. Using a multichannel pipette, 100 μL/well of these cell suspensions are added to 96-well plates (see Note 1).
3. Cells are then incubated overnight at 37°C and 5% CO_2.
4. Cytotoxic drug dilutions should be prepared at 2× their final concentration in cell culture medium, and 100 μL is then added to each well using a multichannel pipette (see Note 2).
5. Cells are incubated for a further 6–7 days at 37°C and 5% CO_2 until the control wells have reached approximately 80–90% confluency (see Note 3).
6. Media is removed from the plates and washed thoroughly with PBS and 100 μL of freshly prepared phosphatase substrate (10 mM p-nitrophenol phosphate in Sodium acetate buffer) is added to each well.
7. The plates are then incubated in the dark at 37°C for 2 h (see Note 4).
8. The enzymatic reaction is stopped by the addition of 50 μL of 1 N NaOH and absorbance read on a plate reader at 405 nm and reference wavelength at 620 nm.
9. The concentration of drug which caused 50% cell kill (IC_{50} of the drug) can be determined from a plot of the percentage survival (relative to the control cells) versus cytotoxic drug concentration.

3.2. Determination of Drug Concentration Required for Pulse Selection

1. Cells are seeded into twelve 25-cm² flasks at $1.5° \times 10^5$ cells per flask and allowed to attach overnight at 37°C.
2. The following day media is removed from the flask and a range of concentrations of appropriate drug added to the

flasks in duplicate. Complete media should be added to two flasks as a 100% survival control. Return flasks to the 37°C incubator for a 4 h incubation, after which the drug is removed, the flasks are rinsed and fed with fresh complete media.

3. The flasks are then incubated for 5–7 days until the cells in the control flasks reach approximately 80% confluency. At this point, remove medium from the flasks and cells are trypsinised and counted in duplicate.

4. The concentration of drug that caused appropriate cell kill can be determined from a plot of the percentage survival relative to the control cells versus cytotoxic drug concentration.

3.3. Pulse-Selection of Lung Cancer Cells

1. Cells at low confluency in 75-cm^2 flasks should be used (see Note 5).

2. Remove culture medium and add drug at the chosen concentration. Return flasks to the incubator for a period of 4 h.

3. After this period, the drug is removed and the flasks rinsed and fed with fresh complete media.

4. The cells are then grown in drug-free media for 6 days, refeeding every 2–3 days.

5. Steps 1–4 should be repeated once a week for 10 weeks (see Note 6).

6. Once the cells have been pulsed for the desired number of pulses, they should be tested for sensitivity to the selecting agent using the method described above for determining IC50 (see Note 7).

7. Stocks of selected cells should be cryopreserved as soon as possible (see Note 8).

3.4. Continuous Selection of Lung Cancer Cells

1. Cells should be grown in 75-cm^2 flasks. The concentration of drug to be used should be determined using the IC50 as a basis (see Note 9).

2. Culture medium is removed and the medium containing drug is added.

3. The cells should be returned to the incubator and re-fed with drug containing medium twice weekly.

4. The drug concentration should be increased regularly as the cells become tolerant of the current concentration (see Note 10).

3.5. Isolation of Clonal Populations Using Cloning Rings

This procedure is suitable for the isolation of anchorage-dependent cells where the CFE of the parent population is low (and therefore possibly the clones) or for subpopulations that can be visualised microscopically as being distinct (see Note 11).

1. Sterilise cloning rings and silicone separately by baking at 120°C overnight.

2. Subculture the cells of interest into a 90-mm Petri dish. There should be a single cell suspension with a low enough density to allow the formation of colonies from a single cell (needs to be determined empirically for cells of interest). Incubate overnight at 37°C and 5% CO_2.

3. Microscopically look for single cells and mark with an indelible marker on the underside of the dish. Where two or more cells have been observed together, these are ignored.

4. Allow the cells to grow, with feeding if necessary, and monitor the selected colonies. When there are about 30–50 cells in the colony of interest, remove the medium from the dish and wash with sterile PBS (see Note 12).

5. With a sterile forceps, dip the flat end of the cloning ring into the sterile silicone grease so that the bottom of the ring is evenly covered. Using the markings on the underside of the dish to locate the colony of interest, place the cloning ring so that the colony is isolated (see Note 13).

6. Add just sufficient trypsin solution to cover the cells in the cloning ring. This will depend on the diameter of the cloning ring, about 30–50 μL. Incubate at 37°C with regular microscopic monitoring until cells have detached. Add 50–100 μL growth medium to the cloning ring and gently pipette up and down.

7. Remove to a well in a 96-well plate.

8. Wash the inside of the cloning ring with 50 μL growth medium to remove any remaining cells and add to the same well of the 96-well plate (see Note 14).

9. Allow the cells to attach overnight at 37°C and refresh with 200 μL fresh medium (see Note 15).

10. The cells are allowed to grow with re-feeding until about 80% confluent. The cells can now be trypsinised into a larger dish, usually either a 48-well or a 24-well dish (see Note 16).

11. When trypsinising from a 96-well plate, wash with PBS as usual. Trypsinise in a minimal volume of trypsin (30–50 μL) as in step 6. Add 100 μL of medium and transfer to another well of 96-well plate or to a larger area plate (48-well or 24-well). Wash out previous well with 50 μL fresh medium. Add sufficient growth medium to larger plates, i.e. 0.5 mL for 48-well plate and 1 mL for 24-well plate (see Note 17).

12. Continue to grow and expand the subpopulation. Once cells in a 24-well plate have reached confluency, the trypsinised cells may be centrifuged at $1,000 \times g$ for 5 min (depending on the cell line treat as with parent initially) to remove residual

trypsin. Expand the new cell line until master stocks of the cells can be frozen down (see Note 18).

Re-cloning often needs to be repeated to ensure successful isolation of a subpopulation, especially where morphologically distinct subpopulations are present in the cloned variant. To do this, the procedure is simply repeated from the beginning using the clonal variant as the starting point.

3.6. Isolation of Clonal Populations by Limiting Dilution

This method is an adaption of the previous method but is suitable only for anchorage-dependent cells with a CFE of greater than 5%.

1. Trypsinise parent cell line to produce a cell density of about 60–120 cells per 12 mL growth medium (or 0.5–1 cell per well). Plate out 100 μL cell suspension into each well of a 96-well plate and allow this to attach overnight at 37°C and 5% CO_2.
2. Monitor and mark wells which contain single cells. Wells containing multiple cells are ignored. Normally, only 5–10 clones will result from each plate (see Note 19).
3. Continue to monitor wells daily and feed if necessary.
4. When the cells are almost 80% confluent, trypsinise as in step 11.

For use of feeder layers or semi-solid media cloning, refer to (11) Chapter on "Cloning Animal Cells."

3.7. Isolation of Clonal Populations by Invasive Behaviour: Variations on a Theme

Much as the drug resistance profile of a subpopulation can be used to isolate out the drug-resistant variants, invasive subpopulations can be isolated out from the parental cell line by carrying out traditional in vitro invasion assays with matrigel and isolating the invasive cell lines from the underside of the insert.

1. Coat transwell inserts (containing 8 μm pores) with 100 μL basement membrane (Matrigel) supported in a 24-well plate and leave overnight at 4°C.
2. Place plate at 37°C for at least 1 h.
3. Trypsinise cells and prepare a 1×10^6 cells/mL suspension.
4. Carefully wash out insert twice, once with serum-free medium and the second time with growth medium, being very careful to not disturb the matrigel.
5. Add 100 μL of the cell suspension followed by 100 μL of growth medium. Add 0.5 mL of growth medium to the well (see Note 20).
6. Incubate for the desired length of time to allow invasive cells to invade through to the underside of the insert. The incubation time can vary depending on the invasive ability of the cell line, e.g. from 12 to 72 h.

7. Aseptically, remove the matrigel and the cells that have not invaded from the top of the insert. Wash out the top of the insert three times with sterile PBS. Rinse the bottom of the insert with PBS by inserting it into a well with PBS. Place insert into a well containing 200 µL trypsin and allow cells to detach. Microscopically check the trypsin to observe single cells in the trypsin. Remove insert and add in 800 µL growth medium to inactivate trypsin and allow cells to grow. Feed with fresh medium after overnight incubation.
8. These cell lines are then expanded as in step 11. Re-cloning of the invasive clones is carried out to segregate the more invasive clones.

4. Notes

1. Plates should be agitated gently to ensure even dispersion of cells over the surface of the given well.
2. A range of drug dilutions should be used to ensure that an accurate IC50 is determined. These assays should be carried out in triplicate.
3. It is important to observe cells microscopically over this time period, since some cell lines will reach confluency sooner than others.
4. Colour development should be monitored during this time; a yellow colour starts to develop as the substrate is metabolised.
5. Cells should be in good condition, low passage and recently fed.
6. If the cells have not recovered sufficiently from the previous pulse, a week should be skipped, but all cell lines must receive the same number of pulses.
7. It is important to test the cells for sensitivity over a number of passages to ensure the stability of resistance. Fold resistance is calculated by dividing the IC50 of the selected cell line by the IC50 of the original cell line.
8. Cells that have been selected with a cytotoxic agent can be sensitive to cryopreservation. It is important to check the viability of frozen stocks and also whether the cells have retained any resistance acquired after cryopreservation.
9. Cells of medium confluency should be used in 75-cm^2 flasks; they should be in good condition, low passage, and recently fed. It is important to note that confluency has an effect on the cells' response to drug treatment; high-density cells will be more resistant than low-density cells.
10. Cells should be monitored regularly for changes in resistance using methods described for determining IC50.

11. The CFE is determined by plating a specific cell number into the well of a plate and counting the number of colonies formed after about 4 days of growth at 37°C and 5% CO_2. CFE = (number of colonies formed/total number of cells seeded) × 100.

12. As a criterion, the colonies to be isolated should be distinct and not encroached upon by other colonies. Also, monitor the appearance of floating cells to ensure that these do not attach near the colonies of interest and contaminate the colony. If there are many floating cells, it may be necessary to wash the cells with PBS several times.

13. It is important to check microscopically at this stage that the colony of interest is isolated. When placing the cloning ring, press down gently with the forceps to ensure that a good seal is formed.

14. At all times during the trypsinisation, it is crucial to ensure that manipulations with pipettes do not move the cloning ring. By marking the position of the cloning ring on the underside of the dish before trypsinising, the position of the cloning ring can be monitored visually.

15. This clonal population should be treated as a separate cell line at this stage.

16. If the cells appear to grow from these single cells and form small colonies but do not continue towards confluency, i.e. they could be contact inhibited, it may be necessary to trypsinise the cells and reset them up in the 96-well to allow the cells to reach a higher density.

17. If the cells appear to be growing slowly, the use of conditioned medium can enhance growth. Conditioned medium contains autocrine factors secreted by the cells to encourage their growth. Retain the medium from 80% confluent cells and centrifuge at $1,000 \times g$ for 5 min. Filter through a 0.2-µm low protein-binding filter and mix in a 50:50 ratio with fresh medium. Once cells have attached to new surface, the conditioned medium can be added to encourage growth.

18. This should be done in the lowest number of passages as possible to keep the new cell line identical to the clones. Also, it is very advisable to keep the cells growing until the viability of the master stocks has been checked.

19. Be very careful that there are no cells growing right at the side of the wells.

20. No gradient is used in this procedure although gradients are an option.

References

1. Parekh, H., Wiesen, K., and Simpkins, H. (1997) Acquisition of Taxol Resistance Via P-Glycoprotein- and Non-P-Glycoprotein-Mediated Mechanisms in Human Ovarian Carcinoma Cells. *Biochem. Pharmacol.* **53**, 461–470.
2. Han, E. K., Gehrke, L., Tahir, S. K., Credo, R. B., Cherian, S. P., Sham, H., Rosenberg, S. H., and Ng, S. (2000) Modulation of Drug Resistance by Alpha-Tubulin in Paclitaxel-Resistant Human Lung Cancer Cell Lines. *Eur. J. Cancer.* **36**, 1565–1571.
3. Clynes, M., Redmond, A., Moran, E., and Gilvarry, U. (1992) Multiple Drug-Resistance in Variant of a Human Non-Small Cell Lung Carcinoma Cell Line, DLKP-A. *Cytotechnology.* **10**, 75–89.
4. Breen, L., Murphy, L., Keenan, J., and Clynes, M. (2008) Development of Taxane Resistance in a Panel of Human Lung Cancer Cell Lines. *Toxicol. In. Vitro.* **22**, 1234–1241.
5. Murphy, L., Henry, M., Meleady, P., Clynes, M., and Keenan, J. (2008) Proteomic Investigation of Taxol and Taxotere Resistance and Invasiveness in a Squamous Lung Carcinoma Cell Line. *Biochim. Biophys. Acta.* **1784**, 1184–1191.
6. Walls, G. A., and Twentyman, P. R. (1985) Cloning of Human Lung Cancer Cells. *Br. J. Cancer.* **52**, 505–513.
7. McBride, S., Meleady, P., Baird, A., Dinsdale, D., and Clynes, M. (1998) Human Lung Carcinoma Cell Line DLKP Contains 3 Distinct Subpopulations with Different Growth and Attachment Properties. *Tumour Biol.* **19**, 88–103.
8. Breen, L., Heenan, M., Amberger-Murphy, V., and Clynes, M. (2007) Investigation of the Role of p53 in Chemotherapy Resistance of Lung Cancer Cell Lines. *Anticancer Res.* **27**, 1361–1364.
9. Heenan, M., O'Driscoll, L., Cleary, I., Connolly, L., and Clynes, M. (1997) Isolation from a Human MDR Lung Cell Line of Multiple Clonal Subpopulations which Exhibit significantly Different Drug Resistance. *Int. J. Cancer.* **71**, 907–915.
10. Chu, Y. W., Yang, P. C., Yang, S. C., Shyu, Y. C., Hendrix, M. J., Wu, R., and Wu, C. W. (1997) Selection of Invasive and Metastatic Subpopulations from a Human Lung Adenocarcinoma Cell Line. *Am. J. Respir. Cell Mol. Biol.* **17**, 353–360.
11. McBride, S., Heenan, M., and Clynes, M. (1998) Cloning Animal Cells, in Animal Cell Culture Techniques (M. Clynes, Ed.) pp 3–4–12, Springer.

Chapter 12

Isolation and Culture of Colon Cancer Cells and Cell Lines

Sharon Glaysher and Ian A. Cree

Abstract

The preparation of cells from heavily contaminated tissue is challenging. It is usually best to avoid such specimens if possible, but for the study of colorectal and some other tumours, it is inevitable that this must be overcome. The best methods seem to use a combination of (1) debridement of necrotic or infected areas of the tumour, (2) enzymatic dissociation in the presence of antibiotics, (3) density centrifugation to remove debris, and (4) extensive washing of the tumour-derived cells prior to their use in cell culture methods. It is possible to obtain cells from up to 80–90% tumours with careful technique and cooperation from the clinical team.

Key words: Colon, Colorectal, Cancer, Antibiotic, Antifungal, Cell culture

1. Introduction

Colorectal cancers are a leading cause of death in both men and women. Despite recent efforts to screen the at risk population for gastrointestinal bleeding, most tumours are large and advanced on presentation. Surgical removal of the affected section of the colon remains the treatment of choice and this usually allows access to a considerable amount of material from the primary tumour. Unfortunately, the colon contains a large flora of microorganisms and the tumour is virtually always ulcerated and heavily contaminated with bacteria and fungi (Fig. 1). The best way we have found to counter this problem is to debride the surface of the tumour vigorously and use relatively high concentrations of antibiotics to kill contaminating microorganisms during an overnight incubation with collagenase which breaks up the tissue and releases cells (1).

This is followed by washing of the cells and density centrifugation to remove debris, and further washing. Finally the cells are suspended in a medium containing more antibiotics

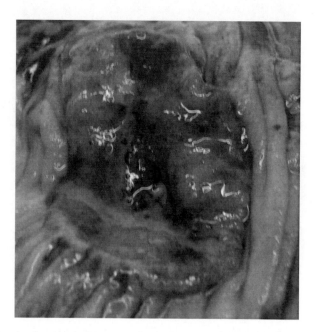

Fig. 1. Opening the colon in this case reveals a typical fungating tumour with an ulcerated surface contaminated with faecal material. Our practice is to ask the pathologist to take a wedge from the anti-mesenteric border for isolation of cells to avoid compromising the resection margins.

than is necessary for 'clean' tumours derived from sterile sites. The cell suspension contains both neoplastic and non-neoplastic cells. It is suitable for use in primary cell culture with selective medium, but can be subjected to magnetic bead or other selection methods to generate a purer cell population for other purposes, such as the generation of new cell lines. In our work, we have used primary cell cultures in an ATP-based tumour chemosensitivity assay (ATP-TCA) to explore the chemosensitivity of colorectal cancer (1, 2) and to develop new combinations for this cancer (2). Similar methods can be used with other tumour types such as those from the upper GI tract, which may also be contaminated (3).

2. Materials

1. Benchtop Centrifuge (e.g. MSE Harrier or similar).
2. 25 and 50 mL sterile plastic disposable universals.
3. Pipette aid for 10–50 mL pipettes and pipettes.
4. Standard automatic pipettes (50–250 µL and 0.25–1.0 mL).
5. Transport/wash medium – DMEM (Sigma) with additions as follows: Penicillin–streptomycin (Sigma), gentamicin

(Sigma), and additional amphotericin B (Sigma) and metronidazole (Sigma).

6. Complete Assay Medium (DCS Innovative Diagnostik Systeme GmbH, Hamburg, Germany) with additional antibiotics as mentioned above.

7. Collagenase H (Sigma) made up from stock in transport medium with antibiotics to 1.0 mg/mL.

8. Waste medium for disposal of infected material (e.g. Virkon).

9. Benchtop centrifuge (e.g. MSE Harrier) – This does not need to be refrigerated.

10. Class II laminar flow hood providing operator and sample protection.

11. Incubator – 5% CO_2 and 37°C.

12. 7-mL disposable bijou plastic universal containers.

13. Cellometer automated cell counter (Nexcelom, Lawrence, USA)

3. Methods

3.1. Preparation of Transport/Wash Medium with Antibiotics

1. In class II hood add 5 mL penicillin–streptomycin (10,000 U/mL penicillin and 10 mg/mL streptomycin) and 5 mL gentamicin (10 mg/mL) to a new 500-mL bottle of DMEM.

2. For potentially contaminated samples, which include virtually all colon cancer samples, also add 5 mL Fungizone (amphotericin B, 250 µg/mL) and 100 µL metronidazole (made up as a 5 mg/mL stock solution).

3. Label the bottle(s) with printed labels indicating name of the reagent, name and address of the manufacturer, date of preparation, lot number and expiry date (see Note 1).

4. Transfer 5 mL of new Transport Medium into a bijou container and place it in the incubator. Check for sterility after 24–48 h (see Note 2).

3.2. Sample Selection and Handling

1. The sample should be obtained from the specimen as rapidly as possible – ideally within the operating theatre (see Note 3).

2. Check that the patient details and specimen receptacle details match.

3. A pathologist or surgeon should normally cut the specimen to ensure that resection margins are respected as these may determine patient care.

4. A sample of 1–2 cm^3 should be cut from the tumour with a sterile scalpel. Take the opportunity to scrape off any obviously contaminated or necrotic material from the surface of the sample at this stage and transfer to a 25-mL labelled plastic universal container with 10 mL transport medium. Shake to remove loose contaminated material and transfer again to a fresh 25-mL labelled universal container with 10 mL transport medium. Repeat until medium is clear.

3.3. Enzymatic Digestion

1. Day 1: Open a database/spreadsheet file for the patient and enter all details from sample submission form. Containers must be appropriately labelled with the sample number at all times. All containers must be kept until the tumour cells are plated out and placed in the incubator. Prepare an excel sheet for the sample and fill in the assay conditions, including batch numbers (see Note 4).

2. Remove the solid tumour sample from its container into a culture dish in the safety cabinet. Using a sterile scalpel blade cut the sample into fine fragments (½–1 mm size) and place them into a universal container with 9 mL transport media (Fig. 2).

3. Add 1.0 mL of preprepared sterile collagenase solution, to achieve a final concentration of 1.0 mg/mL. Ensure that the universal lid is secure and if necessary use parafilm to prevent leakage. Place the universal container horizontally in an incubator at 37°C, 5% CO_2, overnight.

4. Day 2: Remove the universal container, shake it, and replace it in the incubator for 30 min (see Note 5).

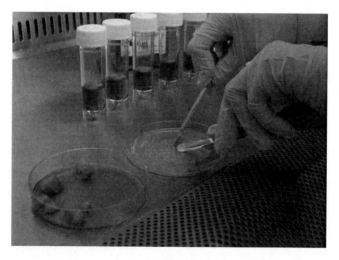

Fig. 2. Cut up the cleaned sample into 0.5–1 mm^3 fragments and transfer to a universal container.

5. Remove the universal container and take it to the safety cabinet. Add transport medium to even out the volumes and balance the centrifuge if more than one universal containers are being processed; then centrifuge at $250 \times g$ for 7 min.

6. Remove the universals from the centrifuge and discard the supernatant into safe disposal solution (see Note 6). Re-suspend the tumour pellet in 10 mL transport medium, then centrifuge again at $250 \times g$ for 7 min.

7. Re-suspend the tumour-derived cells in 10 mL of transport medium.

8. To new universal containers, add 10 mL of Ficoll–Histopaque. Take the re-suspended tumour in transport medium and carefully, drop-by-drop, add it to the surface of the Ficoll. This is easy if the Ficoll universal is held at an angle and the Ficoll is cold. Place the universal containers in the centrifuge at $400 \times g$ for 30 min.

9. Remove the universal containers from the centrifuge and return to the safety cabinet. Remove the layer between the lymphoprep and the wash media and add it to a new, labelled universal container. Add 10 mL transport medium and centrifuge at $250 \times g$ for 7 min.

10. Remove the universal containers from the centrifuge and discard the supernatant; re-suspend the tumour pellet in 10 mL wash media, then centrifuge again at $250 \times g$ for 7 min.

11. Remove the universal containers from the centrifuge and discard the supernatant; re-suspend the tumour pellet in 10 mL Complete Assay Media. This amount may vary depending on the number of cells expected.

3.4. Cell Counting

1. Label an Eppendorf container, add 20 µL of trypan blue, and add 20 µL of the tumour cell suspension. Mix thoroughly, then pipette into one port of an unused chamber on the counting microscope slide (Fig. 3). Count the cells using the Cellometer as indicated by the manufacturer.

2. Dilute the tumour cell suspension with the appropriate amount of CAM to achieve the final concentration desired.

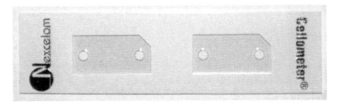

Fig. 3. Cellometer counting chamber – each slide has two counting chambers which can be filled in the hood, and in our hands this has proved much quicker and cleaner than haemocytometer counting methods. For details see http://www.nexcelom.com.

4. Notes

1. The antibiotics can be divided into aliquots and frozen at −20°C for several months.
2. It is wise to check any media made up by incubation for 24 h to ensure that this is sterile – if phenol red is present, this will go yellow if there is contamination and the entire batch should be discarded. Careful batch control is also essential to ensure that older reagents are used first.
3. It is a good practice, and in many countries a requirement, to ensure that the tissue is submitted with a signed consent from the patient for the use of their tissue for research, that the laboratory has appropriate regulatory approvals for human tissue use, and that the study has ethics approval.
4. Batch numbers should be noted as this enables problems to be tracked easily.
5. The fragments should break up with a simple shake of the universal container, but on occasion more vigorous shaking (e.g. using a vortexer) may be necessary.
6. The use of fresh human material carries greater risks of infection than cell lines, and should be a matter of concern. Use of careful technique, gloves, clothing, and safe disposal into a solution that kills viruses and bacteria is essential. We keep a small receptacle of solution in the hood when performing assays with primary cell cultures into which small plastics and other contaminated articles can be placed – most such solutions should be replaced daily.

References

1. Whitehouse, P. A., Knight, L. A., Di Nicolantonio, F., Mercer, S. J., Sharma, S., and Cree, I. A. (2003) Heterogeneity of chemosensitivity of colorectal adenocarcinoma determined by a modified ex vivo ATP-tumor chemosensitivity assay (ATP-TCA), *Anticancer Drugs* 14, 369–375.
2. Whitehouse, P. A., Mercer, S. J., Knight, L. A., Di Nicolantonio, F., O'Callaghan, A., and Cree, I. A. (2003) Combination chemotherapy in advanced gastrointestinal cancers: ex vivo sensitivity to gemcitabine and mitomycin C, *Br J Cancer* 89, 2299–2304.
3. Mercer, S. J., Somers, S. S., Knight, L. A., Whitehouse, P. A., Sharma, S., Di Nicolantonio, F., Glaysher, S., Toh, S., and Cree, I. A. (2003) Heterogeneity of chemosensitivity of esophageal and gastric carcinoma, *Anticancer Drugs* 14, 397–403.

Chapter 13

Isolation and Culture of Melanoma and Naevus Cells and Cell Lines

Julia K. Soo, Alastair D. MacKenzie Ross, and Dorothy C. Bennett

Abstract

Recently developed methods are described for the culture of cells from primary human melanomas and benign or dysplastic naevi (moles). These allow the culture of viable cells from the great majority of such lesions, and the maintenance of what appears to be the predominant population of pigmented cells. These methods should facilitate the study of typical lesional cells, which appear somewhat different from the cell lines that have been established previously under more restrictive conditions: intermediate between these and normal melanocytes. The conditions involve the use of growth-inactivated keratinocytes initially, and a combination of mitogens similar to those routinely used for normal human melanocytes: stem cell factor, endothelin 1, TPA, and cholera toxin.

Key words: Melanoma, Naevus, Culture, Keratinocyte, Radial growth phase, Vertical growth phase, Cell senescence, Stem cell factor, TPA, pH

1. Introduction

Human metastatic melanoma cells have been successfully cultured since at least the 1960s (1), and many laboratories have established numerous such lines, for example those of Herlyn, Parsons, Moore, and Houghton (2–5). These are considered some of the easiest cancer cells to grow, so much so that we will discuss the culture of metastases only briefly. However it appears much more challenging to culture cells from primary melanomas, especially thin ones (2), and far fewer permanent lines have been established from these. Likewise, only a small number of groups have explanted cells from benign melanocytic lesions (naevi or moles), and these cultures have been reported to have a limited lifespan; no permanent lines have been derived (6, 7). We have recently developed methods that efficiently yield viable initial cultures of both primary melanoma cells and naevi, as detailed here. We find, surprisingly, that even under these favourable conditions many primary melanoma cultures, like naevi, grow only for a limited time and then senesce (8). However, around one in seven of the melanoma cultures yield cell lines, which tend to retain the appearance of the early cultures (quite similar to normal melanocytes), and to have only a few chromosomal aberrations. These minimal-deviation lines as well as the primary culture methods should make a useful addition to the resource palette of mainly advanced melanoma lines, allowing the study of more typical primary lesional cells, for the better understanding of the crucial first stages in melanoma oncogenesis.

2. Materials

1. Basic medium for explants ("culture medium"): RPMI 1640 medium (Invitrogen Ltd, Paisley, UK) with penicillin (100,000 U/L), streptomycin sulphate (100 mg/L), glutamine (2 mM) and extra water-soluble phenol red (7.5 µg/mL) (Sigma-Aldrich Co., Poole, UK), pre-gassed with 10% CO_2 (pH 7.0), then supplemented with 10% foetal calf serum (FCS) (Invitrogen).

2. Transport medium (for transporting fresh lesional tissue): As above except without FCS.

3. Medium for XB2 keratinocyte line: Dulbecco's Modified Eagle's medium (DMEM) (Invitrogen) with penicillin (100,000 U/L), streptomycin sulphate (100 mg/L), glutamine or GlutaMAX (4 mM), 10% FCS and 10% CO_2 (pH 7.4). Note that standard DMEM is buffered for use with 10% CO_2.

4. PBSA: Sterile Dulbecco's PBS, solution A (lacking $CaCl_2$ and $MgCl_2$) (Invitrogen).

5. PBSA–BSA: PBSA containing 1 mg/mL BSA (Sigma-Aldrich). Used as a carrier solution for growth factors; both proteins and TPA can otherwise stick to plastic or glass and be lost from dilute solutions.

6. Cholera toxin (Sigma-Aldrich) (CT). A concentrated stock is dissolved in PBSA–BSA at 2 µM (5 mg/30 mL) and filter-sterilized. A 200× working stock is prepared, also in PBSA–BSA, at 40 nM. Both are divided into aliquots and stored at –80°C, or the working stock at 4°C for up to 2 weeks. This is a potent toxin if ingested; handle with care according to the supplier's recommendations and treat CT-containing solutions with concentrated bleach before disposal.

7. 12-O-tetradecanoyl phorbol 13-acetate (Sigma-Aldrich) (TPA), also known as phorbol myristate acetate (PMA), is

dissolved as a concentrated stock in 95% or 100% methanol-free ethanol (not DMSO) (see Note 1) at 2 mM. This is assumed sterile. A 200× working stock is made in PBSA–BSA at 40 μM. Both stocks are aliquotted and stored at –80°C, or the working stock at 4°C for up to 2 weeks. TPA is light-sensitive, so vials should be protected from light and it is recommended to switch out the flow-cabinet light while handling the stock vials. It is also hazardous, as a known tumour promoter that can pass through skin. Handle with care according to the supplier's recommendations and treat TPA-containing solutions with concentrated bleach before disposal.

8. Human stem cell factor (Invitrogen) (SCF), also known as KIT ligand, mast cell growth factor or steel factor (KITL, MGF, SLF) is dissolved in PBSA–BSA at a stock concentration of 5 or 2 μg/mL. The solution is filter-sterilized and aliquots stored at -80°C, or at 4°C for up to 2 weeks.

9. Endothelin 1 (Bachem, Weil am Rhein, Germany) (EDN1) is dissolved in PBSA–BSA at 5 or 2 μM. The solution is filter-sterilized and aliquots stored at –80°C, or at 4°C for up to 2 weeks.

10. Growth medium: Culture medium (#1 above) with TPA (200 nM), CT (200 pM), and SCF (10 ng/mL).

11. Mitomycin C (Sigma-Aldrich). Dissolved at 500 μg/mL in water. Protect from light – wrap vials in foil during use. Can be stored in fridge for up to *about 1 month only*, or indefinitely at –80°C or less. If freezing it, freeze rapidly to avoid precipitation, e.g. drop vials into liquid nitrogen. We prefer to freeze it.

12. Soybean trypsin inhibitor (Sigma-Aldrich). A stock is dissolved at 1 mg/mL in PBSA–BSA, filter-sterilized, divided into aliquots, and stored at –80°C.

13. XB2 is an immortal line of mouse keratinocytes (9), originally obtained from J. Rheinwald, and adapted to growth without fibroblast feeder cells in our laboratory (10).

3. Methods

3.1. Preparation of XB2 Feeder Cells

1. For normal stock culture, XB2 mouse keratinocytes are plated at 3×10^4 cells/mL in DMEM with 10% FCS and 10% CO_2. Medium is changed after 3–4 days. With a healthy culture about $4–5 \times 10^5$ cells/mL can be harvested after 6 days, i.e. doubling time about 1.2 days. It is worth counting the cells and plating evenly, as they grow poorly at densities that are either too low or too high (see Note 2).

2. Cells are subcultured when virtually confluent but below saturation density (at saturation density they attach very firmly

and are hard to detach). For a 10-mL dish: the dish is washed twice (5 mL each) in PBSA containing EDTA (200 μg/mL), and once in the same EDTA solution with trypsin (250 μg/mL) (less trypsin than for fibroblasts, to allow for long exposure). All but 0.75 mL of trypsin–EDTA is removed.

3. The cultures are incubated at 37°C (preferably in air, not CO_2) until cells are completely detached, except no longer than 15–20 min.

4. Cells are resuspended by pipetting, in two washes totalling 10 mL DMEM and 10% FCS; they are counted and replated at 3×10^4 cells/mL. Adequate frozen stocks of growing cells should be prepared soon after receiving the line.

5. For preparation of growth-inactivated feeder cells: several large dishes or flasks are plated as above, e.g. six 10-mL dishes. They are grown as above until nearly confluent.

6. In subdued light, medium is removed and replaced with a half-volume (5 mL on a 10-mL dish) of DMEM+FCS containing 8 μg/mL mitomycin C.

7. Dishes are incubated for 3–3.5 h (timing is important). Medium is aspirated and the cells washed in DMEM.

8. Dishes are incubated in fresh DMEM with 10% FCS for 10 min to ensure elution of the drug.

9. Cells are harvested and subcultured as usual. They can either be re-plated at about 3×10^4 cells/mL for immediate use, or frozen in aliquots of 10^6 or 5×10^5 cells in liquid nitrogen (see Subheading 3.2), and plated at about 5×10^4/mL on thawing.

10. Viability of each batch of feeder cells is checked on thawing (i.e. floating vs. attached cells – none of them should divide); and plating density is adjusted for that batch if necessary.

11. It is also good practice to test batches for residual viable cells, although this should not happen if mitomycin C has been correctly stored and used as above. Plate one aliquot of 10^6 cells on a 10-mL dish and incubate, changing medium twice a week for about 3 weeks before checking carefully for growing colonies of cells.

3.2. Frozen Storage and Thawing of Cells: General Method

We use this method for all types of cultures, but it is given here because melanocytic cells are less viable with some other freezing methods.

1. Sufficient freezing medium is prepared in advance (see step 5 for quantity): standard growth medium for that cell type, with 10% FCS but no growth factors, and with 7.5% dimethyl sulphoxide (DMSO) (see Note 3), chilled to 4°C. Labelled freezing vials are also prepared in advance.

2. Nearly-confluent cells are harvested by appropriate protocol for that cell type, resuspended in their standard medium with serum, and counted.

3. The cell suspension is centrifuged at 4°C at 1,000×g for 5 min, or 7 min for smaller cells like melanocytes.

4. The supernatant is carefully aspirated from the cell pellet.

5. The pellet is resuspended in about 1 mL freezing medium, and the suspension made up to 10^6 cells/mL in more freezing medium and aliquotted into single-use vials. For pigmented melanocytes, speed is important here as DMSO is toxic to these cells.

6. Vials are stored at –80°C overnight (or for no more than a week), and placed in a liquid nitrogen freezer.

7. For recovery, vials are placed immediately into a 37°C waterbath, without getting the screw thread wet (a polystyrene float with holes for vials is helpful).

8. The suspension is pipetted gently and transferred to a 25-mL centrifuge tube. 20 mL of chilled culture medium are gradually added with gentle mixing, the first 1 mL dropwise, to ensure the cells do not lyse. (Important: freezing medium is highly hypertonic and pigmented cells are particularly sensitive here.) Cells are centrifuged at 4°C at 1,000×g for 5–7 min.

9. The supernatant is aspirated and the cells resuspended in complete growth medium at the required volume, then plated and incubated.

3.3. Explantation of Primary Melanoma or Naevus Cultures

1. Ethical permission (US: human subjects approval) is required in most countries for work with human tissue. This can take many months and should be obtained well in advance, but details will vary with country and are not given here. Generally information and permission/consent forms will be needed for the donor patients.

2. Mitomycin-C-inactivated XB2 feeder cells, prepared and thawed as above, are plated at 5×10^4 cells/mL into four wells of a Nunc 24-well plate per specimen, at least 4 h prior to plating of lesional cells, in DMEM with 10% FCS.

3. Sterile 25- and 5-mL screw-topped containers of transport medium (item 2, Subheading 2) are placed on ice and taken to the operating theatre.

4. The excised melanocytic lesion (see Note 4) is placed aseptically into 20 mL transport medium, replaced on ice and taken to a consultant pathologist who assesses the lesion. If the lesion is deemed of adequate size, a sliver of tissue is cut, placed in the 5-mL container of transport medium and taken to the lab. Procedures are sterile hereafter.

5. Using a dissecting microscope in a laminar flow hood, the specimen is placed on a 35 mm tissue culture dish and rinsed using PBSA. Tissue should be kept wet at all times – drying will kill the cells.

6. Fat and reticular dermis are dissected off using sterile forceps and a scalpel with a curved blade and discarded (see Note 5).

7. The specimen is covered with trypsin (250 µg/mL) and EDTA (200 µg/mL) in PBSA in a fresh 35 mm dish with lid. It is either kept overnight at 4°C, which partially or completely separates the epidermis and dermis, or one can proceed to the next step immediately. This step is also used following the overnight incubation if needed.

8. The dish is placed in a 37°C non-gassed incubator for 15 min to complete the separation.

9. Under a dissecting microscope, the epidermis (thin, whitish), is gently pulled away. If the epidermis does not separate easily, the specimen is replaced in the incubator for further trypsinization. The epidermis can be discarded (as it contains normal melanocytes); but we now process it separately in the same way as the dermis (see Note 6).

10. Any remaining reticular dermis is now separated from the papillary dermis (see Note 7) and discarded (unless required as a source of dermal fibroblasts). The trimmed lesional area of papillary dermis is placed on a new 35-mm dish and covered with fresh trypsin–EDTA. It is chopped finely, into a paste, using two apposed, curved-bladed scalpels, to increase surface area, and placed in a 37°C non-gassed incubator in air for 45 min.

11. The tissue and cell suspension is collected with rinsing, transferred into labelled 1.5-mL tubes in a non-refrigerated microcentrifuge, three tubes for the dermal and 1 for the epidermal portion (assuming a small specimen), and centrifuged at $2,000 \times g$ for 5 min.

12. During this time, 2 mL of complete growth medium (item 10, Subheading 2) with 125 µg/mL trypsin inhibitor is prepared.

13. Supernatants are removed; each cell pellet is gently resuspended in 500 µL of growth medium, plated into one well of the prepared feeder cells after removal of their medium, and the plate placed in a humid 37°C incubator with 10% CO_2.

14. The remainder of the lesion is processed routinely for histopathology. Adjacent sections either side of the sample taken for culture are examined and used for the diagnosis assigned to the culture, although the overall lesion diagnosis depends on a series of sections and may be different.

3.4. Maintenance of Primary Lesional Cultures

1. Lesional cells in primary culture are grown in culture medium with CT, TPA and SCF only, as above, except that 20 nM TPA instead of 200 nM is used for VGP melanoma cells, once the diagnosis is available. Medium is changed twice a week (see Notes 8 and 9). Primary lesional cultures are very variable: they may approach confluence quickly, or take some weeks to grow enough for subculture, or may be senescent and not grow noticeably at all, despite attaching (8). Growing cells are often bipolar in shape while senescent cells tend to be larger, with many dendrites and one or more nuclei with a prominent nucleolus (Fig. 1).

2. Fresh feeder cells may be added every 2–3 weeks, if most of them die before the culture is ready for subculture.

3. Once pure melanocytic growth is assured (no fibroblasts), EDN1 (10 µM) can also be added.

4. Cells are subcultured when/if they are approaching confluence. (Cell yield is around 1.5–2×10^5/mL or sometimes more for melanoma cells.) After removing the medium, cells are washed gently once in PBSA, then once in PBSA containing EDTA (200 µg/mL) and trypsin (125 µg/mL: less than

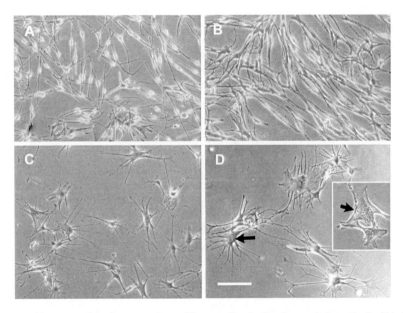

Fig. 1. Appearance of naevus and melanoma cultures. Phase contrast; all cultures photographed within first few passages. *Scale bar* for all panels = 100 µm. All cultures are from dermal (rather than epidermal) portion of a lesion. (**a**) Growing cells from a naevus. Appearance is similar to growing normal human melanocytes (e.g. see ref. 11); small, bipolar cells with 2–4 dendrites. (**b**) Growing cells from an RGP melanoma. Appearance varies; cells in this culture, which yielded an immortal line, are larger than normal melanocytes and with thicker dendrites, but the difference is small. The derived line had the diploid number of chromosomes with minor rearrangements. (**c**) Arrested (senescent) cells from a naevus. Larger and multidendritic. Other arrested naevus cells remain bipolar but larger and flatter (not shown). (**d**) Arrested cells from an RGP melanoma. Very large with many dendrites; bizarre shapes and multinucleated cells (*inset*) are quite common. Nucleoli are often large and prominent (*arrowed*).

for other cell types. Commercial trypsin–EDTA solution can be diluted with EDTA solution if necessary).

5. The trypsin–EDTA is removed immediately and a small fresh amount of the same solution is added (e.g. 0.75 mL for a 10-mL plate, but 0.2 mL for a well of a 24-well plate, to keep cells covered despite the meniscus) (see Note 9).

6. Incubate, preferably in air, until cells completely detached, as seen by gently tilting dish (4–10 min). Do not attempt to detach melanocytes by pipetting or tapping – tends to kill them by breaking their processes.

7. Resuspend in culture medium with 10% FCS; keep suspension on ice while counting cell number. Dilute to 3–5×10^4 cells/mL, add supplements to make complete medium and re-plate (see Note 10). If there are fewer cells than this, they can be plated on to fresh feeder cells. If plating volume is less than about 50× the residual volume of trypsin–EDTA, add soybean trypsin inhibitor (use full amount to neutralise trypsin remaining on dish), or else centrifuge to remove trypsin.

3.5. Culture and Maintenance of Metastatic Melanoma Cells

For completeness, this brief summary of published methods for explantation (2–5) is included. Most groups explant cells from metastatic melanomas by relatively simple means.

1. The sample is aseptically dissected free of other tissues, chopped with two apposed scalpels and directly resuspended and plated in culture medium. No feeder cells are used. The medium is usually a defined medium with serum (such as RPMI 1640 with 10% FCS).

2. Growth factors such as supraphysiological levels of insulin may be added initially but not including TPA, and the factors are generally withdrawn as the culture becomes established. See Hsu et al. (2) for one such combination of growth factors.

3. We find that all established human metastatic melanoma lines can be maintained readily in RPMI 1640 with 10% FCS and 10% CO_2, and can be subcultured and frozen by the methods described above.

4. Notes

1. DMSO at concentrations around 0.5–1% is growth-inhibitory to melanocytes, whereas ethanol is not; we therefore recommend using ethanol as solvent for TPA.

2. The XB2 cells that take longest to detach seem to be the stem cells (small, flat, round, epithelioid, healthy-looking), so

ensure nearly all cells are harvested. It may improve the culture to discard the quickest-detaching fraction every few subcultures, once familiar with the cells' appearance and the time taken to detach. The more-differentiated cells have a scrappy appearance and tend to fragment.

3. Neat DMSO should not be added to medium with cells in it, as it will kill cells locally before mixing. 100% DMSO should be transferred with a blue tip or glass pipette, as it dissolves something toxic out of other plastics. Likewise 100% DMSO should not be stored in polystyrene plastic bottles or tubes. Glass is ok.

4. The minimum size of melanocytic lesion deemed suitable by pathologists at our hospital for this technique is 8 mm. This may vary depending on pathologist experience. Yield of tissue can determine the success of explantation.

5. At this point, the fat and white reticular dermis can clearly be seen under a dissecting microscope; this is cut off and discarded. The pigmented area can be trimmed around and the removed portions discarded, helping to exclude fibroblasts.

6. Initially the epidermis was discarded to exclude normal melanocytes, but a substantial amount of lesion from the epidermal–dermal junction can be lost this way. If keeping the epidermal fraction, it is trimmed as close as possible to the pigmented area before chopping.

7. The lesional area of papillary dermis is pigmented, so it is trimmed as close to the pigmented edge as possible, to avoid fibroblast growth from non-lesional dermis.

8. For about the first week, a Pasteur pipette (3 mL) is used to aspirate spent medium gently, in order to leave any unattached lesional tissue in the well before new medium is added.

9. When changing medium or subculturing cells in small wells, it is important to add new solutions immediately, as surface tension can retract the solution from areas of cells within seconds, killing them. For example one can set the tray in a tilted position, then use a Pasteur pipette in one hand to draw off one solution and a pipette or micropipette in the other hand to add the next. Alternatively, not all the solution may be removed (but for trypsinization, extra washes will then be needed to dilute out serum). Solutions should not be squirted rapidly on to melanocytic cells on plastic, as this can detach them.

10. If growing pigmented cells in flasks, it improves cell viability to gas the empty flask with 10% CO_2 from cylinder (if available), before plating the cell suspension. Gas exchange through the neck of a flask is slow, so this avoids high-pH shock by CO_2 loss before equilibration in the incubator.

Acknowledgments

We are grateful to David Kallenberg for technical assistance, and to Hardev Pandha and Peter Mortimer for their input and advice. Development of the techniques was supported by grants from the British Skin Foundation (722F) and Cancer Research UK (C4704/A8041) supporting JKS, and a St George's Charitable Foundation Fellowship supporting ADMR.

References

1. Romsdahl, M. and Hsu, T. (1967) Establishment and biological properties of human malignant melanoma cell lines grown in vitro *Surg Forum* **18**, 78–9.
2. Hsu, M.-Y., Elder, D. E., and Herlyn, M. (2000) Melanoma: the Wistar melanoma (WM) cell lines. In *Human Cell Culture* (Masters, J. R. W. and Palsson, B., Eds.) pp 259–74, Kluwer Academic Publishers, London.
3. Pope, J. H., Morrison, L., Moss, D. J., Parsons, P. G., and Mary, S. R. (1979) Human malignant melanoma cell lines *Pathology* **11**, 191–5.
4. Semple, T. U., Moore, G. E., Morgan, R. T., Woods, L. K., and Quinn, L. A. (1982) Multiple cell lines from patients with malignant melanoma: morphology, karyology and biochemical analysis *J Natl Cancer Inst* **68**, 365–80.
5. Houghton, A. N., Eisinger, M., Albino, A. P., Cairncross, J. G., and Old, L. J. (1982) Surface antigens of melanocytes and melanomas: markers of melanoma differentiation and melanoma subsets *J Exp Med* **156**, 1755–66.
6. Halaban, R., Ghosh, S., Duray, P., Kirkwood, J. M., and Lerner, A. B. (1986) Human melanocytes cultured from nevi and melanomas *J Invest Dermatol* **87**, 95–101.
7. Mancianti, M. L., Herlyn, M., Weil, D., Jambrosic, J., Rodeck, U., Becker, D., Diamond, L., Clark, W. H., and Koprowski, H. (1988) Growth and phenotypic characteristics of human nevus cells in culture *J Invest Dermatol* **90**, 134–41.
8. Soo, J. K., MacKenzie Ross, A. D., Kallenberg, D. M., Milagre, C., Chong, H., Chow, J., Hill, L., Hoare, S., Collinson, R. S., Keith, W. N., Marais, R., and Bennett, D. C. (Submitted for publication) Malignancy without immortality: cellular immortalization as a late event in melanoma progression *Pigm Cell Melanoma Res.*
9. Rheinwald, J. G. and Green, H. (1975) Formation of a keratinizing epithelium in culture by a cloned cell line derived from a teratoma *Cell* **6**, 317–30.
10. Bennett, D. C., Cooper, P. J., and Hart, I. R. (1987) A line of non-tumorigenic mouse melanocytes, syngeneic with the B16 melanoma and requiring a tumour promoter for growth *Int J Cancer* **39**, 414–8.
11. Bennett, D. C., Bridges, K., and McKay, I. A. (1985) Clonal separation of mature melanocytes from premelanocytes in a diploid human cell strain: spontaneous and induced pigmentation of premelanocytes *J Cell Sci* **77**, 167–83.

Chapter 14

Isolation and Culture of Squamous Cell Carcinoma Lines

Karin J. Purdie, Celine Pourreyron, and Andrew P. South

Abstract

Cutaneous squamous cell carcinoma (SCC) keratinocytes readily grow, expand in culture, and continuously passage, suggesting either spontaneous immortalisation at the early stage of culture or inherent proliferative capacity. One feature of SCC keratinocytes is genomic DNA rearrangement and single-nucleotide polymorphism studies of fresh frozen primary tumour, early and late passage SCC keratinocytes suggest that these rearrangements are stable in culture and retain the parental tumour lesions. SCC keratinocytes are isolated using standard primary culture techniques and become feeder cell independent with little or no observed "crisis" period. SCC keratinocytes readily form tumours in vivo, which retain histological features of the parental tumour, making them an excellent model for the study and development of cancer therapies.

Key words: Keratinocyte, Squamous cell carcinoma, Tissue culture, Single-nucleotide polymorphism, DNA isolation

1. Introduction

The methods described here are based on the technique devised by Rheinwald and Green (1) and previously adapted to successfully cultivate SCC keratinocytes (2). Primary tumour tissue is physically broken down into either single cell suspensions or small explant pieces, and cells are subsequently expanded in culture. All cultures are established on a feeder layer of 3T3 cells, but SCC keratinocytes soon become feeder independent with little or no apparent "crisis" period. The original technique can be used to grow primary keratinocytes from skin, oral cavity, oesophagus, exocervix, and conjunctiva, as well as bladder urothelial cells and cells of the mammary gland (3). Although our technique has been used only to establish cutaneous SCC cell lines, it is conceivable that the method will translate to tumours from these other tissues.

Cell types other than keratinocytes can also be grown out of tumour tissue, in particular, tumour-associated fibroblasts. Human tumour-associated fibroblasts do not spontaneously immortalise and are subject to Hayflick principles of replication capacity (4).

2. Materials

2.1. Isolation and Culture of SCC Keratinocytes

1. Dulbecco's Modified Eagle's Medium (DMEM) with l-glutamine, 4,500 mg/L d-glucose, 110 mg/L sodium pyruvate (Gibco/Invitrogen Ltd, Paisley, UK).
2. F-12 nutrient mixture (Ham's) with l-glutamine (Gibco/Invitrogen Ltd, Paisley, UK).
3. Heat-inactivated fetal bovine serum (FBS, Biosera, Ringmer, UK).
4. Hydrocortisone (1 g, Sigma-Aldrich, Poole, UK).
5. Mouse epidermal growth factor (EGF, 1 mg, AbDSerotec, Oxford, UK).
6. Cholera Toxin (*Vibrio cholerae*) (azide free) (1 mg, Enzo Life Sciences (UK) Ltd Exeter, UK).
7. Insulin from bovine pancreas (500 mg, Sigma-Aldrich, Poole, UK).
8. Human apo-Transferrin (500 mg, Sigma-Aldrich, Poole, UK).
9. Liothyronine: 3,3′,5-Triiodo-l-thyronine sodium salt (100 mg, Sigma-Aldrich, Poole, UK).
10. Versene 1× (EDTA solution, GIBCO/Invitrogen Ltd, Paisley, UK).
11. Trypsin 10× (GIBCO/Invitrogen Ltd, Paisley, UK).
12. Trypsin/EDTA 1× (GIBCO/Invitrogen Ltd, Paisley, UK).
13. Penicillin/Streptomycin solution (PS) used at 1× (100×: 5,000 unit Penicillin/5,000 μg streptomycin/mL, GIBCO/Invitrogen Ltd, Paisley).
14. Fungisone solution used at 1× (100×, GIBCO/Invitrogen Ltd, Paisley, UK).
15. BD Falcon Cell strainer 100 μm pore size (VWR, Lutterworth, UK).
16. Collagenase D (Roche diagnostics, Burgess Hill, UK).
17. NALGENE Cryo 1°C Freezing Container, "Mr. Frosty" (Thermo Fisher Scientific, Loughborough, UK).

2.2. DNA Isolation

1. Nucleon BACC2 kit (Amersham Biosciences UK Ltd, Little Chalfont, UK).
2. 15-mL screw-capped polypropylene centrifuge tubes.
3. Chloroform (HPLC grade).
4. Ethanol 99.7–100% (absolute) AnalaR.
5. TE buffer (10 mM Tris–HCl pH 8.0, 1 mM EDTA pH 8.0).

3. Methods

3.1. Preparation of Keratinocyte Media

1. Transport medium: DMEM supplemented with 10% FBS, 1× Penicillin/Streptomycin, and 1× fungisone (make fresh as required).
2. Keratinocyte medium: 300 mL of DMEM, 100 mL of Ham's F-12 supplemented with 10% FBS, 0.4 μg/mL hydrocortisone, 5 μg/mL insulin, 10 ng/mL EGF, 5 μg/mL transferrin, 8.4 ng/mL cholera toxin, and 13 ng/mL liothyronine (see Note 1).

3.2. Isolation of SCC Keratinocytes (Single Cell Isolation)

1. Tumour tissue is placed in transport medium and moved to the laboratory to be processed as soon as possible. Tissue can be stored at +4°C for up to 4–5 days, but keratinocyte viability declines considerably with time.
2. Tumour tissue is washed in PBS or DMEM and cut into small fragments 1–2 mm in size in a Petri dish containing EDTA/Trypsin to prevent tissue drying out.
3. Minced tumour pieces are transferred to a tube containing trypsin and incubated at 37°C for an hour (or longer depending on the size and amount of tissue; up to 3 h at 37°C or overnight at +4°C). During the incubation period, the tube should be vigorously shaken at 10-min intervals.
4. Pour trypsin and tissue into a sterile Petri dish and disassociate with scalpel, scissors, or needles. Add equal volume of DMEM/10%FBS and pass the suspension through a 100-μm cell strainer. Add more DMEM/10% FBS to the dish to recover all cells and tissue and pass through a 100-μm cell strainer.
5. Centrifuge the cell suspension at $500 \times g$ for 5 min and plate out the recovered cells, ideally at densities of 10^5 to 3×10^5 per 100 mm dish containing keratinocyte medium and feeders, and incubate at 37°C in a 5% CO_2 incubator.
6. Recover the undigested pieces of tissue from the cell strainer and transfer to a small Petri dish with 3–4 mL Collagenase D in DMEM/10%FBS. Leave at 37°C overnight.
7. The following day pass the collagenase/tissue suspension through a 100-μm cell strainer. Add more DMEM/10% FBS

to the dish to recover all cells and tissue and pass through a 100-μm cell strainer.

8. Centrifuge the cell suspension at $500 \times g$ for 5 min and plate out the recovered cells, ideally at densities of 10^5 to 3×10^5 per 100 mm dish containing keratinocyte medium and feeders, and incubate at 37°C in a 5% CO_2 incubator.

9. The remaining tissue can either be used for explant culture or transferred to a dish containing keratinocyte medium and feeders, and incubated at 37°C in a 5% CO_2 incubator.

3.3. Isolation of SCC Keratinocytes (Explant Method)

1. Tumour tissue is placed in transport medium and moved to the laboratory to be processed as soon as possible.

2. Tumour tissue is washed in PBS or DMEM/10% FBS and cut into small fragments 1–2 mm in size in a Petri dish containing a small amount of FBS. More fetal bovine serum is added and the tissue/FBS suspension is transferred to a sterile Petri dish and spread evenly across the surface.

3. The Petri dish containing tissue and FBS is air-dried for approximately 4 min (do not allow the tissue to dry out) to attach the tissue pieces to the plastic.

4. Carefully add keratinocyte culture medium and feeder cells and incubate at 37°C in a 5% CO_2 incubator.

3.4. Culture of SCC Keratinocytes

1. Primary keratinocytes are usually observed 48 h post isolation (see Notes 2 and 3). Media is generally changed every 2–3 days, but if a lot of cells or pieces of tissue are present, then it may be necessary to change more frequently.

2. Cultures are rinsed in EDTA solution to remove feeders and then incubated in trypsin/EDTA solution at 37°C for 20 min or until the cells are visibly rounded and can be easily detached from the culture dish. DMEM/10% FBS is added to inhibit the trypsin, and cells are recovered with the use of a suitable centrifuge at $500 \times g$ for 5 min.

3. The cells are resuspended in keratinocyte medium and seeded at densities of 10^5 to 3×10^5 per 100 mm dish containing keratinocyte medium and feeders, and incubated at 37°C in a 5% CO_2 incubator.

4. To maintain cell stocks for future experiments (see Note 4) cells can be frozen at 10^6 or 2×10^6 cells/mL in 90% FBS/10% dimethylsulphoxide as a cryopreservant. Freezing is best achieved at a rate of −1°C per minute in a −80°C with the aid of a cryo-freezing container before transfer to liquid nitrogen.

5. Frozen vials of cells are thawed rapidly either in a 37°C water bath or under a running hot water tap. The vial is swabbed with 70% ethanol before opening and the cells are washed in

keratinocyte medium with the use of a suitable centrifuge at $500 \times g$ for 5 min before plating as described.

3.5. DNA Isolation for SNP Analysis

1. Non-SCC keratinocytes are also readily isolated from tumour material using this method. In our experience, non-SCC keratinocytes isolated from tumour tissue have limited proliferative capacity and are eventually outgrown by SCC keratinocytes when present. However, this may take a number of passages, and some experiments require very early passage cells. A clear indicator of whether a population of keratinocytes is enriched for tumour cells is by DNA analysis using SNP mapping arrays (5). Mixed populations of cells can be identified through the lack of clear genetic change observed using this method (see Note 5).

2. Cultures are rinsed in EDTA solution to remove feeders and then incubated in trypsin/EDTA solution at 37°C for 20 min or until the cells are visibly rounded and can be easily detached from the culture dish. DMEM/10% FCS is added to inhibit the trypsin, and cells are recovered with the use of a suitable centrifuge at $500 \times g$ for 5 min.

3. The cells are washed twice in PBS to remove all traces of cell culture medium and recovered by centrifugation at $500 \times g$.

4. The DNA is extracted with the Nucleon BACC2 kit according to the manufacturer's instructions. This kit uses a resin-based method to extract the genomic DNA into an aqueous layer while cellular proteins remain bound to the resin. Polypropylene tubes are required because other materials are susceptible to the chloroform used in the protocol.

5. The upper aqueous phase containing the DNA is carefully removed to a clean tube, and the DNA is precipitated by the addition of two volumes of cold absolute ethanol, removed from the tube with a pipette tip and placed directly into TE buffer.

6. To increase DNA yield, the tube containing ethanol may be centrifuged at top speed (minimum $4,000 \times g$) for 5 min and the pellet washed with 70% cold ethanol, air-dried, and re-suspended in TE buffer.

4. Notes

1. Keratinocytes adhere to plastic more efficiently in the absence of EGF: therefore, two keratinocyte media preparations should be made, one with EGF and one without it. Initial isolation and plating should be carried out using medium without EGF. 24–48 h after isolation or plating, medium containing EGF should be used.

2. SCC keratinocytes clearly resemble epithelial cells but can vary in appearance depending on the degree of cell–cell adhesion and propensity to grow in typical epithelioid colonies in the presence of feeder cells. These differences become much less apparent once SCC keratinocytes reach confluence and an epithelial sheet forms. Figure 1 depicts various SCC keratinocyte cultures giving examples of the different cell morphology compared with classical primary normal epidermal keratinocytes grown on a feeder layer.

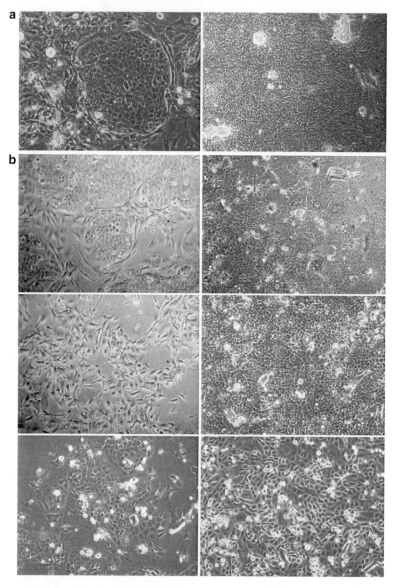

Fig. 1. SCC keratinocyte appearance resembles non-SCC primary keratinocyte culture. (**a**) sub-confluent (with 3T3 feeder cells, *left*) and confluent (*right*) cultures of primary abdominal epidermal keratinocytes. (**b**) Cultures of SCC keratinocytes at different cell densities with or without 3T3 feeder cells.

3. Tumour material contains a variety of cells, and fibroblasts are readily isolated after collagenase treatment. If fibroblasts do become established, they can outgrow SCC keratinocytes, so it is important to regularly observe cultures whilst keratinocytes become established. Fibroblasts can be removed using EDTA solution or short incubation with EDTA/trypsin solution.

4. 75% of the cutaneous SCC cell lines established are tumorigenic in SCID mice using standard approaches (6). The resulting histology varies in the degree of differentiation, but all retain features of primary cutaneous SCC. Figure 2 shows H&E staining of sections from three separate xenograft tumours established with spontaneously immortalised cutaneous SCC lines.

5. As the majority of fresh tumour material is from the margin, normal keratinocyte contamination is common. Mixed populations of cells can be identified through clear genetic change observed using these SNP mapping arrays. Figure 3 depicts examples of heterogeneous DNA SNP profiling. We have found the Nucleon resin-based DNA extraction method gives high yields of pure DNA; however, good results can also be obtained with other commercially available systems such as the Qiagen spin column kits. The Nucleon BACC2 kit is suitable for DNA extraction from 3×10^6 to 1×10^7 cells.

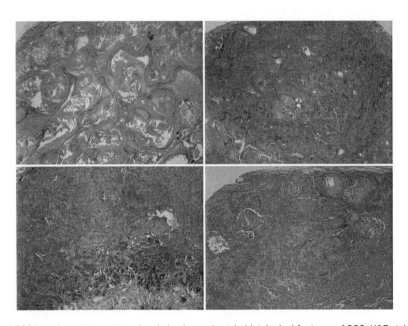

Fig. 2. Cultured SCC keratinocytes are tumorigenic in vivo and retain histological features of SCC. H&E staining of xenograft tumours formed using keratinocytes derived from well (*top left*), moderately (*top right* and *bottom left*), and poorly differentiated SCCs (*bottom right*).

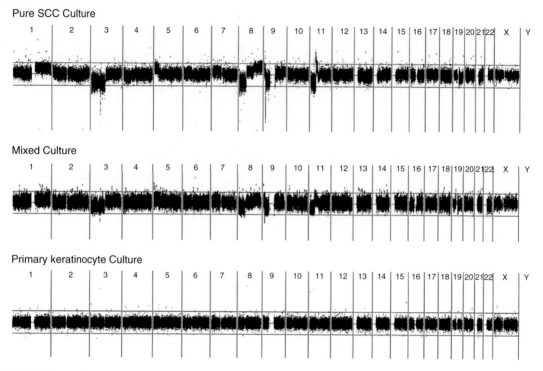

Fig. 3. SNP mapping arrays (5) can be used to distinguish between pure and mixed SCC keratinocyte cultures. Cell culture DNA is compared to non-tumour DNA from the same individual across all chromosomes with deletions shown below the centre line and amplifications above. *Top panel*: pure SCC culture showing clear cancer-specific changes. *Middle panel*: mixed culture of SCC and normal keratinocytes from the same individual with changes less evident. *Bottom panel*: pure culture of normal keratinocytes.

Other methods of DNA analysis such as cytogenetics will also readily identify genomic DNA rearrangements.

Acknowledgments

KJP is funded by Cancer Research UK; CP and APS are funded by the Dystrophic Epidermolysis Bullosa Research Association.

References

1. Rheinwald, J. G., and Green, H. (1975) Serial cultivation of strains of human epidermal keratinocytes: the formation of keratinizing colonies from single cells. *Cell 6*, 331–43.
2. Rheinwald, J. G., and Beckett, M. A. (1981) Tumorigenic keratinocyte lines requiring anchorage and fibroblast support cultures from human squamous cell carcinomas. *Cancer Res 41*, 1657–63.
3. Navsaria, H., Sexton, C., Bouvard, V. & Leigh I.M. (1994) Growth of keratinocytes with a 3T3 feeder layer: basic techniques, in *Keratinocyte Methods* (LEigh, I. M. W., F.M., Ed.) pp 5–12, Cambridge University Press, Cambridge.
4. Hayflick, L., and Moorhead, P. S. (1961) The serial cultivation of human diploid cell strains. *Exp Cell Res 25*, 585–621.

5. Purdie, K. J., Lambert, S. R., Teh, M. T., Chaplin, T., Molloy, G., Raghavan, M., Kelsell, D. P., Leigh, I. M., Harwood, C. A., Proby, C. M., and Young, B. D. (2007) Allelic imbalances and microdeletions affecting the PTPRD gene in cutaneous squamous cell carcinomas detected using single nucleotide polymorphism microarray analysis. *Genes Chromosomes Cancer* 46, 661–9.

6. Song, C., Appleyard, V., Murray, K., Frank, T., Sibbett, W., Cuschieri, A., and Thompson, A. (2007) Thermographic assessment of tumor growth in mouse xenografts. *Int J Cancer 121*, 1055–8.

Chapter 15

Isolation and Culture of Ovarian Cancer Cells and Cell Lines

Christian M. Kurbacher, Cornelia Korn, Susanne Dexel, Ulrike Schween, Jutta A. Kurbacher, Ralf Reichelt, and Petra N. Arenz

Abstract

Ovarian carcinomas show considerable heterogeneity of origin, both in terms of site and tissue. The most important and also most frequent of these tumors arise from the coelomic epithelium and are therefore characterized as epithelial ovarian carcinomas (EOC). EOC is often large and advanced at the time of presentation, so that cells are readily obtainable from surgical specimens or effusions. While the primary tumor may be chemosensitive, they often develop resistance and may do so rapidly. Due to the easy access to tumor cells and its biological behavior, EOC is considered to be an ideal model to investigate principal mechanisms of both antineoplastic drug sensitivity and resistance. Although studies on primary EOC cells are now preferred for many of these investigations, EOC cell line studies remain important too. This chapter gives an overview over major techniques required to establish and maintain primary EOC cell cultures and to initiate and cultivate permanently growing EOC cell lines.

Key words: Ovarian cancer, Chemosensitivity, Chemoresistance, Cell line, ATP, Primary cell culture

1. Introduction

Ovarian cancer is the second-most malignant tumor of the female genital tract and the leading cause of death related to gynecologic malignancies in industrialized countries (1–4). The majority of these tumors, which are also known as epithelial ovarian carcinomas (EOC), arise from the coelomic (or Mullerian) epithelium (3, 5, 6). Less frequently, malignant ovarian tumors may also originate from the oocytes (dysgerminomas, embryonic carcinomas, trophoblast cell carcinomas, and others), the differentiated ovarian stroma (granulosa and theca cell tumors), and the undifferentiated ovarian stroma (all kinds of sarcomas) (7, 8). Carcinosarcomas or so-called mixed Mullerian tumors are also known (3). Due to their underlying embryology, EOC are able to imitate all derivatives of the Mullerian tract and the mesonephros (5, 6). Papillary serous tumors resemble the mucosa of the fallopian tube whereas mucinous tumors may imitate both the endocervical and the intestinal mucosa. Clearly, endometrioid carcinomas are closely related to the endometrium. Clear cell carcinomas imitate the structures of the adrenal cortex and are therefore also known as mesonephroid tumors. Brenner tumors which are mostly benign are of urothelial differentiation. Rare subtypes are squamous cell and small-cell (neuroendocrine) carcinomas. Mixed epithelial tumors are also known. Solid, undifferentiated, or anaplastic carcinomas are mostly of papillary-serous origin (3, 5, 6).

This chapter is mainly focused on EOCs which account for more than 75% of all ovarian malignancies. The incidence of EOC increases by age with a peak observed between 60 and 70 years. Predominant histological subtypes are papillary-serous, mucinous, and endometrioid adenocarcinomas (3, 9). One special aspect of both papillary-serous and mucinous carcinomas is the fact that they can arise from benign lesions (so-called cystadenomas). Approximately one in four serous cystadenomas have the potential of malignant transformation whereas one in ten mucinous cystadenomas may become malignant by time (9). Apart from benign cystadenomas and malignant cystadenocarcinomas, a third subgroup of tumors can be identified, which are known as borderline tumors or tumors of low malignant potential (LMP tumors) (9, 10). These tumors are restricted to the epithelium from which they originated meaning that tumor cells have not yet invaded the organ's stroma harboring local blood and lymph vessels. In contrast to in situ carcinomas of other origin (i.e. the breast or uterine cervix), these tumors are able to metastasize intraperitoneally (and secondarily via extra-ovarian lymph vessels) when located at the ovarian surface (9, 10). Moreover, LMP tumors can also propagate to form invasive carcinomas (10). Although disseminated endometriosis with high proliferative capacity may be regarded as the benign variant of endometrioid carcinomas to some degree, the likelihood of malignant transformation is extremely low in these lesions. Apart from the ovary, papillary-serous carcinomas can also arise from the pelvic peritoneum (3). These tumors, which are biologically identical to papillary-serous ovarian cancers, are classified as peritoneal papillary serous carcinomas (PPSC). Different tumors which share major histomorphological and biological characteristics with EOC are papillary-serous and clear-cell endometrial carcinomas and fallopian tube carcinomas which can show either a papillary-serous or endometrioid differentiation (11–13). Lastly, multifocal papillary-serous carcinomas simultaneously arising from different origins including the ovary, the peritoneum, the fallopian tube, and the endometrium are known to occur (3, 9).

Ovarian malignancies are staged according to both the Féderation Internationale de Gynécologie and d´Obstrétrique (FIGO) and the TNM system (3, 9). The prognosis of patients suffering from EOC decreases by stage. Whereas 85% of patients at FIGO stage I can be cured, the likelihood of cure in stage IV patients is below 5% (3, 9). Due to the relative lack of specific early symptoms, the majority of EOC patients present with advanced disease (i.e. FIGO stage III–IV) at time of primary diagnosis (3, 9). Bulky tumors invading the whole abdominal cavity and even extra-abdominal structures are common in these patients which may also suffer from retroperitoneal lymph node metastases. Malignant ascites which often leads to the diagnosis of ovarian cancer by producing typical symptoms frequently occurs in patients with advanced stage disease (3, 9, 11). Therapy of advanced EOC comprises both surgery and chemotherapy. The residual tumor at the end of primary operation is the most powerful prognostic factor in these individuals (3, 9, 14). Recently, no residual tumor or residual lesions of less than 0.5 cm in diameter are regarded as an optimal surgical result. Optimally debulked patients with FIGO stage III EOC have a more than 50% likelihood of survival which decreases to less than 15% when optimal surgery cannot be achieved (3, 9, 14). In most countries, the primary operation is currently followed by six cycles of intravenous combination chemotherapy, usually including both a platinum analog (cisplatinum or carboplatin) and paclitaxel (3, 9, 15, 16). There is no current evidence supporting the use of other platinum-based combinations or the administration of additional chemotherapy cycles. Intraperitoneal chemotherapy may have advantages over the classical intravenous route in patients with microscopic residual tumor only (17). However, the toxicity of this approach is high and the optimal regimen to be given is still to be defined. Preoperative (neoadjuvant) chemotherapy is another experimental approach, which merits attention especially in patients who are considered bad candidates for optimal resection, but this is currently not considered as the standard due to the lack of conclusive data generated by large-scale prospective controlled trials (3, 18, 19).

Despite its primary chemosensitivity, most patients with advanced stage EOC even including those having had optimal surgery will relapse and ultimately die from their disease. The likelihood of another long-lasting remission following second-line treatment is closely related to the time to relapse after completion of primary therapy (3, 9, 20). Individuals relapsing during or within 6 months after first-line treatment have a particularly poor chance of benefit from a secondary treatment (3, 9, 20, 21). These patients are commonly regarded as being platinum-refractory – whether they are also taxane-refractory is a subject of recent controversy. In contrast, patients relapsing more than 1 year after completion of first-line chemotherapy (who are considered platinum-sensitive) have a good chance of response to platinum-based reinduction chemotherapy. Moreover, secondary radical surgery may be successfully performed in a subset of these individuals (3, 9). Patients relapsing within 7–12 months after completion of first-line treatment represent a third subgroup with so-called intermediate platinum sensitivity. The optimal treatment of these patients is still to be defined. Currently, it is unclear whether they should be subjected to platinum-based or to platinum-free chemotherapy. Additionally, there is no data available supporting the use of secondary debulking surgery in these patients.

EOC is characterized by many features making this disease particularly suitable for preclinical research using cell-based biological and pharmacological tumor models. As with several other solid tumors and hematological malignancies, including breast and colorectal cancer or multiple myeloma, EOC is now widely regarded as a clonal stem cell system (22, 23). Interestingly, EOC has been used for more than 30 years for tumor stem cell research. Additionally, EOC is a tumor with considerable intrinsic chemosensitivity. Nonetheless, EOC has a high potential to develop secondary chemoresistance. Therefore, this disease has a particular importance for preclinical research on cancer chemoresistance and chemosensitivity.

Another important feature making EOC particularly interesting for cell culture research is the easy access of viable tumor cells. Ovarian carcinomas normally present as large bulky tumors. Moreover, EOC patients often suffer from malignant effusions containing lots of disseminated tumor cells (ascites, pleural effusions, and cardiac effusions). Therefore, EOC cells can be easily obtained during tumor-reductive surgery, by surgical or laparoscopic biopsy, or by paracentesis. Generally, ovarian cancer cells are easy to grow under extracorporal conditions. During the last four decades, hundreds of permanent EOC cell lines and numerous variants have been described. Many are now components of institutional and industrial drug screening panels. A number of frequently used parental EOC cell lines are summarized in Tables 1 and 2 (24). It has to be noted, however, that all of the EOC cell lines derived from pretreated patients shown in Table 2 have been initiated prior to the establishment of newer chemotherapeutic agents now routinely used in EOC therapy like carboplatin, paclitaxel, topotecan, gemcitabine, or liposomal doxorubicin. Thus, studies on native tumor cells directly derived from large numbers of clinical tumors which can be tested using short-term assays may become even more important for future drug development and preclinical pharmacological research (25–27).

This chapter provides an overview over basic techniques of isolation and culture of human EOC cells, with a focus on both establishment and maintenance of permanent EOC cell lines and short time EOC cultures for biomedical research. As previously

Table 1
Characteristics of established epithelial ovarian carcinoma cell lines derived from previously untreated patients

Cell line	Histology	Source	References
41M	Adenoca	Ascites	(33)
59M	Endometrioid adenoca	Ascites	(34)
200D	Serous adenoca	Primary	(33)
371M	Mucinous adenoca	Ascites	(35)
A2780	Serous adenoca	Primary	(36)
BUPH:OVSC	Sarcomatoid ca	Solid metastasis	(37)
COLO 110	Serous adenoca	Solid metastasis	(38)
COLO 316	Serous adenoca	Pleural effusion	(38)
COLO 319	Serous adenoca	Ascites	(38)
DO-s	Mucinous adenoca G1	Ascites	(39)
FU-OV-1	Serous adenoca G3	Primary	(40)
HTOA	Serous adenoca G1	Primary	(41)
HOC-1	Serous adenoca G1	Ascites	(42)
HOC-7	Serous adenoca G1	Ascites	(42)
IGROV 1	Adenoca	Primary	(43)
JA-1	Cystadenoca	Primary	(44)
LN1	Mixed Mullerian tumor	Primary	(45)
NOE	Endometrioid adenoca	Primary	(46)
OC 314	Serous adenoca	Ascites	(47)
OTN 11	Serous adenoca G1-2	Ascites	(48)
OTN 14	Mucinous cystadenoca G1	Ascites	(49)
OV-1946	Serous adenoca G3	Ascites	(50)
OVCAR-5	Adenoca	Ascites	(51)
OV-MZ-1	Serous adenoca	Solid tumor	(52)
PE014	Serous adenoca G1	Ascites	(53)
PEA 1	Adenoca G3	Pleural Effusion	(53)
RMG-V	Clear cell adenoca	Primary	(54)
SMOV-2	Clear cell adenoca	Primary	(55)
T014	Serous adenoca G1	Solid metastasis	(53)
TAYA	Clear cell adenoca	Ascites	(56)
TOV-1946	Serous adenoca G3	Primary	(50)
TOV-2223	Serous adenoca G3	Primary	(50)

Adenoca = adenocarcinoma; G1 = well differentiated; G2 = moderately differentiated; G3 = poorly differentiated

Table 2
Characteristics of established epithelial ovarian carcinoma cell lines derived from pretreated patients

Cell line	Histology	Source	Pre-biopsy treatment	References
138D	Serous adenoca	Ascites	CBDCA	(33)
180D	Adenoca	Ascites	DDP	(33)
253D	Serous adenoca	Ascites	CPA/MPA	(33)
Caov-3	Adenoca	Solid tumor	CPA/DOX/FU	(42)
CH1	Papillary adenoca	Ascites	DDP/CBDCA	(34)
COLO 330	Serous adenoca	Ascites	L-PAM/Rx	(38)
OAW 28	Adenoca	Ascites	DDP/L-PAM	(34)
OAW 42	Serous adenoca	Ascites	DDP	(34)
OC 315	Serous adenoca	Ascites	CPA/DDP	(47)
OC 316	Serous adenoca	Pleural effusion	PTC/DDP	(47)
OV-1063	Papillary adenoca	Ascites	CPA/DOX/DDP/HMM	(57)
OV-MZ-1b	Solid adenoca	Ascites	CPA/DOX/DDP, Rx	(52)
OV-MZ-22	Carcinosarcoma	Solid tumor	CPA/EPI/DDP	(58)
OVCAR-2	Adenoca	Ascites	CPA/DDP	(59)
OVCAR-3	Papillary adenoca G3	Ascites	CPA/DOX/DDP	(60)
OVCAR-4	Adenoca	Ascites	CPA/DOX/DDP	(61)
OVISE	Clear cell adenoca	Solid tumor	CPA/DOX/DDP	(62)
OVTOKO	Clear cell adenoca	Solid tumor	CPA/DOX/DDP	(62)
PE01	Serous adenoca G3	Ascites	DDP/FU/CLB	(63)
PE04	Serous adenoca G3	Ascites	DDP/FU/CLB	(63)
PE06	Serous adenoca G3	Ascites	DDP/FU/CLB	(53)
PE016	Serous adenoca G3	Ascites	Rx	(53)
PE023	Serous adenoca G1	Ascites	DDP/CLB	(53)
PEA2	Serous adenoca G3	Ascites	DDP/CLB	(53)
SKOV-3	Adenoca	Ascites	TT	(64)
TR 170	Serous adenoca	Ascites	CPA/DDP	(44)
TR 175	Serous adenoca	Ascites	CLB/CPA	(44)

Adenoca = adenocarcinoma; CBDCA = carboplatin; CPA = cyclophosphamide; DDP = cisplatin; DOX = doxorubicin; EPI = epirubicin; FU = 5-fluorouracil; G1 = well differentiated; G2 = moderately differentiated; G3 = poorly differentiated; HMM = hexamethylmelamine; L-PAM = melphalan; MPA = medroxyprogesterone acetate; PTC = paclitaxel; Rx = radiotherapy

Isolation and Culture of Ovarian Cancer Cells and Cell Lines 167

discussed, EOC is an ideal model for experimental research on cellular pharmacology in primary cultures of native tumor cells tested with various short time assay such as the ATP-based tumor chemosensitivity assay (ATP-TCA), the extreme drug resistance assay (EDR), the fluorescent cytoprint assay (FCA), the histoculture drug response assay (HDRA), different types of clonogenic assays, and many others (28, 29).

2. Materials

2.1. Laboratory Equipment

Most EOC cultures can be grown by using a routine cell culture laboratory providing the following:
Centrifuge (minimum: $400 \times g$).

1. Inverted and phase-contrast microscopes.
2. Class II laminar airflow cabinet.
3. Humidified incubator (37°C and 5% CO_2).
4. Shaker water bath, 37°C.
5. Personal computer.
6. Adjustable automated pipettes (0–200 µL and 200–1,000 µL).
7. Automated pipette pump.
8. Refrigerator (2–8°C).
9. Various freezers: −20°C and −80°C.
10. Dewar container for storage in liquid nitrogen (−196°C) equipped with cryogenic storage racks.

2.2. Laboratory Materials

EOC culture normally requires the following sterile laboratory materials including a variety of glass and plastic ware:

1. Sterile scissors and tweezers.
2. Disposable scalpels (no. 11).
3. Disposable canules (G16–G21).
4. Disposable syringes (1, 2, 5, and 10 mL).
5. Disposable serological pipettes (1, 5, 20, and 25 mL).
6. Autoclavable Pasteur pipettes.
7. Pipette tips at various sizes to be used with automated pipettes.
8. Autoclavable pipette tip boxes.
9. Sterile conical centrifuge tubes (15 mL, 50 mL).
10. Microcentrifuge tubes, snap-capped (1.7 mL).
11. Round bottom polypropylene cryogenic vials (2 mL).

12. Sterile polysterene petri dishes at various diameters.
13. Polysterene angled neck culture flasks with filter caps (25 mL and 75 mL).
14. Disposable cellulose acetate filters (0.2 μm).
15. 70 μm mesh filter gauze.
16. Autoclavable glass tubes, filters, and Erlenmeyer flasks at various sizes.
17. Neubauer Hemocytometer.
18. Autoclavable leukocyte pipettes.

2.3. Cell Culture Media

In the past, various cell culture media (CCM) have been developed and a number of them have been described as being obligatory to maintain specific EOC cell cultures. However, most of the primary EOC cell cultures and the vast majority of established cell lines can be successfully grown in a handful of these media as summarized below. Media should routinely contain phenol red indicator. However, it may be appropriate to use phenol red-free media formulations, in particular, if stimulation experiments with sexual steroids or other hormones are planned (see also Notes 1 and 2). Blank media should be kept at 2–8°C if not otherwise specified by the manufacturer:

1. Dulbecco's modified Eagles' medium (DMEM).
2. McCoy's 5A medium.
3. Roswell Park Memorial Institute medium 1640 (RPMI 1640).
4. Connaught Medical Research Laboratories medium 1066 (CMRL 1066).
5. Hank's F-10 medium.
6. Leibovitz L-15 medium.

2.4. Cell Culture Supplements and Reagents

The following cell culture supplements and reagents should also be available:

1. Hank's balanced salt solution (HBSS), to be stored at room temperature.
2. Dulbecco's phosphate buffered saline (PBS), to be stored at 2–8°C.
3. Fetal bovine serum (FBS), charcoal stripped, to be stored at −20°C.
4. L-Glutamine solution (200 mM), to be stored at −20°C.
5. Bovine insulin solution (10 mg/mL in 25 mM HEPES), to be stored at 2–8°C.
6. Penicillin–Streptomycin (100× concentration); 10,000 IU/mL penicillin and 10 mg/mL streptomycin, to be stored at −20°C.

7. Sterile trypsin-EDTA solution (0.25% in PBS), to be stored at 2–8°C.

8. Trypan-blue dye (0.14%) to be stored at room temperature.

9. Hoechst stain solution (10 mL contains 0.5 μg/mL Hoechst bisbenzamide 33258 fluorochrome stain and thimerosal) to be reconstituted in 10 mL mounting medium (sodium phosphate and citric acid in glycerol), store at 2–8°C.

10. Sterile Ficoll-Hypaque® (Density 1,077 g/mL) to be stored at 2.8°C.

11. Dimethyl sulfoxide (DMSO), to be stored at room temperature.

12. Accumax® Solution (Sigma-Aldrich): Type I collagenase from *Clostridium histolyticum*, DNase, and pronase prepared in Dulbecco's PBS to be stored at –20°C.

13. Lyophilized Tumor Dissociation Enzymes (TDE®; DCS, Hamburg, Germany) which must be reconstituted in 10 mL CCM according to the manufacturer's instructions. Reconstituted TDE® must be stored at –20°C.

2.5. Preparation of Ready-to-Use Cell Culture Media

It is commonly thought that the most appropriate CCM for tumor cells differs from case to case. However, the vast majority of EOC cells grow easily in vitro using the following CCM formulation.

1. Use RPMI 1640, DMEM, or McCoy's 5A as basal medium.

2. Supplement with 10% FBS, 2 mM L-glutamine, 10 μg/mL insulin, 100 IU/mL penicillin and 100 μg/mL of streptomycin.

3. If immediate suppression of B-lymphocyte growth is necessary, use a glutathione-free basal medium (i.e. DMEM) instead of RPMI 1640 or McCoy's 5A (see Subheading 4) and substitute FBS by an appropriate medium replacement (see also Note 1).

Store the ready-to-use CCM at 2–8°C for a maximum for 3 months.

3. Methods

3.1. Collection of Primary Ovarian Carcinoma Cells

As with any other tumor type, strict asepsis is mandatory during all steps of EOC cell collection and processing. As mentioned above, EOC cells are easy to obtain during routine clinical procedures. They can be harvested from either surgical tumor specimens or malignant effusions requiring diagnostic or therapeutic paracentesis. If possible, solid material should be obtained intraoperatively to lower the risk of microbial contamination, provided that histopathological diagnosis and staging will not be compromised. After sampling, solid specimens must be placed into a sterile tightly closable transport vessel containing supplemented DMEM as described above. If microbial infection cannot be excluded, the transport medium should additionally contain 2.5 µg/mL of amphotericin B and 1 µg/mL of metronidazole (30). Malignant effusions must be coagulated with commercial sodium-heparin at 10 IU/mL immediately after paracentesis. EOC cell containing material should be processed optimally, immediately after sampling. If storage or transportation is inevitable, specimens should be kept gently cooled at 8–12°C. Cooling at deeper temperatures should be avoided. Specimens must not be frozen or fixed.

3.2. Preparation of Ovarian Carcinoma Cell Suspensions from Solid Samples

All preparations should be performed under a laminar flow hood. The aim of this step is to obtain EOC cell-enriched suspensions containing both single cells and small organoid particles composed of both tumor cells and stroma, as well. It is not mandatory to achieve pure single cell suspensions (see Note 3).

1. Place EOC specimens in a 10 cm Petri dish to remove necrotic parts, fat, and connective tissue using sterile scissors, tweezers, and scalpels.
2. Add a few milliliters of serum-free DMEM, and then cut the remaining tumor tissue into fragments of 2–5 mm in diameter.
3. Pour the resulting tumor fragments into a 15-mL conical test tube and mix with 10 mL of freshly thawed tumor dissociation reagent (Accumax® or TDE®).
4. Pass the resulting tumor lysate through a 70-µm mesh filter gauze in order to remove undissociated particles.
5. Remove the dissociation enzymes by two 10 min centrifugation steps at $200 \times g$ after adding 10 mL of DMEM.
6. Resuspend the resultant pellet with the appropriate CCM.

Depending on the consistency of the individual EOC material, the enzymatic digestion can be performed either in a 37°C shaker bath or in an incubator for 1–2 h (37°C, humidified 5% CO_2 atmosphere). In extremely fibrotic samples, incubation can be prolonged to 4 h (shaker bath) or overnight (incubator).

3.3. Preparation of Ovarian Carcinoma Cell Suspensions from Malignant Effusions

As mentioned above, EOC patients often present with malignant effusions from which tumors cells can be easily isolated. However, appropriate anticoagulation with sodium-heparin (10 IU/mL effusion) is mandatory prior to further processing.

1. Fill 50-mL conical test tubes with aliquots of heparinized effusions and then centrifuge for 10 min at $200 \times g$ in order to remove all serum components.
2. Resuspend the resulting pellets with 10 mL of DMEM.

3. In specimens containing a high amount of red blood cells, perform an additional purification step. Process these samples for Ficoll-Hypaque® density gradient centrifugation and then resuspend the resulting EOC cell containing interphase in 5 mL CCM.

3.4. Quality Control of Ovarian Carcinoma Cell Suspensions

The resultant tumor cell suspension must be controlled for both quality and viability.

1. Determine the EOC cell content by subsequent Wright-Giemsa or hematoxylin-eosin (HE) staining (usually performed by the collaborating pathologist).
2. Determine the tumor cell number and viability by trypan-blue dye exclusion using a Neubauer hemocytometer.

3.5. Initiation of Primary Ovarian Carcinoma Cell Cultures

Only the general procedures to set up primary cultures of EOC cells are described in this chapter. Techniques of short-time EOC cultures for drug response and resistance assay may differ markedly from these prescriptions and are described in detail elsewhere in this volume (see also Note 4). After successful initiation of primary EOC cultures both short-time assays as well as further propagation in order to establish EOC cell lines is possible.

1. Fill aliquots of 10 mL of the resulting EOC cell suspensions either generated from solid or liquid material into 25-mL canted necked rectangular polypropylene culture flasks.
2. Set up multiple flasks of each tumor cell suspension whenever possible.
3. Close the culture flasks tightly with a filter cap (cap with an integrated 0.22 µm cellulose acetate filter).
4. Transfer the culture flasks to a humidified incubator and incubate at 37°C and 5% CO_2 for 24 h.
5. Control the cultures every day for cell growth and microbial infection under an inverted microscope.
6. Cultivate the primary EOC cultures by using the appropriate CCM as described above. Generally, the CCM used to initiate the EOC culture should be used throughout the whole culture period if not otherwise indicated.
7. Feed the primary EOC cultures by exchanging 50% of the CCM every second or third incubation day depending on the individual cell growth (normally indicated by a yellowish change of the color of the phenol red in the CCM) (see Note 5).

3.6. Subculturing of Epithelial Ovarian Carcinoma Cells Cultures

Both primary EOC cultures and EOC cell lines must be periodically subcultured at subconfluence (i.e. at least 75% of the bottom of the culture flask is covered by cells – tumor cells, fibroblasts,

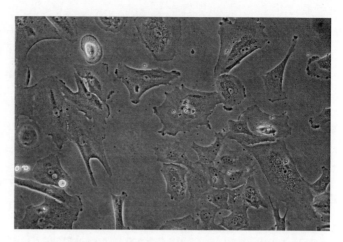

Fig. 1. Microphotograph of a subconfluent ovarian carcinoma cell culture, phase contrast, ×200 original magnification (photo with permission, Sharon Glaysher).

and mesothelial cells) in order to enrich the EOC cell content in the cultures and to remove the fibroblasts by time (Fig. 1).

1. Remove the FBS-containing CCM from the culture flasks completely.
2. Wash twice with serum-free DMEM.
3. Add 10 mL of 0.25% trypsin-EDTA to the culture flasks and then incubate for 10 min at 37°C in a humidified 5% CO_2 incubator.
4. Stop the trypsinization process by adding at least 10 mL of FBS-containing CCM to the culture flask.
5. Resuspend the EOC cells by gently shaking the culture flask or knocking tenderly at the outer surface of its bottom. More vigorous resuspension techniques (i.e. use of cell scrapers) should be avoided (see also Note 6).
6. Pour the whole EOC cell containing supernatant into a 50-mL conical test tube and then wash the cell suspension in the appropriate CCM by two centrifugation steps at $200 \times g$ for 10 min.
7. Transfer the resultant EOC cell containing suspension to new 25-mL culture flasks at a 1:2 or 1:3 ratio depending on the cell density.
8. Add CCM until a total volume of 10–15 mL per culture flask is reached.
9. Allow the cells to attach in the incubator for 24 h at 37°C and 5% CO_2.
10. Cultivate the cells until subconfluence is reached by periodically feeding the cultures as described in the previous paragraph and then perform another passage step.

In cultures derived from malignant effusions, the presence of mesothelial cells may be a particular problem. These cells – in contrast to epithelial cells – are attached to fibrin meshes, which appear during initial culturing. Short-time trypsinization (2 min) may be helpful to completely detach mesothelial cells without removing tumor cells (24).

3.7. Initiation and Maintenance of Ovarian Carcinoma Cell Suspension Cultures

EOC cells normally need adhesion to the inner surface of the cell culture vessel for optimal growth. Anchorage-independent growth which is common in leukemia cells is a rare event in epithelial tumors like EOC. However, some cultures may lose their anchorage-dependence by time and can be grown in suspension. This is most often the case with cultures derived from anaplastic carcinomas and Mullerian mixed tumors, respectively. The procedure of maintaining and subculturing EOC suspension cultures is very similar to those described for leukemia cells (see separate chapters in this volume). Clearly, trypsinization is not necessary. Moreover, one should be aware that nonadherently growing cells normally exhibit a high proliferative capacity requiring a more frequent feeding and/or subculturing. Loss of adherent growth may also be related to mycoplasma infection which, however, mostly results in marked growth reduction (see also Subheading 3.11).

3.8. Establishment and Maintenance of Epithelial Ovarian Carcinoma Cell Lines

Immortalized EOC cell lines can be established by serially subculturing primary EOC cells using the aforementioned cell culture techniques. Both fibroblasts and mesothelial cells will disappear during the subcultivation process (Fig. 2). Generally, immortalization

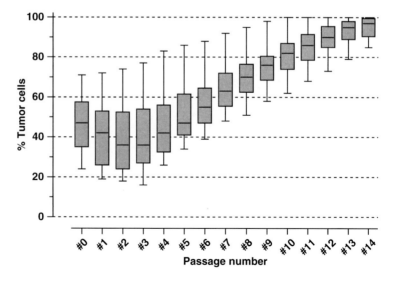

Fig. 2. Correlation between the tumor cell content of 21 primary ovarian carcinoma cell cultures and the number of subsequent passages. The variability with regard to the tumor cell content decreases by time. It has to be noted that the relative tumor cell content drops during the first passages and increases after four subcultures to reach a saturation effect beyond passage #9 (Kurbacher, unpublished data).

of primary EOC cell cultures can be assumed after 10–12 passages (subcultures) provided that the contamination with normal cells does not exceed 15% of the whole cellular content (24).

1. Cultivate the EOC cell lines by using the appropriate CCM.
2. Do not exchange the CCM formulation during the whole culture time if not otherwise indicated.
3. Passage the EOC cell lines at subconfluence by using the techniques previously described.
4. Periodically check for mycoplasma infection by using Hoechst 33258 dye or (preferably) PCR. Commercially available EOC cell lines can be maintained accordingly.

3.9. Cryopreservation of Epithelial Ovarian Carcinoma Cells

If necessary, both native EOC cells and established cell lines can be frozen and stored for later use at any time of the culture process.

1. Resuspend EOC cells by one of the procedures described above.
2. Wash twice by centrifugation at $200 \times g$ for 10 min.
3. Resuspend again in a serum-enriched freezing medium: 75–80% ready-to-use CCM, 15–20% additional FBS, 10% DMSO (or alternatively glycerol).
4. Fill aliquots of 1.5 mL into 2-mL cryotubes and then freeze immediately at –80°C. Slow cooling is not mandatory for most EOC cultures. Snap-freezing normally may give acceptable results in most instances (see Note 7).

Storage at –80°C is possible if liquid nitrogen is not available. However, storage below –150°C may result in a better cell recovery after thawing and should thus be preferred, particularly if cryostorage of more than 3 months is intended.

3.10. Thawing of Cryopreserved Ovarian Carcinoma Cells

Thawing of cryopreserved EOC cells should be performed as fast as possible in order to avoid toxic effects of the cryoprotectants produced during the thawing process. Transfer the cryotubes to a 37°C water bath until complete thawing (i.e. disappearance of all visible ice crystals) (see also Note 8).

1. Immediately pour the suspensions to 15-mL conical test tubes and completely remove the freezing medium by two centrifugation steps at $200 \times g$ for 10 min.
2. Resuspend the frozen-thawed EOC cells in the appropriate CCM and then cultivate as mentioned above.

3.11. Diagnosis and Treatment of Infections

As any other cell culture, EOC cultures are endangered by various microbiological infections. The most important laboratory infections involve fungi, yeasts, bacteria, mycoplasmas, and viruses. Most infections can be easily diagnosed by just observing the content of the culture flasks natively or under the inverted microscope according to the algorithms shown in Figs. 3 and 4. Infections with mycoplasmas which are present in up to 40% of all permanently growing cell cultures worldwide require specific diagnostic means using fluorescence microscopy (DAPI or Hoechst 33258 stain), ELISA, or PCR techniques. Indirect signs of mycoplasma infection can be loss of adherence, diminution of cell growth, and appearance of super-vacuolated cells. Viral infections which mostly remain undetected can be diagnosed the best by using molecular methods like FISH or PCR.

Most laboratory infections are difficult to treat and particularly contaminations with fungi, bacteria, and viruses are often deleterious to the particular culture involved. Since most germs associated with laboratory infections stem from the laboratory staff by itself, strict asepsis is the most effective way to avoid microbial contaminations (see Note 9). In most cases, contaminated culture flasks should be

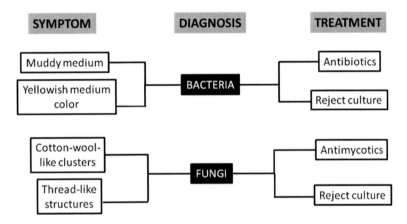

Fig. 3. Diagnosis of laboratory infections I: directly visible symptoms.

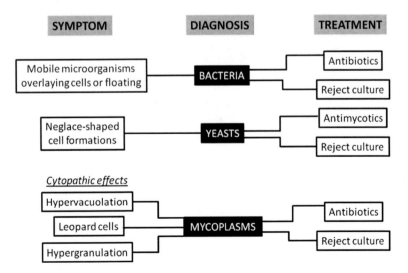

Fig. 4. Diagnosis of laboratory infections II: symptoms which can be diagnosed by light microscopy.

immediately discarded in order to prevent the infection of other cultures grown in the same incubator. If infections occur in very rare or valuable cultures, a treatment attempt can be made by using the appropriate compounds among a variety of commercially available antibiotics and/or antimycotics. The involved culture flasks should be immediately transferred to a separate incubator and kept there until the unequivocal confirmation of successful decontamination. In most cases, the attempt of decontamination (in particular with fungi and bacteria) remains unsuccessful so that extinction of the involved culture is the best way to handle laboratory infections in most instances.

4. Notes

1. A lot of cell culture researchers believe that a specific CCM formulation is mandatory to achieve optimal results with human cancer cell cultures. This resulted in thousands of publications in this field which suggest that tumor cell culture is a rather complicated field of research. However, the majority of human ovarian carcinoma cultures (both primary cultures and cell lines) can be easily grown in vitro and to our own experience, only a handful of basal media as mentioned in Subheading 2.3 is sufficient to grow most EOC cultures with considerable success. Depending on the investigations planned, it may be preferable to use glutathione-free formulations (such as DMEM) instead of RPMI-1640 or McCoy's 5A.

2. Phenol red exhibits a weak estrogen-like effect which may disturb stimulation experiments with sexual steroids or other hormonal compounds (30– 32). In these instances, the use of phenol red-free media may be preferable. However, the omission of the indicator requires a particularly attentive handling of the cell cultures.

3. Generally, preparation of pure single cell suspensions forseen for EOC primary cultures is impossible. Moreover, the presence of stroma cells may well facilitate the growth of EOC cells at the beginning of the culture period by enriching the medium with growth factors and cytokines (so-called feeder effect).

4. Short-time ovarian EOC cell cultures have been investigated in depth by using both clonogenic and nonclonogenic assays (including the ATP-based tumor chemosensitivity assay, the MTT assay, or the fluorescent cytoprint assay). Using short-time EOC cultures instead of established cell lines may be advantageous since primary EOC cells are much closer related to the clinical situation. However, the use of primary cultures is limited to the exhaustible number of available tumor cells.

Therefore, the use of permanently growing EOC cell lines is preferable in studies which require a large number of repeated experiments.

5. After initiation of most primary EOC cultures, cells need up to 1 week to grow out sufficiently. During this initial phase of incubation, there is normally no need to exchange the medium or to subcultivate.

6. Generally, EOC cells are not tightly attached to the inner surface of the culture flask. After trypsinization, they can be resuspended easily by gentle techniques. The use of more vigorous resuspension methods (cell scrapers or vigorous shaking) may exhibit deleterious effects to cells by disrupting the integrity of the cell membranes or leading to a loss of membrane-bound receptors.

7. In the past, a lot of slow freezing protocols have been developed to cryopreserve mammalian cells. Since EOC cells appear to be rather cryoresistant to our personal experience, snap-freezing seems to be appropriate in most cases.

8. If a 37°C water bath is not available, frozen EOC cells can also be thawed at body temperature by simply warming the cryotubes in the fist.

9. One of the most important problems in the culture of EOC cells is the prevention of microbial infections. Strict aseptic conditions are therefore a must for all methodological steps of cultivation and storage of EOC cells. Although the majority of contaminated cultures are destined to be lost, anti-infectious treatment may be considered in special instances. Recent publications have shown that even primary cultures derived from colorectal cancer specimens which have to be considered as potentially infected can be successfully decontaminated by the addition of both amphotericin B and metronidazole without markedly deteriorating the proliferative capacity in vitro (30).

References

1. Tortolero-Luna, G., and Mitchel, M. F. (1995) The epidemiology of ovarian cancer. *J. Cell. Biochem. Suppl.* **23**, 200–207.
2. Holschneider, C. H., and Berek, J. S. (2000) Ovarian cancer: epidemiology, biology and prognostic factors. *Semin. Oncol.* **19**, 3–10.
3. Cannistra, S. A. (2004) Cancer of the ovary. *N. Engl. J. Med.* **351**, 2519–2529
4. Permuth-Wey, J., and Sellers, T. A. (2009) The epidemiology of ovarian cancer. *Methods Mol. Biol.* **472**, 402–437.
5. Krigman, H., Bentley, R., and Robboy, S. J. (1994) Pathology of epithelial ovarian tumors. *Clin. Obstet. Gynecol.* **37**, 475–491.
6. Soslow, R. A. (2008) Histologic subtypes of ovarian carcinoma: an overview. *Int. J. Gynecol. Pathol.* **27**, 161–174.
7. Williams, S. D., Gershenson, D. M., Horowitz, C. J., and Silva E. (2003) Ovarian germ cell tumors, in *Principles and Practice of Gynecologic Oncology,* 3rd ed. (Hoskins, W. J., Perez, C. A., Young, R. C., eds.)

8. Hartmann, L. C., Young, R. C., and Podratz, K. C. (2003) Ovarian sec cord-stromal tumors, in *Principles and Practice of Gynecologic Oncology*, 3rd ed. (Hoskins, W. J., Perez, C. A., Young, R. C., eds.) Lippincott Williams & Wilkins, Philadelphia, pp. 1074–1097.
9. Ozols, R. F., Rubin, S. C., Thomas, G. M., and Robboy, S. J. (2003) Epithelial ovarian cancer, in *Principles and Practice of Gynecologic Oncology*, 3rd ed. (Hoskins, W. J., Perez, C. A., Young, R. C., eds.) Lippincott Williams & Wilkins, Philadelphia, pp. 980–1057.
10. Tropé, C., Davidson, B., Paulsen, T., Abeler, V. M., and Kaern, J. (2009) Diagnosis and treatment of ovarian borderline neoplasms "state of the art". *Eur. J. Gynaecol. Oncol.* **30**, 471–482.
11. Markman, M., Zaino, R. J., Fleming, P. A., and Barakat, R. R. (2003) Carcinoma of the fallopian tube, in *Principles and Practice of Gynecologic Oncology*, 3rd ed. (Hoskins, W. J., Perez, C. A., Young, R. C., eds.) Lippincott Williams & Wilkins, Philadelphia, pp. 1098–1115.
12. Naumann, R. W. (2008) Uterine papillary serous carcinoma: state of the state. *Curr. Oncol. Rep.* **10**, 505–511.
13. Salvador, S., Gilks, B., Köbel, M., Huntsman, D., Rosen, B., and Miller, D. (2009) The fallopian tube: primary site of most pelvic high-grade serous carcinomas. *Int. J. Gynecol. Cancer.* **19**, 58–64.
14. Pomel, C., Jeyarajah, A., Oram, D., Shepherd, J., Milliken, D., Dauplat, J., and Reynolds, K. (2007) Cytoreductice surgery in ovarian cancer. *Cancer Imaging* **17**, 210–215.
15. McGuire, W. P., Hoskins, W. J., Brady, M. F., Kucera, P. R., Partridge, E. E., Look, K. Y., Clarke-Person, D. L., and Davidson, M. (1996) Cyclophosphamide and cisplatin compared with paclitaxel and cisplatin in patients with stage III and stage IV ovarian cancer. *N. Engl. J. Med.* **334**, 1–6.
16. Du Bois, A., Lück, H. J., Meier, W., Adams, H. P., Möbus, V., Costa, S., Bauknecht, T., Warm, M., Schröder, W., Olbricht, S., Nitz, U., Jackisch, C., Emons, G., and Wagner U for the Arbeitsgemeinschaft Gynäkologische Onkologie Ovarian Cancer Study Group (2003) A randomized clinical trial of cisplatin/paclitaxel versus carboplatin/paclitaxel as first-line treatment of ovarian cancer. *J. Natl. Cancer Inst.* **95**, 1320–1329.
17. Armstrong, D. K., Bundy, B., Wenzel, L., Huang, H. Q., Baergen, R., Lele, S., Copeland, L. J., Walker, J. L., and Burger, R. A., for the Gynecologic Oncology Group. Intraperitoneal cisplatin and paclitaxel in ovarian cancer. *N. Engl. J. Med.* **354**, 34–43.
18. Vergote, I. B., De Wever, I., Decloedt, J., Tjalma, W., Van Gamberen, M., and van Dam, P. (2000) Neoadjuvant chemotherapy versus primary debulking surgery in advanced ovarian cancer. *Semin. Oncol.* **27**, 31–36.
19. Pölcher, M., Mahner, S., Ortmann, O., Hilfrich, J., Diedrich, K., Breitbach, G. P., Höss, C., Leutner, C., Braun, M., Möbus, V., Karbe, I., Stimmler, P., Rudlowski, C., Schwarz, J., and Kuhn, W. (2009) Neoadjuvant chemotherapy with carboplatin and docetaxel in advanced ovarian cancer – a prospective multicenter phase II trial (PRIMOVAR). *Oncol. Rep.* **22**, 605–613.
20. Markman, M., Rothman, R., Hakes, T., Reichman, B., Hoskins, W., Rubin, S., Jones, W., Almadrones, L., and Lewis, J. L., Jr. (1991) Second-line platinum therapy in patients with ovarian cancer previously treated with cisplatin. *J. Clin. Oncol.* **9**, 389–393.
21. Markman, M. (2008) Pharmaceutical management of ovarian cancer: current status. *Drugs* **68**, 771–789.
22. Alvero, A. B., Chen, R., Fu, H. H., Montagna, M., Schwartz, P. E., Rutherford, T., Silasi, D. A., Steffensen, K. D., Waldstrom, D., Visintin, I., and Mor, G. (2009) Molecular phenotyping of human ovarian cancer stem cells unravels the mechanisms of repair and chemoresistance. *Cell Cycle* **8**, 158–166.
23. Fong, M. Y., and Kakar, S. S. (2010) The role of cancer stem cells and the side population in epithelial ovarian cancer. *Histol. Histopathol.* **25**, 113–120.
24. Langdon, S. P., and Lawrie, S. S. (2000) Establishment of ovarian cancer cell lines. *Methods Mol. Med.* **39**, 155–159.
25. Andreotti, P. E., Linder, D., Hartmann, D. M., Cree, I. A., Pazzagli, M., and Bruckner, H. W. (1994) TCA-100 tumour chemosensitivity assay: differences in sensitivity between cultured tumour cell lines and clinical studies. *J. Biolumin. Chemilumin.* **9**, 373–378.
26. Andreotti, P. E., Cree, I. A., Kurbacher, C. M., Hartmann, D. M., Linder D, Harel G., et al. (1995) Chemosensitivity testing of human tumors using a microplate adenosine luminescence assay: clinical correlation for cisplatin resistance of ovarian carcinoma. *Cancer Res.* **55**, 5276–5282.
27. Cree, I. A., and Kurbacher, C. M. (1999) ATP based tumour chemosensitivity testing: assisting new agent development. *Anti-Cancer Drugs* **10**, 431–435.

28. Blumenthal, R. D. (2005) An overview of chemosensitivity testing. *Methods Mol. Med.* **110**, 3–18.
29. Kurbacher, C. M., and Cree, I. A. (2005) Chemosensitivity testing using microplate adenosine triphosphate-based luminescence measurements. *Methods Mol. Med.* **110**, 101–120.
30. Whitehouse, P. A., Knight, L.A., Di Nicolantonio, F., Mercer, S. J., Sharma, S., and Cree, I. A. (2003) Combination chemotherapy in advanced gastrointestinal cancers: ex vivo sensitivity to gemcitabine and mitomycin C. *Br. J. Cancer* **89**, 2299–2304.
31. Simon, W. E., Albrecht, M., Hänsel, M., Dietel, M., and Hölzel, F. (1983) Cell lines derived from human ovarian carcinomas: growth stimulation by gonadotropic and steroid hormones. *J. Natl. Cancer Inst.* **70**, 839–845.
32. Kurbacher, C. M., Jäger, W., Kurbacher, J. A., Bittl, A., Wildt, L., and Lang, N. (1995) Influence of human luteinizing hormone on cell growth and CA 125 secretion of primary epithelial carcinomas in vitro. *Tumor Biol.* **16**, 374–384.
33. Wilson, A. P., Dent, M., Pelovic, T., Hubbold, L., and Radford, H. (1996) Characterisation of seven human ovarian tumour cell lines. *Br. J. Cancer* **74**, 722–727.
34. Hills, C. A., Kelland, L. R., Abel, G., Siracky, J., Wilson, A. P., and Harrap, K. P. (1989) Biological properties of ten human ovarian carcinoma cell lines: calibration in vitro against four platinum complexes. *Br. J. Cancer* **59**, 527–534.
35. Sato, S., Kobayashi, Y., Okuma, Y., Kondo, H., Kanaishi, Y., Saito, K., and Kiguchi, K. (2002) establishment and characterization of a cell-line originated from human mucinous carcinoma of the ovary. *Hum. Cell*, **15**, 171–177.
36. Hamilton, T. C., Winker, M. A., Louie, K. G., Batist, G., Behrens, B. C., et al. (1985) Augmentation of adriamycin, melphalan, and cisplatin cytotoxicity in drug-resistant and -sensitive human ovarian carcinoma cell lines by buthionine sulfoximine mediated glutathione depletion. *Biochem. Pharmacol.* **34**, 2583–2586.
37. Guo, H. F., Feng, J., Liu, G., Cui, H., Ye, X., Yao, Y., and Fu, T. (2005) Establishment and characterization of a human ovarian sarcomatoid carcinoma cell line BUPH:OVSC.
38. Woods, L. K., Morgan, L. T., Quinn, L. A., Moore, G. E., Semple, T. U., and Stedman, K. E. (1979) Comparison of four new cell lines from patients with adenocarcinoma of the ovary. *Cancer Res.* **39**, 4449–4459.
39. Briers, T. W., Strootbants, P., Vandeputte, T. M., Nouwen, E. J., Conraads, M. V., Eestermans, G., et al. (1989) Establishment and characterization of a human ovarian neoplastic cell line DO-s. *Cancer Res.*, **49**, 5153–5161.
40. Emoto M., Oshima K, Ishiguro, M., Iwasaki, H., Kawarabayashi, T., and Kichuchi, M. (1999) Establishment and characterization of a serous papillary adenocarcinoma cell line of the human ovary in serum-free culture. *Pathol. Res. Pract.* **195**, 237–242.
41. Ishiwata, I., Ishiwata, C., Soma, M., Nozawa, S., and Ishikawa, H. (1987) Characterization of newly established human ovarian carcinoma cell line – special reference of the effects of cis-platinum on cellular proliferation and release of CA 125. *Gynecol. Oncol.* **26**, 340–354.
42. Buick, R. N., Pullano, R., and Trent, J. M. (1985) Comparative properties of five human ovarian adenocarcinoma cell lines. *Cancer Res.*, **45**, 3668–3676.
43. Benard, J., Da Silva, J., De Bois, M.-C., Boyer, P., Duvillard, P., Chiric, E., et al. (1985) Characterization of a human ovarian adenocarcinoma line, IGROV 1, in tissue culture and nude mice. *Cancer Res.* **45**, 4970–4979.
44. Hill, B. T., Wheelan, R. D., Gibby, E. M., Sheer, D., Hosking, L. K., Shellard, S. A., and Rupniak, H. T. (1987) Establishment and characterization of three new human ovarian carcinoma cell lines and initial evaluation of their potential in experimental chemotherapy studies. *Int. J. Cancer* **39**, 219–225.
45. Becker, J. L., Papenhausen, P. R., and Widen, R. H. (1997) Cytogenetic, morphologic and oncogene analysis of a cell line derived from a heterologous mixed mullerian tumor of the ovary. *In Vitro Cell Dev. Biol. Anim.* **33**, 325–331.
46. Umezu, T., Kajiyama, H., Terauchi, M., Shibata, K., Ino, K., Nawa, A., and Kikkawa, F. (2007) Establishment of a new cell line of endometrioid carcinoma of the ovary and its chemosensitivity. *Hum. Cell* **20**, 71–76.
47. Alama, A., Barbieri, F., Favre, A., Cagnoli, M., Noviello, E., Pedullà, F. et al. (1996) Establishment and characterization of three new cell lines derived from the ascites of ovarian carcinomas. *Gynecol. Oncol.* **62**, 82–88.
48. Poels, L. G., Jap, P. H., Ramaekers, F. F., Scheres, J. M., Thomas, C. M., Voojs, O. G. et al. (1989) Characterization of a hormone-producing ovarian carcinoma cell line. *Gynecol. Oncol.* **32**, 203–214.
49. Van Niekerk, C. C., Poels, L. G., Jap, P. K. H., Smeets, D. C. F. M., Thomas, C. M. G., Ramaekers, F. C. S., et al. (1988) Characterization

of a human ovarian carcinoma cell line, OTN 14, derived from a mucinous cycstadenocarcinoma. *Int. J. Cancer* **42**, 104–111.

50. Ouellet, V., Zietarska, M., Portelance, L., Lafontaine, J., Madore, J., Puiffe, J. et al. (2008) Characterization of three new serous epithelial ovarian cancer cell lines. *BMC Cancer* **8**, 152.

51. Louie, K. G., Hamilton, T. C., Winker, M. A., Behrens, B. C., Tsuruo, T., Klecker, R. W., et al. (1986) Adriamycin accumulation and metabolism in adriamycin-sensitive and -resistant human ovarian cancer cell lines. *Biochem. Pharmacol.* **65**, 467–472.

52. Möbus, V., Gerharz, C. D., Press, U., Moll, R., Beck, T., Mellin, W., et al. (1992) Morphological, immunohistochemical and biochemical characterization of 6 newly established human ovarian carcinoma cell lines. *Int. J. Cancer* **52**, 76–84.

53. Langdon, S. P., Lawrie, S. S., Hay, F. G., Hawkes, M. M., McDonald, A., Hayward, I. P., et al. (1988) Characterization and properties of nine human ovarian adenocarcinoma cell lines. *Cancer Res.* **48**, 6166–6172.

54. Aoki, D., Suzuki, N., Susumu, N., Noda, T., Suzuki, A., Tamada, Y., et al. (2005) Establishment and characterization of the RMV-G cell line from human ovarian clear cell adenocarcinoma. *Hum Cell.* **18**, 143–146.

55. Yonamine, K., Hayashi, K. and Iida, T. (1999) Establishment and characterization of human ovarian clear cell adenocarcinoma cell line (SMOV-2), and its cytotoxicity by anticancer agents. *Hum. Cell* **12**, 139–148.

56. Saga, Y., Suzuki, M., Machida, S., Ohwada, M., and Sato, I. (2002) Establishment of a new cell line (TAYA) of clear cell adenocarcinoma of the ovary and its radiosensitivity. *Oncology* **62**, 180–184.

57. Horowitz, A. T., Treves, A. J., Voss, R., Okon, E., Fuks, Z., Davidson, L. et al. (1985) A new human ovarian carcinoma cell line: establishment and analysis of tumor-associated markers. *Oncology* **42**, 332–337.

58. Möbus, V. J., Gerharz, C. D., Weikel, W., Merk, O., Dreher, L., Kreienberg, R., Moll, R. (2001) Characterization of a human carcinosarcoma cell line of the ovary established after in vivo change of histologic differentiation. *Gynecol. Oncol.* **83**, 523–532.

59. Pirker, R., FitzGerald, D. J., Hamilton, T. C., Ozols, R. F., Laird, W., Frenkel, A. E., et al. (1985) Characterization of immunotoxins active against ovarian carcinoma cell lines. *J. Clin. Invest.* **76**, 1261–1267.

60. Hamilton, T. C., Young, R. C., McKoy, W. M., Grotzinger, K. R., Green, J. A., Chu, E. W., et al. (1983) Characterization of a human ovarian carcinoma cell line (NIH:OVCAR-3) with androgen and estrogen receptors. *Cancer Res.* **43**, 5379–5389.

61. Louie, K. G., Behrens, B. C., Kinsella, T. J., Hamilton, T. C., Grotzinger, K. R., McKoy, W. M., et al. (1985) Radiation survival parameters of antineoplastic drug-sensitive and -resistant human ovarian carcinoma cell lines and their modification by buthionine sulfoximine. *Cancer Res.* **45**, 2110–2115.

62. Gorai, I., Nakazawa, T., Miyagi, E., Hirahara, F., Nagashima, Y., Minaguchi, H. (1995) Establishment and characterization of two human ovarian clear cell adenocarcinoma line from metastatic lesions with different properties. *Gynecol. Oncol.* **57**, 33–46.

63. Wolf, C. R., Hayward, I. P., Lawrie, S. S., Buckton, K., McIntyre, M. A., Adams, D. J. (1987) Cellular heterogeneity and drug resistance in two ovarian adenocarcinoma cell lines derived from a single patient. *Int. J. Cancer* **39**, 695–702.

64. Fogh, J, Fogh, J. M., and Orfeo, T. (1977) One hundred and twenty-seven cultured tumour cell lines producing tumours in nude mice. *J. Natl. Cancer Inst.* **59**, 221–226.

Chapter 16

Establishment and Culture of Leukemia–Lymphoma Cell Lines

Hans G. Drexler

Abstract

The advent of continuous human leukemia–lymphoma cell lines as a rich resource of abundant, accessible, and manipulable living cells has contributed significantly to a better understanding of the pathophysiology of hematopoietic tumors. The first leukemia–lymphoma cell lines were established in 1963 and since then large numbers of new cell lines have been described. The major advantages of continuous leukemia–lymphoma cell lines are the unlimited supply and worldwide availability of identical cell material and the infinite viable storability in liquid nitrogen. These cell lines are characterized generally by monoclonal origin and differentiation arrest, sustained proliferation in vitro under preservation of most cellular features, and by specific genetic alterations. Here some of the more promising techniques for establishing new leukemia–lymphoma cell lines and the basic principles for culturing these cells are described. Several clinical and cell culture parameters might have some influence on the success rate, e.g., choice of culture medium and culture conditions, specimen site of the primary cells, and status of the patient at the time of sample collection.

Key words: Cell lines, Cryopreservation, Culture, Establishment, Immortalization, Leukemia, Lymphoma, Myeloma

1. Introduction

1.1. Importance of Leukemia–Lymphoma Cell Lines

The availability of continuous human leukemia–lymphoma (LL) cell lines as a rich resource of abundant, accessible, and manipulable living cells has contributed significantly to a better understanding of the pathophysiology of hematopoietic tumors (1). The first malignant hematopoietic cell lines, i.e. Burkitt's lymphoma-derived lines, were established in 1963 (2). Since then large numbers of cell lines have been described, although not all of them have been characterized in full detail (3). The major

Table 1
Advantages and characteristics of leukemia–lymphoma cell lines

Unlimited supply of cell material
Worldwide availability of identical cell material
Indefinite storability in liquid nitrogen and recoverability
Monoclonal origin
Differentiation arrest at a discrete maturation stage
Sustained proliferation in culture
Stability of most features in long-term culture
Specific genetic alterations

advantages and common characteristics of these cell lines are listed in Table 1. The spectrum of malignant hematopoietic cell lines is not only restricted to the various type of leukemia but also includes the lymphomas and myelomas. The following refers generally also to lymphoma- and myeloma-derived cell lines unless indicated otherwise.

Truly malignant cell lines must be discerned from Epstein-Barr virus (EBV)-immortalized normal cells, so-called B-lymphoblastoid cell lines (B-LCL). Some types of malignant cell lines are indeed EBV$^+$, and some EBV$^+$ normal cell lines may carry also genetic aberrations (1).

LL cell lines are now ubiquitously used and have become nearly irreplaceable tools in a multitude of research areas within the various scientific disciplines, e.g., immunology, cytogenetics, molecular biology, pharmacology, toxicology, virology, just to name a few. Nevertheless, it should be kept in mind that once cells are removed from their native environment and grown in cell culture, they are uncoupled from critical extracellular cues and may, but mostly not, undergo phenotypic drift. During this process, transcriptional control mechanisms may change in ways that are difficult to account for. Therefore, in vitro-derived results must always be interpreted with caution and confirmed in the context of the cells' native environment.

1.2. Basic Principles of Leukemia–Lymphoma Cell Line Establishment

It is very difficult to establish LL cell lines; clearly the majority of attempts fail. The simple seeding of the malignant cells directly into suspension cultures is the most commonly used approach in attempts to establish cell lines. While the success rate for the establishment of continuous cell lines is overall low, it depends on the subtype of leukemia or lymphoma and varies strongly between the different reports in the literature.

It has been suggested that certain types of malignant hematopoietic cells that were derived from patients at relapse or from cases with poor prognostic features have an enhanced growth potential in vitro in comparison to samples obtained at presentation or from patients with good prognostic parameters. Indeed, a review of the literature showed that in aggregate the success rate for B-cell precursor (BCP)-cell lines established from patients at diagnosis was 6%, whereas the success rate was 29% for samples at relapse (4). Furthermore, there seems to be a higher success rate in cases which have certain primary chromosomal aberrations or gene mutations, e.g. T-cell lines harboring an (8;14)(q24;q11) translocation and cell lines carrying alterations of the *P53*, *P15INK4B*, or *P16INK4* genes (3, 5).

It has been reported that a hypoxic environment (e.g. 5% O_2) is more suitable to the culture and establishment of leukemic T- and BCP-cell lines than the standard incubation in 5–10% CO_2 in air (6). According to this method, cells are cultured in wells with a feeder layer consisting of complete media, human serum, and agar supplemented with insulin-like growth factor. For the BCP-cell lines, the monocyte toxin l-leucine methyl ester and insulin were applied. While the success rate of cell line establishment was reported to be high, some of these cell lines have extremely long doubling times (10–14 days), which limit their usefulness. Furthermore, no other group has so far confirmed the reproducibility of this method.

The reasons for the frequent failure to establish cell lines remain unclear. The major causes appear to be culture deterioration with cessation of multiplication of the neoplastic cells and overgrowth by normal fibroblasts, macrophages, or lymphoblastoid cells. While the lymphoblastoid cells may give rise to a continuous cell line (a nonmalignant EBV⁺ B-LCL), human fibroblast and macrophage cultures are commonly not immortalized.

Despite the fact that the proliferation of malignant hematopoietic cells in vivo seems to be independent of the normal regulatory mechanisms, these cells usually fail to proliferate autonomously in vitro even for short periods of time. In vivo, at least initially, these cells seem to require one or probably several hematopoietic growth factors for proliferation. The addition of regulatory proteins, e.g. recombinant hematopoietic growth factors or conditioned medium (CM) secreted by certain tumor cell lines (which often contains various factors), is a culturing technique that appears to increase the frequency of success by overcoming the "crisis" period in which the neoplastic cells cease proliferating in vitro. These molecules enable the cells from the majority of patients to multiply for about 2–4 weeks. Out of these short-term cultures containing surviving cells, continuous cell lines that are derived from the malignant cells can be established. For cell line immortalization, malignant cells must acquire a selective advantage in the cell culture environment. However, the mechanisms involved in the establishment of an LL cell line-permissive milieu are largely unknown.

Taken together, the efficiency of cell line establishment is still rather low and the deliberate establishment of LL cell lines remains by and large an unpredictable random process. Clearly, difficulties in establishing continuous cell lines may be due to the inappropriate selection of nutrients and growth factors for these cells. Thus, a suitable microenvironment for hematopoietic cells, either malignant or normal, cannot yet be created in vitro. Hence, this is the area with the greatest challenge in the culture of LL cells. A systematic investigation to define the optimal cell numbers, culture conditions, growth factor combinations and other supplements, matrix characteristics, and a multitude of different parameters has not been reported yet. Further work is required to achieve significant improvements in the success rates of LL cell line immortalization. A breakthrough will, likely, benefit leukemia and lymphoma research in a myriad of experimental questions. In the following, some of the most commonly used techniques for establishing LL cell lines are described.

2. Materials

2.1. Establishment and Culture of Cell Lines

1. Ficoll–Hypaque solution: density 1.077 g/L; store at 4°C in the dark.
2. Cell culture medium: all are commercially available; the most commonly used are RPMI 1640, Iscove's Modified Dulbecco's Medium, α-MEM, or McCoy's 5A store at 4°C in the dark; prepare aliquots of complete medium (maximally 100–200 mL) in a separate sterile glass bottle (to be used only for one given cell line) which should be kept at room temperature in the culture laboratory to detect quickly and easily any microbiological contamination.
3. Fetal bovine serum (FBS): inactivate toxic components in the FBS prior to use in a 56°C waterbath for 45 min; store aliquots of maximally 50 mL in plastic tubes at –20 to –30°C.
4. Neubauer hematocytometer and Trypan blue solution.
5. Recombinant growth factors: for example, erythropoietin (EPO), granulocyte-macrophage colony-stimulating factor (GM-CSF), granulocyte-CSF (G-CSF), interleukin-2 (IL-2), IL-3, IL-6, stem cell factor (SCF), or thrombopoietin (TPO).
6. Conditioned medium (CM) from tumor cell line cultures: for example, the human bladder carcinoma cell line 5637 is known to produce and secrete large quantities of various

growth factors (7); aliquot the CM in 30 or 50 mL tubes and store at −20°C; test the 5637 CM in proliferation assays using an indicator cell line or determine the exact growth factor concentration with ELISAs.

2.2. Freezing and Storage of Cell Lines

1. Dimethylsulfoxide (DMSO).
2. Freezing ampoules (plastic cryo vials).
3. Controlled rate freezer.
4. Cryo Freezing Container (Nalgene Cryo 1°C Freezing Container).

3. Methods

3.1. Establishment of Cell Lines

3.1.1. Acquisition of Cells (see Notes 1 and 2)

1. Collect heparinized or otherwise anticoagulated specimens of peripheral blood or bone marrow or other samples in sterile tubes (see Notes 3 and 4). Place lymph nodes and other solid tissues in sterile containers.
2. Specimens should ideally be processed as soon as possible after receipt, but may be stored overnight at room temperature. Peripheral blood and bone marrow should remain undiluted, solid tissues should be placed in culture medium.
3. Cryopreserved samples can also be used for attempts to establish cell lines. But it appears to be of advantage to isolate the mononuclear cells prior to cryopreservation by Ficoll–Hypaque density gradient centrifugation (see Subheading 3.1.2).

3.1.2. Isolation of Cells (see Note 5)

1. Cut solid tissue specimens (e.g. lymph nodes) with scissors and force the particles through a fine metal mesh. Suspend the cells in 50–100 mL culture medium (possibly also more depending on the size of the tissue specimen). The most commonly used media are listed in Note 6.
2. Dilute the blood and bone marrow samples 1:2 with culture medium. Isolation of cells from a leukapheresis collection requires dilution of the sample with culture medium at 1:4.
3. Pipette the Ficoll–Hypaque density gradient solution (density 1.077 g/L) into a 15 or 30 mL conical centrifuge tube. Slowly layer the mixture of medium and sample over the Ficoll–Hypaque solution. Use equal volume of sample mixture and Ficoll–Hypaque solution (see Note 7).
4. Centrifuge for 20–30 min at $450 \times g$ at room temperature (with the centrifuge brakes turned off). A layer of mononuclear cells should be visible on top of the Ficoll–Hypaque phase as they have a lower density than the Ficoll–Hypaque solution.

The anucleated erythrocytes and the polynucleated granulocytes are concentrated as pellet below the Ficoll–Hypaque layer.

5. Using a sterile Pasteur pipette, transfer the interface layer containing the mononuclear cells to a centrifuge tube.

6. Wash the cells by adding culture medium plus 10% FBS (add about five times the volume of the mononuclear cell solution) and centrifuge for 5–10 min at $200 \times g$ at room temperature (see Note 8).

7. Discard the supernatant, resuspend the cell pellet in culture medium plus 10% FBS and repeat the washing procedure.

8. Finally, resuspend the cells in 20 mL of culture medium with 20% FBS. Count the cells and determine their viability (see Notes 9 and 10).

3.1.3. Culture Conditions

1. Adjust the cell suspension of the original patient cells to a concentration of $2–5 \times 10^6$/mL in culture medium (see Note 6) with 20% FBS plus additional 10% conditioned medium (CM) of cell line 5637 (see Note 11) or with an appropriate concentration of purified or recombinant growth factors (see Note 12). See Note 13, for further additives such as antibiotics and others.

2. Place 5–10 mL of the cell suspension in the complete culture medium in an 80 cm² plastic culture flask (see Note 14). If 24-well plates are used, add 1–2 mL cell suspension into each well. Add 100–200 µL of cell suspension into wells of 96-well flat-bottomed microplates (see Note 15).

3. Place the cells in a humidified incubator at 37°C and 5% CO_2 in air (see Note 16).

4. Expand the cells by exchanging half of the spent culture volume with culture medium plus 20% FBS plus 10% 5637 CM (or with appropriate concentrations of recombinant growth factors) once a week. After a few hours, some cells become adherent (see Note 17).

5. During the first weeks, the neoplastic cells may appear to proliferate actively. If the medium becomes acidic quickly (yellow in the case of RPMI 1640 medium), change half of the volume of medium at 2–3 days interval (seldom daily). If the number of the cells increases rapidly, readjust the cells weekly to a concentration of at least 1×10^6/mL in fresh complete medium by dilution or subdivision into new flasks or wells of the plate. The neoplastic cells from the majority of patients undergo as many as four doublings in 2 weeks, but after 2–3 weeks most malignant cells cease proliferating. Following a lag time of 2–4 weeks (crisis period), a small percentage of cells from the total population may still proliferate actively and may continue to grow forming a cell line.

6. If the malignant cells continue to proliferate for more than 2 months, there is a high possibility of generating a new cell line. Then, the task of characterizing the proliferating cells should be begun as soon as possible (8). Prior to the characterization of the cells, freeze ampoules of the proliferating cells containing a minimum of 3×10^6 cells/ampoule in liquid nitrogen to avoid loss of the cells due to occasional contaminations or other accidents.

7. Limiting dilution of the cells in 96-well plates leads to the generation of monoclonal cell lines (see Note 18).

3.1.4. Documentation of Cells

It is absolutely mandatory to freeze aliquots of the original cells (see Subheading 3.2) and to store them in appropriate locations for later documentation, authentication, and comparisons. Once the cultured cells start to proliferate, ampoules should be frozen at regular intervals during the initial period of the culture expansion. The morphology of the primary cell lines and also later of the cells from the resulting cell line should be documented on stained cytospin slide preparation (see Note 19).

3.2. Freezing and Storage of Cell Lines

3.2.1. Freezing of Cells

It is generally assumed that cell lines can be cryopreserved at $-196°C$ in liquid nitrogen for more than 10 years, if not indefinitely, without any significant changes in their biological features. The viable cell lines can be recovered at any time when needed. Prior to freezing in liquid nitrogen, the cells are suspended traditionally in the appropriate medium containing 20% FBS and 10% DMSO which can lower the freezing point to protect the frozen cells from damage caused by ice crystals (see Note 20). It appears that no single suspending medium and procedure will be perfect for processing and cryogenic storage of all cell cultures. However, the procedures described here are suitable for most cell lines and are compatible with prolonged preservation of viability and other characteristics of the cell lines.

1. Harvest the cells of cell lines by transfer of the flask contents to 30 or 50 mL centrifuge tubes. The cells should be harvested in their logarithmic growth phase and when they are at their best of health (see Note 21). Sufficient numbers of primary cells should be frozen for later documentation and authentication (see Subheading 3.1.4); immediately upon isolation, primary cells can be frozen after thorough washing steps (see Subheading 3.1.2).

2. Determine the total number of viable cells by counting the cell density as well as the viability of the cells using the Trypan blue dye exclusion method (see Note 10).

3. Prepare a sufficient volume of freezing solution that consists of 70% culture medium, 20% FBS, and 10% DMSO and that should be cooled on wet ice (see Note 22).

4. Centrifuge the cell suspensions at $200 \times g$ for 10 min, discard the supernatant.

5. Adjust the cells to a concentration of $5–10 \times 10^6$/mL for cell lines and $10–50 \times 10^6$/mL for primary material using freshly prepared freezing solution (see Note 23).

6. Distribute the cells into freezing ampoules (plastic cryovials) with 1 mL per ampoule, thus containing at least 5×10^6 cells (see Note 24).

7. Using a computer-controlled rate freezer (cryofreezing system) a cooling rate of 1°C per minute can be achieved, from room temperature down to –25°C. When the temperature reaches –25°C, the cooling rate is increased to 5–10°C per minute. Upon reaching –100°C, the ampoules can be transferred quickly to a liquid nitrogen container (see Note 25).

3.2.2. Cryopreservation in Liquid Nitrogen

Permanent storage of ampoules with primary cells or cells from cell lines should be in the liquid phase of the liquid nitrogen. However, long-term storage is also possible in the vapor phase of the liquid nitrogen tank. Long-term preservation of cell lines or primary cells at –80°C (beyond 1 week) cannot be recommended as the cells will die under these conditions within months or even weeks.

3.3. Thawing, Expansion, and Maintenance of Cell Lines

Commonly, cell lines are stored frozen in liquid nitrogen. Frozen cells must be thawed carefully to minimize cell loss.

3.3.1. Thawing of Cell Lines

1. The thawing-out solution consists of 80% culture medium and 20% FBS of room temperature or 37°C.

2. Remove the frozen ampoule from liquid nitrogen, thaw the cells rapidly in a 37°C waterbath by gently shaking the ampoule in the water (see Note 26).

3. Wipe the ampoule with a tissue pre-wetted with 70% ethanol before the vial is opened. As soon as the cell suspension is thawed, transfer the cell suspension into a centrifuge tube.

4. Dilute the cell suspension slowly by adding 10–20 mL culture medium plus 20% FBS to the tube, shake the tube gently (see Note 27).

5. Centrifuge the cells at $200 \times g$ for 10 min. Discard the supernatant.

6. Wash the cells again using another 10–20 mL culture medium plus 20% FBS. During the centrifugation, determine the total cell number and the percentage of viable cells in the improved Neubauer hematocytometer with Trypan blue vital staining (see Note 10).

7. Finally, resuspend the washed cells in the desired culture medium with the recommended FBS concentration at the optimal cell density.

3.3.2. Expansion of Cell Lines

After thawing, cells may be resuspended in complete medium at a general cell concentration of ca. $0.2–1.0 \times 10^6$ cells/mL (see Note 28). Suspension cell lines may be cultured in flasks or 24-well plates (see Note 29).

1. Incubate the cells in a humidified 37°C incubator with 5% CO_2 in air. Loosen the top of the flask slightly to allow for free gaseous exchange into and out of the flask.

2. Feed the cells by exchanging half of the culture volume with culture medium plus FBS at 2–3 days intervals (see Note 30). Remove gently half of the spent medium from the flask and then add the same volume of new complete medium into the flask.

3. If the cells proliferate actively, the culture medium will soon change color due to a pH change caused by cellular metabolism. In this case, it is necessary to change the medium more frequently. Should the cells have doubled, subdivide the cells from the original flask into a second flask by diluting the suspension 1:2 with new medium.

4. When changing the medium, it is important to calculate the total cell number by determining the density as well as the viability of the cells using Trypan blue dye exclusion (see Note 10).

5. A careful documentation of all manipulations, macroscopic and microscopic observations, intentional and accidental changes in the cellular conditions, and data on cell density, viability, and total cell number at different time points is mandatory.

3.3.3. Maintenance of Cell Lines

The cell lines can be maintained as long as required (see Note 31). The cells can be harvested at any time for different uses. If the cells proliferate more quickly than needed, keep the cell growth at a slower pace by decreasing the FBS to a lower percentage in the medium, changing the medium at longer intervals, or discarding a certain amount of the cells (up to 75%) during the exchange of medium. There is, however, always the risk that a subclone with a growth advantage and somewhat different genotypic and/or phenotypic features than the parental cell line will overgrow the culture during long-term maintenance. A telltale example is growth factor-dependent cell lines that during extended culture (1–2 months) may become completely growth factor-independent. Hence, it appears to be best to establish a master and a working bank of any given cell line with a sufficient number of frozen back-ups for quick and easy withdrawal so that cultures do not need to be kept for extended periods of times between experiments. A reasonable time frame for continuous uninterrupted cultivation of one cell line is in the range of 1–2 months.

As after longer usage culture flasks or plates will often contain a certain amount of unused ingredients of the medium, metabolized molecules and cell detritus, it is recommendable that the plastic culture vessel be changed once every 1–2 months.

3.4. Further Considerations

3.4.1. General Considerations

Although in vivo the malignant cells enjoy a selective growth advantage over normal hematopoietic cells, in vitro LL cells are so difficult to grow and to maintain that attempts to establish cell lines meet much more often with failure than with success. Although currently there is no one single cell culture system which assures consistent establishment of cell lines, several methods for immortalizing neoplastic cells have been developed. The technique of seeding cells in suspension cultures as described in this chapter is certainly the most often used. Other methods recommended by several researchers have their advantages and might meet with success in some attempts.

Growth of LL cells in soft agar or methylcellulose offers the advantage that the colonies formed are well fixed and can be easily removed from the supporting medium for further culture in other environments. Thus, a cell line might be established by passaging single colonies.

Some lymphocytic leukemia cells can be immortalized using transforming viruses. EBV can promote growth of malignant B-cell lines from some patients with mature B-cell malignancies. But the EBV can also transform normal B-lymphocytes. Human T-cell leukemia virus (HTLV)-I allows the growth of malignant T-cells by inducing the IL-2 receptor. Considering the fact that EBV and HTLV can also transform normal cells, it is necessary to ascertain the malignant origin of the established cell lines by means of karyotype and molecular genetic analysis (8).

The growth of malignant and normal hematopoietic cells in vitro and in vivo is the result of complex interactions between growth factors and their respective receptors. The addition of some growth factors into the culture medium can support the proliferation of the neoplastic cells and induce the formation of cell lines.

A number of completely synthetic media, including several media designed specifically for unique types of LL cells, have been used by researchers for establishment and maintenance of cell lines in suspension cultures. It appears that no single medium is well suited for the growth of all types of neoplastic hematopoietic cells. Should one medium fail to support cell growth, it may become necessary to try another kind of medium.

FBS is the standard supplement in the suspension culture system of cell lines. It is commonly used in concentrations of 5–20%. Cell lines that do not require FBS in the culture medium have been described; however, they appear to be rather rare (3, 9). Prior to usage, it is recommended that batches of serum be pretested for their ability to support vigorous cell growth and for viral (in particular for bovine viral diarrhea virus), mycoplasmal, and other bacterial contamination. If possible, a large supply of the FBS from a pretested batch that supports cell growth well and has no contamination should be purchased and stored at –20°C for future use. Alternatives are newborn calf serum (usually at only 25% of the price of the expensive FBS) or serum-free media. However, not all cell lines will grow in newborn calf serum as well as in FBS. Although serum-free media do not provide financial advantages, they do allow for certain experimental manipulations which are not possible with FBS as one can control all the substances to which the cells are exposed. FBS is known to contain many unidentified ingredients at highly variable concentrations.

3.4.2. Culture Environment

A minimal amount of oxygen is essential for the growth of most types of cells in suspension cultures. The majority of cell lines grow well when they are incubated in a humidified 37°C incubator with 5% CO_2 in air. However, the partial pressure of oxygen (pO_2) in normal body fluids is significantly less than that of air. Studies have shown that the pO_2 in human bone marrow is 2–5%. This is considerably lower than the pO_2 (15–20%) existing in the typical cell culture incubator maintained at 5% CO_2 in air. It has been reported that growth of cultured cells could be improved by reducing the percentage of oxygen in the gaseous phase to between 1 and 10%. Growing LL cells under low oxygen conditions of 6% CO_2, 5% O_2, and 89% N_2 may be a useful method for the establishment of cell lines (6), but is logistically much more difficult due to the necessity for specialized incubators and a rather large consumption of N_2.

3.4.3. Common Cell Culture Problems

In attempts to establish cell lines, overgrowth of fibroblasts and normal EBV+ lymphoblastoid cells is the most common problem. Should the nutrients in the medium become exhausted too quickly, the adherent cells should be removed by passaging the suspension cells into new flasks containing fresh medium. The overgrowth of EBV+ B-lymphocytes can become visible as early as 2 weeks after seeding of the new culture. The EBV+ B-cells look small and have irregular contours with some short villi. They proliferate preferentially in big floating clusters or colonies (10). The colonies can be picked out with a Pasteur pipette and the EBV genome should be detected as soon as possible.

It is always necessary to freeze aliquots of the fresh primary cells in liquid nitrogen before culture. If a cell line should subsequently become established, the original cells can be used as a control for characterization of the established cell line (8). In case of failure to immortalize a cell line, the frozen primary back-up cells can be used for further attempts (a number of cell lines have been established from cryopreserved material).

Maintenance of cell lines requires careful attention. Every cell line appears to have its optimal growth environment; parameters such as culture medium, cell density, nutrition supplements, and pH all play a major role. If cell growth becomes suboptimal or cells inexplicably die during culture, some of the following problems should be considered: suitability of the culture medium or the growth supplements for this particular cell line; proper functioning of the incubator at the appropriate temperature, humidity, and CO_2 levels; and selection of the adequate cell density. Some cell lines clearly grow better in 24-well plates than in culture flasks.

Contamination with *Mycoplasma*, other bacteria, fungus, and viruses and cross-contamination with other "foreign" cells are the most common problems encountered in the maintenance of cell lines (11, 12). Therefore, analyses at regular intervals must be undertaken to ensure a contamination-free environment for cell growth (Fig. 1). In particular, *Mycoplasma* contamination should be examined regularly, e.g., using the polymerase chain reaction. All solutions and utensils coming into contact with cells must be sterilized prior to use; sterile techniques and good laboratory practices must be followed strictly (13). Although antibiotics can be added to the culture medium to prevent bacterial infection, they do not usually inhibit virus, fungus, or *Mycoplasma* infection. Therefore, antibiotics such as penicillin and streptomycin are not necessary, if care is taken regarding cell culture techniques (see Note 13). Because contamination can cause the loss of valuable cell lines, it is important to cryopreserve a sufficient amount of cell material from each cell line. To prevent cellular contamination and misidentification, it is mandatory to use a separate bottle of medium for each cell line. Furthermore, cell culturists should not deal with and feed more than one cell line at the same time.

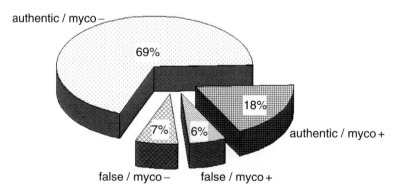

Fig. 1. Percentages of false and mycoplasma-contaminated LL cell lines. Shown is an analysis of 598 LL cell lines for which data on these two parameters are available: parameters "authentic versus false" refer to cross-contamination of cell lines; parameters "Myco– versus Myco+" refer to the mycoplasma contamination status. Authentic/myco–: 411 cell lines (69%); authentic/myco+: 108 (18%); false/myco–: 41 (7%); false/myco+: 38 (6%). Updated from ref. 12.

Table 2
Most important requirements for a new leukemia cell line

Immortality of cells
Verification of neoplasticity
Authentication of derivation
Scientific significance
Characterization of cells
Availability to scientific community

3.4.4. Time Considerations

Generation of a cell line may require 2–6 months. However, a cell culture should not be considered a continuous cell line until the cells have been passaged and expanded for at least half a year, better 1 year. Expansion of most cell lines after thawing takes 2–3 weeks. Depending on the parameters analyzed, 2–4 months might be needed for a thorough characterization of the established cell line.

3.4.5. Availability of Cell Lines

The most important requirements for a new cell line are listed in Table 2 (8). The cell line must be immortalized, i.e., it should be able to grow indefinitely. To differentiate clearly between normal and malignant cell lines, the neoplasticity of the cell line should be verified, e.g., by the analysis of numerical and structural chromosomal alterations. The derivation from the assumed patient must be proven by unequivocal authentication procedures, e.g., by forensic-type DNA fingerprinting. The scientific significance of a new cell line can best be documented by its full characterization to uncover its most relevant features.

Finally, it is of utmost importance that new cell lines are not only published but also made available to outside investigators. On one hand, scientific authenticity requires reproducibility. On the other hand, a cell line is certainly much more useful, if it is transferred to other colleagues inside and outside the own research field. The best option is the deposition of a cell line in a public cell line bank which is able to perform adequate quality and identity controls, to further characterize the cells, to store sufficiently large master and working stocks, and to distribute the cells non-commercially to requesting scientists.

4. Notes

1. The most commonly used specimens are peripheral blood or bone marrow samples as these are relatively easily obtained from patients. Other liquid specimens (such as pleural effusion,

ascites, cerebrospinal fluid, etc.) and solid tissue samples (lymph node, tonsil, spleen, etc.) are less often used but can be processed similar to peripheral blood or bone marrow samples. Solid tissues require a prior step whereby the tissue is dissociated mechanically. All solutions and utensils must be sterile. Work in a laminar flow cabinet (class 2 biological safety cabinet) under sterile conditions is recommended (for specifics on the biological safety, see Note 2).

2. During work with human blood or other tissues including primary LL cells, established cell lines, and pathogenic and infectious agents, biosafety practices must be followed rigorously. Some cell lines may be virally infected: for example, EBV and human herpesvirus-8 (HHV-8) are assigned to biological safety risk category 2; human T-cell leukemia virus (HTLV)-I/-II, and human immunodeficiency virus (HIV) fall into risk category 3. Fresh primary material may contain hepatitis viruses B or C (both in risk category 3). A standard code of practice including safety and legal considerations has been published with the UKCCCR (United Kingdom Co-Ordinating Committee on Cancer Research) Guidelines for the Use of Cell Lines in Cancer Research (13).

3. A review of 601 new LL cell lines (9) for which sufficient data were available produced the following statistical findings with regard to the specimen site from where primary cells had been obtained: peripheral blood, 46%; bone marrow, 25%; pleural effusion, 13%; ascites, 6%; bulk tumor/tumor tissue, 5%; lymph node, 3%; cerebrospinal fluid, 1%; pericardial effusion, 1%; and others (liver, meninges, spleen, and tonsils), <1%. Only data from well-characterized cell lines were counted, without sister cell lines or subclones (applies also for Notes 4 and 6).

4. According to a review of 547 new LL cell lines for which data were available (9), it seems that cells obtained from patients at later stages of the malignancy are more likely to be immortalized than at earlier stages; treatment status of patient when the sample was taken: at relapse/refractory/terminal, 54%; at diagnosis/presentation, 34%; at blast crisis/leukemic transformation, 10%; and during therapy, 2%.

5. It appears to be of advantage to prepare single-cell suspensions from the patient-derived samples instead of placing the specimens directly into culture medium. In the case of peripheral blood or bone marrow, all erythrocytes, granulocytes, and thrombocytes that are end-stage cells without the possibility to proliferate will die within days and the resulting metabolic byproducts may be toxic, hampering any potential proliferation of the blast cells.

6. An analysis of 621 new LL cell lines with available data (9) showed that the following media were most commonly used for the establishment of cell lines: RPMI 1640, 72%; Iscove's Modified Dulbecco's Medium, 14%; McCoy's 5A, 5%; α-MEM, 4%; and others (Cosmedium, Dulbecco's MEM, Eagle's MEM, Fischer's, Ham's F10, Ham's F12, L-15, Opti-MEM, SFM-Pro, X-Vivo 10, and YU-KLS), 4%.

7. Do not disturb the Ficoll–Hypaque/sample interface. It is helpful to hold the centrifuge tube at a 45°angle.

8. The washing steps are performed to remove the acidic heparin or other anticoagulants which may harm the cells, to remove the patient's serum which may inhibit the cell growth, and to remove the Ficoll–Hypaque which is hypertonic for the cells. Two washes will generally suffice. FBS which should have been inactivated prior to use in a 56°C waterbath for 30–45 min is included to prevent cells from adhering to one another.

9. In general, more than 90% of the mononuclear cells can be recovered by this procedure. Cell yields depend on the number of malignant cells in the specimen and are highly variable from patient to patient. On average, each milliliter of leukemia bone marrow yields $15-30 \times 10^6$ mononuclear cells and each milliliter of peripheral blood yields $1-2 \times 10^6$ mononuclear cells, the latter depending obviously on the white blood cell count. Several hundred millions of mononuclear cells can usually be recovered from lymph nodes.

10. The cell count and cellular viability can be determined by Trypan blue dye exclusion. Viable cells will not take up the Trypan blue stain. Make a 1:10 dilution of the cell suspension in Trypan blue by adding 10 µL of the cell suspension to 90 µL of the ready-for-use Trypan blue solution in a test tube and mixing well. Cells should be analyzed immediately as viable cells may begin to take up the dye also after about 5–10 min. Fill the counting chamber of an improved Neubauer hematocytometer. The chamber should not be under- or overfilled. Using a microscope, count the number of blue stained (nonviable) and unstained (viable) cells in the 1-mm middle square and the four 1-mm corner squares. Cells touching the top and left lines of the square (but not those touching the bottom and right lines) are counted as well. Calculate the average number of viable and nonviable cells per square ($1 \text{ mm} \times 1 \text{ mm} = 1 \text{ mm}^2$). The square is 0.1 mm deep, hence the volume is 0.1 mm^3 ($=0.1$ µL). The number of cells per mL (cell density) and the total number of cells in the cell suspension and their percentage of viability can be calculated as follows (1) cell density: average number of cells/1-mm square × dilution factor × 10,000 = cells/mL; (2) total number

of cells: cell count/mL × volume of cell suspension; and (3) % cell viability: total number of viable cells × 100 divided by total number of viable + nonviable cells. There should ideally be about 100 cells per square. Otherwise a lower dilution of the cell suspension (e.g., 1:2 or 1:5) or a higher dilution (e.g. 1:20) should be used.

11. 5637-conditioned medium: The adherent bladder carcinoma cell line 5637 is known to produce and secrete large quantities of various growth factors, including G-CSF, GM-CSF, SCF, and others (but not IL-3) (7). CM should be made in large volumes to guarantee continuity over a period of time, e.g., 6 months to 1 year. At initiation of culture, seed out at ca. 2×10^6 5637 cells/80 cm^2 in 10 mL of complete medium. Incubate at 37°C in 5% CO_2 in a humidified atmosphere, with the flask tops slightly loosened, until the cell monolayer becomes almost confluent. Replace the medium and grow for a further 4 days. Collect this medium and store it at 4°C. Add fresh complete medium to the 5637 culture. Repeat this process at 3–5 days interval, pooling the culture supernatants. After 1–2 weeks, split the confluent 5637 cultures 1:4 to 1:5 using trypsin/EDTA for 5 min and reseed the cells in fresh flasks. 5637 cells have a doubling time of about 24 h, leading to a cell harvest of ca. $8–10 \times 10^6$ cells/80 cm^2 (14). Spin down the pooled conditioned medium at $200 \times g$ for 5–10 min before filtering through sterile filters of 0.2 µm to exclude any contaminating nonadherent cells. Aliquot into 30 or 50 mL tubes and store at –20°C. Test the 5637 CM in proliferation assays at 5, 10, and 20% v/v, using specific growth factor-dependent cell lines, e.g., cell line M-07e (14), to determine the optimum volume necessary for maximal stimulation. Alternatively, the exact growth factor concentrations in the 5637 CM can be determined with enzyme-linked immunoassays (7).

12. The most effective growth factors in terms of induction of proliferation appear to be GM-CSF, IL-3, and SCF which may be used singly or in combination; a concentration of 10 ng/mL would provide a surplus of growth factors.

13. Most cell culturists add antibiotics (commonly penicillin–streptomycin) to their cell cultures, as ready-for-use commercially available 100× solutions. While this does not appear to be harmful, it is also clearly not necessary, if proper cell culture techniques (good culturing practices) are used. L-Glutamine (at 2 nM) and 2-mercaptoethanol (at 5×10^{-5} M) may also be added, but their possible effects, if any at all, on the success rate of establishment are questionable. We strongly discourage the routine addition of antimycoplasmal reagents as this practice would lead very quickly to the development of resistant mycoplasma strains (15).

14. Most cell lines have been established using the "direct method" of placing cells into liquid culture in a humidified incubator at 37°C and 5% CO_2 in air. Alternative approaches are inoculation in semisolid media (methylcellulose and soft agar), initial heterotransplantation and serial passage in immunodeficient mice with subsequent adaptation to in vitro culture, or culture on temporary feeder layer (e.g. on fibroblasts).

15. There are various types and sizes of plastic culture vessels: flasks (e.g. 25, 80, 175 cm^2) or plates (with 12, 24, or 96 wells). The number of flasks or wells used depends on the number of the primary cells available. As many of the malignant cells as possible should be used in attempts to establish a cell line. In theory, a cell line starts from one single cell. Thus, the more attempts, the higher the chances.

16. Alternatively (but seldom an option as this requires a special type of incubator and large quantities of N_2), incubate the cells in a humidified 37°C incubator with 6% CO_2, 5% O_2, and 89% N_2.

17. These adherent cells appear to be the source of colony-stimulating factors for both normal and malignant cells. During the first 2 weeks, it is not necessary to remove the adherent cells from the culture unless there is a specific reason to do so, for example because of the addition of a specific recombinant growth factor to the medium to obtain a unique type of cell line. After 2 weeks, if the suspension cells grow very rapidly, the adherent cells can be removed simply by transferring the suspension cells into new culture vessels to reduce the potential for overgrowth of fibroblasts and normal lymphoblastoid cells.

18. After prolonged culture in vitro, the cell line will become anyway oligoclonal or monoclonal due to the outgrowth of selected cell clones. In most cases, it is not absolutely necessary to subclone the cell line by limiting dilution. In some types of cell lines, e.g. immature T- and BCP-cell lines, it might be very difficult or virtually impossible to "clone" the cells.

19. Harvest cells, count and adjust the concentration to $0.5–1.0 \times 10^6$ cells/mL. Spin 50, 70, and 90 µL of cell suspension onto glass microscope slides in a Shandon cytocentrifuge at 500 rpm for 5 min and air dry at room temperature overnight (slides should not be left unstained for longer than 1 week). Select the best preparations under the microscope: the cells should be neither too packed nor too isolated for an appropriate morphological examination. Stain with the May-Grünwald and Giemsa solutions as follows (1) fix for 5 min in 100% methanol; transfer for 5 min to May-Grünwald stain which has been diluted 1:2 with Weise buffer, and then rinse with Weise buffer; (2) transfer for 20 min to Giemsa stain

which has been diluted 1:10 with Weise buffer; (3) finally rinse thoroughly with Weise buffer. Let the stained preparations dry overnight. Mount the stained cells with a coverslip using Entellan and examine the preparation under the microscope.

20. Glycerol has been used as an alternative to DMSO, but it is a bit cumbersome due to its high viscosity.

21. Freezing and storing the cells will not improve the status and quality of the cell culture prior to the freezing; at best, the status quo will be preserved, but most often will be diminished to various degrees.

22. It is not necessary to sterilize the DMSO solution as pure DMSO is lethal to bacteria.

23. The freezing solution should be added quickly to the cells and then mixed thoroughly. Long-time exposure to DMSO at room temperature can trigger significant cellular changes such as activation and so-called "induction of differentiation." Keeping cells in DMSO-containing media on ice could minimize the effect of DMSO on the cells.

24. Seal the ampoules which have been properly labeled with the name of the cell line and the date of freezing; keep a written record. More cells per ampoule can be frozen if needed, depending on the cell type.

25. Alternatively, if only a few ampoules are frozen, the sealed ampoules can be placed in a plastic box with an inset for ampoules (Nalgene Cryo 1°C Freezing Container); the box is half-filled with isopropanol and stored in a –80°C freezer for at least 4 h. With this method, a 1°C per minute cooling rate can be achieved as well.

26. It is important that the frozen cell solution be thawed in about 1 min. Rapid warming is necessary so that the frozen cells pass quickly through the temperature zone between –50 and 0°C where most cell damage is believed to occur. Slow thawing will harm the cells by the formation of ice crystals in the cells causing hypertonicity and breakage of cellular organelles.

27. Cells frozen with DMSO are usually dehydrated. During the washing steps with medium, water will diffuse into the cells. Diluting the suspension slowly is supposed to reduce the loss of electrolytes, to counteract extreme pH changes, and to prevent denaturation of cellular proteins.

28. It appears that most cell lines grow better at higher cell concentrations than at lower ones. Some cell lines (e.g., some BCP-cell lines) prefer a concentration higher than $1.0 \times 10^6/$mL. Usually, the optimal concentration of a cell line for expansion must be explored empirically. If after 2–3 days of

culture, the cells do not grow well (perhaps due to the presence of many dead cells), it might be useful to concentrate the cells and to culture them at a higher cell density. It is recommended that the cells be resuspended first in medium containing 20% FBS; should the cells start to multiply and resume their expected growth activity, the percentage of FBS can be decreased stepwise.

29. General recommendations are 5–10 mL suspension into an 80 cm² flask or 1–2 mL suspension into each well of a 24-well plate. For "difficult" cell lines, it may be advantageous to suspend some cells in a flask and another aliquot in a 24-well plate (or even a 96-well microplate). There are distinct differences between flask and plate regarding exposure to CO_2, accessibility to microscopic observation, and possibilities of manipulation.

30. Before changing the medium, set the flasks upright in the incubator for at least 30 min to let the cells sink to the bottom of the flask. One of the advantages of plates is the fact that cells are always concentrated at the bottom of the wells.

31. There is a fundamental difference between "expansion" and "maintenance" of a cell line. Some cell lines will deteriorate over long-time culture under maintenance conditions. In such cases, it might then be better to freeze and rethaw the cells when needed.

References

1. Drexler, H.G., Matsuo, Y., and MacLeod, R.A.F. (2000) Continuous hematopoietic cell lines as model systems for leukemia-lymphoma research *Leukemia Res* **24**, 881–911.
2. Drexler, H.G. and Minowada, J. (2000) Human leukemia-lymphoma cell lines: Historical perspective, state of the art and future prospects, in *Human Cell Culture, Vol. III - Cancer Cell Lines Part 3: Leukemias and Lymphomas* (Masters, J.R.W. and Palsson, B.O., eds.), Kluwer Academic Pub, Dordrecht, pp. 1–18.
3. Drexler, H.G. (ed.) (2000) *The Leukemia-Lymphoma Cell Line FactsBook*. Academic Press, San Diego, CA.
4. Matsuo, Y. and Drexler, H.G. (1998) Establishment and characterization of human B cell precursor-leukemia cell lines *Leukemia Res* **22**, 567–579.
5. Drexler, H.G., Fombonne, S., Matsuo, Y., Hu, Z.B., Hamaguchi, H., and Uphoff, C.C. (2000) p53 alterations in human leukemia-lymphoma cell lines: In vitro artifact or prerequisite for cell immortalization? *Leukemia* **14**, 198–206.
6. Smith, S.D., McFall, P., Morgan, R., Link, M., Hecht, F., Cleary, M., and Sklar, J. (1989) Long-term growth of malignant thymocytes in vitro *Blood* **73**, 2182–2187.
7. Quentmeier, H., Zaborski, M., and Drexler, H.G. (1997) The human bladder carcinoma cell line 5637 constitutively secretes functional cytokines *Leukemia Res* **21**, 343–350.
8. Drexler, H.G. and Matsuo, Y. (1999) Guidelines for the characterization and publication of human malignant hematopoietic cell lines. *Leukemia* **13**, 835–842.
9. Drexler, H.G. (2010) *Guide to Leukemia-Lymphoma Cell Lines*. 2nd Edition Braunschweig.
10. Drexler, H.G. and Matsuo, Y. (2000) Malignant hematopoietic cell lines: In vitro models for the study of multiple myeloma and plasma cell leukemia. *Leukemia Res* **24**, 681–703.
11. Drexler, H.G., Dirks, W.G., and MacLeod, R.A.F. (1999) False human hematopoietic

cell lines: Cross-contaminations and misinterpretations. *Leukemia* **13**, 1601–1607.

12. Drexler, H.G., Uphoff, C.C., Dirks, W.G., and MacLeod, R.A.F. (2002) Mix-ups and mycoplasma: The enemies within. *Leukemia Res* **26**, 329–33.

13. UKCCCR (2000) UKCCCR guidelines for the use of cell lines in cancer research *Br J Cancer* **82**, 1495–1509.

14. Drexler, H.G., Dirks, W., MacLeod, R.A.F., Nagel, S., Quentmeier, H., Steube, K.G., and Uphoff C.C. (eds.) (2010) *DSMZ Catalogue of Human and Animal Cell Lines.* http://www.cellines.de.

15. Drexler, H.G. and Uphoff, C.C. (2002) Mycoplasma contamination of cell cultures: Incidence, sources, effects, detection, elimination, prevention *Cytotechnology* **39**, 23–38.

Chapter 17

Isolation of Inflammatory Cells from Human Tumours

Marta E. Polak

Abstract

Inflammatory cells are present in many tumours, and understanding their function is of increasing importance, particularly to studies of tumour immunology. The tumour-infiltrating leukocytes encompass a variety of cell types, e.g. T lymphocytes, macrophages, dendritic cells, NK cells, and mast cells. Choice of the isolation method greatly depends on the tumour type and the leukocyte subset of interest, but the protocol usually includes tissue disaggregation and cell enrichment. We recommend density centrifugation for initial enrichment, followed by specific magnetic bead negative or positive panning with leukocyte and tumour cell selective antibodies.

Key words: Tumour-infiltrating lymphocytes, Dendritic cells, Antigen-presenting cells, Isolation, Density gradient

1. Introduction

The presence of tumour-infiltrating inflammatory cells has been reported in many human tumours, including ovarian, colon, prostate, and lung carcinomas, and skin cancer (1–10). The tumour-infiltrating leukocytes encompass a variety of cell types, e.g. T lymphocytes, macrophages, dendritic cells, NK cells, and mast cells. Infiltration of tumours by inflammatory cells can be associated both with positive and negative prognosis, and it is of interest to investigate in detail their biology and function. Choice of the isolation method greatly depends on the tumour type and the leukocyte subset, but the protocol must include steps for disaggregation of tumour tissue and for enrichment of cells in interest population. As tumour-infiltrating lymphocytes create only a small fraction of tumour mass, we recommend to use the enrichment on the density gradient as the initial step to fractionate distinctive populations and debulk the leukocytes of dead cells and tissue debris.

If high purity of isolated cell population is a priority, this initial step can be followed by positive or negative separation (panning) for a particular population of tumour infiltrating leukocytes with magnetic beads. Conjugation of leukocyte receptors with antibodies can cause cell activation and affect functional assays. To avoid it, cells can be used directly after the separation on density gradient, or alternatively the separation on density gradient can be repeated for the enriched cell population only. Negative panning to deplete the enriched population of selected cell types is also useful.

The phenotype and function of isolated cells can be subsequently tested by a variety of methods, allowing investigation of their biological role or monitoring of an immunotherapy.

2. Materials

1. Wash medium: RPMI supplemented with 2 mM L-glutamine and 1% Penicillin/Streptiomycin (Invitrogen, UK).
2. Heat-inactivated foetal bovine serum FBS (Invitrogen, UK).
3. Heat-inactivated human AB serum (Sigma, UK).
4. PBS (Sigma, UK).
5. Bovine serum albumin (BSA) (Sigma, UK).
6. Collagenase – working solution 1.0 mg/mL (Sigma, UK).
7. OptiPrep™ (Axis-Shield, UK).
8. Ficoll-Histopaque Lymphoprep (Sigma, UK).
9. Scalpel blades.
10. Filters, 70-μm nylon (BD, Falcon, UK).
11. MACS beads (Miltenyi Biotec, UK) (see Note 7).
12. MACS buffer: PBS +0.5% BSA, 2 mM EDTA, filtered and degassed O/N.
13. FcR block (Miltenyi Biotec, UK).
14. Fungazone (Sigma, UK).

3. Methods

3.1. Enzymatic Disaggregation of Tumour Tissue

1. Remove the solid tumour specimen from its container into a culture dish in the safety cabinet. Using a sterile scalpel blade cut the specimen into fine sections (½–1 mm size) and place them into a universal container with 9 mL of culture medium with penicillin and streptomycin (see Note 1).

2. Add 1 mL of pre-prepared sterile collagenase solution, mix well, and leave in the incubator at 37°C, 5% CO_2, overnight (see Note 2).

3. The next day, remove the universal container, vortex it, and replace it in the incubator for 30 min.

4. In the safety cabinet, add an equal volume of cell culture medium, mix well, and centrifuge at $400 \times g$ for 7 min. After discarding the supernatant, repeat the wash with fresh 10 mL of culture medium.

5. In a fresh container, prepare 10 mL of Lymphoprep. Re-suspend the pellet in 10 mL of culture medium and carefully overlay it onto lymphoprep. Centrifuge the cells at $600 \times g$ for 30 min, without using the brake (see Notes 2 and 3).

6. Collect the live cells from the interface and wash twice with culture medium.

7. Remove the universal containers from the centrifuge and discard the supernatant; re-suspend the tumour pellet in 10 mL of wash media and then centrifuge again at $400 \times g$ for 7 min.

8. Re-suspend the cell pellet in 10 mL of culture medium and count the cells in a haemocytometer.

9. Leave the cells on ice for further preparation (see Note 4).

3.2. Separation of Leukocytes by Density Gradient

The principle of the separation is that the differing sizes and densities of cells extracted from disaggregated tumour tissue allow separation of cell populations on density gradient. All protocols presented below use OptiPrep™ as the density gradient medium. OptiPrep™, a 60% (w/v) solution of iodixanol in water, density = 1.32 g/mL, is less toxic to cells than the standard Histopaque Ficoll. Cells of monocytic lineage, e.g. dendritic cells, monocytes, and macrophages have lower density than lymphocytes, and they separate on a 1.065 g/cm³ density layer (11.5% ioxidanol), while the lymphocytes separate on 15% Ioxidanol layer, density = 1.085 g/cm³. Additionally, many tumour cell types are significantly bigger than leukocytes, which may enable tumour cell separation in a separate band, the density of which needs to be established by experiment for each tumour type studied. The three protocols presented below enable enrichment of one (Subheading 3.2.1), two (Subheading 3.2.2), and three (Subheading 3.2.3) distinct populations from mixed cell population from disaggregated tumour tissue. The volumes of reagents and the sizes of separating containers should be chosen dependently initially and expected in the positive fraction cell count (see Notes 2 and 5).

3.2.1. Separation of Antigen-Presenting Cells from Mixed Cell Suspension, Density Barrier 1.065 g/cm³

1. Pre-cool 4 vol. of RPMI and 1 vol. of OptiPrep.
2. Mix 1 vol. of RPMI and 1 vol. of OptiPrep (Opti-mix).
3. Gently but thoroughly re-suspend the cell pellet in remaining 3 vol. of RPMI.

4. Add Opti-mix into cell suspension and mix gently.
5. Transfer the cell suspension into appropriate separation container (see Note 5).
6. Overlay the cell suspension with 1 vol. of FBS to prevent aggregation of cells on the fluid–air interface.
7. Centrifuge at $600 \times g$ for 20 min, at 4°C if possible, allow the rotor to decelerate *without* using the brake (see Note 3).
8. Harvest the dendritic cell fraction from the top of the 11.5% OptiPrep layer (below the FBS).
9. Wash the cells twice in culture medium, $400 \times g$, 7 min.
10. Count cells in a haemocytometer (see Note 6).
11. Leave the cells on ice for further preparations.
12. Further enrichment of antigen-presenting cells from the low-density fraction using specific antibody bound magnetic beads might be necessary owing to the contamination with tumour cells.

3.2.2. Separation of Antigen-Presenting Cells and T lymphocytes from Tumours

1. Pre-cool OptiPrep and RPMI (see Note 4).
2. Prepare working suspension (WS; 11.5% Ioxidanol) mixing 4.2 vol. of RPMI with 1 vol. of OptiPrep.
3. Aliquot equal volumes of the WS into 30-mL universal containers.
4. Gently but thoroughly resuspend the cell pellet in 3 vol. of pre-cooled RPMI.
5. Add 1 vol. of pre-cooled OptiPrep into cell suspension and mix gently.
6. Using Pasteur pipette *gently underlay* the cell suspension under aliquoted WS.
7. Overlay the prepared gradient with one full Pasteur pipette of FBS–RPMI.
8. Centrifuge at $600 \times g$ for 15 min at room temperature.
9. Allow the rotor to decelerate *without using the brake*.
10. Harvest the Monocytes and DCs from the upper band (over the WS layer) and T cells from the lower band (below the WS layer).
11. Wash the cells in wash media, at $400 \times g$ for 7 min.
12. Count cells using a haemocytometer (see Note 6).
13. Further enrichment of antigen-presenting cells from the low-density fraction using specific antibody bound magnetic beads might be necessary owing to the contamination with tumour cells.

3.2.3. Separation of Distinct Populations of Antigen-Presenting Cells, Lymphocytes, and Tumour Cells from Squamous Cells Carcinoma

1. Pre-cool OptiPrep and RPMI.
2. Prepare working suspensions 1 and 2 (WS1: 10% Ioxidanol, WS2: 11.5% Ioxidanol) mixing 5 vol. of RPMI with 1 vol. of OptiPrep for WS1 and 4.2 vol. of RPMI with 1 vol. of OptiPrep for WS2.
3. Aliquot equal volumes of the WS1 into 30-mL universals.
4. Using Pasteur pipette *gently underlay* the WS2 under aliquoted WS1.
5. Gently but thoroughly re-suspend the cell pellet in 3 vol. of pre-cooled RPMI.
6. Add 1 vol. of pre-cooled OptiPrep into cell suspension and mix gently.
7. Using Pasteur pipette *gently underlay* the cell suspension under aliquoted WS1 and WS2.
8. Overlay the prepared gradient with one full Pasteur pipette of FBS–RPMI.
9. Centrifuge at $600 \times g$ for 15 min at room temperature.
10. Allow the rotor to decelerate *without using the brake*.
11. Harvest the tumour cells from the lowest density band (directly under the FBS layer), monocytes and DCs from the middle band (between the WS1 and WS2 layer), and T cells from the lowest band (below the WS layer).
12. Wash the cells in wash media, at $400 \times g$ for 7 min.
13. Count cells in a haemocytometer (see Note 6).
14. Further enrichment of separated cells from each band using specific antibody bound magnetic beads might be necessary owing to the contamination with tumour cells.

3.3. Magnetic Separation of Antigen-Presenting Cells

Many subsets of lymphocytes express cluster of differentiation (CD) antigens specific only to this subsets. Using antigen-specific magnetic bead conjugated antibodies, it is possible to obtain more than 98% pure population of cells. The choice of antigen depends solely on investigator preferences (see Note 7). If the microbead-conjugated antibody for the antigen of choice does not exist, it is possible to separate antigen-positive cell using indirect technique in which cells are labeled using antibodies against the cell of interest and captured using a common anti-immunoglobulin or protein A conjugated bead.

1. Count cells, wash in MACS buffer after any preparations, and spin at $400 \times g$ for 7 min.
2. Re-suspend cells in 200 µL of MACS buffer for each of 2×10^8 cells. For other cell quantities, see Note 7.

3. Add 100 μL of FcR blocking buffer per each 2×10^8 cells.
4. Add 100 μL of cell marker-specific antibody of choice.
5. Mix thoroughly the cell suspension with a sterile disposable pipette.
6. Incubate in the fridge for 15 min: incubation time and temperature are critical.
7. Wash cells with 10× volume of staining buffer.
8. Equilibrate the MS column with 3×0.5 mL of staining buffer.
9. Re-suspend the pellet in 500 μL of staining buffer per each 10^8 cells and mix thoroughly to obtain a single cell suspension. For other cell quantities, see Note 8.
10. Apply gently the cell suspension onto a pre-wetted column.
11. Let the cell suspension pass through the column, collecting the effluent as the negative fraction.
12. Wash the column three times with 0.5 mL of MACS buffer and collect the effluent as the negative fraction.
13. Remove the column from the magnet and place it on a suitable collection tube.
14. Apply 1 mL of staining buffer onto the MS column and flush the positive cells out of the column firmly with enclosed plunger.
15. Count cells in a haemocytometer (see Note 6).
16. Spin cell suspension down at $400 \times g$ for 7 min.
17. Re-suspend cells in culture media or any other media required.
18. The cell population purity can be assessed using flow cytometry or cytospin technique prior to further experiments.

4. Notes

1. For specimens which are likely to be contaminated with fungi, supplement the medium with fungazone.
2. The method of disaggregation of tumour tissue can be modified by the investigator, depending on tumour type, specimen size, and tissue properties. Some tumours may require enzymatic digestion using stronger enzymes, e.g. trypsin, and to limit cell loss during preparation, smaller samples (i.e. biopsies) may benefit from avoiding separation on density gradient, and cell separation using magnetic beads only.
3. Centrifuging without using the brake is a prerequisite to ensure good sample separation.

4. To ensure maximum cell viability, it is recommended to perform all the steps at low temperature, i.e. handle the samples on ice and centrifuge at +4°C in a temperature-controlled centrifuge if available.

5. Since the tumour-infiltrating leukocytes are infrequent, it is likely that the numbers of purified cells are very low. All the procedures must be carried out very carefully to avoid cell loss during the preparation. It is advisable that the volumes of the containers are carefully chosen depending on the sample size, as the overloading of the density gradient can affect the separation and using too big container may render collection of cells from the interphase very difficult. For an initial cell count of approximately 5×10^6, the volumes are 4 mL of cell culture medium and 1 mL of Optiprep, and the separation container is a 15-mL Falcon tube. If the initial cell count is low, e.g. 5×10^5 to 2×10^6 cells, the recommended separation container is a 5-mL tube, and the volumes need to be decreased by half. For some very large tumour cells preparation (cell count between 2×10^7 and more), the separation can be done in a bigger volume (8 mL of medium and 2 mL of OptiPrep) of separation mix and in bigger containers, e.g. 30-mL universal container, or in multiple tubes.

6. If the initial size of the specimen was small, and the separation was carried out in a 5-mL tube, perform all counts in small volumes, e.g. 250 µL.

7. The markers most commonly used for myeloid antigen-presenting cells separation are CD11c for myeloid dendritic cells, CD1a for Langerhans' cells, CD123 for plasmacytoid dendritic cells, CD14 for monocytes, and CD68 for macrophages. The general markers for T lymphocytes are CD3, CD8, and CD4, and for B lymphocytes CD19. For the full list of available microbead-conjugated antibodies, please refer to Miltenyi Biotec Web site, http://www.miltenyibiotec.com.

8. For cell numbers less than 2×10^8, use volumes as given. For higher cell number, adjust the reagent volumes proportionally.

References

1. Chaux, P., Moutet, M., Faivre, J., Martin, F., and Martin, M. (1996) Inflammatory cells infiltrating human colorectal carcinomas express HLA class II but not B7-1 and B7-2 costimulatory molecules of the T-cell activation, *Lab Invest* **74**, 975–983.
2. Hartmann, E., Wollenberg, B., Rothenfusser, S., Wagner, M., Wellisch, D., Mack, B., Giese, T., Gires, O., Endres, S., and Hartmann, G. (2003) Identification and functional analysis of tumor-infiltrating plasmacytoid dendritic cells in head and neck cancer, *Cancer Res* **63**, 6478–6487.
3. Makitie, T., Summanen, P., Tarkkanen, A., and Kivela, T. (2001) Tumor-infiltrating macrophages (CD68(+) cells) and prognosis in malignant uveal melanoma, *Invest Ophthalmol Vis Sci* **42**, 1414–1421.

4. Schwaab, T., Weiss, J. E., Schned, A. R., and Barth, R. J., Jr. (2001) Dendritic cell infiltration in colon cancer, *J Immunother* **24**, 130–137.
5. Polak, M. E., Borthwick, N. J., Johnson, P., Hungerford, J. L., Higgins, B., Di Palma, S., Jager, M. J., and Cree, I. A. (2007) Presence and phenotype of dendritic cells in uveal melanoma, *Br J Ophthalmol* **91**, 971–976.
6. Polak, M. E., Johnson, P., Di Palma, S., Higgins, B., Hurren, J., Borthwick, N. J., Jager, M. J., McCormick, D., and Cree, I. A. (2005) Presence and maturity of dendritic cells in melanoma lymph node metastases, *J Pathol* **207**, 83–90.
7. Huang, S. J., Hijnen, D., Murphy, G. F., Kupper, T. S., Calarese, A. W., Mollet, I. G., Schanbacher, C. F., Miller, D. M., Schmults, C. D., and Clark, R. A. (2009) Imiquimod Enhances IFN-gamma Production and Effector Function of T Cells Infiltrating Human Squamous Cell Carcinomas of the Skin, *J Invest Dermatol*.
8. Leffers, N., Gooden, M. J., de Jong, R. A., Hoogeboom, B. N., ten Hoor, K. A., Hollema, H., Boezen, H. M., van der Zee, A. G., Daemen, T., and Nijman, H. W. (2009) Prognostic significance of tumor-infiltrating T-lymphocytes in primary and metastatic lesions of advanced stage ovarian cancer, *Cancer Immunol Immunother* **58**, 449–459.
9. Ruffini, E., Asioli, S., Filosso, P. L., Lyberis, P., Bruna, M. C., Macri, L., Daniele, L., and Oliaro, A. (2009) Clinical significance of tumor-infiltrating lymphocytes in lung neoplasms, *Ann Thorac Surg* **87**, 365–371; discussion 371–362.
10. Sugihara, A. Q., Rolle, C. E., and Lesniak, M. S. (2009) Regulatory T cells actively infiltrate metastatic brain tumors, *Int J Oncol* **34**, 1533–1540.

Chapter 18

Isolation of Endothelial Cells from Human Tumors

Elisabeth Naschberger, Vera S. Schellerer, Tilman T. Rau, Roland S. Croner, and Michael Stürzl

Abstract

Antiangiogenic drugs have been used successfully for the treatment of colorectal cancer (CRC) and several other tumor types. Until recently, viable tumor endothelial cells (TEC) and normal endothelial cells of uninvolved colon tissue (NEC) from the same patient have not been available to optimize treatment strategies in vitro. Here, we describe a protocol for the isolation of TEC and NEC. These cells were isolated at a very high purity via magnetic cell sorting of tissue samples obtained from surgical specimens of patients suffering from CRC. Isolated TEC and NEC expressed CD31, CD105, VE-cadherin, VCAM-1, ICAM-1, and E-selectin, formed capillaries in basal membrane extract, and were able to take up acetylated LDL. They were negative for podoplanin, CD45, CD68, and CK-20, indicating blood vessel endothelial lineage. Expression of vWF was more pronounced in NEC cultures, whereas vWF was absent or only slightly expressed in all TEC cultures in vitro. Lower intracellular concentrations of vWF were also detected in TEC as compared to NEC at the tissue level. The latter finding demonstrated that differential features of TEC and NEC in vivo are stably perpetuated in culture. The isolated endothelial cell cultures may provide a useful in vitro model system to elucidate epigenetic effects on angiogenesis in cancer and to optimize antiangiogenic therapy.

Key words: Endothelial cells, Angiogenesis, Colon, Colorectal carcinoma, Tumor, Collagenase II, MACS, CD31

1. Introduction

The formation of new blood vessels from preexisting vasculature (angiogenesis) is key to progression and metastasis of solid tumors (1–4). Without adequate vascular supply, solid tumors do not grow beyond a size of 1–2 mm. For further progression, angiogenesis is needed to compensate the lack of nutrients and oxygen (5).

Angiogenesis is initiated by proangiogenic factors, such as vascular endothelial growth factor and basic fibroblast growth factor, which are released by the tumor cells and activate resting endothelial cells (EC) to migrate, proliferate, differentiate, and finally to form new blood vessels (5–7). The rapid proliferation of tumor endothelial cells (TEC) and the persistent proangiogenic stimulation lead to the formation of incomplete and irregular vessels with fenestrations, irregular blood flow, and increased permeability as compared to normal vasculature (2, 8, 9). Evidence exists that TEC are different from normal endothelial cells (NEC) as indicated by the expression of specific "tumor endothelial markers" and cytogenetic abnormalities (10, 11).

Colorectal cancer (CRC) is one of the leading cancers in Western countries with 500,000 deaths worldwide per year (12). Almost a third of all patients suffer from metastatic disease at first diagnosis (13). Nearly 50% of all diagnosed CRC patients will develop metastasis with mostly fatal prognosis in the following (13). For more than a century, surgery has played the major role in curative treatment of CRC. The presently applied treatment of choice for advanced disease is surgery combined with adjuvant chemotherapy (12). Of note, a significant response to antiangiogenic therapy has been observed when patients with advanced disease are treated with the anti-VEGF antibody bevacizumab in combination with chemotherapeutics (14). Since this first report, antiangiogenic strategies are increasingly considered as a relevant therapeutic option for the therapy of CRC. However, further improvement of treatment protocols by drug optimization and patient selection is required. This demands in vitro model systems that allow optimization of antiangiogenic treatment strategies in CRC.

Here, we describe a protocol for the isolation of pure, viable EC from human CRC and healthy colon tissue. These cells may allow optimization of antiangiogenic therapy. In addition, the comparative molecular analysis of TEC and NEC may identify further molecular targets for improved antiangiogenic treatment of CRC (15).

2. Materials

Our group was the first to succeed in the isolation of pure microvascular endothelial cells from CRC (15). Here, we describe a detailed protocol of the isolation process.

2.1. Tissue Collection

1. 1× Hank's Balanced Salt Solution (1× HBSS; PAA, Pasching, Austria) supplemented with 1× penicillin/streptomycin (PAA) and 250 μg/mL amphotericin B (PAA).

2. 50 mL Falcon tubes, sterile capped (BD Biosciences, Heidelberg, Germany).

2.2. Generation of a Single Cell Suspension

1. Collagenase II (Biochrom, Berlin, Germany): dissolve 1 g of collagenase II in a suitable volume of 1× HBSS without antibiotics to obtain a final stock solution of 34,200 U/mL; sterile filtered, aliquoted, and stored at −20°C.
2. EBM-2-MV (Lonza, Cologne, Germany).
3. Bovine skin gelatin, type B 1.5% (Sigma-Aldrich, Munich, Germany) dissolved in 1× PBS (Biochrom), stirred for 1 h at 65°C, autoclaved, and sterile filtered; storage at 4°C for 6 weeks.
4. Scalpel No. 23 (Feather Safety Razor, Osaka, Japan) and forceps.
5. Cell strainer 100 μm (BD Biosciences).
6. Syringes, 10 mL (BD Biosciences).
7. Dynal sample mixer, model MX1 (Invitrogen, Karlsruhe, Germany).
8. 15-mL Falcon tubes (BD Biosciences).

2.3. Enrichment of Tumor and Normal Endothelial Cells by MACS

1. CD31 MicroBead kit (Miltenyi Biotec, Bergisch Gladbach, Germany).
2. FcR Blocking Reagent (Miltenyi Biotec).
3. MACS buffer: 1× PBS pH 7.2, 0.5% bovine serum albumin, 2 mM EDTA; sterile filtered and stored at 4°C; degas before use.
4. MACS separation columns (Miltenyi Biotec).
5. MACS preseparation filter (Miltenyi Biotec).
6. MACS separation unit (separator and stand, Miltenyi Biotec).
7. Accutase (PAA).
8. EBM-2-MV (Lonza).
9. 1.5% gelatin type B (Sigma) dissolved in 1× PBS (Biochrom), stirred for 1 h at 65°C, autoclaved, and sterile filtered; storage at 4°C for 6 weeks.
10. 1× PBS (Biochrom).

3. Methods

In the following sections, the methodological procedures are described in detail. Specific tips that may improve the outcome are described in Subheading 4.

3.1. Tissue Collection

1. Cut pieces of 0.5 up to 3 g (depending on tumor size) from the center of the malignancy to be used for the isolation of tumor endothelial cells (TEC). Morphologically visible

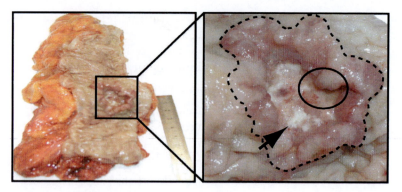

Fig. 1. Viable tumor tissue can be obtained from the center of malignancy by cutting nonhypoxic, nonnecrotic areas from colorectal carcinoma. Pieces of up to 3 g were cut from a nonhypoxic, nonnecrotic area from the central part of the tumor (area of resection marked by *circle*; invasive margin marked by *dashed line*; necrotic area shown by *arrow*). Tissue for the isolation of normal endothelial cells has been removed at a distance of about 10 cm from the tumor margin (*dashed line*). The figure is partly reproduced with permission from Schellerer et al. (15).

hypoxic or necrotic parts of the tumor should be avoided (Fig. 1). Uncompromised healthy colon tissue is harvested at a safety distance of at minimum 10 cm apart from the tumor and is used for the isolation of corresponding normal endothelial cells (NEC). Place the resected specimen immediately in a 50 mL falcon tube with ice-cold HBSS (see Note 1).

2. From now on, work under a sterile laminar flow for the whole procedure.

3. Wash the specimen by transferring it using sterile forceps from a 50 mL falcon with fresh ice-cold 1× HBSS to a new falcon with 1× HBSS four times consecutively.

3.2. Generation of a Single Cell Suspension

1. Remove the specimen from the falcon tube and place it in a 10-cm cell culture dish. Mince the tissue into approximately 10-mm^3 pieces using a fresh, sterile scalpel (see Note 2).

2. Put the minced tissue into a 15-mL falcon tube filled with 3 mL EBM-2 supplemented with 0.5% FBS (=EBM-2-Low). Put approx. 0.2 g of tissue in one falcon tube (see Note 3).

3. According to the weight of the tissue pieces in each falcon tube, calculate the required amount of collagenase II (50 μl enzyme per 0.1 g tissue; 17,100 U/g; see Note 4).

4. Pipet the calculated amount of collagenase II into each falcon tube and add EBM-2-Low up to a total volume of 5 ml.

5. Put the falcon tubes in the Dynal sample mixer at 37°C for 1 h at 5% CO_2 with the lowest speed available.

6. Put a cell strainer on a 50 mL falcon tube and pour the digested tissue through the strainer. Wash the filter from the inside and outside using 3 mL EBM-2-Low each time.

7. Centrifuge with $500 \times g$ for 5 min at 20°C and discard the supernatant.

8. Resuspend the cell pellet in 5 mL EBM-2-MV and cultivate the cells until 70–80% confluence in a T-25 cell culture flask precoated with 1.5% gelatin for at least 2 h. Wash the cells twice with 1× PBS the next day and add fresh medium. Afterward renew the medium every second day.

3.3. Enrichment of Tumor and Normal Endothelial Cells by MACS

1. Usually, 5–7 days after initial seeding, the cultures reach 70–80% confluence and the cells can be positively selected for CD31 by MACS (see Note 5 and Fig. 2). Repeat MACS selection until all nonendothelial cells are removed (Fig. 2).

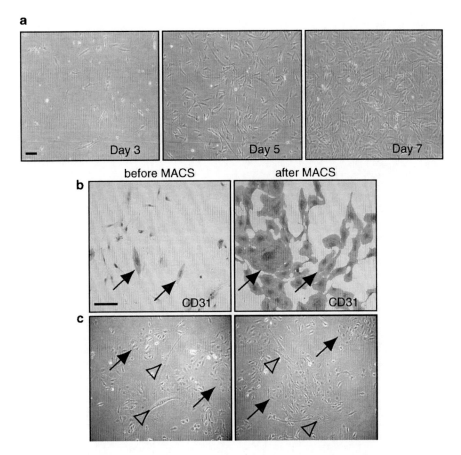

Fig. 2. Tumor endothelial cells can be enriched from a mixture of cells arising after initial cultivation of digested colorectal carcinoma tissue using CD31-MACS. (**a**) Tissues were enzymatically digested by Collagenase II and mechanically dispersed. The resulting single cell suspension was seeded in gelatin-coated culture flasks. After 5–7 days phenotypically different cells arose. The scale bar corresponds to 100 μm. (**b**) Endothelial cells isolated from healthy colon and colorectal cancers were positively selected using CD31-MicroBeads. The yield of endothelial cells was monitored by immunocytochemical staining of the CD31 surface antigen before and after MACS. Positive staining is visualized by *pink color* (*arrows*). Counterstaining was performed with hematoxylin (*blue color*). Scale bar corresponds to 100 μm. (**c**) Clones of endothelial cells can be observed arising between nonendothelial cells after some enrichment rounds of the endothelial cells by CD31-MACS (*arrows*, endothelial cells; *arrowheads*, nonendothelial cells). The figure is partly reproduced with permission from Schellerer et al. (15).

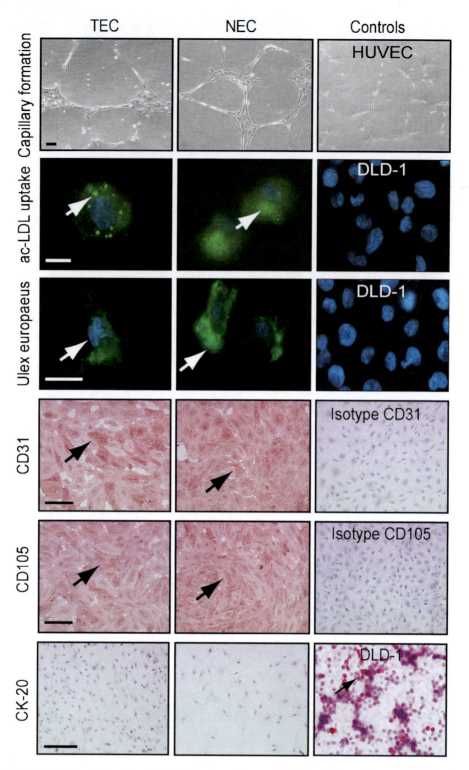

Fig. 3. The isolated tumor and normal endothelial cell cultures show endothelial cell lineage. Pure endothelial cell populations (>93%) of tumor (TEC) and normal endothelial cells (NEC) were characterized by different assays. For capillary formation, TEC and NEC were seeded onto an extract of extracellular basement membrane and were stimulated with angiogenic growth factors. HUVEC were used as a positive control. For the uptake of acetylated low-density lipoprotein (ac-LDL), TEC and NEC were incubated with Alexa488-labeled acetylated LDL (*green fluorescent signal*, *arrows*) for 1 h and were counterstained by DAPI (*blue fluorescent signal*). TEC and NEC cultures were incubated with FITC-conjugated *Ulex europaeus*-lectin for 1 h and the signal (*green*, *arrow*) was visualized using epifluorescence microscopy.

2. Wash the cells with 1× PBS and add 1–2 mL accutase until cells are completely detached; rinse the dish with EBM-2-MV and collect the cells in a falcon tube.

3. Follow the manufacturer's instructions for a positive selection for endothelial cells using the CD31 MicroBead Kit as detailed below (see Note 6).

4. Put a preseparation filter on a 15-mL falcon tube and prewet it using 0.5 mL EBM-2-MV; from now on, work all the time on ice and use precooled solutions!

5. Centrifuge the cells at 4°C for 3 min with 300 ×g and discard the supernatant.

6. Dissolve the pellet in 60 µl MACS-buffer, add 20 µL FcR blocking reagent, vortex briefly, and add 20 µL CD31 MicroBeads (see Note 7); incubation for 15 min at 4°C.

7. Add 1 mL MACS buffer to the cells and centrifuge at 4°C for 3 min with 300 ×g; discard the supernatant.

8. During this centrifugation step, fix a suitable column (see Note 7) in the MACS separator and prewet the column with 0.5 mL MACS buffer.

9. Dissolve the pellet after centrifugation in 1 mL MACS buffer and pipet 2× 0.5 mL of the cell suspension on the column.

10. Wash the column with 3× 0.5 mL MACS buffer.

11. Elute the cells from the first column using the stamp with 1 mL MACS buffer on a second (prewet before with 0.5 mL MACS buffer) column.

12. Wash the column two times with 0.5 mL MACS buffer and once with 0.5 mL EBM-2-MV

13. Elute the cells using the stamp with 1 mL EBM-2-MV in a suitable cell culture dish ($1–2 \times 10^4$ cells per cm^2) precoated with 1.5% gelatin (see Note 8).

14. Cultivate the cells until confluence with EBM-2-MV medium renewal every second day.

15. In case of significant (>10%) contamination by nonendothelial cells, perform additional cycles of magnetic cell sorting (see Note 9). Contamination with nonendothelial cells can be analyzed by staining an aliquot of the cells for CD31, and subsequent immunocytochemical or FACS analysis (Figs. 2b and 3).

Fig. 3. (continued) Counterstaining was performed with DAPI (*blue fluorescence*). Negative controls of ac-LDL and *U. europaeus* binding have been performed with the colorectal carcinoma cell line DLD-1. Immunocytochemical stainig of TEC and NEC was performed using CD31, CD105 and CK-20 antibodies. Positive staining is visualized by *pink colour* (*arrows*). Counterstaining was performed with hematoxylin (*blue colour*). For CD31 and CD105 isotype staining controls are displayed. DLD-1 cells were stained as a positive control for the CK-20 immunostaining. The *scale bars* correspond to 100 µm (capillary formation, CD31, CD105), 50 µm (CK-20), 10 µm (ac-LDL - uptake, *U. europaeus*). The figure is partly reproduced with permission from Schellerer et al. (15).

16. The first confluent T-25 flask of pure TEC/NEC is designated as passage 0 (see Notes 10 and 11). One passage is defined by a split ratio of 1:4.

4. Notes

1. The elapsed time from resection of the specimen to the start of the isolation procedure is critical for the success of isolation. Keep this time as short as possible. In our laboratory, this does not take more than 30 min.
2. Avoid quenching, pressing, or tearing of the specimen while cutting. This affects the viability of the endothelial cells. Try to use the scalpel like a rocking tool.
3. Weigh the falcon tubes filled with EBM-2-Low before putting the tissue pieces inside and note the weight. Weigh the falcon tubes again after putting the tissue pieces inside and calculate the difference in weight.
4. The protocol was developed for the digestion of colorectal carcinoma. If you intend to use tissue from another tumor, you might have to standardize newly the concentration of the collagenase to be used to obtain a single cell suspension.
5. A MACS separation column with suitable size has to be selected for the procedure. Therefore, count an aliquot of the cells, calculate the total cell number, and choose your column according to the guidelines of Miltenyi Biotec (http://www.miltenyibiotec.com).
6. A MACS separation unit is chosen according to the size of the column and the number of columns that you intend to use simultaneously. Instructions for the selection of a suitable MACS separator can be found on the home page of Miltenyi Biotec (http://www.miltenyibiotec.com).
7. Here, the volumes/concentrations are given for a MS separation column and up to 1×10^7 cells.
8. If you want to make use of the negative cell fraction, keep the flow-through, centrifuge and cultivate the CD31-negative cells.
9. A CD31 antibody (clone JC70A, Dako, Hamburg, Germany) in combination with a mouse APAAP detection system (Dako) is used to evaluate the purity of the cultures by immunocytochemistry in our laboratory (Figs. 2b and 3).
10. The initially established CD31-positive cell cultures may also be characterized for additional endothelial cell markers to

show endothelial cell origin. We used in addition CD105, vWF, VE-cadherin, ICAM-1, VCAM-1, E-selectin, CD45, CD68, CK-20, and podoplanin. Moreover, the cells were analyzed for their ability to form capillary structures, bind *Ulex europaeus*, and take up acetylated-LDL (Fig. 3) (15).

11. In general, viable tumor or normal endothelial cells can be obtained from approx. 30% of the specimens undergoing the isolation and selection procedure.

Acknowledgments

We thank Christina von Kleinsorgen for excellent technical support and all patients that were participating in this study. We also thank the Institute of Pathology for kindly preparing the specimens and providing the histology reports. This work was supported by grants from the ELAN-program of the University Medical Center Erlangen to EN/RSC (# 05.06.05.1), a grant from the Doktor Robert Pfleger-Stiftung Bamberg to VSS/EN and from the interdisciplinary center of clinical research (IZKF) of the University Medical Center Erlangen to MS/EN.

References

1. Quesada, A. R., Munoz-Chapuli, R., and Medina, M. A. (2006) Anti-angiogenic drugs: from bench to clinical trials, *Med Res Rev.* **26**, 483–530
2. Jain, R. K. (2005) Normalization of tumor vasculature: an emerging concept in antiangiogenic therapy, *Science* **307**, 58–62.
3. Folkman, J. (1971) Tumor angiogenesis: therapeutic implications, *N Engl J Med* **285**, 1182–1186.
4. Folkman, J. (2006) Angiogenesis, *Annu Rev Med* **57**, 1–18.
5. Carmeliet, P. (2005) VEGF as a key mediator of angiogenesis in cancer, *Oncology* **69 Suppl 3**, 4–10.
6. Zanetta, L., Marcus, S. G., Vasile, J., Dobryansky, M., Cohen, H., Eng, K., Shamamian, P., and Mignatti, P. (2000) Expression of Von Willebrand factor, an endothelial cell marker, is up-regulated by angiogenesis factors: a potential method for objective assessment of tumor angiogenesis, *Int J Cancer* **85**, 281–288.
7. Klagsbrun, M. (1992) Mediators of angiogenesis: the biological significance of basic fibroblast growth factor (bFGF)-heparin and heparan sulfate interactions, *Semin Cancer Biol* **3**, 81–87.
8. Jain, R. K. (2001) Normalizing tumor vasculature with anti-angiogenic therapy: A new paradigm for combination therapy, *Nat Med* **7**, 987–989.
9. Jain, R. K. (2003) Molecular regulation of vessel maturation, *Nat Med* **9**, 685–693.
10. St Croix, B., Rago, C., Velculescu, V., Traverso, G., Romans, K. E., Montgomery, E., Lal, A., Riggins, G. J., Lengauer, C., Vogelstein, B., and Kinzler, K. W. (2000) Genes expressed in human tumor endothelium, *Science* **289**, 1197–1202.
11. Hida, K., and Klagsbrun, M. (2005) A new perspective on tumor endothelial cells: unexpected chromosome and centrosome abnormalities, *Cancer Res* **65**, 2507–2510.
12. Saunders, M., and Iveson, T. (2006) Management of advanced colorectal cancer: state of the art, *Br J Cancer* **95**, 131–138.
13. Hurwitz, H., and Kabbinavar, F. (2005) Bevacizumab combined with standard fluoropyrimidine-based chemotherapy regimens to treat colorectal cancer, *Oncology* **69 Suppl 3**, 17–24.
14. Hurwitz, H., Fehrenbacher, L., Novotny, W., Cartwright, T., Hainsworth, J., Heim, W., Berlin, J., Baron, A., Griffing, S.,

Holmgren, E., Ferrara, N., Fyfe, G., Rogers, B., Ross, R., and Kabbinavar, F. (2004) Bevacizumab plus irinotecan, fluorouracil, and leucovorin for metastatic colorectal cancer, *N Engl J Med* **350**, 2335–2342.

15. Schellerer, V. S., Croner, R. S., Weinländer, K., Hohenberger, W., Stürzl, M., and Naschberger, E. (2007) Endothelial cells of human colorectal cancer and healthy colon reveal phenotypic differences in culture, *Lab Invest* **87**, 1159–1170.

Chapter 19

Cellular Chemosensitivity Assays: An Overview

Venil N. Sumantran

Abstract

Data on cell viability have long been obtained from in vitro cytotoxicity assays. Today, there is a focus on markers of cell death, and the MTT cell survival assay is widely used for measuring cytotoxic potential of a compound. However, a comprehensive evaluation of cytotoxicity requires additional assays which measure short and long-term cytotoxicity. Assays which measure the cytostatic effects of compounds are not less important, particularly for newer anticancer agents. This overview discusses the advantages and disadvantages of different non-clonogenic assays for measuring short and medium-term cytotoxicity. It also discusses clonogenic assays, which accurately measure long-term cytostatic effects of drugs and toxic agents. For certain compounds and cell types, the advent of high throughput, multiparameter, cytotoxicity assays, and gene expression assays have made it possible to predict cytotoxic potency in vivo.

Key words: Cytotoxicity, Non-clonogenic, Clonogenic, High throughput, Microfluidic

1. Introduction

The need for better in vitro methods for measuring and predicting cytotoxicity has never been greater. This is true because toxicity is the main reason for the high rate of failure (40–50%) of pharmaceutical drugs. Further, combinatorial chemistry, high-throughput screening (HTS) technologies, and the interest in natural products have produced huge libraries of new, complex, and highly diverse chemical compounds which must be screened for in vitro cytotoxicity. The European Centre for the Validation of Alternative Methods (ECVAM) suggests that at least 30,000 new chemicals will have to be tested for potential cytotoxicity within the next 15 years. In this light, the choice of the appropriate in vitro cytotoxicity assay becomes crucial.

Indeed, evidence shows that the degree of cytotoxicity of a substance can vary, and strongly depends on the assays used to estimate it (1). Therefore, this chapter describes and discusses the advantages and disadvantages of different assays for detecting cytotoxic and cytostatic effects of compounds. A cytotoxic compound causes a short-term loss of cell viability by triggering cell death or causing a large decrease in cell survival. In contrast, a cytostatic compound affects long-term cell survival or cell proliferation without affecting viable cell number in the near term. There are several types of non-clonogenic assays which measure acute/short-term cytostatic and cytotoxic effects. Both types of non-clonogenic assays are explained in detail below. However, long-term cytostatic changes (growth inhibition due to cell cycle arrest), are best measured by clonogenic assays, which are also discussed below. Finally, we discuss how HTS, cell-based microfluidic devices, and DNA microarrays are revolutionizing the field of in vitro cytotoxicity testing.

2. Non-clonogenic Cytotoxicity Assays

2.1. Enzyme Release Assays

The loss of cell membrane integrity due to cell death or membrane damage results in release of certain soluble, cytosolic enzymes. Therefore, acute cytotoxicity can be measured by measuring activity of these enzymes released into the culture supernatant. The level of enzyme activity correlates with the amount of cell death/membrane damage, and provides an accurate measure of the cytotoxicity induced by the test substance. These assays do not provide insight into the mechanism underlying the observed cytotoxicity.

2.1.1. Lactate Dehydrogenase Release Assay

The Lactate dehydrogenase (LDH) release assay was developed in the 1980s as a rapid and sensitive assay for assaying cytotoxicity in immune cells (2). Since the results of LDH release assays compared favorably with those from conventional ^{51}Cr release assays, the latter is no longer in use (3). Due to the ubiquitous nature of LDH, this assay is now widely used for measuring acute cytotoxicity of any chemical in other cell types.

Commercial kits are available which measure LDH activity by a coupled two-step reaction. In the first step, LDH catalyzes the reduction of NAD^+ to NADH and H^+ by oxidation of lactate to pyruvate. In the second step, the diaphorase enzyme uses the newly-formed NADH and H^+ to reduce a tetrazolium salt (INT) to a colored formazan, which absorbs maximally at 490–520 nm. This two-step reaction increases the sensitivity of the assay. Adequate controls are required to account for false positives due to the presence of phenol red and endogenous LDH activity present in the serum within the culture medium.

2.1.2. Glucose 6-Phosphate Dehydrogenase Release Assay

An alternative method monitors the release of the cytosolic enzyme glucose 6-phosphate dehydrogenase (G6PD) from damaged cells into the surrounding medium (4). This assay is also marketed as a kit. G6PD activity is assayed by a coupled enzymatic assay, wherein oxidation of glucose 6-phosphate by G6PD generates NADPH, which in turn leads to the reduction of resazurin by diaphorase to yield a fluorescent product-resorfurin. The assay is rapid and can detect activity originating from around 500 cells. According to the manufacturers, the G6PD release assay should give lower background signals than that observed with LDH release assays, because levels of G6PD activity in common cell culture sera are typically lower than levels of LDH activity.

2.1.3. Glyceraldehyde-3-Phosphate Dehydrogenase Release Assay

The aCella-TOX kit is based on a new and highly sensitive method using Coupled Luminescent technology for the detection of cytotoxicity. This assay quantitatively measures the release of Glyceraldehyde-3-Phosphate Dehydrogenase (GAPDH) from primary cells, cell lines, and bacteria. The release of GAPDH is coupled to the activity of the enzyme 3-Phosphoglyceric Phosphokinase (PGK) to produce ATP, which is then detected via the luciferase–luciferin bioluminescence technology. Thus, increased levels of ATP correlate with increased GAPDH release and increased cytotoxicity (5).

This patented method claims much higher sensitivity than the LDH enzyme release assay, because the coupled luminescent signal-amplification system yields a strong signal even for small amounts of GAPDH released. This method has been tested with many modes of cytolysis, including T cell cytotoxicity, cytolysis induced by complement (6), pore-forming agents, antibiotic-mediated lysis of bacteria, and detergent mediated or mechanical lysis (3). Since the assay is nondestructive, one can also measure cell viability and even gene expression in the same culture plate. Extra culture supernatants can also be removed from the original plate and assayed for kinetic studies.

In addition to enzyme release assays which measure acute cytotoxicity, there are other non-clonogenic assays which specifically measure acute effects of compounds on cell viability or cell survival. Assays in each category are explained below.

2.2. Cell Viability Assays

2.2.1. Trypan Blue and Alamar Blue Assays

Early assays measured decreases in viable cell number, as a direct measure of cytotoxicity. The Trypan Blue dye was and is commonly used as it is excluded by living cells but stains dead cells. Indeed, cell counting by the Trypan Blue exclusion method, is still used as a confirmatory test for measuring changes in viable cell number caused by a drug/toxin (7). A more recent vital dye based assay for cell viability uses Alamar blue. Alamar blue contains a redox indicator which exhibits both fluorescence and colorimetric changes in the oxidation-reduction range of cellular metabolism. Thus, growing cells cause a reduction of Alamar blue, while growth inhibition causes dye oxidation. Since these redox changes of Alamar blue are stable, one can conduct long term assays and kinetic studies of drug induced changes on cell viability. In addition, Alamar blue is easy to use, has low toxicity, and can be used with suspension/adherent cell cultures (8).

2.2.2. Sulforhodamine B Assay

A decrease in total cell protein caused by a compound can also used as a parameter of cell viability. Sulforhodamine B (SRB) is an anionic aminoxanthene dye which forms an electrostatic complex with basic amino acid residues of proteins, and provides a sensitive linear response (9). The color development is rapid and stable and is measured at absorbances between 560 and 580 nm. The sensitivity of the SRB assay compares favorably with that of several fluorescence assays. This assay is also useful in quantitating clonogenicity, and is well suited to high-volume, automated drug screening (10).

Assays measuring a loss in cell viability are insufficient since decreases in viable cell number/total cell protein can be due to decreased cell proliferation, decreased cell survival, or increased cell death. Therefore, the next section describes the major assays used to quantitate cell survival. These assays are primarily colorimetric methods which measure a single intracellular end-point.

2.3. Cell Survival Assays

2.3.1. Neutral Red Uptake Assay

The neutral red uptake-cytotoxicity assay is a cell survival/viability assay based on the ability of viable cells to incorporate the neutral red (NR) dye. NR is a weak cationic supravital dye that accumulates within lysosomes of viable cells. Toxic substances cause decreased uptake of NR which can be quantitated spectrophotometrically. Cytotoxicity is expressed as a concentration dependent reduction of the uptake of NR after chemical exposure, and serves as a sensitive indicator of both cell integrity and growth inhibition (11). Indeed, the neutral red uptake (NRU) assay proved as sensitive as a mouse whole genome array for estimating the differential cytotoxic potential of three types of cigarettes with varying tar content (12). However, the NRU assay cannot differentiate between cytotoxic and cytostatic drugs or compounds.

2.3.2. Applications of the NRU Assay

2.3.2.1. Phototoxicity

The NRU assay has been standardized to specifically detect and measure phototoxic chemicals in Balb/c 3T3 mouse fibroblast cell cultures. Thus, cells are pretreated with the test chemicals, and then exposed to the dark/UV-A wavelengths for 50 min. Subsequently, NR is added to cells for 24 h, and NR uptake is measured. Cytotoxic chemicals are those which cause reduced NR uptake in the dark, whereas the phototoxic chemicals decrease NR uptake only in UV-A exposed cultures (13).

2.3.2.2. Cosmetic Testing

The NRU assay accurately quantitated and predicted cytotoxicity caused by cosmetics and shampoos. The results are in agreement with in vivo data from Draize rabbit eye irritation tests of cosmetic products, and could finally lead to the termination of such animal based tests for cytotoxicity testing of these products (14).

2.4. Notes on Cytotoxicity Assays Based on Altered Cell Permeability

2.4.1. Advantages

The ease, sensitivity, and rapidity with which these cytotoxicity assays (Enzyme release assays, assays for uptake of vital dyes, and NR) can be performed have made them popular. For example, Putnam et al. (15) used eight assays with different endpoints to evaluate short and long-term cytotoxicity in CHO cells exposed to cigarette smoke condensate. Assays which measured membrane integrity (LDH release) were most sensitive for detecting short-term effects (1 h), whereas the NRU or protein binding assays were most sensitive for detecting longer-term damage (12–24 h) (15). The LDH release and the NRU assays can also be used to evaluate paraptosis, a caspase-independent and non-apoptotic mechanism of cell death characterized by cell swelling, mitochondrial changes, and cytoplasmic vacuolization (16).

2.4.2. Disadvantages

A disadvantage of cytotoxicity assays based on altered cell permeability, is that the initial sites of cellular damage caused by most toxic agents is intracellular. Therefore, cells may be irreversibly damaged and committed to die, while the plasma membrane is still intact. Thus, these permeability based assays can *underestimate* cellular damage when compared to other methods. Despite this fact, many permeability assays are widely used as accepted methods for the measurement of cytotoxicity (17).

2.4.3. MTT Assay

The MTT assay measures the mitochondrial function and is most often used to detect loss of cell survival/cell viability due to a drug or toxin. Other colorimetric assays to measure cell survival include the XTT and WST-1 assays. However, these metabolic assays cannot differentiate between cytotoxic and cytostatic drugs or compounds, and may not be adequately sensitive when working with low cell numbers (18–20).

2.4.4. ATP Assay

Measuring cytotoxicity by quantitation of intracellular concentrations of adenosine triphosphate (ATP) as a measure of cell survival, has now gained wide acceptance for evaluating medium, long-term cytotoxic effects of chemicals (48–72 h in vitro). The assay is based on bioluminescent detection of cellular ATP (21) and is extremely sensitive, being able to measure ATP levels in a single adherent or non-adherent mammalian cell. The large dynamic range and long signal duration are additional advantages of this assay compared to the MTT assay (20, 22). However, the ATP quantitation assay is also incapable of differentiating between cytotoxic and cytostatic drugs, i.e. changes in ATP levels could be due to changes in cell survival, viable cell number, or cell death.

2.4.5. Applications of the ATP Cell Survival Assay

The ATP Assay could predict the chemosensitivity of platinum-resistant epithelial ovarian cancer tumors to a panel of other drugs with an accuracy of 85% (23). This assay also measured intrinsic radiosensitivity of cervical cancer cells with results similar to conventional clonogenic assays (24).

2.5. Notes on Cell Survival Assays

2.5.1. Advantages

The ease, sensitivity, rapidity, and low cost have made the MTT method as one of the most widely used assays for measuring acute cytotoxic effects of compounds. Nowadays, the ATP Cell survival assay is increasingly being used in high-throughput mode to compare cytotoxicity of large numbers of drugs on one/more cell types simultaneously.

2.5.2. Disadvantages

There are two disadvantages associated with the use of metabolic assays such as the MTT, XTT, WST-1, and the ATP cell survival Assay. Firstly, like the permeability assays mentioned above, these assays can underestimate cellular damage and cell death because these methods work best for detecting the later stages of apoptosis when the metabolic activity of the cells is severely reduced. Nevertheless, these assays are useful for quantitating cytotoxicity in short-term cell cultures (a 24–96 h period) and the ATP assay overcomes this problem by using a longer (6 day) incubation period (21). Secondly, cell survival assays are of limited value for measuring cell-mediated cytotoxicity. This is because most effector cells become activated upon binding to the target cells. This activation can result in increased formazan production by the effector cell, which tends to mask the decreased formazan production that results from target cell death (25).

2.6. Fluorometric Assays for Measuring Cytotoxic and Cytostatic Effects

The colorimetric MTT-based non-clonogenic assays to measure cytotoxicity, cell viability, and cell survival discussed above, are increasingly being replaced by fluorometric assays with higher sensitivities and dynamic range. In comparison with the MTT and LDH release assays, a fluorescence-based oxygen uptake assay proved to be the most sensitive method for detecting changes in mitochondrial integrity due to known toxicants in several tumor cell lines (26). Another fluorometric microplate-based assay measures cytotoxicity based on hydrolysis of a fluorescein diacetate (FDA) probe by esterases in intact cells. This method has the advantage of being able to detect cytotoxic and/or cytostatic effects of different compounds in vitro, and can be used on cell lines, and fresh tumor cells from patients (27). Altered cell adhesion due to cellular damage can also be a useful parameter for evaluating short-term cytotoxicity. Indeed, measuring loss of monolayer adherence by fluorescent methods, proved as sensitive as assays measuring organelle-specific damage, for investigating cytotoxic effects of combined photodynamic therapy with chemotherapeutic drugs in different cell lines (28).

2.7. Multiplex Cytotoxicity Assays

An important principle is that the most sensitive cytotoxicity assay for a given agent depends on the cellular site at which it causes direct damage in the target cell at a given time point. Therefore, careful selection of the appropriate non-clonogenic, cytotoxicity assay(s) can provide valuable data on the potency of short-term effects of a cytotoxic agent. Since each of the various assays listed above measures a single intracellular end-point, one can perform multiple assays to allow simultaneous measurements of several endpoints to estimate cell damage with greater accuracy. This approach is also logical since a drug/chemical may affect cell size, morphology, membrane integrity, and/or organelle function. This trend is reflected in the availability of multiplex assay kits for high content screening of cells. These kits simultaneously quantify hallmark indicators of cytotoxicity, such as viable cell number, nuclear size and morphology, cell membrane permeability, lysosome number, and/or integrity of the mitochondrial membrane. Indeed, the measurement of several endpoints in a cytotoxicity assay is particularly useful while screening mixtures of natural compounds or drug extracts, because it increases the chance that potential bioactive/cytotoxic compounds are discovered during screening. For example, a bioassay for simultaneous measurement of metabolic activity, membrane integrity, and lysosomal activity found three fungal secondary metabolites that affected different intracellular targets (29).

Having described the non-clonogenic assays which measure cytotoxic and/or cytostatic effects of different compounds in vitro, we now focus on assays which measure cell death, a common mechanism underlying the cytotoxicity of large numbers of drugs, chemicals, and toxins.

2.8. Cell Death Assays

Apoptosis is a distinctive, coordinated mode of genetically "programmed" cell death which is often energy-dependent, and involves the activation of a group of cysteine proteases called "caspases". On the other hand, necrosis is a toxic, energy-independent mode of cell death and often involves direct damage to cell membranes. Since many morphological and biochemical features of apoptosis and necrosis can overlap, it is crucial to perform at least two or more distinct assays to confirm that cell death is occurring via apoptosis. Ideally, one assay should detect early apoptotic events (initiation) and the second assay should quantitate a later (execution) event in apoptosis (30).

2.9. Assays for Early Apoptosis

2.9.1. Detection of Caspases

Detection of caspase activation can be used as an early marker of apoptosis. This is done by Western Blot or ELISA using polyclonal/monoclonal antibodies against both the inactive pro-caspases and active caspases. Measuring caspase activity by its action on a fluorescent or luciferin labeled substrate is a more definitive and sensitive method (usually requiring 1×10^5 cells) (31, 32).

However, this method does not permit determination of the cell type undergoing apoptosis. Moreover, the specificity of the assay can be compromised due to overlapping substrate preferences of members of the caspase family.

2.9.2. Membrane Alterations

A good method for early detection of apoptosis is monitoring the externalization of phosphatidylserine residues on the outer plasma membrane of individual apoptotic cells. This is done by detection of flourescence-tagged Annexin V (33). The advantages of this method are high sensitivity and the ability to confirm the activity of initiator caspases. The disadvantage is that the membranes of necrotic cells can also be labeled with Annexin V. However, this problem can be solved by performing a control to demonstrate the membrane integrity of the Annexin V positive cells. This control is based on the fact that cells in *early* stages of apoptosis *retain* membrane integrity. Therefore, while both apoptotic and necrotic cells would be Annexin V positive (phosphatidylserine-positive), only apoptotic cells can exclude nucleic acid dyes such as propidium iodide or trypan blue, whereas necrotic cells lacking membrane integrity, will take up these specific dyes.

2.10. Assay for Late Apoptosis

2.10.1. DNA Fragmentation

When the DNA from a cell homogenate is visualized, apoptotic cells show the presence of a characteristic "DNA ladder" on agarose gels. This is due to the programmed degradation of nuclear DNA by endonucleases. This methodology is simple, but requires large cell numbers (at least 1×10^6 cells). False positives can occur since necrotic cells also generate DNA fragments, and because DNA fragmentation can occur during preparation of the cell homogenate (34).

2.10.2. DNA-Histone Cell Death ELISA

During apoptosis, endogenous endonucleases cleave double-stranded DNA at the accessible internucleosomal linker region and generate nucleosomes. In contrast to linker DNA, the DNA of nucleosomes is tightly complexed with core histones, and is thus protected from cleavage by endonucleases. After induction of apoptosis, the cytoplasm of the apoptotic cell is enriched with nucleosomes (DNA–histone complexes), because DNA degradation occurs several hours *before* plasma membrane breakdown. Levels of nucleosomes can be quantitated by an ELISA for detection of the DNA–histone complexes (35).

In this ELISA, the cells are treated with/without the test compound, lysed, and then centrifuged to separate low molecular weight DNA from high molecular weight (nuclear) DNA. The low molecular weight DNA containing the nucleosomes in the supernatant is then quantitated by a "sandwich enzyme immunoassay" using sequential mouse monoclonal antibodies directed against species specific histones, followed by an antiDNA-antibody conjugated to the peroxidase enzyme. After removal of unbound antibodies, the amount of peroxidase activity retained in the immunocomplexes is assayed with an appropriate substrate (for peroxidase) and levels of the product are quantitated colorimetrically/photometrically. The levels of peroxidase activity obtained are directly proportional to the degree of apoptosis induced by the test compound. The "units" of peroxidase activity can be converted to equivalent cell numbers using an internal standard wherein increasing cell numbers are treated with a known inducer of apoptosis under fixed conditions, and run through the ELISA.

This method is quite sensitive (10^2–10^4 cells/test required) and gives quantitative evidence of DNA fragmentation compared to the qualitative method of visualizing "DNA ladders". In addition, this method has the advantage of distinguishing between apoptosis and necrosis, because the nucleosomes leak out of necrotic cells – but remain cytosolic in apoptotic cells. Therefore, one can check for possible necrosis by assaying samples of conditioned media (CM) from cells treated with/without test compound, *prior* to cell lysis. If the ELISA detects significant levels of nucleosomes in these CM samples, it indicates that the test compound has induced necrosis.

2.10.3. (TUNEL) Terminal deoxynucleotidyl transferase dUTP Nick End-Labeling assay

The TUNEL method is also capable of detecting nuclear DNA fragmentation in apoptotic cells. Here, the endonuclease cleavage products are enzymatically end-labelled at the 3'-end with labeled dUTP, using the enzyme terminal transferase (36). The labeled dUTP is then detected by light/fluorescence microscopy, or flow cytometry. This assay is very sensitive, allowing quantitation of DNA damage in a single cell to a few hundred cells by flow cytometry. However, false positives can arise from necrotic cells and cells in the process of DNA repair and gene transcription.

2.10.4. Mitochondrial Assays

Mitochondrial assays allow unequivocal detection of apoptotic cell death.

2.10.4.1. Mitochondrial Membrane Potential

Laser scanning confocal microscopy with appropriate fluorescent dyes can be used to track mitochondrial permeability transition (MPT), and the depolarization of the inner mitochondrial membrane. A more definitive method is based on the collapsed electrochemical gradient across the mitochondrial outer membranes of apoptotic cells. This phenomenon is detected with a fluorescent cationic dye which aggregates and accumulates within viable mitochondria, to emit a specific fluorescence. However, in apoptotic cells the dye diffuses into the cytoplasm and emits a fluorescence which differs from that of the aggregated form of the dye (37).

2.10.4.2. Cytochrome c Release Assays

Cytochrome c release from mitochondria is a confirmatory assay for apoptotic cell death. Cytochrome c can be assayed using fluorescence and electron microscopy in living or fixed cells (38). However, cytochrome c is unstable after release into the cytoplasm (39). Therefore, a positive control should be used to ensure that the assay conditions can reliably detect cytosolic cytochrome c.

2.11. Clonogenic Cell Survival Assay

Although the non-clonogenic assays for cytotoxicity, cell viability, cell survival, and cell death described above, can measure the potency of a cytotoxic agent, these short-term assays can *underestimate* cytotoxicity in comparison with long-term assays for cell growth or cloning efficiency. Conversely, non-clonogenic cytotoxicity assays can sometimes *overestimate* cytotoxicity by not accounting for reversible damage or regrowth of cells resistant to the drug/cytotoxic agent. Usage of multiparameter, non-clonogenic, cytotoxicity assays can reduce these errors, but cannot eliminate them. For these reasons, it may be advisable to include clonogenic cell survival assays in studies of in vitro cytotoxicity of cell lines when feasible. The clonogenic cell survival assay (CSA) measures the long-term cytostatic effects of a drug/cytotoxic agent, by measuring the proliferative ability of a single cell to form a clone and produce a viable colony. In one early study, when compared to results from cell-labeling index, dye exclusion, and metabolic assays, the CSA gave the most reliable, dose-dependent index of cell lethality (40). These early observations can be explained by the finding that DNA damage correlated directly with reduced cloning efficiency and was associated with the appearance of apoptotic markers in certain tumor cell lines (41).

2.11.1. Applications of the Clonogenic Cell Survival Assay

The clonogenic cell survival assay is still widely used for testing and predicting cytotoxicity of anticancer drugs (42) although the proportion of primary tumors of a given type that can be successfully tested is limited by the same factors that lead to inefficient production of cell lines from many tumor types. For example, this assay helped explain the role of extracellular matrix proteins such as fibronectin, in tumor cell survival after irradiation (43). Recently, the clonogenic assay proved extremely reliable for differentiating degrees of in vitro toxicity of carbon-based nanoparticles between different tumor cell lines. Furthermore, it was possible to distinguish between effects on cell viability and cell proliferation by including colony size as an endpoint in the assay (44).

A recent study suggests that the clonogenic cell survival (CSA) assay is the "gold standard" because it measures the sum of all modes of cell death, and also accounts for delayed growth arrest (45). Indeed, the CSA is still widely considered the single most reliable in vitro cell line assay for measuring potency of a cytotoxic drug. This observation is supported by gene microarray studies which suggest that a drug's cytotoxic potency strongly correlates with its ability to reduce clonogenic potential. This maybe because a drug's cytotoxic potency depends on its ability to inhibit genes regulating cell-survival, the cell cycle, DNA replication/DNA repair, and oxygen levels (46). Understandably, many of these same genes would also regulate a cell's clonogenic potential, because they control a cell's long-term survival and proliferative potential after exposure to a drug/compound.

2.12. Notes on the Clonogenic Cell Survival Assay

The clonogenic cell survival assay does have limitations. Although it proved as accurate and sensitive as fluorescence based viability assays (47), the clonogenic cell survival assay lacks the dynamic range of newer fluorescent methods or the ATP assay (48). The conventional CSA also cannot measure impact of cell–cell communication on cell proliferation, because cells are plated at low densities to form colonies. In addition, this assay is not applicable when a substance decreases growth without inhibiting DNA synthesis and/or cell cycle progression. This assay is also inappropriate for testing agents, which inhibit growth solely by causing cytoskeletal damage (49), or by inducing apoptosis (50).

2.13. Choice of Non-clonogenic Versus Clonogenic Assays

Initial screens for measuring the cytotoxic effects of a chemical/drug usually employ an assay to measure acute loss of cell viability or cell survival in cell lines. Thus, the enzyme release assays (LDH/G6PD/GAPDH), and cell survival assays (MTT, NRU, and ATP) are useful for this purpose. If the test chemical/drug reproducibly decreases cell viability/survival, one can determine whether the compound induces cell death by performing specific assays for early and late apoptosis (as explained above). If the test compound does not alter the rate of cell survival/cell death, it is cytostatic rather than cytotoxic. One can then check if the chemical/drug decreases long-term cell proliferation by using the clonogenic cell survival assay. Results of the clonogenic assay can be confirmed by uptake of bromodeoxyuridine (BRDU) or radiolabelled-Thymidine, to quantitate cell proliferation. If data from these assays suggest that the test chemical/drug induces growth arrest, then fluorescence-activated cell sorting (FACS) analysis of cells with labeled DNA can be done to determine which phase of the cell cycle is arrested by the test compound.

Since a chemical can have both cytotoxic and cytostatic effects, it may be useful to run multiple non-clonogenic assays to measure short-term (acute) cytotoxicity, and the clonogenic cell survival assay to detect possible long-term/reversible growth arrest in the target cells exposed to the test compound.

2.14. High-Throughput Cytotoxicity Assays

Non-clonogenic cytotoxicity assays with single and multiple endpoints are now routinely done in high-throughput format for rapid screening of drug toxicity. This section discusses how HTS has revolutionized toxicity testing (51).

A common single parameter used in HTS of compounds, is the ATP cell viability assay. Besides measuring kinetics of toxicity caused by different agents, this assay could also detect species or cell type specific cytotoxicity (52). According to this study, the ATP assay can give data with qualitative and quantitative significance comparable to that obtained from animal studies if conducted in multiple cell lines with a dynamic dose range. Multiparameter, high throughput assays provide additional high quality and quantity of cytotoxicity data. The parameters most often measured in HTS multiplex cytotoxicity studies include viable cell number, nuclear, and mitochondrial changes. Thus, a HTS study measuring multiple intracellular end points for in vitro cytotoxicity across the therapeutic range of drug concentrations in HepG2 cells, could predict the human hepatotoxic potential of some of these drugs. Mutiparameter HTS also been used to reevaluate toxicity of drugs which lacked promise in conventional assays (53). Validated in vitro three-dimensional (3D) cell cultures of other human cell types (such as corneal, gingival, oral, and skin epithelium), are available and can provide valuable cytotoxicity data in high-throughput format.

2.14.1. Cell-Based Microfluidic Devices

The major technological advances in HTS involve the advent of microfluidic devices which have been developed for conducting analytical or biochemical processes on a very small scale. These microscale perfusion devices (also known as "lab-on-a-chip,") consist of microscope slide/credit card-sized units containing chambers interconnected by channels. Fluid flow through the chips is controlled by a micropump. The cell-based microfluidic devices are also described as "cell chips," "cell biochips," or "micro-bioreactors." These devices are the new tools for rapid screening for drug toxicity. A device may contain one cell type in one/more chambers, or different cell types in different chambers. Primary animal or human cells, or cell lines which grow in an adherent or non-adherent manner, can be used in microfluidic devices. Newer devices also permit cells to be cultured in 3D, stratified, multilayered, or aggregated cultures. Thus far, assay detection methods for cell-based high-throughput assays primarily involve electrochemical and optical detection methods (54).

2.14.2. Applications of Microfluidic Devices in Cytotoxicity Testing

Microscale perfusion devices have also been developed for measuring cytotoxicity in specific cell types. The first studies focused on hepatocytes. Thus, Sivaraman et al. (55) developed a microperfused system of 3D rat liver cells and demonstrated that this system retained functions similar to the in vivo tissue. In another study, drugs and their active metabolites were screened in miniaturized 3D arrays of hepatocytes, in order to determine the IC (50) values for nine compounds and their secondary metabolites (56). More recently, cell-based microfluidic devices have been developed for other cell types. Thus, a renal microchip was developed using the MDCK kidney cell line as an in vitro model for chronic toxicity testing of chemicals (57). Primary human keratinocytes (skin cells) have been used for cytotoxicity testing in a microfluidic device (58). Stem cells have also been used in perfused micro-bioreactors for toxicity testing. Thus Cui et al. (59) observed significant differences in the toxicity responses of human bone marrow cells cultured in 2-dimensional (2D) versus 3D formats and concluded that their 3D micro-bioreactor platform was "an efficient and standardized alternative testing method" for toxicity testing. Yang et al. (54) also concluded that 3D cell cultures are essential for obtaining cytotoxicity data which is comparable to the in vivo response.

Microfluidic bioreactors are being further modified in order to yield cytotoxicity data which is physiologically relevant. Thus, micro-bioreactors with miniaturized cultures and sensor technologies, permit real-time monitoring of cell viability and function with noninvasive detection methods (54). Importantly, such micro-bioreactors with continuous perfusion can be used to study *chronic* toxicity in long-term cultures. The next step is analysis of systemic toxicity by investigating interactions between different cell types. For example, Viravaidya et al. (60) developed monolayer cultures of liver and lung cell lines in separate chambers, connected by a common perfused fluid. This was done in order to attempt to replicate the toxic effects of a chemical on the human lung. In the experiment, liver cells were exposed to the chemical and the media containing the chemical and/or its metabolites, were transported by fluid flow to the lung cells wherein their toxic effect was assessed. In 2004, a novel in vitro system called the integrated discrete multiple organ cell culture (IdMOC) system, was developed to measure the comparative cytotoxicity of tamoxifen towards normal human cells (from five major organs) versus MCF-7 adenocarcinoma breast cancer cells (61). Similar devices using cell types from other organs can be set up to create an in vitro ADMET (absorption, distribution, metabolism, excretion/toxicity) assay system (60).

In summary, multiparameter, high-throughput, microfluidic devices using 3D human cell cultures exposed to therapeutically relevant concentrations of a drug, can provide highly reliable in vitro estimates of a drug's potency/cytotoxicity in humans. The advantages of the unique cell environment provided by microfluidic devices include the small volume of liquid within the system which more closely replicates the in vivo extracellular volume, the proximity of different types of cells, the continuous perfusion of media, and the controlled delivery and recirculation of reagents. However, the problems of cell shear stress, bubble formation, incompatibility with samples dissolved in organic solvents, and inter-assay reproducibility need to be addressed before these devices can be widely used for toxicity testing (62). Thus, these high-throughput methods also face challenges for assay validation and acceptance, similar to earlier cell based methods, and may not replace the use of animal based tests required by the regulatory authorities. Nevertheless, the improved predictions on drug potency and in vitro cytotoxicity data from studies using high-throughput microfluidic devices can be used to design better animal studies and/or studies which use fewer animals.

2.14.3. DNA Microarrays and Cytotoxicity Testing

In recent years, DNA microarrays have been used to evaluate expression of genes regulating drug metabolism and toxicity (63). Notably, oligonucleotide microarrays with up to 25,000 genes have examined the induction of genes regulating absorption, distribution, metabolism, and excretion (ADME) in tissue necropsies from animal models (64). With respect to expression of drug metabolizing enzymes, the dynamic range and sensitivity of DNA microarrays is comparable to northern blotting analysis and variability of the data is less than the inter-animal variability (65). The use of gene expression microarrays for assessment of the potency of cytotoxic drugs has already been explained (see Subheading 2.11.1). Gene expression microarrays have also been applied to measure differential cytotoxicity of closely related drugs. For e.g., a DNA microarray of 60 genes, could distinguish between the patterns of gene expression of two classes of retinoid synergists with different effects on apoptosis of HL60 cells. This in vitro data can be used to develop more effective and less toxic retinoids, which must then be confirmed in the appropriate in vivo models (66). More recently, it has been possible to show that primary cell cytotoxicity is predictable from the expression of genes involved in known resistance/sensitivity pathways, measured by polymerase chain reaction (PCR) arrays (67).

3. Conclusion

In conclusion, all of the assays described have their own advantages and disadvantages. If drug cytotoxicity is assayed with multiple parameters (e.g. nuclear damage, mitochondrial potential), then these data can give more valuable information on the mechanism of cell damage caused by the drug and its cytotoxic potency. New technologies are emerging, particularly in vitro HTS and cell-based microfluidic devices. Microfluidic devices are providing data, which more closely approximate the in vivo cytotoxic potency of a test compound and helps to narrow the gap between in vitro and in vivo data on drug toxicity. Finally, for a more comprehensive analysis of cytotoxicity, the data on cytotoxic potency of a compound (from non-clonogenic and the clonogenic cell survival assays in appropriate target cells), can be correlated with gene expression data to obtain information on genes regulating crucial genes in signaling pathways regulating cell survival, cell proliferation, drug bioavailability (i.e. ADME), and drug sensitivity/resistance.

References

1. Fellows, M.D., and O'Donovan, M.R. (2007) Cytotoxicity in cultured mammalian cells is a function of the method used to estimate it. *Mutagenesis* **22**, 275–280.
2. Decker, T., and Lohmann-Matthes, M.L. (1988) A quick and simple method for the quantitation of lactate dehydrogenase release in measurements of cellular cytotoxicity and tumor necrosis factor (TNF) activity. *J. Immunol. Methods* **115**, 61–69.
3. Korzeniewski, C., and Callewaert, D.M. (1983) An enzyme-release assay for natural cytotoxicity. *J. Immunol. Methods* **64**, 313–320.
4. Batchelor, R.H., and Zhou, M. (2004) Use of cellular glucose-6-phosphate dehydrogenase for cell quantitation, applications in cytotoxicity and apoptosis assays. *Anal. Biochem.* **329**, 35–42.
5. Corey M.J., and Kinders, R.J. Methods and compositions for coupled luminescent assays. United States Patent 6,811,990, issued November 2, 2004.
6. Corey, M.J., et al, (1997) A Very Sensitive Coupled Luminescent Assay for Cytoxicity and Complement-Mediated Lysis. *J. Immunol. Methods* **207**, 43–51.
7. Tolnai, S.A. (1975) A method for viable cell count. *Meth. Cell Science* **1**, 37–38.
8. Ahmed, S.A., Gogal, R.M., Jr, and Walsh, J.E. (1994) A new rapid and simple non-radioactive assay to monitor and determine the proliferation of lymphocytes: an alternative to [3H]thymidine incorporation assay. *J. Immunol. Methods.* **170**, 211–24.
9. Perez, R.P., Godwin, A.K., Handel, L.M., and Hamilton, T.C. (1993) A comparison of clonogenic, microtetrazolium and sulforhodamine B assays for determination of cisplatin cytotoxicity in human ovarian carcinoma cell lines. *Eur. J. Cancer* **29A**, 395–399.
10. Skehan, P., Storeng, R., Scudiero, D., Monks, A., McMahon, J., Vistica, D., Warren J.T., Bokesch, H., Kenney, S., and Boyd, M.R. (1990) A New colorimetric cytotoxicity assay for anticancer-drug screening. *J. Natl. Cancer Inst.* **82**, 1107–1112.
11. Borenfreund, E., and Puerner, J.A. Toxicity determined in vitro by morphological alterations and neutral red absorption. (1985) *Toxicol. Lett.* **24**, 119–124.
12. Lu, B., Kerepesi, L., Wisse, L., Hitchman, K., and Meng, Q.R. (2007) Cytotoxicity and gene expression profiles in cell cultures exposed to whole smoke from three types of cigarettes. *Toxicol. Sci.* **98**, 469–478.
13. Lasarow, R.M., Isseroff, R.R., and Gomez, E.C. (1992) Quantitative in vitro assessment of phototoxicity by a fibroblast-neutral red assay. *J. Invest. Dermatol.* **98**, 725–729.
14. Xingfen, Y., Wengai, Z., Ying, Y., Xikun, X., Xiaoping, X., and Xiaohua, T. (2007) Preliminary study on neutral red uptake assay as an alternative method for eye irritation test. AATEX 14, Special Issue, 509–514. Proc. 6th World Congress on Alternatives & Animal Use in the Life Sciences August 21–25, Tokyo, Japan
15. Putnam, K.P., Bombick, D.W., and Doolittle, D.J. (2002) Evaluation of eight in vitro assays for assessing the cytotoxicity of cigarette smoke condensate. *Toxicol. In Vitro.* **16**, 599–607.
16. Sperandio, S., Poksay, K., de Belle, I., Lafuente, M.J., Liu, B., Nasir, J., and Bredesen, D.E. (2004) Paraptosis: mediation by MAP kinases and inhibition by AIP-1/Alix. *Cell Death Differ.* **11**, 1066–1075.
17. Roche Applied Science: Apoptosis, Cell Death, and Cell Proliferation Manual: 3rd edition, page 59.
18. Mosmann, T. (1983) Rapid colorimetric assay for cellular growth and survival: application to proliferation and cytotoxicity assays. *J Immunol. Methods,* **65**, 55–63.

19. Scudiero, D.A., Shoemaker, R.H., Paul, K.D., Monks, A., Tierney, S., Nofziger, T.H., Currens, M.J., Seniff, D., and Boyd, M.R. (1988) Evaluation of a soluble tetrazolium/formazan assay for cell growth and drug sensitivity in culture using human and other tumor cell lines. *Cancer Res.* **48**, 4827–4833.

20. Crouch, S.P., Kozlowski, R., Slater, K.J., and Fletcher, J. (1993) The use of ATP bioluminescence as a measure of cell proliferation and cytotoxicity. *J. Immunol. Methods.* **160**, 81–88.

21. Petty, R.D., Sutherland, L.A., Hunter, E.M., and Cree, I.A. (1995) Comparison of MTT and ATP-based assays for the measurement of viable cell number. *J. Biolumin. Chemilumin.* **10**, 29–34.

22. Ulukaya, E., Ozdikicioglu, F., Oral, A.Y., and Demirci, M. (2008) The MTT assay yields a relatively lower result of growth inhibition than the ATP assay depending on the chemotherapeutic drugs tested. *Toxicol. In Vitro* **22**, 232–239.

23. Ng, T.Y., Ngan, H.Y., Cheng, D.K., and Wong, L.C. (2000) Clinical applicability of the ATP cell viability assay as a predictor of chemoresponse in platinum-resistant epithelial ovarian cancer using nonsurgical tumor cell samples. *Gynecol. Oncol.* **76**, 405–408.

24. Tam, K.F., Ng, T.Y., Liu, S.S., Tsang, P.C.K., Kwong, P.W.K., and Ngan, H.Y.S. (2005) Potential application of the ATP cell viability assay in the measurement of intrinsic radiosensitivity in cervical cancer. *Gynecol. Oncol.* **96**, 765–770.

25. Roche Applied Science: Apoptosis, Cell Death, and Cell Proliferation Manual: 3rd edition, page 98.

26. Hynes, J., Hill, R., and Papkovsky, D.B. (2006) The use of a fluorescence-based oxygen uptake assay in the analysis of cytotoxicity. *Toxicol. In Vitro* **5**, 785–792.

27. Lindhagen, E., Nygren, P., and Larsson, R. (2008) The fluorometric microculture cytotoxicity assay. *Nat. Protoc.* **3**, 1364–1369.

28. Mickuviene, I., Kirveliene, V., and Juodka, B. (2004) Experimental survey of non-clonogenic viability assays for adherent cells in vitro. *Toxicol. In Vitro* **18**, 639–648.

29. Ivanova, L., and Uhlig, S. (2008) A bioassay for the simultaneous measurement of metabolic activity, membrane integrity, and lysosomal activity in cell cultures. *Anal. Biochem.* **379**, 16–19.

30. Elmore, S. (2007) Apoptosis: A Review of Programmed Cell Death. *Toxicologic Pathology* **35**, 495–516.

31. Gurtu, V., Kain, S.R., and Zhang, G. (1997) Fluorometric and colorimetric detection of caspase activity associated with apoptosis. *Anal Biochem*, **251**, 98–102.

32. Grabarek, J., Amstad, P., and Darzynkiewicz, Z. (2002). Use of fluorescently labeled caspase inhibitors as affinity labels to detect activated caspases. *Hum Cell*, **15**, 1–12.

33. Bossy-Wetzel, E., and Green, D.R. (2000) Detection of apoptosis by Annexin V labeling. *Methods Enzymol*, **322**, 15–18.

34. Wyllie, A.H. (1980) Glucocorticoid-induced thymocyte apoptosis is associated with endogenous endonuclease activation. *Nature* **284**, 555–556.

35. Rubin, R.L., Joslin, F.G., and Tan, E.M. (1983) An improved ELISA for anti-native DNA by elimination of interference by anti-histone antibodies. *J. Immunol. Methods* **63**, 359–366.

36. Kressel, M., and Groscurth, P. (1994) Distinction of apoptotic and necrotic cell death by in situ labelling of fragmented DNA. *Cell Tissue Res*, **278**, 549–556.

37. Poot, M., and Pierce, R.H. (1999) Detection of changes in mitochondrial function during apoptosis by simultaneous staining with multiple fluorescent dyes and correlated multiparameter flow cytometry. *Cytometry*, **35**, 311–317.

38. Scorrano, L., Ashiya, M., Buttle, K., Weiler, S., Oakes, S.A., Mannella, C.A., and Korsmeyer, S.J. (2002) A distinct pathway remodels mitochondrial cristae and mobilizes cytochrome c during apoptosis. *Dev. Cell*, **2**, 55–59.

39. Goldstein, J.C., Waterhouse, N.J., Juin, P., Evan, G.I., and Green, D.R. (2000). The coordinate release of cytochrome *c* during apoptosis is rapid, complete and kinetically invariant. *Nat. Cell Biol.* **2**, 156–162.

40. Roper, P.R., and Drewinko, B. (1976) Comparison of in vitro methods to determine drug-induced cell lethality. *Cancer Res*, **36** (7 PT 1), 2182–2188.

41. Jarvis, W.D., Fornari, F.A., Traylor, R.S., Martin, H.A., Kramer, L.B., Erukulla, R.K., Bittman, R., and Grant, S. (1996) Induction of apoptosis and potentiation of ceramide-mediated cytotoxicity by sphingoid bases in human myeloid leukemia cells. *J. Biol. Chem.* **271**, 8275–8284.

42. Yalkinoglu, A.O., Schlehofer, J.R., and zur Hausen, H. (1990). Inhibition of N-methyl-N'-nitro-N-nitrosoguanidine-induced methotrexate and adriamycin resistance in CHO cells by adeno-associated virus type 2. *Int. J. Cancer.* **45**, 1195–1203.

43. Cordes, N., and Meineke, V. (2003) Cell adhesion-mediated radioresistance (CAM-RR): Extracellular matrix-dependent improvement of cell survival in human tumor and normal cells in vitro. *Strahlenther. Onkol.* **179**, 337–344.

44. Herzog, E., Casey, A., Lyng, F.M., Chambers, G., Byrne, H.J., and Davoren, M. (2007) A new approach to the toxicity testing of carbon-based nanomaterials: the clonogenic assay. *Toxicol. Lett.* **174**, 49–60.

45. Mirzayans, R., Andrais, B., Scott, A., Tessier, A., and Murray, D. (2007) A sensitive assay for the evaluation of cytotoxicity and its pharmacologic modulation in human solid tumor-derived cell lines exposed to cancer-therapeutic agents. *J. Pharm. Pharm. Sci.* **10**, 298s–311s.

46. Stroncek, D.F., Jin, P., Wang, E., and Jett, B. (2007) Potency analysis of cellular therapies: the emerging role of molecular assays. *J. Transl. Med.* 5, 24–29.

47. Duerst, R.E., and Frantz, C.N. (1985) A sensitive assay of cytotoxicity applicable to mixed cell populations. *J. Immunol. Methods* **82**, 39–46.

48. Frgala, T., Kalous, O., Proffitt, R.T., and Reynolds, C.P. (2007) A fluorescence microplate cytotoxicity assay with a 4-log dynamic range that identifies synergistic drug combinations. *Mol. Cancer Ther.* **6**, 886–897.

49. Aragon, V., Chao, K., and Dreyfus, L.A. (1997) Effect of cytolethal distending toxin on F-actin assembly and cell division in Chinese hamster ovary cells. *Infect Immun.* **65**, 3774–3780.

50. Sacks, P.G., Harris, D., and Chou, T.C. (1995) Modulation of growth and proliferation in squamous cell carcinoma by retinoic acid: a rationale for combination therapy with chemotherapeutic agents. *Int. J. Cancer* **61**, 409–415.

51. Schoonen, W.G., Walter, M.A., Westerink, W.M., and Horbach, G.J. (2009) High-throughput screening for analysis of *in vitro* toxicity. *Mol. Clin. Environ. Toxicol.* **99**, 401–452.

52. Xia, M., Huang. R., Witt, K.L., Southall, N., Fostel, J., Cho, M.H., Jadhav, A., Smith, C.S., Inglese, J., Portier, C.J., Tice, R.R., and Austin, C.P. (2008) Compound cytotoxicity profiling using quantitative high-throughput screening. *Environ. Health Perspect.* **116**, 284–291.

53. O'Brien, P.J., Irwin, W., Diaz, D., Howard-Cofield, E., Krejsa, C.M., Slaughter, M.R., Gao, B., Kaludercic, N., Angeline, A., Bernardi, P., Brain, P., and Hougham, C. (2006) High concordance of drug-induced human hepatotoxicity with in vitro cytotoxicity measured in a novel cell-based model using high content screening. *Arch. Toxicol*, **80**, 580–604.

54. Yang, S.T., Zhang, X., and Wen, Y. (2008) Microbioreactors for high-throughput cytotoxicity assays. *Curr. Opin. Drug. Discov. Devel.* **1**, 111–127.

55. Sivaraman, A., Leach, J.K., and Townsend, S., Hogan, B.J., Stolz, D.B., Fry, R., Samson, L.D., Tannenbaum, S.R., and Griffith, L.G. (2005) A microscale in vitro physiological model of the liver: predictive screens for drug metabolism and enzyme induction. *Curr. Drug Metab.* **6**, 569–591.

56. Lee, M.Y., Kumar, R.A., Sukumaran, S.M., Hogg, M.G., Clark, D.S., and Dordick, J.S. (2008) Three-dimensional cellular microarray for high-throughput toxicology assays. *Proc. Natl. Acad. Sci. U. S. A.* **105**, 59–63.

57. Baudoin, R., Griscom, L., and Monge, M., et al. (2007) Development of a renal microchip for in vitro distal tubule models. *Biotechnol. Prog.* **23**, 1245–1253.

58. Walker, G.M., Monteiro-Riviere, N., Rouse, J. and O'Neill, A.T. (2007) A linear dilution microfluidic device for cytotoxicity assays. *Lab Chip.* **7**, 226–232.

59. Cui, Z. F., Xu, X., Trainor, N., Triffitt, J. T., Urban, J. P. G., and Tirlapur, U. K. (2007) Application of multiple parallel perfused microbioreactors and three-dimensional stem cell culture for toxicity testing. *Toxicol. In Vitro.* **21**, 1318–1324.

60. Viravaidya, K., Sin, A. and Shuler, M.L. (2004) Development of a microscale cell culture analog to probe naphthalene toxicity. *Biotechnol. Prog.* **20**, 316–323.

61. Li, A.P., Bode, C. and Sakai, Y. (2004) A novel in vitro system, the integrated discrete multiple organ cell culture (IdMOC) system, for the evaluation of human drug toxicity: comparative cytotoxicity of tamoxifen towards normal human cells from five major organs and MCF-7 adenocarcinoma breast cancer cells. *Chem. Biol. Interact.* **150**, 129–136.

62. Anderson, E.J. and Knothe-Tate, M.L. (2007) Open access to novel dual flow chamber technology for in vitro cell mechanotransduction, toxicity and pharmacokinetic studies. *BioMedical Engineering Online*, **6**, 46–57.

63. Gerhold, D., Lu, M., Xu, J., Austin, C., Caskey, C.T., and Rushmore, T. (2001) Monitoring expression of genes involved in drug metabolism and toxicology using

DNA microarrays. *Physiol. Genomics.* **5**, 161–170.

64. Slatter, J.G., Cheng, O., Cornwell, P.D., de Souza, A., Rockett, J., Rushmore, T., Hartley, D., Evers, R., He, Y., Dai, X., Hu, R., Caguyong, M., Roberts, C.J., Castle, J., and Ulrich, R.G. (2006) Microarray-based compendium of hepatic gene expression profiles for prototypical ADME gene-inducing compounds in rats and mice in vivo. *Xenobiotica* **36**, 902–937.

65. Bartosiewicz, M., Trounstine, M., Barker, D., Johnston, R., and Buckpitt, A. (2000) Development of a toxicological gene array and quantitative assessment of this technology. *Arch. Biochem. Biophys.* **376**, 66–73.

66. Ishida, S., Shigemoto-Mogami, Y., Kagechika. H., Shudo, K., Ozawa, S., Sawada, J., Ohno, Y., and Inoue, K. (2003) Clinical potential of subclasses of retinoid synergists revealed by gene expression profiling. *Mol. Cancer. Ther.* **2**, 49–58.

67. Glaysher, S., Yiannakis, D., Gabriel, F.G., Johnson, P., Polak, M.E., Knight, L.A., Goldthorpe, Z., Peregrin, K., Gyi, M., Modi, P., Rahamim, J., Smith, M.E., Amer, K., Addis, B., Poole, M., Narayanan, A., Gulliford, T.J., Andreotti, P.E., and Cree, I.A. (2009) Resistance gene expression determines the in vitro chemosensitivity of non-small cell lung cancer (NSCLC). BMC Cancer **9**, 300.

Chapter 20

Cell Sensitivity Assays: The MTT Assay

Johan van Meerloo, Gertjan J.L. Kaspers, and Jacqueline Cloos

Abstract

The MTT (3-[4,5-dimethylthiazol-2-yl]-2,5 diphenyl tetrazolium bromide) assay is based on the conversion of MTT into formazan crystals by living cells, which determines mitochondrial activity. Since for most cell populations the total mitochondrial activity is related to the number of viable cells, this assay is broadly used to measure the in vitro cytotoxic effects of drugs on cell lines or primary patient cells. In this chapter the protocol of the assay is described including important considerations relevant for each step of the assay as well as its limitations and possible applications.

Key words: MTT, 3-[4,5-Dimethylthiazol-2-yl]-2,5 diphenyl tetrazolium bromide, Viability assay, IC_{50}, LC_{50}, Drug sensitivity assay, Cytotoxicity assay

---

1. Introduction

The general purpose of the MTT (3-[4,5-dimethylthiazol-2-yl]-2,5 diphenyl tetrazolium bromide) assay is to measure viable cells in relatively high throughput (96-well plates) without the need for elaborate cell counting. Therefore the most common use is to determine cytotoxicity of several drugs at different concentrations. The principle of the MTT assay is that for most viable cells mitochondrial activity is constant and thereby an increase or decrease in the number of viable cells is linearly related to mitochondrial activity. The mitochondrial activity of the cells is reflected by the conversion of the tetrazolium salt MTT into formazan crystals, which can be solubilised for homogenous measurement. Thus, any increase or decrease in viable cell number can be detected by measuring formazan concentration reflected in optical density (OD) using a plate reader at 540 and 720 nm. For drug sensitivity measurements the OD values of wells with cells incubated with drugs are compared to the OD of wells with cells not exposed to drugs.

The MTT assay is suitable for the measurement of drug sensitivity in established cell lines as well as primary cells. For dividing cells (usually cell lines) the decrease in cell number reflects cell growth inhibition and the drug sensitivity is then usually specified as the concentration of the drug that is required to achieve 50% growth inhibition as compared to the growth of the untreated control (50% inhibitory concentration, IC_{50}). For primary (nondividing) cells, drug sensitivity is measured as enhanced cell kill of treated cells as compared to the loss of cells already commonly seen in untreated cells (50% lethal concentration, LC_{50}). For some cell types such as fresh acute myeloid leukemia (AML) cells the control median cell survival is 100%, which is no problem as long as results for treated cells are compared to the controls.

For primary pediatric acute lymphoblastic leukemia (ALL) cells, the in vitro MTT assay has been extensively applied to predict drug sensitivity in vivo. In these studies, a relatively good correlation was observed between in vitro sensitivity and clinical outcome (1–4). Despite these and other positive results, the MTT assay is not commonly used as predictive test for the clinic, but rather as an in vitro tool to determine potential antitumor activity of new drugs and/or combinations of drugs (5). In addition, the MTT assay is a suitable tool to study resistance to drugs (6, 7) (see Table 1 for applications).

In order to set up the MTT assay for cells and/or drugs that have not been tested before, several considerations are of crucial importance and will be discussed in this chapter.

Table 1
Possible applications for MTT assay

Application	References
Study of cross-resistance between related and unrelated drugs	(2, 3, 5–7)
Measuring drug sensitivity ex vivo to predict clinical outcome	(1–4, 7, 8)
Sensitivity testing of new drugs	(4–6)
Drug screening on cell lines and/or patient samples	(3–5, 7, 8)
Testing drug combinations on cell lines and/or patient material	(1, 2)

2. Materials

2.1. Cells and Controls

Class 2B biocabinet suitable for drug experiments. Incubator with 5% CO_2 at 37°C. Microplates – 96 well (Greiner Bio-one, Alphen a/d Rijn, The Netherlands). For suspension cells, we use round bottom wells or flat bottom wells, while for attached cells only flat bottoms can be used (see Subheading 3.1 for the different loading volume per plate) (see Note 1).

2.2. Solutions and Solvents

2.2.1. MTT Solution

Dissolve 500 mg MTT powder (Sigma, St. Louis, USA) in 10 mL PBS. Stir with a magnetic stirrer for approximately 1 h in the dark. Filter sterilize the solution with a 0.22 μm filter (Millipore, Carrigtwohill, Ireland) and store in 10-mL aliquots at −20°C.

Warning: MTT is toxic and harmful. MTT is light sensitive, hence protect from light.

2.2.2. Acidified Isopropanol

To dissolve the formazan crystals, different solutions can be used such as methanol, ethanol, and DMSO. Our lab has the best experience with acidified isopropanol. To make this solution, add 50 mL 2 M HCl to 2.5 L isopropanol. Store the solution at least a month at room temperature before use. When the isopropanol is not acidified correctly the suspension will become cloudy.

2.3. Equipment

1. Plateshaker (IKA schuttler MST4, Janke & Kunkerel, Staufen, Germany).
2. Pipettes 0.001–1 mL, single channel and 0.01–0.3, multichannel.
3. Class 2B hood.
4. Benchtop centrifuge.
5. Microplate reader (Anthos-Elisa-reader 2001, Labtec, Heerhugowaard, The Netherlands).
6. O_2 incubator.

3. Methods

3.1. Plate Setup

Generally cells are plated in triplicates to minimize the variability of the results. The volume of cells depends on the type of plate used. For round bottom and flat bottom plates, 80 μL and 120 μL are used, respectively. Each plate should contain control wells (without drugs) and blank wells (without cells). For some drugs that also show absorbance at the given wavelenghts, an additional control is required of wells with medium (without cells) including the range of drug used. The number of plates needed depends on the specific experiment. A common MTT assay experiment requires a testing plate for the OD of the drug, a testing plate to determine the growth curve for the starting amount of cells seeded per well, and a broad dilution range to determine the dilution range for the experiments. However, for a standardized drug sensitivity assay of patient material, a single plate often suffices. For cell lines, it is recommended to include a day 0 plate in order to accurately determine the extent of growth (in the course of the experiment for the control cells). The outer wells are not used for the experiment due to evaporation and are filled with phosphate buffered saline (PBS) to keep the evaporation of the plate to the minimum, some drugs can also have influence on neighbouring wells; this has to be investigated before starting the experiment (see Note 2).

3.2. Drug Incubation

There are two methods of drug exposure (1); if the drug is unstable or cannot be stored in medium at $-20°C$, the plates have to be prepared fresh by making the drug dilution stocks and adding the total amount of drugs in 20 µL for round bottom plates and 30 µL of drug solution to flat bottom wells to each well just before adding the cells. This renders the total volume of cell and drug suspension to 100 µL for round bottom and 150 µL for flat bottom 96-well plates. (2) For testing a stable drug, plates can be prepared with the 30 µL drug concentrations and can be stored in $-20°C$ for later use. When the cells are added to the wells containing the drug suspension, this will mix the drugs through the cells (see Note 2).

3.3. Culture Period

For primary leukemic cells, the plates are incubated for 4 days to determine the optimal effect for most standard drugs. For cell lines, the plates have to be incubated during log phase, which is usually 72–96 h in an incubator with 5% CO_2 at $37°C$ (see Note 1).

3.4. MTT Incubation

After the appropriate incubation time, add 1:10 volume of MTT solution (5 mg/mL); e.g. 10 µL for round bottom and 15 µL for flat bottom 96-well plates. Unused MTT can be frozen and reused. Shake plates for 5 min on a plateshaker by slowly increasing the shaking speed to a maximum of 900 shakes/min. Then incubate the plate for another 4–6 h at $37°C$ in a CO_2 incubator, depending on the cell type.

3.5. Dissolution of Formazan Crystals

150 µL of acidified isopropanol is added to each well and resuspended until all crystals have been dissolved. Mix each well thoroughly using a multichannel pipette. Between each row, rinse tips of multichannel pipette with isopropanol and discard the used isopropanol. Blow out tips thoroughly before mixing the next rows. Start with the control wells, before mixing the rows with drugs.

3.6. OD Measurement

The OD is measured at 540 and 720 nm in order to get a more exact measurement by correcting for background noise. The 720 nm OD background will be subtracted from the 540 nm OD total signal. The measured data are copied into an excel sheet and with the use of the following formula, the percentage of living cells can be determined: The average OD of the blank control wells (without cells and if the drug has no specific OD without drug as well) is subtracted from the average OD of the control wells (cells but no drugs) and the wells containing the drugs. Some drugs may interfere with the OD measurement and then the average OD of the wells containing drugs, but without cells is subtracted from the average OD of the control wells. The leukemic cell survival is calculated by: (OD treated well [–blank])/(mean OD control well [–blank]) × 100. The LC_{50} (the drug concentration which results in 50% leukemic cell survival) can be calculated. For more reliable results the experiment should be done in triplicate and in duplicate in case of primary cells and when a range of drug concentrations is being tested.

3.7. Example Results

After adding acidified isopropanol and dissolving the formazan crystals thoroughly the plate has to rest for 10 min before measuring. In Fig. 1, a plate is shown after dissolving the formazan crystals with acidified isopropanol showing the blank culture medium wells, untreated cell control wells, and three drugs for which the cells are either sensitive (C), intermediate sensitive (E), and resistant (D) in these dose ranges. To be able

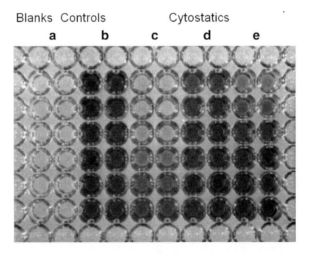

Fig. 1. A 96-well plate after formazan crystals are dissolved in acidified isopropanol. (**a**) Blanks control wells, (**b**) untreated cell control wells, (**c**) cell line with drug C with a dose–response curve from 100% cell death to no response on cell growth, (**d**) cell line with drug D with a dose–response curve showing no growth inhibition, (**e**) cell line with drug E with a dose–response curve with dose-dependent modest growth inhibition at high drug concentrations. Outer wells are not used because of possible evaporation.

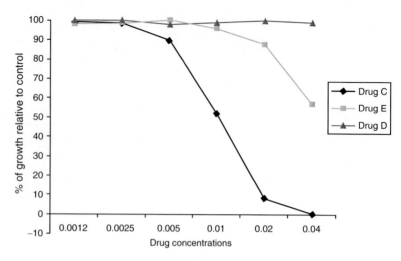

Fig. 2. Dose–response curve of cell lines C, D, and E. (C) cell line with drug C with a dose–response curve from control growth without drugs (set at 100%) to complete cell kill at the highest drug concentrations, (D) cell line with drug D with a dose–response curve showing no growth inhibition, (E) cell line with drug E with a dose–response curve showing dose-dependent modest growth inhibition (IC_{50} is not reached).

to determine IC_{50} values of drugs D and E, different dose ranges are required.

When measured in a plate reader, these measurements can then be used to determine the IC_{50} for cell lines and LC_{50} for patient material. The results can be plotted into a graph as shown in Fig. 2. The wells containing untreated cells are set as 100% and the dose–response curves of the three different drugs are depicted. It is clearly shown in the picture that for drugs D and E, the IC_{50} is not reached.

4. Notes

1. The cell concentration that is plated may vary for different cell lines and primary cells. For leukemic cell lines, we use concentrations in the range of $3.0–4.0 \times 10^3$/well, while primary AML cells are plated in $0.08–0.12 \times 10^6$ cells/well and primary ALL cells in 0.16×10^6 cells/well. For cells that have not been used before in this assay, several cell concentration experiments have to be done before starting the drug experiments as recommended below.

2. Cells are plated in duplicate or triplicate, and in each plate, 4–6 control wells are included that contain the same amount of cells as the experimental wells and which will not be exposed to drugs. Additional 4–6 wells are needed for

measuring the blanks, which contain the culture medium only. Some drugs can have an OD value of their own that can influence the measurement when diluted, and therefore a dilution of the drugs has to be measured without the adding cells. If there is a high difference in OD between the different drug dilutions, extra drug dilution control wells have to be added to the plate setup. The absorbance value for the blanks should be between 0.001 and 0.1 OD units as measured on a Microplate reader using filters for 540 and 720 nm. In addition, the absorbance range for untreated cells should typically be between 0.75 and 1.25 OD units (see Notes 3 and 4).

3. To determine whether cells are suitable for use in the MTT assay, several parameters need to be verified:

   (a) *Linearity of viable cell number with OD level.* To determine this, several cell concentrations are plated and related to OD values. For cell types where the mitochondrial activity of cells is not constant, the MTT assay cannot be used in to measure the influence of the drugs.

   (b) *Cell concentrations.* The optimal cell concentration to be plated is dependent on the basic level of mitochondrial activity and the rate of proliferation. In order to establish this, several concentrations of cells should be plated in, for instance, seven plates, and measured daily to determine the growth curve of the cell line to prevent overgrowth, which will influence the experiment. The starting OD value of day 0 should not exceed 0.125. By constructing a growth curve, the logarithmic cell phase is determined at which the cells duplicate. At a certain time point the cell growth will plateau due to exhaustion of the medium, contact inhibition, and exceeding of the maximal OD value that can accurately be measured. The most optimal concentration of plating is when cells have almost no lag phase and the assay should not proceed after the log phase.

   (c) *Cell culture.* Cells should be cultured for a relevant time period to be able to demonstrate the effect of the drug. For cell lines this should only be during the log phase, while for primary cells the MTT assay has to be completed before all untreated cells are dead. Only for cell lines, a day 0 plate is used to precisely measure the activity of the starting cell dilution at day 0 without the drugs' effects. Moreover, when using primary leukemic cells, in this assay, the blast count at day 4 should be $\geq$70% leukemic cells; this will be a blast count of $\geq$80–90% for ALL (7, 8) and $\geq$70–75% for AML at day 0. Samples can be purified with immunomagnetic beads directed against contaminating nonmalignant cells (8).

4. Drug incubations.

    (a) *Control medium.* In order to account for possible influences of the dissolvent of the drugs on the background OD, the control medium should contain the concentrations of dissolvent of drugs.

    (b) *Stability of the drugs.* To determine if cytostatic drugs have to be added fresh or if the drugs can be stored in plates at −20°C, drugs should be added into the 96-well plates and stored at −20°C in advance of the experiments. Possible differences in the dose–response curves will imply whether the plates with drugs can be stored or that the drugs need to be added fresh at the day of plating.

    (c) *Drug combinations* of drug A + drug B can be measured in several ways of which two are the most common (1) One fixed concentration of drug A can be used (usually in a concentration between the IC_{20} and IC_{50}) in combination with a dose range of drug B. (2) Mix both drugs at IC_{50} as highest dose and then make a dose–response curve by diluting this mixture further to obtain a combined dose–response curve. Synergism or antagonism can then be investigated, for instance, by the use of "calcusyn" software (Biosoft, Cambridge, UK) which can deal with both methods.

    (d) *Reproducibility* is important for an accurate measurement and so the use of at least a duplicate row setup of each drug, and if possible a triplicate row setup is preferred within each plate. For cell lines we always perform the assay in triplicate while for primary material this is commonly not possible.

    (e) *Plate set-up.* It is our experience that some drugs can affect surrounding cells possibly by aerosols; therefore, the wells with drugs have to be separated by PBS to make sure that the drugs will not affect the adjacent wells. The plate set-up also needs to include 4–6 control wells with the used culture medium, which are used to measure the background of culture medium.

Acknowledgments

This work was supported by KiKa "Stichting Kinderen Kankervrij" – Dutch Children Cancer-free foundation and VONK "VUmc Onderzoek Naar Kinderkanker" foundation.

References

1. Kaspers, G.J., Pieters, R., Van Zantwijk, CH., Van Wering, E.R., Van Der Does-Van Den Berg, A., Veerman, A.J. (1998) Prednisolone resistance in childhood acute lymphoblastic leukemia: vitro-vivo correlations and cross-resistance to other drugs. *Blood.* **92**, 259–266.

2. Salomons, G.S., Smets, L.A., Verwijs-Janssen, M., Hart, A.A., Haarman, E.G., Kaspers, G.J., Wering, E.V., Der Does-Van Den Berg, A.V., Kamps, W.A. (1999) Bcl-2 family members in childhood acute lymphoblastic leukemia: relationships with features at presentation, in vitro and in vivo drug response and long-term clinical outcome. *Leukemia.* **13**, 1574–1580.

3. Kaspers G.J., Veerman A.J., Pieters R, Van Zantwijk CH., Smets L.A., Van Wering E.R., Van Der Does-Van Den Berg A. (1997) In vitro cellular drug resistance and prognosis in newly diagnosed childhood acute lymphoblastic leukemia. *Blood.* **90**, 2723–2729.

4. Kaspers, G.J., Wijnands, J.J., Hartmann, R., Huismans, L., Loonen, A.H., Stackelberg, A., Henze, G., Pieters, R., Hählen, K., Van Wering, E.R., Veerman, A.J. (2005) Immunophenotypic cell lineage and in vitro cellular drug resistance in childhood relapsed acute lymphoblastic leukaemia. *Eur J Cancer.* **41**, 1300–1303.

5. Hubeek, I., Peters, G.J., Broekhuizen, R., Zwaan, C.M., Kaaijk, P., van Wering, E.S., Gibson, B.E., Creutzig, U., Janka-Schaub, G.E., den Boer, M.L., Pieters, R., Kaspers, G.J. (2006) In vitro sensitivity and cross-resistance to deoxynucleoside analogs in childhood acute leukemia. *Haematologica.* **91**, 17–23.

6. Oerlemans, R., Franke, N.E., Assaraf, Y.G., Cloos, J., van Zantwijk, I., Berkers, C.R., Scheffer, G.L., Debipersad, K., Vojtekova, K., Lemos, C., van der Heijden, J.W., Ylstra, B., Peters, G.J., Kaspers, G.J., Dijkmans, B.A., Scheper, R.J., Jansen, G. (2008) Molecular basis of bortezomib resistance: proteasome subunit beta5 (PSMB5) gene mutation and overexpression of PSMB5 protein. *Blood.* **112**, 2489–2499.

7. Klumper E., Pieters R., Veerman A.J., Huismans D.R., Loonen A.H., Hählen K., Kaspers G.J., van Wering E.R., Hartmann R., Henze G. (1995) In vitro cellular drug resistance in children with relapsed/refractory acute lymphoblastic leukemia. *Blood.* **86**, 3861–3868.

8. Kaspers G.J., Veerman A.J., Pieters R, Broekema G.J., Huismans D.R., Kazemier K.M., Loonen A.H., Rottier M.A., van Zantwijk CH., Hählen K. (1994) Mononuclear cells contaminating acute lymphoblastic leukaemic samples tested for cellular drug resistance using the methyl-thiazol-tetrazolium assay. *Br J Cancer.* **70**, 1047–1052.

Chapter 21

Cell Sensitivity Assays: The ATP-based Tumor Chemosensitivity Assay

Sharon Glaysher and Ian A. Cree

Abstract

The ATP-based tumor chemosensitivity assay (ATP–TCA) is a standardised system which can be adapted to a variety of uses with both cell lines and primary cell cultures. It has a strong track record in drug development, mechanistic studies of chemoresistance and as an aid to clinical decision-making. The method starts with the extraction of cells in suspension from continuous cell culture (for cell lines), malignant effusions or biopsy material. Enzymatic digestion is used to obtain cells from tumor tissue. The assay uses a serum-free medium and polypropylene plates to prevent the growth of non-neoplastic cells over a 6-day incubation period followed by detergent-based extraction of cellular ATP for measurement by luciferin–luciferase assay in a luminometer. The assay results are usually shown as percentage inhibition at each concentration tested, and can be used with suitable software to examine synergy between different anticancer agents.

Key words: ATP, Combination, Cell line, Chemosensitivity, Chemoresistance, Primary cell culture, Selective medium

1. Introduction

Many types of tumor cell survival assays have been developed for assessing the sensitivity or resistance to drug treatment. Nearly all use the same basic method involving tumor cell isolation, drug exposure and assessment of cell survival. Their differences lie in sample processing and in the techniques used to measure the assay endpoint, as reviewed in Chapter 19 of this volume.

The ATP-based tumor chemosensitivity assay (ATP–TCA) was developed in the early 1990s (1, 2) as an answer to the problems encountered with other assay types, which lacked sensitivity and were insufficiently reproducible for potential clinical use.

It encompasses two advances and has been standardised (2). Firstly, the endpoint is highly reproducible and very sensitive – ATP measured by recombinant luciferase and luciferin is able to measure the presence of single cells, and its concentration is linear with cell number up to 10^8 cells, though there are choices to be made whether sensitivity or dynamic range are the most important issues when designing the reagent. Secondly, in kit-form the assay uses a serum-free medium which was formulated so that it does not support the growth of lymphoid cells, which are commonly present in large numbers in tumor tissue (2). The use of an adherence-free culture method in polypropylene microplates with this serum-free medium inhibits the growth of fibroblasts and other non-neoplastic cells which require this. Over the course of the 6-day incubation period, non-neoplastic cells die off, but they are there long enough to condition the medium, and this seems to be important in maintaining the phenotype of the neoplastic cells, including their chemosensitivity. The assay shows inter- and intra-plate variation of less than 15% in expert hands. The assay has been used extensively by a number of groups, mainly in Germany and the UK, with primary cell cultures to study the chemosensitivity of human tumors and shows good correlation with clinical outcome in a number of tumor types – notably breast (3) and ovarian cancers (2, 4–6), and melanoma (7).

The ATP based tumor chemosensitivity assay (ATP–TCA) has proved to be a useful tool for cell based research and drug development work (8), leading to the development of several novel chemotherapy combinations with clinical utility, notably mitoxantrone+paclitaxel (9), treosulfan+gemcitabine (10–13) and liposomal doxorubicin with vinorelbine (14). Although in theory, cell survival assays like the ATP–TCA are not obviously suited to look at cytostatic effects, they do seem to be useful. The ATP–TCA has been used to study the effects of targeted agents such as gefitinib (15) and imatinib (16), as well as inhibitors of drug resistance (17). Latterly, it has proven useful for the development of novel drugs with uncertain mechanisms of action, such as the *N*-bisphosphonates (18).

The ATP–TCA can be used with both cell lines (in which case, the medium and type of plate can be varied) as well as human tumor-derived cells in primary cell culture. The comparison of results from the two can be interesting – cell lines rarely show similar chemosensitivity to primary cell cultures unless they are re-adapted to a serum-free and adherence-free existence (19).

Use of the ATP–TCA data to study mechanisms of resistance and sensitivity has advantages over clinical data in that drugs can be tested alone or in combination that would never be used clinically. By harvesting cells before and after assay, or from medium only and drug-treated wells, it is possible to show strong correlation between expression of genes involved in chemoresistance and in re-growth after short-term culture (20). Comparison of these observations with those from tumor biopsies taken before and after treatment show good correlation (21). In fact, we have now shown that the profile of expression of genes involved in sensitivity or resistance to particular cytotoxic agents is predictive of chemosensitivity in the ATP–TCA (22).

In those tumor types where neoadjuvant therapy and screening has yet to diminish average tumor size to a point at which few tumor cells are available for such studies, the ATP–TCA can also function as an aid for oncologists in determining drug sensitivity of tumors from individual patients. However, the practicalities of providing such a service are difficult to overcome and cellular chemosensitivity assays are probably best regarded as useful research tools with both cell lines and tumor-derived cells in primary cell culture.

2. Materials

2.1. Cell Culture

1. ATP–TCA kit (DCS Innovative Diagnostik-Systeme GmbH, Hamburg, Germany) or alternative using a serum free based DMEM (Sigma, UK) media supplemented with antibiotics.
2. Polypropylene round bottom 96-well cell culture plates individually wrapped sterile with lid(Thermo Fisher).
3. 50 mL reservoirs sterile (Fisher).
4. Pipettes (ranging between 10–100 µl) Multichannel pipette (12 channel) (Finnpipette: Fisher, UK, Ovation: Vistalab Technologies, US).

2.2. Bioluminescence

1. Tumor cell extraction reagent (TCER) this needs to match luciferin–luciferase reagent and will either be included in the chemosensitivity assay kit or available from the same company to which your luciferin–luciferase is sourced.
2. ATP standard-stock solution stored at –20°C make up standard fresh each time.
3. Luminometer or CCD camera imaging device capable of reading multiwell plates (e.g. Berthold Detection Systems, Pforzheim, Germany – Orion 1/2 or MPL1/2 ranges).
4. Luciferin–Luciferase Reagent (Included in kit from DCS Innovative Diagnostik-Systeme, Promega -CellTiter-Glo® Luminescent Cell Viability Assay, or other commercially available alternatives).
5. White opaque 96-well flat bottom plates (Thermo Fisher).

3. Methods

3.1. Cell Culture Microplate Preparation

1. Make up drug combinations at 800% the test dose concentration (TDC) in complete assay medium (CAM) or alternative (see Note 1).
2. Label 96-microwell polypropylene plates (see Note 2) with maximum inhibitor (MI), media only (MO) and relevant drug names (Fig. 1).

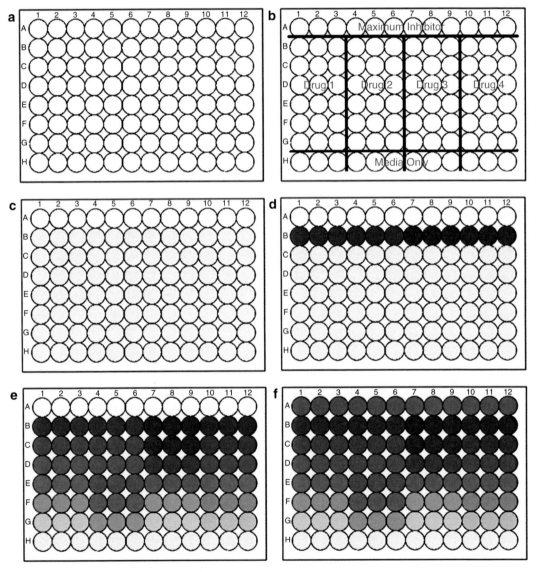

Fig. 1. Preparation of ATP–TCA plates before addition of cells. (**a**) Polypropylene 96-well round bottom plate sterile, with lid. (**b**) Shows the positioning of the reagents and design for labelling. (**c**) Add CAM to rows B–H Columns 1–12. (**d**) Add drugs to Row B in triplicate. (**e**) Perform dilution series from rows B–G mixing between each dilution. (**f**) Add MI to row A.

3. Pipette 9 mL of complete assay media (CAM) into a plastic reservoir (see Note 3). Using a multi-channel pipette, add 100 μL of CAM into each well except the row of MI wells (wells B1–H12).

4. Add 100 μL of MI solution to the MI row of wells (wells A1–A12) (see Note 4).

5. Pipette 100 μL of the appropriate drug into the appropriate 200% TDC wells in the microplate. Using a multichannel pipette, mix the contents of wells B1–B12, then perform a 100 μL serial dilution down the plate until you reach wells G1–G12 (6.25% TDC wells) mixing contents in each well as you go. Upon reaching 6.25% TDC wells, extract 100 μL and discard. Note that nothing has been added to the MO well (see Note 5).

6. Using a multi-channel pipette, add 100 μL of the tumor cell suspension to all 96-wells on the plate (see Note 6). Cells should be resuspended in CAM to a concentration of 200,000 cells/mL for solid tumor samples, 100,000 cells/mL for ascites samples or between 10,000–50,000 cells/mL for cell lines (depending on cell doubling time).

7. The plate is then placed into a cleaned plastic box containing dampened tissue. The lid must not be closed, allowing free exchange of gases while the dampened tissue keeps the humidity high in this little microenvironment preventing media evaporation (see Note 7).

8. Place in an incubator at 37°C 5% CO_2 for 6 days before reading the plate (see Note 8).

3.2. ATP-Standard Curve

1. Switch on luminometer at least 30 min prior to use (see Note 9).

2. Remove luciferin–luciferase (Lu–Lu) reagent and dilution buffer from fridge and allow this to reach room temperature (see Note 10).

3. Switch on computer and start the luminometer program (e.g. Simplicity 2.1). Select the 96-well raw data plate format and highlight the 9 wells that will contain the ATP. This means the luminometer will read only these wells and not all 96-wells.

4. Label one bijou "ATP" and cover with silver foil and a second bijou with "dilution buffer".

5. Pipette 4 mL and 1 mL of dilution buffer into the "ATP" and "dilution buffer" bijous respectively.

6. Pipette 50 μL of dilution buffer from its corresponding bijou into 9 wells of one row of a 96-well white polystyrene microplate (see Note 11).

7. Remove one ATP stock aliquot (concentration) from the freezer and cover in foil. Leave to defrost; this will take roughly 2 min. Do not defrost by holding in your hand as this can adversely affect the activity of the ATP (see Note 12).

8. Pipette 10 µL of ATP from the stock aliquot into the "ATP" labelled bijou. Replace lid and invert to mix.

9. Pipette 25 µL from the "ATP" bijou into the first well of the white plate containing the dilution buffer.

10. Mix and perform a serial dilution with 25 µL before discarding the final 25 µL from well 9 (see Note 13).

11. Add 50 µL Lu–Lu to each of the 9 wells taking care not to touch the wells containing ATP and contaminating the Lu–Lu reagent.

12. Tap plate gently on work surface to settle the reagents and place into the luminometer.

13. Start the program to commence reading (see Note 14).

14. Save data generated, this may be automatically transferred to an excel worksheet where it can be saved (see Note 15).

15. Reset the luminometer program ready to read experimental plates.

3.3. Preparation of Cell Extract and ATP Measurement by Luciferin–Luciferase

1. Always start by carrying out an ATP standard curve (see Subheading 3.2).

2. Remove the plates to be read from the incubator and observe wells under an inverted microscope, making note of any irregular wells.

3. Using a multi-channel pipette add 50 µl of tumor cell extraction reagent (TCER) to each well. Leave for 10–15 min.

4. Take a white polypropylene 96-well plate and label it appropriately. Using a multi-channel pipette, mix and transfer 50 µL from each well of the cell plate to into an appropriate well in the white plate. The pipette tips should be changed after every drug triplicate (i.e. every three wells).

5. Using a multi-channel pipette add 50 µL of luciferin–luciferase to each well of the white plate (see Note 16).

6. Tap plate gently on work surface to settle the reagents and place into the luminometer.

7. Start the program to commence reading.

8. Once reading is completed, Label data with experiment details and produce graphs of inhibition versus concentration tested (Fig. 2) (see Notes 17 and 18).

9. Reset the luminometer program ready for the next plate.

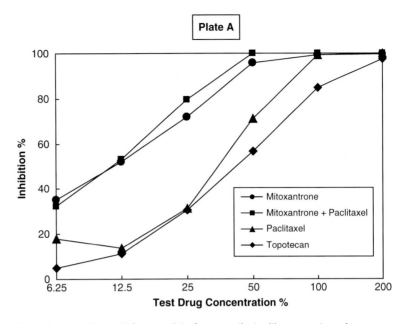

Fig. 2. An example result for one plate from a patient with recurrent ovarian cancer, using cells harvested from ascites. While the combination of mitoxantrone and paclitaxel shows the greatest degree of inhibition, there is relatively little sensitivity to paclitaxel. Toptecan shows the least activity in this example.

4. Notes

1. Care should be taken when storing drugs: (23) store these as directed by manufacturer. Small aliquots should be made where possible to prevent repeated freeze/thaw from those drugs stored below 0°C. This also prevents cross contamination of drugs, which can be very costly. Make 800% TDC stock concentrations in 2–5 mL of CAM or DMEM based media according to the kit insert or previously published data (23). TDCs are derived for each drug tested from pharmacokinetic data. In most cases, tumor concentrations achieved are not available and TDCs must be estimated from plasma concentrations, making due allowance for differences in protein binding of drugs with assay media compared with plasma. Assay media should be supportive for the cell type used (e.g. lymphoma cells with RPMI 1640 (Sigma), rather than a glutamine-free medium). To enhance the buffering capacity of the media, HEPES can be added if not already present.
2. Ensure that only polypropylene round bottom plates are used. The polypropylene surface promotes adherence-free growth preventing attachment of cells forcing them to form

spheroids mimicking a more natural cellular environment. When using tumor derived cells this also helps in the prevention of fibroblast colonisation. The use of polypropylene is also important to prevent adsorption of drugs which can occur with other plastics (24, 25). Positioning of the MI and MO controls within the plate is also important. When using drugs which may form aerosols this format helps limit crosstalk between wells preventing cellular inhibition from a 200% TDC well inhibiting a medium only well. This is also true for when reading plates: luminescent well cross-talk is limited when this plate design is followed. Ensure when labelling plates that both the lid and plate are labelled with sample details, plate number and date of setup. This will ensure that you will never lose track of multiple plates when lids are removed for assay preparation.

3. If preparing multiple plates pipette 9 mL of CAM per plate into the reservoir.

4. When preparing multiple plates use a multichannel pipette and reservoir when pipetting MI to prevent errors and to shorten assay preparation time.

5. When pipetting drugs into the corresponding 200% TDC wells in the assay plate make sure to change the pipette tip after each triplicate to prevent contamination between wells. When performing the serial dilution, be extra vigilant to prevent pipetting into the media only row.

6. When adding the cells to the plates mix the cells thoroughly in the reservoir prior to pipetting. When using tumor derived cells take extra care to mix the cells in the reservoir again after each three transfers to prevent sedimentation of cells in small aggregates. When pipetting cells remember to check the pipette tips to make sure cell aggregates haven't blocked the end of the tips and caused inaccurate volumes to be dispensed. It is also important not to touch the contents of the well with your tips when dispensing cells to prevent cross-contamination with drugs.

7. To prevent infection from unknown organisms in the local water supply make sure the tissue (thick paper towel) placed in the plate box is moistened with deionised or sterile water.

8. During the 6 day incubation period plates should be checked regularly for infection and cell overgrowth. The phenol red within the CAM will change from a pink colouration to yellow if either of these instances occurs. Using an inverted microscope, wells can be viewed individually to assess whether the colour chance is due to the depletion of nutrients by cells or by a bacterial or yeast infection.

9. At least once a month the luminometer machine's calibration should be assessed using the instrument's luminescence test plate. This alongside the ATP Standard curve confirms the luminometer machine and the luciferin–luciferase reagent are performing at optimal conditions.

10. Depending on the standard room temperatures of your laboratory equilibration of the luciferin–luciferase reagents could take up to 30 min. and so this should be removed from the fridge at the same time the luminometer is switched on. Make sure that this is kept out of direct light: it may be wise to wrap the container in foil. DO NOT be tempted to help the luciferin–luciferase reagent to warm: it is extremely temperature sensitive.

11. Standard curve white microplates can be re-used on separate occasions using different rows each time (8 rows) preventing excess plastic waste.

12. Keeping ATP covered while not in use will reduce its degradation from light.

13. The dilution series should include between five and ten dilutions within a range to incorporate expected ATP levels in your experiments.

14. Make sure that your luminometer is set to read at 1 s intervals.

15. Check the data to ensure that the luciferin–luciferase reagent is still active. This is important to check if reagents are stored for longer periods of time.

16. Make note of any wells that appear to be infected or overgrown as this may help in analysis when anomalous results can be rectified by the elimination of these wells.

17. Do not mix the luciferin–luciferase reagent and cell lysate mixture by pipetting. This can create bubbles that interfere with luminescence readings as well as cross contamination of wells.

18. Save data indicating sample name, plate number (if more than one) and date. Highlight any of those wells where infection or overgrowth may have occurred. Label data with drug names from corresponding plate making note of the order in which the drugs were used. This is important for identification as when returning to results at a later period of time which is without this basic information can render the data meaningless. Interpretation of data requires knowledge of the drugs used and their likely effects in the ATP–TCA: do not be tempted to provide data for clinical use unless you have considerable experience and appropriate training.

References

1. Hunter, E. M., Sutherland, L. A., Cree, I. A., Dewar, J. A., Preece, P. E., Wood, R. A., Linder, D., and Andreotti, P. E. (1993) Heterogeneity of chemosensitivity in human breast carcinoma: use of an adenosine triphosphate (ATP) chemiluminescence assay, *Eur J Surg Oncol* **19**, 242–249.
2. Andreotti, P. E., Cree, I. A., Kurbacher, C. M., Hartmann, D. M., Linder, D., Harel, G., Gleiberman, I., Caruso, P. A., Ricks, S. H., Untch, M., Sartori, D.C. and Bruckner H. (1995) Chemosensitivity testing of human tumors using a microplate adenosine triphosphate luminescence assay: clinical correlation for cisplatin resistance of ovarian carcinoma, *Cancer Res* **55**, 5276–5282.
3. Cree, I. A., Kurbacher, C. M., Untch, M., Sutherland, L. A., Hunter, E. M., Subedi, A. M., James, E. A., Dewar, J. A., Preece, P. E., Andreotti, P. E., and Bruckner, H. W. (1996) Correlation of the clinical response to chemotherapy in breast cancer with ex vivo chemosensitivity, *Anticancer Drugs* **7**, 630–635.
4. Sharma, S., Neale, M. H., Di Nicolantonio, F., Knight, L. A., Whitehouse, P. A., Mercer, S. J., Higgins, B. R., Lamont, A., Osborne, R., Hindley, A. C., Kurbacher, C. M., and Cree, I. A. (2003) Outcome of ATP-based tumor chemosensitivity assay directed chemotherapy in heavily pre-treated recurrent ovarian carcinoma, *BMC Cancer* **3**, 19.
5. Cree, I. A., Kurbacher, C. M., Lamont, A., Hindley, A. C., and Love, S. (2007) A prospective randomized controlled trial of tumour chemosensitivity assay directed chemotherapy versus physician's choice in patients with recurrent platinum-resistant ovarian cancer, *Anticancer Drugs* **18**, 1093–1101.
6. Konecny, G., Crohns, C., Pegram, M., Felber, M., Lude, S., Kurbacher, C., Cree, I. A., Hepp, H., and Untch, M. (2000) Correlation of drug response with the ATP tumorchemosensitivity assay in primary FIGO stage III ovarian cancer, *Gynecol Oncol* **77**, 258–263.
7. Ugurel, S., Schadendorf, D., Pfohler, C., Neuber, K., Thoelke, A., Ulrich, J., Hauschild, A., Spieth, K., Kaatz, M., Rittgen, W., Delorme, S., Tilgen, W., and Reinhold, U. (2006) In vitro drug sensitivity predicts response and survival after individualized sensitivity-directed chemotherapy in metastatic melanoma: a multicenter phase II trial of the Dermatologic Cooperative Oncology Group, *Clin Cancer Res* **12**, 5454–5463.
8. Kurbacher, C. M., and Cree, I. A. (2005) Chemosensitivity testing using microplate adenosine triphosphate-based luminescence measurements, *Methods Mol Med* **110**, 101–120.
9. Kurbacher, C. M., Bruckner, H. W., Cree, I. A., Kurbacher, J. A., Wilhelm, L., Poch, G., Indefrei, D., Mallmann, P., and Andreotti, P. E. (1997) Mitoxantrone combined with paclitaxel as salvage therapy for platinum-refractory ovarian cancer: laboratory study and clinical pilot trial, *Clin Cancer Res* **3**, 1527–1533.
10. Myatt, N., Cree, I. A., Kurbacher, C. M., Foss, A. J., Hungerford, J. L., and Plowman, P. N. (1997) The ex vivo chemosensitivity profile of choroidal melanoma, *Anticancer Drugs* **8**, 756–762.
11. Neale, M. H., Myatt, N., Cree, I. A., Kurbacher, C. M., Foss, A. J., Hungerford, J. L., and Plowman, P. N. (1999) Combination chemotherapy for choroidal melanoma: ex vivo sensitivity to treosulfan with gemcitabine or cytosine arabinoside, *Br J Cancer* **79**, 1487–1493.
12. Corrie, P. G., Shaw, J., Spanswick, V. J., Sehmbi, R., Jonson, A., Mayer, A., Bulusu, R., Hartley, J. A., and Cree, I. A. (2005) Phase I trial combining gemcitabine and treosulfan in advanced cutaneous and uveal melanoma patients, *Br J Cancer* **92**, 1997–2003.
13. Pfohler, C., Cree, I. A., Ugurel, S., Kuwert, C., Haass, N., Neuber, K., Hengge, U., Corrie, P. G., Zutt, M., Tilgen, W., and Reinhold, U. (2003) Treosulfan and gemcitabine in metastatic uveal melanoma patients: results of a multicenter feasibility study, *Anticancer Drugs* **14**, 337–340.
14. Di Nicolantonio, F., Neale, M. H., Knight, L. A., Lamont, A., Skailes, G. E., Osborne, R. J., Allerton, R., Kurbacher, C. M., and Cree, I. A. (2002) Use of an ATP-based chemosensitivity assay to design new combinations of high-concentration doxorubicin with other drugs for recurrent ovarian cancer, *Anticancer Drugs* **13**, 625–630.
15. Knight, L. A., Di Nicolantonio, F., Whitehouse, P., Mercer, S., Sharma, S., Glaysher, S., Johnson, P., and Cree, I. A. (2004) The in vitro effect of gefitinib ('Iressa') alone and in combination with cytotoxic chemotherapy on human solid tumours, *BMC Cancer* **4**, 83.
16. Knight, L. A., Di Nicolantonio, F., Whitehouse, P. A., Mercer, S. J., Sharma, S., Glaysher, S., Hungerford, J. L., Hurren, J., Lamont, A., and Cree, I. A. (2006) The effect of imatinib mesylate (Glivec) on human tumor-derived cells, *Anticancer Drugs* **17**, 649–655.

17. Di Nicolantonio, F., Knight, L. A., Glaysher, S., Whitehouse, P. A., Mercer, S. J., Sharma, S., Mills, L., Prin, A., Johnson, P., Charlton, P. A., Norris, D., and Cree, I. A. (2004) Ex vivo reversal of chemoresistance by tariquidar (XR9576), *Anticancer Drugs* **15**, 861–869.
18. Knight, L. A., Kurbacher, C. M., Glaysher, S., Fernando, A., Reichelt, R., Dexel, S., Reinhold, U., and Cree, I. A. (2009) Activity of mevalonate pathway inhibitors against breast and ovarian cancers in the ATP-based tumour chemosensitivity assay, *BMC Cancer* **9**, 38.
19. Fernando, A., Glaysher, S., Conroy, M., Pekalski, M., Smith, J., Knight, L. A., Di Nicolantonio, F., and Cree, I. A. (2006) Effect of culture conditions on the chemosensitivity of ovarian cancer cell lines, *Anticancer Drugs* **17**, 913–919.
20. Di Nicolantonio, F., Mercer, S. J., Knight, L. A., Gabriel, F. G., Whitehouse, P. A., Sharma, S., Fernando, A., Glaysher, S., Di Palma, S., Johnson, P., Somers, S. S., Toh, S., Higgins, B., Lamont, A., Gulliford, T., Hurren, J., Yiangou, C., and Cree, I. A. (2005) Cancer cell adaptation to chemotherapy, *BMC Cancer* **5**, 78.
21. Mercer, S. J., Di Nicolantonio, F., Knight, L. A., Gabriel, F. G., Whitehouse, P. A., Sharma, S., Fernando, A., Bhandari, P., Somers, S. S., Toh, S. K., and Cree, I. A. (2005) Rapid up-regulation of cyclooxygenase-2 by 5-fluorouracil in human solid tumors, *Anticancer Drugs* **16**, 495–500.
22. Glaysher, S., Yiannakis, D., Gabriel, F. G., Johnson, P., Polak, M. E., Knight, L. A., Goldthorpe, Z., Peregrin, K., Gyi, M., Modi, P., Rahamim, J., Smith, M. E., Amer, K., Addis, B., Poole, M., Narayanan, A., Gulliford, T. J., Andreotti, P. E., and Cree, I. A. (2009) Resistance gene expression determines the in vitro chemosensitivity of non-small cell lung cancer (NSCLC), *BMC Cancer* **9**, 300.
23. Hunter, E. M., Sutherland, L. A., Cree, I. A., Subedi, A. M., Hartmann, D., Linder, D., and Andreotti, P. E. (1994) The influence of storage on cytotoxic drug activity in an ATP-based chemosensitivity assay, *Anti-cancer drugs* **5**, 171–176.
24. Palmgren, J. J., Monkkonen, J., Korjamo, T., Hassinen, A., and Auriola, S. (2006) Drug adsorption to plastic containers and retention of drugs in cultured cells under in vitro conditions, *Eur J Pharm Biopharm* **64**, 369–378.
25. Di, S., Li-Fen, H., and Jessie, L. S. A. (1996) Binding of taxol to plastic and glass containers and protein under in vitro conditions, *Journal of Pharmaceutical Sciences* **85**, 29–31.

Chapter 22

Differential Staining Cytotoxicity Assay: A Review

Larry M. Weisenthal

Abstract

Differential Staining Cytotoxicity (DiSC) assay is the prototype for a closely related family of assays based on the concept of total cell kill, or, in other words, cell death occurring in the entire population of tumor cells. It is probably the most versatile of the cell-death end points, in that it (1) can be applied to both solid and hematologic neoplasms, (2) can be applied to specimens in which it is not possible to obtain a pure population of highly enriched tumor cells, and (3) can be applied to a wide variety of drugs, ranging from traditional cytotoxic agents to biological response modifiers with activity mediated through tumor-infiltrating effector cells, to "targeted" kinase inhibitors, and to antivascular agents, such as bevacizumab and pazopanib. The basic principles of the assay are to culture three-dimensional fresh tumor cell clusters in anchorage-independent conditions. At the conclusion of the culture period, Fast Green dye is added to the microwells, the contents of which are then sedimented onto permanent Cytospin centrifuge slides and then counterstained with hematoxylin–eosin or Wright–Giemsa. "Living" cells stain with the cytologic stain in question and can be identified as either normal or neoplastic, based on standard morphologic criteria. "Dead" cells stain blue-green. Nonviable endothelial cells appear as strikingly hyperchromatic, blue-black, and often refractile objects, which may be readily distinguished from other types of dead cells. This assay has been biologically and clinically validated in a number of ways, as described in this chapter.

Key words: DiSC assay, Staining, Cytotoxicity, Leukemia, Lymphoma, Carcinoma, Chemosensitivity

1. Introduction

Differential Staining Cytotoxicity Assay (DiSC) is the prototype for a closely related family of assays based on the concept of total cell kill, or, in other words, cell death occurring in the entire population of tumor cells (as opposed to only in a small fraction of the tumor cells, such as the proliferating fraction or clonogenic fraction) (1–6). It is probably the most versatile of the cell-death end points, in that it (1) can be applied to both solid and hematologic neoplasms (2), can be applied to specimens in which it is not possible to obtain a pure population of highly enriched tumor cells, and (3) can be applied to a wide variety of drugs, ranging from traditional cytotoxic agents to biological response modifiers with activity mediated through tumor-infiltrating effector cells, to "targeted" kinase inhibitors, and to antivascular agents, such as bevacizumab and pazopanib.

The basic technology concepts for cell-death assays are straightforward. A fresh specimen is obtained from a viable neoplasm. The specimen is most often a surgical specimen from a viable solid tumor. Less often, it is a malignant effusion, bone marrow, or peripheral blood specimen containing "tumor" cells (a word used to describe cells from either a solid or hematologic neoplasm). These cells are isolated and then cultured in the continuous presence or absence of drugs, most often for 3–7 days. At the end of the culture period, a measurement is made of cell injury, which correlates directly with cell death. While there is evidence that the majority of available anticancer drugs may work through a mechanism of causing sufficient damage to trigger so-called programmed cell death, or apoptosis (7–10), the DiSC assay and related cell-death end points are capable of detecting cell death mediated through both apoptotic and nonapoptotic mechanisms.

Although there are methods for specifically measuring apoptosis, per se, there are practical difficulties in applying these methods to mixed (and clumpy) populations of tumor cells and normal cells. One problem is that the peak times for a given apoptotic signal may be different between different specimens and different drugs. Thus, more general measurements of cell death have been applied, after a sufficiently long period of time has passed to allow for drug-induced cell death to occur. What is being measured is not a specific death signal, but rather the loss of a cell viability signal. These include (a) delayed loss of cell membrane integrity (which has been found to be a useful surrogate for apoptosis), as measured by differential staining in the DiSC assay method, which allows selective drug effects against tumor cells to be recognized in a mixed population of tumor and normal cells (11, 12), (b) loss of mitochondrial Krebs cycle activity, as measured in the MTT assay (13), (c) loss of cellular ATP, as measured in the ATP assay (14–16), and (d) loss of cell membrane esterase activity and cell membrane integrity, as measured by the fluorescein diacetate assay (17–19).

It is very important to realize that all of the above four end points can and do, in most cases, produce valid and reliable measurements of cell death, which correlate very well with each other on direct comparisons of the different methods (13, 18–28). We have performed direct correlations between the DiSC and MTT assays in approximately 6,500 fresh human tumor specimens, testing an average of 15 drugs per specimen at two different concentrations. Thus, we have approximately 180,000 direct comparisons between DiSC (membrane integrity) and MTT (mitochondrial Krebs cycle activity) end points in fresh human tumor specimens.

The overall correlation coefficient between these end points in specimens containing >60% tumor cells is 0.85 (These data do not include assays on 5FU, which, for biological reasons, may be somewhat more accurately tested in the MTT assay than in the DiSC assay. These data also do not include assays for paclitaxel and docetaxel, which, for different biological reasons, are better tested in the DiSC assay than in the MTT assay).

The above studies, demonstrating the comparability of results with the multiple different cell death end points, are important for the following reason. For perfectly understandable reasons, clinical studies correlating assay results with clinical outcome are very difficult to perform. The literature in this field may be characterized as including a great many small studies, but not very large studies. Additionally, different investigators have favored different cell-death end points, depending on the laboratory and clinical situation.

For example, the DiSC assay is labor-intensive and requires expertise in recognizing and counting tumor cells using a microscope, but it may be applied to specimens containing a heterogeneous mixture of tumor cells and normal cells. MTT, ATP, and FDA end points use semiautomated instrument readouts, but can only be applied to specimens which are relatively homogeneous for tumor cells. In addition, there are a number of additional reasons why one type of cell-death end point may be advantageous in a given tumor specimen and why laboratories may apply several different cell-death end points in the testing of a single specimen.

It should be noted that, historically, the DiSC assay studies of the early 1980s (11, 29, 30) provided the prototype for later studies of the other cell-death end points. When the MTT end point was first introduced in the late 1980s, the first published studies compared the MTT results to the DiSC results, with culture conditions and drug exposures being otherwise identical (13, 21, 23, 31). Many laboratories have preferred the MTT end point, and later the ATP and FDA end points, which, as in the case of the MTT end point, were compared to the DiSC end point, in early publications (19, 27, 28), because of the difficulty in preparing and scoring the DiSC assay microscope slides. What is important is that each of the above cell-death end points does give essentially the same results (except in the case of isolated drugs, such as taxanes and 5FU). Thus, it is entirely reasonable and proper to consider as a whole the clinical validation data which have been published using the above four end points.

The second point to understand is that cell-death assays are not intended to be scale models of chemotherapy in the patient. The DiSC assay was designed to address the major practical problems with the popular clonogenic assays of the late-70s/early-80s. Chief among these problems were (1) low evaluability rates and (2) uncertainty of what was being measured in individual assays (true tumor cell colonies, arising from clonogenic cell growth versus artifactual colonies arising from cell aggregation). Unlike the case with the clonogenic assays, there was no attempt to model in vivo pharmacokinetics (i.e., no attempt to utilize clinically achievable drug concentrations or to determine something analogous to an antibacterial minimal inhibitory concentration). Instead, the assay conditions were rigorously fixed, with respect to culture media and drug exposure time (the latter being, most typically, 96 h). Drugs were first tested in training sets to determine the drug concentration which gave the widest scatter of results (mathematically defined as the greatest standard deviation). The hypothesis to be tested with clinical correlations was a very simple one – that above-average drug effects in the assays would correlate with above-average drug effects in the patient, as measured by both response rates and patient survival. This hypothesis has now been proven to be correct, in a broad spectrum of neoplasms, and with a broad spectrum of drugs.

2. Historical Perspective: Development of the DiSC Assay

Two important approaches to measure total cell kill have been (1) measuring the structural integrity of cells and (2) measuring the metabolic integrity of cells (e.g., MTT, ATP). The former end point is measured in the DiSC assay.

Microscopic inspection of the structural integrity of tumor cells can be made somewhat more objective by the principle of dye exclusion. In 1917, Pappenheimer (32) described experiments in which he exposed freshly excised thymic lymphocytes to various toxic agents and then added trypan blue. Trypan blue penetrated the incompetent membranes of dead cells and stained the dead cells blue, while the living cells remained unstained and clear.

Richter and MacDowel in 1933 (33) used this method to study the effects of chemical agents on mouse leukemia cells. Schrek in 1936 (34) applied this method to study the effects of ricin, hyperthermia, and snake venom on murine solid tumor cells and also on cells dissociated from normal tissues. Schrek in the 1950s through 1980s used a related assay based upon measuring the structural integrity of cells by means of phase contrast microscopy to study drug and radiation sensitivities of normal and neoplastic cells from human lymphatic neoplasms (35–51). Shortly before his death at age 86, Schrek reviewed his work and concluded that 23% of CLL patients have cells which are radiation-resistant in this assay and that these patients also tend to exhibit clinically drug-resistant disease (50, 52–55). In my opinion, Schrek's work is one of the great overlooked achievements in clinical cancer research in the past 50 years. In his published work of 40–50 years ago, there are obvious clues to the existence of practicable methodologies for drug resistance testing in CLL and NHL, which should have long ago been developed to replace the empiric clinical trial in the study of hematologic neoplasms.

In the late 1970s, Durkin reported a perfect correlation between the results of a 2-day trypan blue assay in non-Hodgkin's lymphoma and the response of eight patients to clinical chemotherapy (56).

My own interest in the trypan blue assay dated to work I did as a graduate student in Raymond Ruddon's laboratory at the University of Michigan in 1970. I studied the activity of a series of investigational purine derivatives obtained from Dr. A.C. Sartorelli in an established line of Burkitt lymphoma (P3J) cells (Fig. 1). The assay system consisted of measuring the growth curve of the cells by means of a Coulter Counter. Some of the

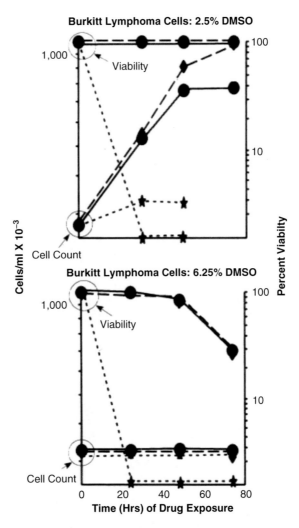

Fig. 1. Direct comparisons of different cell death end points in 96-h microcluster culture of fresh human tumor specimens tested with an assortment of drugs.

purine derivatives were soluble only in 6.25% DMSO. The cells proliferated in 2.5% DMSO, but not in 6.25% DMSO where they, in fact, slowly died. I could not use the Coulter Counter in this situation, but instead measured Trypan Blue uptake. In spite of the lack of cell proliferation, I was still able to determine structure–activity relationships in the Trypan Blue assay. In 1978, I was a Clinical Associate in the Medicine Branch of the US NCI. One of my colleagues was Dan Von Hoff, who at the time was carrying out early studies with the Human Tumor Colony Assay (HTCA), which he had just learned from Sydney Salmon of Arizona (57). Dan told me that leukemia and lymphoma did not grow all that well in the culture system. I remembered my previous studies with the Burkitt lymphoma cells in the high DMSO and how I was able to obtain results with cells that would not grow by testing them in a Trypan Blue assay. The problem with fresh tumors, as opposed to cell lines, was that the former tissues often contained a heterogeneous mixture of normal and neoplastic cells. To circumvent this and other problems, my laboratory devised a system in which dead cells were first stained green with Fast Green dye (which stains dead cells in the same fashion as Trypan Blue) and then concentrated (along with an internal standard of acetaldehyde-fixed duck red blood cells) onto a permanent microscope slide using a cytocentrifuge (Fig. 2). Finally, "living" cells were counterstained with hematoxylin–eosin (solid tumors) or Wright–Giemsa (hematologic neoplasms), which permitted identification of surviving cells as either tumor cells or

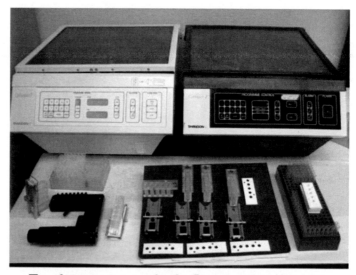

Fig. 2. Equipment needed for the DiSC assay: a cytospin is essential.

normal cells (1, 11, 12, 48–50, 58, 59). The interpretation of the assay is quite subjective and can be labor-intensive, in the case of suboptimum specimens and/or slides, and requires the recognition and scoring of slides by a highly skilled, dedicated, and experienced cytopathologist or technologist (Fig. 3).

We applied this assay to a variety of fresh human tumor specimens and found good correlations between assay results and clinical response (1, 11, 12, 60). We also reported that the assay results in vitro accurately reflected known, disease-specific patterns of clinical drug resistance (1, 11, 12, 61–63) and clinical radiation resistance (64, 65), that specimens from previously treated patients were significantly more resistant in the assay than specimens from previously untreated patients (1, 12, 66–68), and that serial assays on specimens for individual patients showed increasing drug resistance in the presence of intervening chemotherapy, but no significant change in the absence of intervening therapy (12). We further reported that clinically achievable concentrations of verapamil and lidocaine had the capacity to reverse drug resistance in some specimens of fresh hematologic neoplasms (67, 69) and later confirmed this finding in the clinic (68, 70). We also reported results which showed that the assay system had some unique advantages for the study of biologic response modifiers (63, 71, 72).

Other, independent investigators have studied the assay system. The DiSC assay has been studied most extensively in hematologic neoplasms (1, 12, 19, 23, 27–29, 69, 73–100), but the more limited studies in solid tumors (1, 11, 64, 67, 95, 101–106) have been entirely consistent with the findings in hematologics. Overall, there have been greater than 1,200 published correlations between assay results and clinical response, with the largest number of correlations coming from the long-term, focused work in chronic lymphocytic leukemia by Andrew Bosanquet and collaborators at the former Bath Cancer Research Unit. Patients treated with drugs which were "sensitive" in the assays responded to chemotherapy in 81% of cases. Patients treated with drugs which were "resistant" in the assays responded in 11% of cases. Specificity for drug resistance was 95%, and sensitivity for drug resistance was 63%. The most important way to look at these results is to note that patients treated with drugs which were "sensitive" in the assays had a 7.4-fold greater likelihood of clinical response than patients treated with assay-"resistant" drugs (5, 6).

In every published study, treatment with assay "positive" drugs was associated with a higher than otherwise expected response rate (i.e., higher than the response rate seen in the entire patient population studied) and treatment with assay negative drugs was associated with a lower than otherwise expected response rate. In a number of studies, assay results were strongly

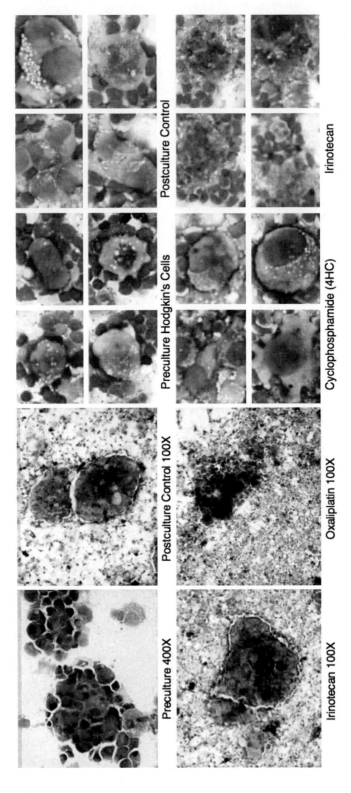

Fig. 3. Representative appearance of DiSC assay slides, which are permanent, and from which archival DNA and RNA may be isolated. *Left panel* – colonic cancer (from liver biopsy), *right panel* – Hodgkin's Disease (from node biopsy). Example results from the DiSC assay to show the appearance of cells.

correlated with patient survival (78, 84, 91, 93, 100, 102, 106). Although most of these were small studies, the number of different institutions and consistency of results provide support for the clinical and biological validity of the DiSC assay in human neoplasms.

3. Acquired Drug Resistance, DiSC Assay

With cell proliferation assays (e.g., colony formation assays, thymidine uptake assays), there are negligible published data to indicate that assays on fresh tumors from treated patients are demonstrably more drug resistant than assays on tumors from untreated patients. In the early 1980s, we reported that DiSC assays carried out on cells from previously treated CLL and ALL patients showed significantly greater resistance to nitrogen mustard, melphalan, and doxorubicin (CLL) and to dexamethasone, vincristine, cytarabine, and doxorubicin (ALL) than assays carried out on untreated patients (1, 12, 68, 106).

There are now supporting data from other laboratories to indicate that the DiSC assay clearly demonstrates greater resistance when cells from previously treated patients are tested (77, 99). This finding is of major importance for drug resistance research because it suggests that the DiSC and related assays (107) are perhaps the most relevant model systems available to study the causes and circumvention of clinically acquired drug resistance.

Table 1 shows mean percent of control tumor cell survivals for a number of human neoplasms in which acquired clinical drug resistance is known to occur. Examples of this are (1) cisplatin in non-small cell lung cancer, (2) cisplatin/carboplatin in ovarian cancer, (3) fluorouracil (5FU) in colon cancer, (4) cyclophosphamide in breast cancer, and (5) cyclophosphamide in non-Hodgkin's lymphoma. Shown separately are mean percent of control tumor cell survivals for specimens from previously untreated patients and for specimens from previously treated patients who were biopsied within 6 months of having received their most recent chemotherapy. In the case of breast cancer, cases of neoadjuvant chemotherapy were censored, and only specimens (from both treated and untreated patients) that were obtained from outside of the breast were included to reduce confounding variables of tumors in the process of responding to neoadjuvant therapy as well as possible second primary tumors. These data show that, with both DiSC and MTT end points, there is significantly greater in vitro drug resistance in specimens obtained from populations of patients known to have greater clinical drug resistance. These data provide additional evidence for the biologic validity of cell-death assays in a variety of human tumors.

Table 1
Comparison between assay results in specimens from previously untreated versus previously treated patients

Neoplasm/drug	Untreated mean % control cell survival (n) Top = DiSC Bottom = MTT	Treated mean % control cell survival (n) Top = DiSC Bottom = MTT	P2 Untreated versus treated Top = DiSC Bottom = MTT
NSCLC (AdenoCa) Cisplatin	47% (n=198) 54% (n=161)	58% (n=71) 64% (n=49)	=0.0041 =0.022
Ovarian Cisplatin	22% (n=193) 23% (n=185)	38% (n=310) 39% (n=280)	<0.0001 <0.0001
Colon 5FU	56% (n=88) 49% (n=148)	68% (n=56) 58% (n=94)	=0.0010 =0.0002
Breast Nonbreast specimens; non-neoadjuvant prior therapy; Cyclophosphamide (4HC)	49% (n=118) 51% (n=95)	60% (n=175) 58% (n=144)	=0.0017 =0.052
Non-Hodgkin's lymphoma; Cyclophosphamide (4HC)	29% (n=106) 48% (n=86)	43% (n=144) 59% (n=72)	=0.0046 =0.030

4. Drug Development Using the DiSC Assay

As noted previously, there are no published studies which convincingly show the ability of any in vitro assay system based on cell proliferation to correctly model known disease-specific activity patterns of known anticancer agents. In contrast, we have reported that drug activity in the DiSC assay tends to mirror known patterns of clinical drug resistance (1, 11, 12, 61, 62, 66, 68). The above studies showed that hematologic neoplasms tend to be much less drug resistant than solid tumors, small cell lung cancer is less resistant than non-small cell lung cancer, dexamethasone and vincristine are much more active in ALL than in ANLL, cytarabine is dramatically more active in hematologic neoplasms than in solid tumors, cisplatin is more active in solid tumors than in hematologic neoplasms, and so on. The figure below shows comparisons between cisplatin (tested at 3.3 μg/mL) and oxaliplatin (tested at 10 μg/mL) in 96-h incubation DiSC assays (Fig. 4). Particularly striking are the much greater activity for oxaliplatin than cisplatin in NHL/CLL, the much greater activity of cisplatin than oxaliplatin in ovarian cancer, sarcomas, and non-small cell lung cancer, and the greater activity of oxaliplatin in endometrial adenocarcinoma than oxaliplatin in

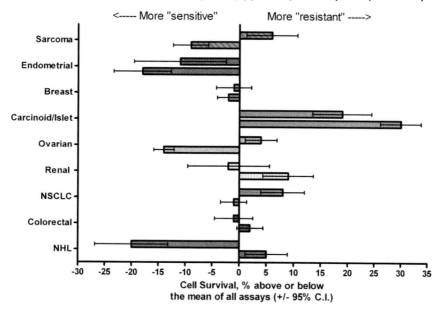

Fig. 4. Disease-specific activity of platinums in the DiSC assay.

ovarian cancer. These data serve as models for the way in which these assays could be used to direct attention to the most promising target diseases in phase II trials.

As a further example, we asked the question if a particularly troublesome subtype of breast cancer could be considered as a distinct disease from the standpoint of being a chemotherapy target. Prognosis of triple negative (ER–/PR–/Her2–) breast cancer (TNBC) is greatly improved by complete response to chemotherapy, and it is therefore essential to improve complete remission rate in this disease. We screened all of the "standard" breast cancer drugs and found that only cisplatin had exceptional activity in TNBC, with activity of cisplatin in TNBC being equivalent to that in previously untreated, poorly differentiated ovarian cancer (Table 2). Of the above three breast cancer markers, ER had the greatest impact (ER– tumors being much more platinum sensitive than ER+ tumors). Her2 negativity had a more modest impact, but ER–/Her2– tumors were more platinum sensitive than ER–/Her2+ tumors. PR did not have a major impact beyond the impact of ER.

In addition, the assay has produced patterns of both radioresistance (61, 64, 65) and cytokine sensitivity (63, 71, 108), which are compatible with known disease-specific activity patterns. Again, these data serve as models for the way in which these assays could be used to direct attention to the most promising target diseases for new drugs in phase II trials and the most promising drugs to target specific diseases, in situations where multiple candidate drugs exist.

Table 2
Cisplatin activity in breast and ovarian cancer sub-types

Control cell survival following 96-h exposure to cisplatin (%)

Tumor type	Control cell survival after cisplatin 3.3 mg/ml (95% C.I.)	Control cell survival after cisplatin 1.65 mg/ml (95% C.I.)	No. of different fresh tumor specimens tested	P2 (comparison with entire breast cancer cisplatin database, n=650)	
				(3.3 mg/ml)	(1.65 mg/ml)
Breast, ER+	46 (41–51)	75 (71–79)	79	N.S.	N.S.
Breast, ER–	22 (16–28)	57 (49–65)	44	<0.0001	<0.0001
Breast, Her2+	42 (32–53)	76 (70–82)	30	N.S.	N.S.
Breast, Her2–	34 (28–40)	64 (59–69)	73	=0.0009	=0.0006
Breast, ER– Her2+	33 (16–50)	71 (58–84)	11	N.S.	N.S.
Breast, ER– Her2–	17 (12–22)	50 (41–59)	28	<0.0001	<0.0001
Ovarian, poorly-differentiated, untreated	17 (13–21)	44 (38–50)	90	<0.0001	<0.0001
Ovarian, poorly-differentiated, relapsed within 6 months to treatment	31 (27–35)	65 (60–70)	93	<0.0001	=0.0008
Ovarian, poorly-differentiated, relapsed greater than 6 months	21 (16–26)	53 (47–59)	61	<0.0001	<0.0001

5. Biologic Response Modifiers in the DiSC Assay

Because of the unique capability of the DiSC assay to detect specific cell killing in individual cell populations present in a mixed population of tumor and normal cells, I suggested the application of this assay to the study of biological response modifiers in 1983 (11). Subsequently, Lewensohn and colleagues (109) and Lepri (110) applied the assay to the study of interferons in multiple myeloma, and my laboratory used a modification of the assay to study a variety of biologic response modifiers in a variety of fresh human neoplasms (63, 68, 71, 72, 108). This assay is much more sensitive and specific than the classic chromium release assay and shows promise for the study of immunologic mechanisms (72, 109) and as a predictive test for clinical BRM therapy (63, 72).

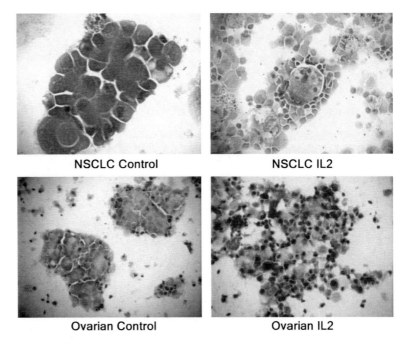

Fig. 5. Activity of IL2 in individual cases of NSCLC and ovarian cancer.

Figure 5 shows 7-day assays performed on cells from patients with non-small cell lung cancer (adenocarcinoma) and ovarian cancer, respectively. It is very rare for Interleukin-2 (IL2) to have a notable effect in these neoplasms (63, 71). In the first of the above cases (the NSCLC), the patient had stage-4 disease and achieved a virtual complete remission with assay-directed therapy consisting of vinorelbine + gefitinib + tamoxifen. The patient was then placed on maintenance therapy with low-dose, subcutaneous IL2 and remained in remission for 8 months following discontinuation of chemotherapy, when she developed leptomeningeal disease, in the absence of a systemic recurrence. The ovarian cancer patient had stage 3B, platinum-resistant disease and had received 5 prior chemotherapy regimens. She was also treated with low-dose, subcutaneous IL2. Within 6 weeks, her CA-125 dropped from 128 to 52 and thereafter remained stable, on IL-2 maintenance, for a total of 11 months, until she again developed progressive disease. Metabolic end points cannot be used for biological response modifiers, in which drug activity is mediated through normal lymphocytes (which often transform to metabolically robust cells in culture, particularly when stimulated by BRMs, e.g., see the NSCLC slides, above) and/or macrophages, which can also have considerable metabolic activity.

Shown below (Fig. 6, reprinted from reference (72), with permission) are DiSC/BRM assays performed on (1) drug-resistant forms of cancer (chiefly colon, lung, and renal) and (2) drug-sensitive forms of cancer (ovarian and breast). Agents tested were the nonspecific macrophage-activating agent ImuVert (left panel)

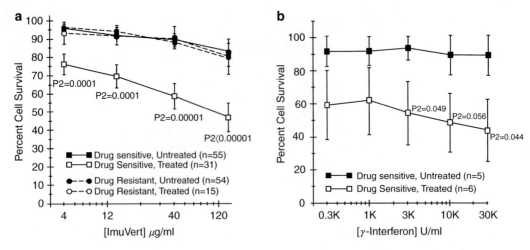

Fig. 6. Activity of macrophage-activating BRMs in drug-resistant and drug-sensitive tumors. (**a**) ImuVert, (**b**) γ-interferon.

and interferon gamma (right panel). Both agents were significantly more active in previously treated, "drug-sensitive" neoplasms than in untreated neoplasms of both types and from treated, "drug-resistant" neoplasms. We hypothesized that this difference was owing to "in situ vaccination" from the chemotherapy-induced release of tumor antigens as a consequence of effective chemotherapy. This, we speculated, would serve to prime tumor infiltrating effector cells so that they would respond to additional stimulation with macrophage-activating BRMs with potent, specific antitumor effects.

At a later time, other investigators performed a clinical trial in advanced ovarian cancer, in which patients were randomized to receive cisplatin + cyclophosphamide alone versus the same regimen with the addition of subcutaneous interferon (IFN) gamma, 0.1 mg, on 3 days per week, every other week (i.e., week on/week off), and this showed improved progression-free survival and a trend for improved overall survival (*111*) (Fig. 7). These authors cited our study in their discussion, as far as providing a mechanism for the positive results they obtained. Their study was terminated before scheduled completion, because of the substitution of paclitaxel for cyclophosphamide in standard therapy.

Subsequent to this, there was a larger, international trial (*112*) in which patients were randomized to carboplatin/paclitaxel alone versus the same regimen with the addition of low-dose, subcutaneous gamma interferon, 0.1 mg, for 3 days per week, every week (i.e., continuously). This study was closed early, because of an *inferior* survival in the gamma interferon arm! However, a close inspection of the two trials reveals the following differences: (a) In the 2nd (IFN-gamma inferior) study, the total dose of IFN-gamma was twice that in the 1st (IFN-gamma superior) study. (b) The hematologic toxicity was much greater in the IFN-gamma arm in the 2nd study, resulting in significantly less

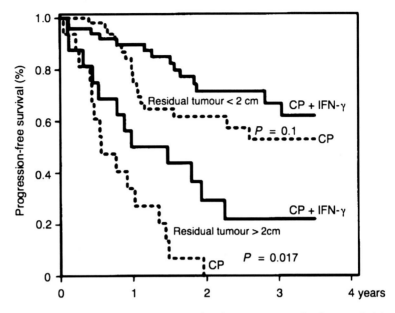

Fig. 7. Impact of subcutaneous gamma interferon on progression-free survival in advanced ovarian cancer, reprinted from (111), with permission.

chemotherapy actually received, significantly greater myelosuppression and probably significantly more use of hematopoetic growth factors, which have been shown to accelerate cancer growth, and probably greater immunosuppression. (c) The greater toxicity in the IFN-gamma arm in the 2nd study may have compromised the ability to administer 2nd line, "salvage" therapy. (d) The overall survival of the cisplatin/cyclophosphamide/IFN-gamma arm in the first study compared favorably with the carboplatin/paclitaxel/no IFN-gamma arm of the second study, even though the second study took place many years after the first, during which time overall results of first-line ovarian cancer therapy had improved substantially worldwide.

While my hypothesis of "in situ vaccination" remains unsettled, there is no doubt that occasional individual patients do benefit substantially from BRM therapy, and the DiSC assay affords a potential mechanism for identifying such patients and for future development of more effective BRM therapy.

6. Anti-angiogenic Agents, DiSC Assay

We have recently reported a completely unexpected finding: dead microvascular, microcapillary cells stain strikingly differently than dead tumor cells and dead normal cells which are not capillary cells (113). As virtually all specimens of solid tumors and hematologic neoplasms contain copious endothelial and other microvascular cells, this makes it possible to test fresh human specimens for the activity of antiangiogenic agents which have the capacity to kill tumor-associated microvascular cells. We used this discovery to develop a unique assay, which I believe to offer the only existing means to study the effectiveness of putative

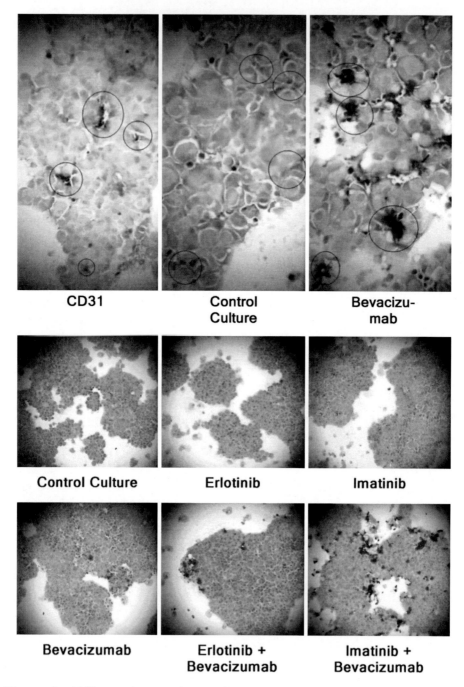

Fig. 8. Microvascular viability assay in pancreatic carcinoid tumor (upper three panels) and in breast cancer (lower six panels).

antiangiogenic agents directly against microcluster cultures of human tumors and hematologic neoplasms (which typically contain endothelial cells in apposition to lymphoma or leukemia cells, even in peripheral blood specimens).

The upper three panels of Fig. 8 show the appearance of CD31 (endothelial cell antigen)-stained cells, viable (pink-stained) microvascular cells in control cultures (circled in the middle of the upper three panels), and nonviable (dense, refractile, blue-black stained) microvascular cells in the bevacizumab-exposed cultures. Bevacizumab removes VEGF from the culture medium, producing death of microvascular cells in VEGF-dependent tumors (112). The lower six panels show an example of where erlotinib had a minor effect on enhancing bevacizumab activity, while imatinib had a dramatic and striking effect to enhance bevacizumab activity. We have extended this work, in a search to identify exploitable synergy between putative antiangiogenic agents (113, 114).

7. DiSC Assay for "Targeted" Antikinase Agents

There is an explosion of small molecule inhibitors of protein kinases being introduced into clinical trials and approved for cancer treatment. These are all expensive drugs, with limited effectiveness, in most cases, against the most common neoplasms. There is a great need to individualize therapy with this class of agents. The presence and/or absence of molecular markers has been of some utility, but the ability to test these agents in cell culture systems, both alone and in combination with other agents, would be of great potential benefit. These protein kinase inhibitors are particularly interesting, in that they may have antiangiogenic activity, as well as direct antitumor activity. This DiSC assay for cytotoxicity, combined with recognition of antimicrovascular effects (see above), provides a potentially attractive means for testing this class of agents.

We have been routinely testing and reporting out results for gefitinib as "sensitive," "intermediate," and "resistant" since 2003, where "intermediate" is a percent cell death within plus or minus one half standard deviation from the median of reference, database assays; "sensitive" is 1/2 standard deviation or greater above the mean cell death, and "resistant" is 1/2 standard deviation or greater below the mean cell death.

Figure 9 shows the overall (not progression-free) survival for previously treated patients with non-small cell lung cancer (115), as a function of gefitinib results prospectively reported to the referring clinicians and as a function of whether or not they ever received subsequent treatment with gefitinib or erlotinib following

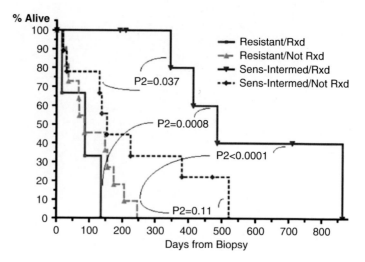

Fig. 9. Overall survival of previously treated patients with NSCLC, as a function of prospectively reported assay results and subsequent management with (Rxd) or without (Not Rxd) gefitinib/erlotinib.

the assay. Although the numbers are small (only 32 patients), the relationships were both significant and striking, as was the excellent survival of gefitinib sensitive-intermediate patients who actually received subsequent therapy with gefitinib (recalling that these were all patients with progressive disease following first-line chemotherapy). Assays for gefitinib-activating EGFR mutations only identify a small subset of patients with projected benefit and do not have the potential ability to identify cases where gefitinib (and other protein kinases) may synergize with other cytotoxic or antiangiogenic agents. The data in Fig. 9 suggest that cell culture assays may have the potential to identify a larger number of patients who may benefit from specific forms of targeted antikinase therapy, alone or in combination with other agents.

8. Limitations of the DiSC Assay

In addition to the general limitations of all in vitro assays (26), the DiSC assay has practical and theoretical limitations (26). The most important technical requirement for performance of the DiSC assay is the availability of talented, knowledgeable, experienced, and dedicated slide readers. Recognizing and accurately scoring up to scores of microscope slides per assay and completing several assays per day, on a continuing basis, obviously requires such qualities, and a laboratory which lacks personnel with these qualities will not have success with the DiSC assay.

I believe that the DiSC assay is applicable to most of the anticancer drugs available today, from traditional cytotoxics to biologic response modifiers, with indirect antitumor activity mediated by immune effector cells, to antiangiogenic agents, and to "targeted" protein kinase inhibitors. There are, however, several exceptions. We do not obtain a cell-death signal with the DiSC assay after 96 h of exposure to methotrexate or pemetrexed. There is also an inconsistent cell-death signal with directly cytotoxic monoclonal antibodies, such as cetuximab, trastuzumab, and rituximab. Finally, although we get both a good signal and reasonably good data with the DiSC assay in the case of fluoropyrimidines, such as 5FU, I feel that the MTT is a uniquely valuable end point for this latter class of agents, as I think that an important effect of 5FU is to inhibit mitochondrial DNA synthesis and MTT is a more specific mitochondrial probe than is the DiSC (and other) end points. On the other hand, I feel that the DiSC end point is more reliable for taxanes, for different biological reasons. With most agents, however, DiSC, MTT, ATP, and fluorescein diacetate results are virtually interchangeable, when all end points are tested properly, in reasonably "pure" populations of tumor cells.

Because of the stringent demands of DiSC assay slide scoring, it is entirely understandable that many investigators would prefer a semiautomated assay, such as the MTT assay or other similar assay, such as those based on fluorescein metabolism and trapping or ATP content. However, there are problems which routinely emerge when the metabolic assays are applied to nonuniform cell populations.

A fundamentally important point is that it is the composition of the cell culture at the end of the assay that determines the suitability for the metabolic end points, not the cell composition at the time when the assay is initially plated. In both solid and hematologic neoplasms, we have found that there can be important changes in the cell composition of cultures during the period of cell culture. If the percentage of malignant cells is going down, then these cells are, by definition, dying faster (or, in rare cases, proliferating slower than the normal cells). This also implies that the surviving tumor cells, as representatives of a dying cell population, may have a reduced rate of cell metabolism, relative to the normal, "healthier" cells, and thus the total contribution of normal cells to cell metabolism may be even greater than their cell numbers would suggest. Thus, assay results may be seriously altered by the presence of even comparatively small numbers of "contaminating" normal cells.

I feel most strongly that the lack of quality controls with these assay systems will inevitably lead to the frequent contamination of the medical literature with results which are avoidably erroneous and misleading. It is sobering to recall that it was the lack of suitable quality controls which led to the publication of erroneous results and the unfortunate and avoidable discrediting to the "human tumor colony assay" (116). This state of affairs led directly to the discrediting of the entire field of predictive cell culture assays, which persists in the USA to the present day. In science, as in other endeavors, one ignores history at one's own peril.

In our laboratory, we have, since 1992, been routinely testing all tumors with at least two (and most often 3) of the following end points: DiSC, MTT, resazurin (117), and/or ATP (most typically testing the cell cultures with the first three of these end points, largely reserving the ATP assay for solid tumor assays in which the cell yield is very low, but in which there are >80% tumor cells, and fewer than 20% normal cells, at the conclusion of the culture period). When we have similar results with multiple end points, I have greater confidence in using the results to inform clinically important decisions.

9. Conclusion

The virtual abandonment of research into fresh tumor cell culture assays in the late 1980s was, in my opinion, the greatest lost opportunity in all of clinical cancer research. I predict that there will be increasing recognition that cell culture assays provide many striking advantages, compared to "molecular" approaches, and that the next 10 years will see a much needed renaissance regarding the development and application of fresh tumor, cell culture assays in cancer research and treatment.

Post-script: All of the DiSC assay studies described herein were performed on true, fresh-tumor primary cell cultures, as floating cell clusters, or, in the case of hematologic neoplasms, as discohesive floating cells, in anchorage-independent conditions. In addition to this body of work, there are also several published studies from the US National Cancer Institute, all of which utilized subcultured (i.e., passaged) cells, grown in predominately monolayer cultures. These studies are highly controversial and complicated to explain. A detailed critique of these NCI studies is available at: http://weisenthal.org/chapt12.htm.

References

1. Weisenthal, L.M., Shoemaker, R.H., Marsden, J.A., Dill, P.L., Baker, J.A. and Moran, E.M. (1984) In vitro chemosensitivity assay based on the concept of total tumor cell kill. *Recent Results Cancer Res.* **94**, 161–73.
2. Weisenthal, L.M. and Lippman, M.E. (1985) Clonogenic and nonclonogenic in vitro chemosensitivity assays. *Cancer Treat. Rep.* **69**, 615–32.
3. Weisenthal, L.M. (1993) Cell culture assays for hematologic neoplasms based on the concept of total tumor cell kill. In: *Drug Resistance in Leukemia and Lymphoma*, G.J.L. Kaspers, Pieters, R., Twentyman, P.R., Weisenthal, L.M., and Veerman, A.J.P., Editors., Harwood Academic Publishers: Langhorne, PA. p. 415–432.
4. Weisenthal, L.M. (1994) Clinical correlations for cell culture assays based on the concept of total tumor cell kill. *Contrib. Gynecol. Obstet.* **19**, 82–90.
5. Bosanquet, A.G., Nygren, P., and Weisenthal, L.M. (2008) Individualized tumor response testing in leukemia and lymphoma. In:

Innovative Leukemia and Lymphoma Therapy, G.J.L. Kaspers, Coiffier, B., Heinrich,M.C., and Estey,E.H., Editors, Informa Healthcare: New York. p. 23–43.
6. Weisenthal, L.M. and Nygren, P. (2002) Current Status of Cell Culture Drug Resistance Testing (CCDRT). http://weisenthal.org/oncol_t.htm.
7. Hickman, J.A. (1992) Apoptosis induced by anticancer drugs. *Cancer Metastasis Rev.* **11**, 121–39.
8. Zunino, F., Perego, P., Pilotti, S., Pratesi, G., Supino, R. and Arcamone, F. (1997) Role of apoptotic response in cellular resistance to cytotoxic agents. *Pharmacol. Ther.* **76**, 177–85.
9. Jaffrezou, J.P., Bettaïeb, A., Levade, T. and Laurent, G. (1998) Antitumor agent-induced apoptosis in myeloid leukemia cells: a controlled suicide. *Leuk. Lymphoma.* **29**, 453–63.
10. Castejon, R., Yebra, M., Citores, M.J., Villarreal, M., García-Marco, J.A. and Vargas, J.A. (2009) Drug induction apoptosis assay as predictive value of chemotherapy response in patients with B-cell chronic lymphocytic leukemia. *Leuk. Lymphoma* **50**, 593–603.
11. Weisenthal, L.M., Marsden, J.A., Dill, P.L. and Macaluso, C.K. (1983) A novel dye exclusion method for testing in vitro chemosensitivity of human tumors. *Cancer Res.* **43**, 749–57.
12. Weisenthal, L.M., Dill, P.L., Finklestein, J.Z., Duarte, T.E., Baker, J.A. and Moran, E.M. (1986) Laboratory detection of primary and acquired drug resistance in human lymphatic neoplasms. *Cancer Treat. Rep.* **70**, 1283–95.
13. Carmichael, J., DeGraff, W.G., Gazdar, A.F., Minna, J.D. and Mitchell, J.B. (1987) Evaluation of a tetrazolium-based semiautomated colorimetric assay: assessment of chemosensitivity testing. *Cancer Res.* **47**, 936–42.
14. Kangas, L., Gronroos, M. and Nieminen A.L. (1984) Bioluminescence of cellular ATP: a new method for evaluating cytotoxic agents in vitro. *Med. Biol.* **62**, 338–43.
15. Garewal, H.S., Ahmann, F.R., Schifman, R.B. and Celniker, A. (1986) ATP assay: ability to distinguish cytostatic from cytocidal anticancer drug effects. *J. Natl. Cancer Inst.* **77**, 1039–45.
16. Sevin, B.U., Peng, Z.L., Perras, J.P., Ganjei, P., Penalver, M. and Averette, H.E. (1988) Application of an ATP-bioluminescence assay in human tumor chemosensitivity testing. *Gynecol. Oncol.* **31**, 191–204.
17. Rotman, B., Teplitz, C., Dickinson, K. and Cozzolino, J.P. (1988) Individual human tumors in short-term micro-organ cultures: chemosensitivity testing by fluorescent cytoprinting. *In Vitro Cell. Dev. Biol.* **24**, 1137–46.
18. Larsson, R., Nygren, P., Ekberg, M. and Slater, L. (1990) Chemotherapeutic drug sensitivity testing of human leukemia cells in vitro using a semiautomated fluorometric assay. *Leukemia* **4**, 567–71.
19. Nygren, P., Kristensen, J., Jonsson, B., Sundström, C., Lönnerholm, G., Kreuger, A. and Larsson, R. (1992) Feasibility of the fluorometric microculture cytotoxicity assay (FMCA) for cytotoxic drug sensitivity testing of tumor cells from patients with acute lymphoblastic leukemia. *Leukemia* **6**, 1121–8.
20. Twentyman, P.R., Fox, N.E. and Rees, J.K. (1989) Chemosensitivity testing of fresh leukaemia cells using the MTT colorimetric assay. *Br. J. Haematol.* **71**,19–24.
21. Pieters, R., Huismans, D.R., Leyva, A. and Veerman, A.J. (1989) Comparison of the rapid automated MTT-assay with a dye exclusion assay for chemosensitivity testing in childhood leukaemia. *Br. J. Cancer* **59**, 217–20.
22. Pieters, R., Huismans, D.R., Leyva, A. and Veerman, A.J. (1989) Sensitivity to purine analogues in childhood leukemia assessed by the automated MTT-assay. *Adv. Exp. Med. Biol.* **253A**, 447–54.
23. Kirkpatrick, D.L., Duke, M. and Goh, T.S. (1990) Chemosensitivity testing of fresh human leukemia cells using both a dye exclusion assay and a tetrazolium dye (MTT) assay. *Leuk. Res.* **14**, 459–66.
24. Tsai, C.M., Ihde, D.C., Kadoyama, C., Venzon, D. and Gazdar, A.F. (1990) Correlation of in vitro drug sensitivity testing of long-term small cell lung cancer cell lines with response and survival. *Eur. J. Cancer* **26**, 1148–52.
25. Dmitrovsky, E., Seifter, E.J., Gazdar, A.F., Tsai, C.M., Edison, M., Brantley, P., Veach, S.R., Batist, G., Ihde, D.C. and Mulshine, J.L. (1990) A phase II trial of carboplatin (CBDCA) in small-cell and non-small-cell lung cancer with correlation to in vitro analysis of cytotoxicity. *Am. J. Clin. Oncol.* **13**, 285–9.
26. Hanson, J.A., Bentley, D.P., Bean, E.A., Nute, S.R. and Moore, J.L. (1991) In vitro chemosensitivity testing in chronic lymphocytic leukaemia patients. *Leuk. Res.* **15**, 565–9.
27. Rhedin, A.S., Tidefelt, U., Jönsson, K., Lundin, A. and Paul, C. (1993) Comparison of a bioluminescence assay with differential staining cytotoxicity for cytostatic drug testing in vitro in human leukemic cells. *Leuk. Res.* **17**, 271–6.
28. Nygren, P., Hagberg, H., Glimelius, B., Sundström, C., Kristensen, J., Christiansen, I.

and Larsson, R. (1994) In vitro drug sensitivity testing of tumor cells from patients with non-Hodgkin's lymphoma using the fluorometric microculture cytotoxicity assay. *Ann. Oncol.* **5** Suppl 1, 127–31.
29. Weisenthal, L.M. and Marsden, J.A. (1981) A novel dye exclusion assay for predicting response to cancer chemotherapy. *Proc. Am. Assoc. Cancer Res.* **22**, 155.
30. Weisenthal, L., Marsden, J.A., Malefatto, J. and Dill, P.L. (1981) Predicting response to cancer chemotherapy with novel dye exclusion assay. In: *XIIth International Congress of Chemotherapy.* Florence, Italy.
31. Carmichael, J., DeGraff, W.G., Gazdar, A.F., Minna, J.D. and Mitchell, J.B. (1987) Evaluation of a tetrazolium-based semiautomated colorimetric assay: assessment of radiosensitivity. *Cancer Res.* **47**, 943–6.
32. Pappenheimer, A.M. (1917) Experimental Studies Upon Lymphocytes: I. The Reactions of Lymphocytes under Various Experimental Conditions. *J. Exp. Med.* **25**, 633–650.
33. Richter, M.N. and Macdowell, E.C. (1933) Studies on Mouse Leukemia: Vii. The Relation of Cell Death to the Potency of Inoculated Cell Suspensions. *J. Exp. Med.* **57**, 1–20.
34. Schrek, R. (1936) A method for counting the viable cells in normal and in malignant cell suspensions. *Am. J. Cancer* **28**, 389–392.
35. Schrek, R. and Ott, J.N.Jr. (1952) Study of the death of irradiated and non-irradiated cells by time-lapse cinemicrography. *AMA Arch. Pathol.* **53**, 363–78.
36. Vycital, R.O., Schrek, R. and Clarke, T.H. (1953) Unstained cell counts as a method of evaluating cancerocidal agents. *J. Lab. Clin. Med.* **42**, 326–34.
37. Schrek, R., Leithold, S.L. and Friedman, I.A. (1957) In vitro sensitivity of human leukemic cells to x-rays. *Proc. Soc. Exp. Biol. Med.* **94**, 250–3.
38. Schrek, R., Friedman, I.A. and Leithold, L. (1958) Variability of the in vitro sensitivity of human leukemic lymphocytes to x-rays and chemotherapeutic agents. *J. Natl. Cancer. Inst.* **20**, 1037–50.
39. Schrek, R., Leithold, S.L., Friedman, I.A. and Best, W.R. (1962) Clinical evaluation of an in vitro test for radiosensitivity of leukemic lymphocytes. *Blood* **20**, 432–42.
40. Schrek, R. (1964) Prednisolone Sensitivity and Cytology of Viable Lymphocytes as Tests for Chronic Lymphocytic Leukemia. *J. Natl. Cancer Inst.* **33**, 837–47.
41. Schrek, R. (1965) In vitro methods for measuring viability and vitality of lymphocytes exposed to 45 degree, 47 degree, and 50 degree C. *Cryobiology* **2**, 122–8.
42. Dolowy, W.C., Elrod, L.M., Ammeraal, R.N. and Schrek, R. (1967) Toxicity of L-asparaginase to resistant and susceptible lymphoma cells in vitro. *Proc. Soc. Exp. Biol. Med.* **125**, 598–601.
43. Schrek, R. and Dolowy, W.C. (1971) In vitro test for sensitivity of leukemic cells to L-asparaginase. *Cancer Res.* **31**, 523–6.
44. Schrek, R. (1975) *Sensitivity of leukaemic lymphocytes to microtubular reagents.* Br. J. Exp. Pathol. **56**, 280–5.
45. Schrek, R. and Stefani, S.S. (1976) Cytarabine: cytocidal effect on normal and leukemic lymphocytes. Synergism with x-rays and comparison with mechlorethamine. *Exp. Mol. Pathol.* **24**, 84–90.
46. Knospe, W.H., Gregory, S.A., Trobaugh, F.E. Jr., Stedronsky, J.A. and Schrek, R. (1977) Chronic lymphocytic leukemia: correlation of clinical course and therapeutic response with in vitro testing and morphology of lymphocytes. *Am. J. Hematol.* **2**, 73–101.
47. Schrek, R. (1979) Utility and efficiency of viable cell counts. Cancer Res. **39**, 4288.
48. Schrek, R. and Stefani, S.S. (1980) Effects of hyperthermia on radiosensitivity of normal and leukaemic lymphocytes. *Br. J. Exp. Pathol.* **61**, 256–60.
49. Schrek, R. and Stefani, S.S. (1981) Toxicity of microtubular drugs to leukemic lymphocytes. *Exp. Mol. Pathol.* **34**, 369–78.
50. Schrek, R., (1988) Chronic lymphocytic leukemic patients, resistant to chemotherapy. *Med. Hypotheses* **26**, 227–8.
51. Schrek, R., Best, W.R. and Stefani, S. (1988) Relationship between in vitro and in vivo radiosensitivity of lymphocytes in chronic lymphocytic leukemia. *Acta Haematol.* **80**, 129–33.
52. Schrek, R. (1988) Intractable chronic lymphocytic leukemia. J. *Natl. Cancer. Inst.* **80**, 604.
53. Schrek, R. (1990) Essential in vitro test before treatment of patients with intractable chronic lymphocytic leukemia. *Acta Haematol.* **84**, 104–5.
54. Schrek, R. (1990) Differences between responsive and intractable chronic lymphocytic leukemia. *Med. Hypotheses* **31**, 81–2.
55. Schrek, R. (1991) Intractable chronic lymphocytic leukemia and interferon. *Med. Hypotheses.* **35**, 182–3.
56. Durkin, W.J., Ghanta, V.K., Balch, C.M., Davis, D.W. and Hiramoto, R.N. (1979) A methodological approach to the prediction of anticancer drug effect in humans. *Cancer Res.* **39**, 402–7.

57. Salmon, S.E., Hamburger, A.W., Soehnlen, B., Durie, B.G., Alberts, D.S. and Moon, T.E. (1978) Quantitation of differential sensitivity of human-tumor stem cells to anticancer drugs. *N. Engl. J. Med.* **298**, 1321–7.
58. Weisenthal, L.M., Dill, P.L., Kurnick, N.B. and Lippman, M.E. (1983) Comparison of dye exclusion assays with a clonogenic assay in the determination of drug-induced cytotoxicity. *Cancer Res.* **43**, 258–64.
59. Weisenthal, L.M., Lalude, A.O. and Miller J.B. (1983) In vitro chemosensitivity of human bladder cancer. *Cancer* **51**, 1490–6.
60. Wilbur, D.W., Camacho, E.S., Hilliard, D.A., Dill, P.L. and Weisenthal, L.M. (1992) Chemotherapy of non-small cell lung carcinoma guided by an in vitro drug resistance assay measuring total tumour cell kill. *Br. J. Cancer* **65**, 27–32.
61. Weisenthal, L.M., (1991) Predictive assays for drug and radiation resistance. In: *Human Cancer in Primary Culture: A Handbook*, J.M. Masters, Editor. Kluwer Academic Publishers: Dordrecht, The Netherlands.
62. Weisenthal, L.M., Dill, P. and Birkhofer, M. (1991) Accurate identification of disease-specific activity of antineoplastic agents with an in vitro fresh tumor assay measuring killing of largely non-dividing cells. *Proc. Am. Assoc. Cancer Res.* **32**, 384.
63. Weisenthal, L.M. and Dill, P. (1992) In vitro effects of interleukin-2 (IL2) on fresh human tumor cell cultures measured by the DISC assay. *Proc. Am. Assoc. Cancer Res.* **33**, A3313.
64. Kurohara, W., Colman, M., Nagourney, R.A., Weisenthal, L.M., Swingle, K. and Redpath, J.L. (1989) Radiation response of cells from human tumor biopsies as assessed by a dye exclusion technique: a possible predictive assay. *Int. J. Radiat. Biol.* **56**, 767–70.
65. Weisenthal, L.M., Dill, P.L., and Swingle, K.F. (1989) Clinical radiation sensitivity profiles of human neoplasms are reproduced by a short-term in vitro (DiSC) assay measuring cytotoxicity in the total (largely nondividing) tumor cell population following ultra- high dose, single-fraction radiation. *Proc. Am. Assoc. Cancer Res.* **30**, 401.
66. Weisenthal, C.L., Meade, R.C., Owenby, J. and Irwin, R.I. (1961) An investigation of kutapressin as a hemostatic agent in transurethral surgery of the prostate. *J. Urol.* **86**, 346–9.
67. Weisenthal, L.M., Su, Y.Z., Duarte, T.E. and Nagourney, R.A. (1988) Non-clonogenic, in vitro assays for predicting sensitivity to cancer chemotherapy. *Prog. Clin. Biol. Res.* **276**, 75–92.
68. Weisenthal, L.M., Nagourney, R.A., Kern, D.H., Boullier, B., Bosanquet, A.G., Dill, and M. P.L., J.C., and Moran, E.M. (1989) Approach to the clinical circumvention of drug resistance utilizing a non-clonogenic in vitro assay measuring the effects of drugs, radiation, and interleukin-II on largely non-dividing cells. In: *Strategies in Cancer Medical Therapy: Biological Bases and Clinical Implications.* Pavia, Italy: Edimes.
69. Weisenthal, L.M., Su, Y.Z., Duarte, T.E., Dill, P.L. and Nagourney, R.A. (1987) Perturbation of in vitro drug resistance in human lymphatic neoplasms by combinations of putative inhibitors of protein kinase C. *Cancer Treat. Rep.* **71**, 1239–43.
70. Moran, E., Nagourney, R.A., Ottenheimer, E.J., Mahutte, K., and Weisenthal, L.M. (1988) Reversal of acquired drug resistance with lidocaine and verapamil: a phase I study. *Proc. Am. Assoc. Cancer Res.* **29**, 218.
71. Weisenthal, L.M. (1991) Effect of prior chemotherapy on biologic response modifier activity. *J. Natl. Cancer Inst.* **83**, 790–3.
72. Weisenthal, L.M., Dill, P.L. and Pearson, F.C. (1991) Effect of prior cancer chemotherapy on human tumor-specific cytotoxicity in vitro in response to immunopotentiating biologic response modifiers. *J. Natl. Cancer Inst.* **83**, 37–42.
73. Bosanquet, A.G., Bird, M.C., Price, W.J. and Gilby, E.D. (1983) An assessment of a short-term tumour chemosensitivity assay in chronic lymphocytic leukaemia. *Br. J. Cancer* **47**, 781–9.
74. Bird, M.C., Bosanquet, A.G. and Gilby, E.D. (1985) In vitro determination of tumour chemosensitivity in haematological malignancies. *Hematol. Oncol.* **3**, 1–10.
75. Bird, M.C., Bosanquet, A.G., Forskitt, S. and Gilby, E.D. (1986) Semi-micro adaptation of a 4–day differential staining cytotoxicity (DiSC) assay for determining the in-vitro chemosensitivity of haematological malignancies. *Leuk. Res.* **10**, 445–9.
76. Bird, M.C., Godwin, V.A., Antrobus, J.H. and Bosanquet, A.G. (1987) Comparison of in vitro drug sensitivity by the differential staining cytotoxicity (DiSC) and colony-forming assays. *Br. J. Cancer* **55**, 429–31.
77. Bird, M.C., Bosanquet, A.G., Forskitt, S. and Gilby, E.D. (1988) Long-term comparison of results of a drug sensitivity assay in vitro with patient response in lymphatic neoplasms. *Cancer* **61**, 1104–9.
78. Bosanquet, A.G. (1991) Correlations between therapeutic response of leukaemias and in-vitro drug-sensitivity assay. *Lancet* **337**, 711–4.

79. Bosanquet, A.G. (1993) In vitro drug sensitivity testing for the individual patient: an ideal adjunct to current methods of treatment choice. *Clin. Oncol. (R. Coll. Radiol.)* **5**, 195–7.

80. Bosanquet, A.G., McCann, S.R., Crotty, G.M., Mills, M.J. and Catovsky, D. (1995) Methylprednisolone in advanced chronic lymphocytic leukaemia: rationale for, and effectiveness of treatment suggested by DiSC assay. *Acta Haematol.* **93**, 73–9.

81. Bosanquet, A.G. and Bell, P.B. (1996) Enhanced ex vivo drug sensitivity testing of chronic lymphocytic leukaemia using refined DiSC assay methodology. *Leuk. Res.* **20**, 143–53.

82. Bosanquet, A.G., et al. 1997 Ex vivo cytotoxic drug evaluation by DiSC assay to expedite identification of clinical targets: results with 8-chloro-cAMP. *Br J Cancer* **76**(4): p. 511–8.

83. Bosanquet, A.G., Copplestone, J.A., Johnson, S.A., Smith, A.G., Povey, S.J., Orchard, J.A. and Oscier, D.G. (1999) Response to cladribine in previously treated patients with chronic lymphocytic leukaemia identified by ex vivo assessment of drug sensitivity by DiSC assay. *Br. J. Haematol.* **106**, 474–6.

84. Bosanquet, A.G., Johnson, S.A. and Richards, S.M. (1999) Prognosis for fludarabine therapy of chronic lymphocytic leukaemia based on ex vivo drug response by DiSC assay. *Br. J. Haematol.* **106**, 71–7.

85. Mason, J.M., Drummond, M.F., Bosanquet, A.G. and Sheldon, T.A. (1999) The DiSC assay. A cost-effective guide to treatment for chronic lymphocytic leukemia? *Int. J. Technol. Assess. Health Care* **15**, 173–84.

86. Bosanquet, A.G. and Bosanquet, M.I. (2000) Ex vivo assessment of drug response by differential staining cytotoxicity (DiSC) assay suggests a biological basis for equality of chemotherapy irrespective of age for patients with chronic lymphocytic leukaemia. *Leukemia* **14**, 712–5.

87. Bosanquet, A.G., Burlton, A.R. and Bell, P.B. (2002) Parameters affecting the ex vivo cytotoxic drug sensitivity of human hematopoietic cells. *J. Exp. Ther. Oncol.* **2**, 53–63.

88. Thornton, P.D., Matutes, E., Bosanquet, A.G., Lakhani, A.K., Grech, H., Ropner, J.E., Joshi, R., Mackie, P.H., Douglas, I.D., Bowcock, S.J. and Catovsky, D. (2003) High dose methylprednisolone can induce remissions in CLL patients with p53 abnormalities. *Ann. Hematol.* **82**, 759–65.

89. Bosanquet, A.G. and Bell, P.B. (2004) Ex vivo therapeutic index by drug sensitivity assay using fresh human normal and tumor cells. *J. Exp. Ther. Oncol.* **4**, 145–54.

90. Tidefelt, U., Sundman-Engberg, B. and Paul, C. (1988) Effects of verapamil on uptake and in vitro toxicity of anthracyclines in human leukemic blast cells. *Eur. J. Haematol.* **40**, 385–95.

91. Tidefelt, U., Sundman-Engberg, B., Rhedin, A.S. and Paul, C. (1989) In vitro drug testing in patients with acute leukemia with incubations mimicking in vivo intracellular drug concentrations. *Eur. J. Haematol.* **43**, 374–84.

92. Lathan, B., von Tettau, M., Verpoort, K. and Diehl, V. (1990) Pretherapeutic drug testing in acute leukemias for prediction of individual prognosis. *Haematol. Blood Transfus.* **33**, 295–8.

93. Staib, P., Lathan, B., Schinköthe, T., Wiedenmann, S., Pantke, B., Dimski, T., Voliotis, D. and Diehl, V. (1999) Prognosis in adult AML is precisely predicted by the DISC-assay using the chemosensitivity-index Ci. *Adv. Exp. Med. Biol.* **457**, 437–44.

94. Kirkpatrick, D.L., Chen, L. and Goh, T.S. (1992) Modification of the DiSC assay by the incorporation of monoclonal antibody staining. *Leuk. Res.* **16**, 1097–103.

95. Beksac, M., Kansu, E., Kars, A., Ibrahimoglu, Z. and Firat, D. (1988) A rapid drug sensitivity assay for neoplasmatic cells. *Med. Oncol. Tumor Pharmacother.* **5**, 253–7.

96. Bosanquet AG, Raper, S.L, Durant, J., Scadding, S.M., Graham, D., Oscier, G.D., Richards, S.M., and Catovsky, D. (2006) Comparison of ex vivo drug sensitivity by TRAC assay and patient response in the UK LRF CLL4 trial. *Haematologica/Hematol. J.* **91**(suppl 1), 100.

97. Bosanquet, A., Raper, S.L., Durant, J., Scadding, S.M., Graham, D., Oscier, G.D., Richards, S.M., and Catovsky, D. on Behalf of the NCRI CLL Working Group. (2006) Drug sensitivity by TRAC (DiSC) assay as a prognostic factor for patient response in untreated CLL: results from the UK LRF CLL4 trial. *Blood* **108**, 94a.

98. Else, M., Smith, A.G., Raper, S.L., Kim S. and Cocks, S.C. and Catovsky, D. (2007) An association between drug sensitivity by TRAC (DiSC) assay and quality of life in the UK LRF CLL4 trial. 12th International Workshop on CLL, September 2007. www.caltri.org/pdf/ElseIWCLL07.pdf.

99. Bosanquet, A.G., Richards, S.M., Wade, R., Else, M., Matutes, E., Dyer, M.J., Rassam, S.M., Durant, J., Scadding, S.M., Raper, S.L., Dearden, C.E. and Catovsky, D. (2009) Drug cross-resistance and therapy-induced resistance in chronic lymphocytic leukaemia by an enhanced method of individualised tumour response testing. *Br. J. Haematol.* **146**, 384–95.

100. Staib, P., Staltmeier, E., Neurohr, K., Cornely, O., Reiser, M., Schinköthe, T. (2005) Prediction of individual response to chemotherapy in patients with acute myeloid leukaemia using the chemosensitivity index Ci. *Br. J. Haematol.* **128**, 783–91.

101. Carstensen, H., and Tholander, B. (1985) Chemosensitivity of ovarian carcinoma: In vitro/in vivo correlations using the dye exclusion assay of Weisenthal. In: *Proceedings: 3rd European Conference on Clinical Oncology.* Stockholm, Sweden.

102. Nagourney, R.A., Brewer, C.A., Radecki, S., Kidder, W.A., Sommers, B.L., Evans, S.S., Minor, D.R. and DiSaia, P.J. (2003) Phase II trial of gemcitabine plus cisplatin repeating doublet therapy in previously treated, relapsed ovarian cancer patients. *Gynecol. Oncol.* **88**, 35–9.

103. Carstensen, H. (1983) [Predictive testing of cytostatics in ovarian cancer using Weisenthal's method]. *Lakartidningen* **80**, 2812–6.

104. Nagourney, R.A., Sommers, B.L., Harper, S.M., Radecki, S. and Evans, S.S. (2003) Ex vivo analysis of topotecan: advancing the application of laboratory-based clinical therapeutics. *Br. J. Cancer* **89**, 1789–95.

105. Brewer, C.A., Blessing, J.A., Nagourney, R.A., Morgan, M. and Hanjani, P. (2006) Cisplatin plus gemcitabine in platinum-refractory ovarian or primary peritoneal cancer: a phase II study of the Gynecologic Oncology Group. *Gynecol. Oncol.* **103**, 446–50.

106. Weisenthal LM, W.C., Smith ME, Sanchez CG, and Berglund RF. (2003) Platinum resistance determined by cell culture drug resistance testing (CCDRT) predicts for patient survival in ovarian cancer. http://weisenthal.org/w_ovarian_cp_toc.html.

107. Pieters, R., Huismans, D.R., Loonen, A.H., Hählen, K., van der Does-van den Berg, A., van Wering, E.R. and Veerman, A.J. (1991) Relation of cellular drug resistance to long-term clinical outcome in childhood acute lymphoblastic leukaemia. *Lancet* **338**, 399–403.

108. Weisenthal, L.M., Dill, P.L., and Pearson, F.C. (1990) Tumor and patient-specific activity of biologic response modifiers (ImuVert, tumor necrosis factor, alpha-interferon) in fresh specimens of human neoplasms detected by a sensitive and specific in vitro assay. *Proc. Am. Assoc. Cancer Res.* **31**, 299.

109. Einhorn, S., Fernberg, J.O., Grandér, D. and Lewensohn, R. (1988) Interferon exerts a cytotoxic effect on primary human myeloma cells. *Eur. J. Cancer Clin. Oncol.* **24**, 1505–10.

110. Lepri, E., Barzi, A., Menconi, E., Portuesi, M.G., Liberati, M. (1991) In vitro synergistic activity of PDN-IFN alpha and NM + IFN alpha combinations on fresh bone-marrow samples from multiple myeloma patients. *Hematol. Oncol.* **9**, 79–86.

111. Windbichler, G.H., Hausmaninger, H., Stummvoll, W., Graf, A.H., Kainz, C., Lahodny, J., Denison, U., Müller-Holzner, E. and Marth, C. (2000) Interferon-gamma in the first-line therapy of ovarian cancer: a randomized phase III trial. *Br. J. Cancer* **82**, 1138–44.

112. Alberts, D.S., Marth, C., Alvarez, R.D., Johnson, G., Bidzinski, M., Kardatzke, D.R., Bradford, W.Z., Loutit, J., Kirn, D.H., Clouser, M.C., Markman, M. for the GRACES Clinical Trial Consortium. (2008) Randomized phase 3 trial of interferon gamma-1b plus standard carboplatin/paclitaxel versus carboplatin/paclitaxel alone for first-line treatment of advanced ovarian and primary peritoneal carcinomas: results from a prospectively designed analysis of progression-free survival. *Gynecol. Oncol.* **109**, 174–81.

113. Weisenthal, L.M., Patel, N. and Rueff-Weisenthal, C. (2008) Cell culture detection of microvascular cell death in clinical specimens of human neoplasms and peripheral blood. *J. Intern. Med.* **264**, 275–87.

114. Weisenthal, L., Lee, D.J., and Patel, N. (2008) Antivascular activity of lapatinib and bevacizumab in primary microcluster cultures of breast cancer and other human neoplasms, In: *ASCO 2008 Breast Cancer Symposium.* Washington, D.C. Abstract # 166.

115. Weisenthal, L. (2006) Gefitinib-induced cell death in short term fresh tumor cultures predicts for long term patient survival in previously-treated non-small cell lung cancer. *J. Clin. Oncol..* **24(18S)**, 17117.

116. Selby, P., Buick, R.N. and Tannock, I. (1983) A critical appraisal of the "human tumor stem-cell assay". *N. Engl. J. Med.* **308**, 129–34.

117. Al-Nasiry, S., Geusens, N., Hanssens, M., Luyten, C. and Pijnenborg, R. (2007) The use of Alamar Blue assay for quantitative analysis of viability, migration and invasion of choriocarcinoma cells. *Hum. Reprod.* **22**, 1304–9.

Chapter 23

Real-Time Cytotoxicity Assays

Donald Wlodkowic, Shannon Faley, Zbigniew Darzynkiewicz, and Jonathan M. Cooper

Abstract

Validation of new therapeutic targets calls for the advance in innovative assays that probe both spatial and temporal relationships in signaling networks. Cell death assays have already found a widespread use in pharmacological profiling of anticancer drugs. Such assays are, however, predominantly restricted to end point DEAD/LIVE parameter that provides only a snapshot of inherently stochastic process such as tumor cell death. Development of new methods that can offer kinetic real-time analysis would be highly advantageous for the pharmacological screening and predictive toxicology.

In the present work we outline innovative protocols for the real-time analysis of tumor cell death, based on propidium iodide (PI) and SYTOX Green probes. These can be readily adapted to both flow cytometry and time-lapse fluorescence imaging. Considering vast time savings and kinetic data acquisition such assays have the potential to be applied in a number of areas including accelerated anticancer drug discovery and high-throughput screening routines.

Key words: Cytotoxicity, Real-time assays, Antitumor drugs, Flow cytometry, Time-lapse microscopy

1. Introduction

Tumor cell death serves as a useful end point in pharmacological profiling of anticancer drugs (1). Despite the large variety of techniques that have been developed so far to detect cell death, technological innovations can make the deployment of these assays more effective (2). The permeability of plasma membrane to charged fluorescent probes is an accepted marker that distinguishes LIVE from DEAD. Since it is generally assumed that such probes are inherently cytotoxic their use is mostly restricted to end point assays (3). The major drawback of such analysis is, however, capturing only a snapshot of the incidence of cell death which is inherently a stochastic process. Therefore, development of new methods that can provide kinetic quantification of drug induced cytotoxicity would be highly advantageous for the pharmacological screening and predictive toxicology (3). In such assays, markers applied supravitally, should have minimal effects on the structure, function, and survival of cells (4).

Recently, we have provided new evidence that many plasma membrane integrity markers such as propidium iodide (PI), SYTOX Green, SYTOX Red, and YO-PRO 1 can be used to dynamically probe and quantify cytotoxicity in real time (2, 3). Such assays meet the following criteria of dynamic and high-throughput analysis: (1) the straightforward staining and adaptability for automated dispensing; (2) the lack of side-effects on cellular viability, proliferation or cell migration; and (3) the lack of interference with the assay readout (2, 3).

Reduction of sample processing achieved with these protocols is important for the preservation of fragile apoptotic cells. Our data indicate that such simple bioassays can be readily adapted for novel microfluidic chip-based (Lab-on-a-Chip) platforms with minimal protocol modifications (5, 6).

2. Materials

2.1. Dynamic Detection of Cell Death Using Flow Cytometry

1. Cell suspension ($1–5 \times 10^5$ cells/ml).
2. 1× PBS.
3. 1 mg/ml PI stock solution in PBS. Store protected from light at +4°C. Stable for over 12 months. Caution: PI is a DNA binding molecule and thus can be considered as a potential carcinogen. Always handle with care and use protective gloves.
4. *Optional: 1 mM SYTOX Green stock solution in DMSO. Store protected from light at –20°C. Stable for over 12 months. Caution: although there are no reports on SYTOX Green toxicity, appropriate precautions should always be applied when handling SYTOX Green solutions.*
5. *Optional: 10 μM SYTOX Green working solution in PBS (prepare fresh as required).*
6. 1.5-ml Eppendorf tubes.
7. 12×75 mm polystyrene FACS tubes.

2.2. Real-Time Detection of Cell Death Using Time-Lapse Imaging

1. Cell suspension ($1–5 \times 10^5$) or cell monolayer.
2. 1× PBS.
3. 1 mg/ml PI stock solution in PBS. Store protected from light at +4°C. Stable for over 12 months. Caution: PI is a DNA binding molecule and thus can be considered as a potential carcinogen. Always handle with care and use protective gloves.

4. *Optional: 1 mM SYTOX Green stock solution in DMSO. Store protected from light at −20°C. Stable for over 12 months. Caution: although there are no reports on SYTOX Green toxicity, appropriate precautions should always be applied when handling SYTOX Green solutions.*

5. *Optional: 100 µM SYTOX Green working solution in PBS (prepare fresh as required).*

6. 1.5-ml Eppendorf tubes.

7. Optical grade cell culture plates or cell culture chambers.

3. Methods

3.1. Dynamic Detection of Cell Death Using Flow Cytometry

The near real-time detection of cell death is based on a continuous presence of the fluorescent probe in the culture medium and performing sequential specimen sampling by flow cytometry (2, 3). As presence of the fluorescent dye has no impact on cellular viability, proliferation or cell migration, method presented here is a single-step and time saving assay (2, 3). Elimination of washing steps enhances preservation of fragile apoptotic cells in an intact state without compromising assay sensitivity (1 3).

1. Seed cells at a desired concentration in 24-well culture plates (see Note 1).

2. Add drug into the culture as appropriate (Fig. 1a).

3. Add 1 µL of 1 mg/ml PI stock solution (final concentration 1 µg/ml; see Note 2).
   – *Optional: instead of PI use 1.5 µL of 100 µM SYTOX Green working solution (final concentration 150 nM, see Notes 3 and 4).*

4. Culture cells in the presence of PI or SYTOX Green and collect sample aliquots into 12×75 mm Falcon FACS at desired time-points (see Note 5).

5. Analyze on a flow cytometer with 488 nm excitation line (Argon-ion laser or blue solid-state laser) with emissions collected at 530 nm (SYTOX Green) or 575–610 nm (PI). Adjust the logarithmic amplification scale to distinguish between viable cells (bright PI^- and $SYTOX^-$) from late apoptotic and/or necrotic cells with compromised plasma membranes ($PI^+/SYTOX^+$) (Fig. 1b; see Notes 5–7).

3.2. Real-Time Detection of Cell Death Using Time-Lapse Imaging

The principle of this assay is similar to the previously described protocol for flow cytometry. The main advantage of this protocol is a true real-time detection of cell death based on a time-lapse fluorescent microscopy (3, 5, 6).

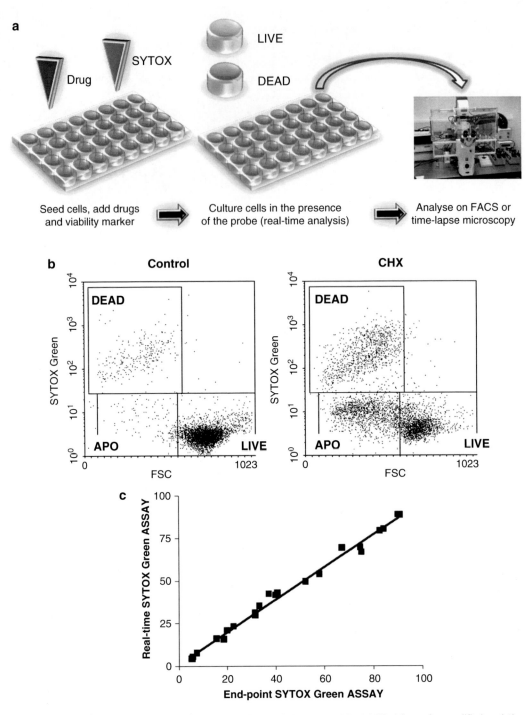

Fig. 1. Dynamic analysis of cytotoxicity using simplified real-time protocols: (**a**) Workflow of a modified real-time (no-wash) protocol. Note that viability marker SYTOX Green is continuously present in the culture medium as opposed to a standard, end-point staining procedure. (**b**) Viability marker SYTOX Green was applied to dynamically track drug-induced cytotoxicity using flow cytometry. Human promyelocytic leukemia HL60 cells were exposed to a pro-apoptotic drug cycloheximide (CHX; 50 μg/ml) for 24 h in the continuous presence of SYTOX Green (100 nM). Fluorescent probe was excited using 488 nm Argon-ion laser. SYTOX Green fluorescence signal was logarithmically amplified using 530 nm band-pass filter. Debris was excluded electronically. Analysis based on bivariate dot plots FSC vs. SYTOX Green is shown.

1. Seed cells at a desired concentration in optical grade culture plates.
2. Add drug into the culture as appropriate.
3. Add 1 µL of 1 mg/ml PI stock solution (final concentration 1 µg/ml; see Note 2).
   - *Optional: instead of PI use 1.5 µL of 100 µM SYTOX Green working solution (final concentration 150 nM, see Notes 3 and 4).*
4. Position cell carrier on a time-lapse microscope stage.
5. Culture cells in the presence of PI or SYTOX Green and collect time-lapse images at desired time-points (see Note 8).

4. Notes

1. Cell seeding densities should be empirically adjusted to a particular type of cell line and/or primary cell culture.
2. Continuous presence of PI in the culture medium does not affect cellular viability, proliferation or cell migration. Our results indicate that human promyelocytic HL60 cells remain viable and reproductively competent even when challenged with PI concentrations up to 5 µg/ml for up to 72 h (3). Similar results were obtained on a panel of diverse tumor cell lines (suspension: U937, HL60, K562, MOLT-4, and Jurkat; adherent: U2OS, Saos2, MDA-MB-231, and 3T3). Importantly, presence of the probe does not appear to affect cell cycle and long-term cell proliferation as estimated using (methyl-^{3}H)-thymidine incorporation and Trypan Blue assays (3). We recommend, however, initial titration of PI to find an optimal concentration for a particular cell line.
3. A green fluorescent SYTOX Green (Ex max: 504/Em max: 523 nm) probe can be continently substituted for PI. Remaining fluorescent channels can be utilized e.g. for multiparameter analysis of apoptotic markers such as calcium flux, caspase activation, or externalization of phosphatidyl serine residues (5, 6).

Fig. 1. (continued) LIVE – viable cells, APO – apoptotic cells, DEAD – late apoptotic/necrotic cells. (c) Comparison between percentages of cell death estimated using standard SYTOX Green end-point vs. new kinetic protocol. Human promyelocytic leukemia HL60 cells were exposed to a range of pro-apoptotic drugs cycloheximide (CHX), campthothecin (CAM), and staurosporine (STS) for 24 h. Data were acquired using BD FACS Calibur flow cytometer equipped with 488 nm excitation line and 530 nm band-pass filter. Note excellent agreement between results obtained with both assays ($R^2 \geq 0.98$ for $p < 0.05$ in Pearson and Lee linear correlation test).

4. Similar to PI, SYTOX Green does not display any side effects on cellular viability, proliferation, or cell migration when used in concentrations up to 1 mM (3).

5. Cell suspension can be collected and analyzed without any centrifugation and washing steps (3). The continuous labeling procedure not only provided similar results to a standard end point staining protocol, but also allows for a straightforward adaptation for high-throughput screening (HTS) (3).

6. Cells cultured in the presence of PI or SYTOX Green exhibit an overall increase of background florescence as compared to end point protocols. Adjust the logarithmic amplification scale to distinguish between viable cells (bright PI^- and $SYTOX^-$) from late apoptotic and/or necrotic cells with compromised plasma membranes (PI^+/$SYTOX^+$) as depicted in Fig. 1b (3).

7. Flow cytometry allows quantitative measurements of laser light scatter characteristics that reflect morphological features of cells (1). Cell shrinkage due to the dehydration can be detected at early stages of apoptosis as a decrease in intensity of forward light scatter (FSC) signal (1). Depending on a cell line model and stimuli being used, analysis based on FSC and SYTOX bivariate dotplots can provide additional information about apoptotic cells (APO – FSC^{low}/$SYTOX^-$) as depicted in Fig. 1b. It should be noted, however, that observable changes in light scattering are not a reliable marker of apoptosis and should be always confirmed by other dedicated assays (1).

8. A wide variety of optical grade cell carries can be exploited that include optical culture plates, cell culture chambers, and microfluidic chip-based devices (3, 5, 6). Program time-lapse protocol to collect images at desired time intervals. PI and SYTOX Green display substantial resistance to photobleaching. Therefore specimens can be repeatedly imaged for extended periods of time (3, 5, 6). No phototoxic reactions have been observed so far, but we recommend a careful assessment of selected fluorescent probes for a particular experimental protocol and biological specimen being used (3).

Acknowledgments

Supported by BBSRC, EPSRC and Scottish Funding Council, funded under RASOR (DW, SF, JMC) and NCI CA RO1 28 704 (ZD). Views and opinions described in this chapter were not influenced by any conflicting commercial interests.

References

1. Darzynkiewicz, Z., Juan, G., Li, X., Gorczyca, W., Murakami, T. and Traganos, F. (1997) Cytometry in cell necrobiology: analysis of apoptosis and accidental cell death (necrosis). *Cytometry.* **27**, 1–20.
2. Wlodkowic, D., Skommer, J., Faley, S., Darzynkiewicz, Z., Cooper, J.M. (2009) Dynamic analysis of apoptosis using cyanine SYTO probes: from classical to microfluidic cytometry. *Exp Cell Res.* **315**, 1706–14.
3. Wlodkowic, D., Skommer, J., McGuinness, D., Faley, S., Kolch, W., Darzynkiewicz, Z., Cooper, J.M. (2009) Chip-based dynamic real-time quantification of drug-induced cytotoxicity in human tumor cells. *Anal Chem.* **Jul 2** (Epub ahead of print), DOI: 10.1021/ac9010217.
4. Wlodkowic D and Darzynkiewicz Z (2008) Please do not disturb: Destruction of chromatin structure by supravital nucleic acid probes revealed by a novel assay of DNA-histone interaction. *Cytometry A.* **10**, 877–879.
5. Wlodkowic, D., Faley, S., Zagnoni, M., Wikswo, J.P., Cooper, J.M. (2009) Microfluidic single-cell array cytometry for the analysis of tumor apoptosis. *Anal Chem.* **81**, 5517–23.
6. Faley S, Copland M, Wlodkowic D, Kolch W, Seale KT, Wikswo JP, Cooper JM (2009) Microfluidic single-cell arrays to interrogate signalling dynamics of individual, patient-derived hematopoietic stem cells. *Lab Chip. (in press)*, DOI: 10.1039/b902083g.

Chapter 24

Purification of Annexin V and Its Use in the Detection of Apoptotic Cells

Katy M. Coxon, James Duggan, M. Francesca Cordeiro, and Stephen E. Moss

Abstract

Cell death by apoptosis has been studied for many years using fluorescently labeled annexin V. Annexin V shows high affinity for the phosphatidylserine that becomes enriched in the outer leaflet of the plasma membrane during apoptosis, but not necrosis, allowing differentiation between the two types of cell death. In this chapter we detail two methods for the purification of annexin V. The first is an untagged recombinant protein using a three step Fast Protein Liquid Chromatography (FPLC) method, and the second using a single step purification protocol via a glutathione S-transferase (GST) tag. Labeling of the resulting annexin V with a fluorescent dye to allow visualization of the protein is also explained. Finally, two methods are described in which a fluorescently labeled derivative of annexin V is used to detect apoptosis, namely the in vitro method of fluorescence-activated cell sorting (FACS) where fluorescent annexin V is used to differentiate apoptotic and necrotic cells within a population; and detection of apoptosing retinal cells (DARC) allowing the identification of apoptotic cells in the retina in vivo.

Key words: Annexin V, Apoptosis, Protein purification, FACS, DARC

1. Introduction

Apoptosis or programmed cell death plays an important role in development, homeostasis, and disease progression. Cells instructed to undergo apoptosis progress through a regulated sequential process of cellular events resulting in death without the release of cytotoxic agents, thus preventing secondary death of the surrounding cells. Initiation of the apoptotic cascade promotes the activation of cellular proteases (caspases) and endonucleases and results in mitochondrial dysfunction. If the pro-apoptotic signal is not reversed, the cell is then committed to the process and DNA fragmentation, extensive protein degradation, and cytoskeletal reorganization is observed. Furthermore, during the early phase of apoptosis the negatively charged membrane phospholipid, phosphatidylserine (PS), usually found in the inner leaflet of the plasma membrane, is externalized (1, 2).

Annexin V (P08758 Swiss-Prot), a 36 kDa protein, is known to have a high affinity for PS in a Ca^{2+}-dependent manner (3). This affinity to PS via calcium has allowed the development of a number of methods using fluorescently labeled annexin V for the detection of apoptotic cells, both in vitro and in vivo. These include fluorescence microscopy based methods, fluorescence-activated cell sorting (FACS) (4) and *d*etection of *a*poptosing *r*etinal *c*ells (DARC) (5).

In addition to the annexin V-dependent protocols for the detection of apoptotic cells, a number of annexin V-independent methods have been established. These include TUNEL, DNA laddering, assays of caspase activity, and visualization of apoptotic bodies.

Detection and quantification of apoptosis in vitro has been performed in labs for many years; however, more recently, methods of detection in vivo, using fluorescence or radiolabeled annexin V have been developed. The ability to detect and quantify apoptotic events in animal models and humans has attracted considerable attention in fields such as ophthalmology, cardiology, and oncology by potentially offering significant clinical value in the diagnosis and treatment of glaucoma (5–7) and myocardial infarction (8, 9) and to monitor therapeutic efficacy during cancer treatment (10–12). The following protocol describes two methods by which annexin V can be used to detect apoptotic cells, one in vitro and the other in vivo.

Two methods for the purification of annexin V are described below. The first method addresses the purification of an untagged annexin V and requires access to Fast Protein Liquid Chromatography (FPLC) instruments to generate accurate salt gradients and to monitor protein elution throughout the process. This method tends to generate a very pure annexin V prep due to the number of different chromatography steps used. The second method describes the purification of a glutathione S-transferase (GST)-tagged annexin V, which can be performed manually, rapidly, and without the use of any specialized equipment. This method is relatively straightforward and lends itself to those who have no or little experience in protein purification, but it does tend to generate a more contaminated final product resulting from the use of a single chromatography step.

2. Materials

Before beginning the purification, labeling and use of annexin V to detect apoptotic cells see Notes 1–4.

2.1. Untagged Annexin V

2.1.1. Bacterial Growth and Induction of Untagged Annexin V Expression

1. *E. coli* strain BL21 harboring an isopropyl β-D-1-thiogalactopyranoside (IPTG) inducible expression vector containing the cDNA encoding annexin V.
2. Luria-Bertani (LB) Media: to 700 mL distilled H_2O (dH_2O) add 10 g bacto-tryptone, 5 g yeast extract, and 10 g NaCl; adjust pH to 7.5 with NaOH and make up to 1 L with dH_2O. Autoclave in two 1 L conical flasks sealed with a bung and covered with foil.
3. Ampicillin: prepare at a concentration of 100 mg/mL in water. Store at −20°C (see Note 5).
4. IPTG: prepare a 1 M solution in water and sterilize using a 0.22 µm syringe filter. Store at −20°C.
5. Lysis buffer: 50 mM Tris–HCl and 3 mM $CaCl_2$ at pH 7.4. Store at 4°C.

2.1.2. Harvesting and Lysis of Bacterial Cultures

1. Lipid binding buffer: 50 mM Tris–HCl and 3 mM $CaCl_2$ at pH 7.4. Store at 4°C.
2. 0.22 µm syringe filter: Millipore, Watford, UK.

2.1.3. Lipid Binding Enrichment of Annexin V

3. Lipid binding buffer: 50 mM Tris–HCl and 3 mM $CaCl_2$ at pH 7.4. Store at 4°C.
4. Lipid elution Buffer: 50 mM Tris–HCl and 60 mM EGTA at pH 7.4. Store at 4°C.

2.1.4. Analysis of Fractions by SDS-PAGE

1. 2× SDS-PAGE sample buffer: 4% (w/v) SDS, 10% (v/v) 2-mercaptoethanol, 20% (v/v) glycerol, 0.004% (v/v) bromophenol blue, and 0.125 M Tris–HCl at pH 6.8. Store in aliquots at −20°C.
2. 12% SDS-PAGE gel:
3. 1× SDS-PAGE running buffer: 25 mM Tris–HCl, 190 mM glycine, and 0.1% (w/v) SDS at pH 8.3. Store at room temperature. Can be made up as a 10× solution.
4. Prestained molecular weight protein marker: Dual Colour Precision Plus Prestained Protein Standards from BioRad, Hemel Hempstead, Hertfordshire.
5. Coomassie blue: 50% (v/v) methanol, 10% (v/v) acetic acid, and 0.05% (w/v) coomassie brilliant blue R-250. Filter sterilize before use and store at room temperature.

2.1.5. Desalting

1. 5 mL HiTrap Desalting column from GE Healthcare, Buckinghamshire, UK.
2. Centrifugal concentrator, 10,000 Da molecular weight cut off (mwco) from Sartorius Stedim Biotech, Surrey, UK.
3. Ion exchange binding buffer: 50 mM Tris–HCl and 20 mM NaCl at pH 7.4. Store at 4°C.

2.1.6. Ion Exchange Chromatography

1. POROS HQ 50 μm column from Applied Biosystems, Foster City, CA, USA.
2. Ion exchange binding buffer: 50 mM Tris–HCl and 20 mM NaCl at pH 7.4. Store at 4°C.
3. Ion exchange elution buffer: 50 mM Tris–HCl and 1 M NaCl at pH 7.4. Store at 4°C.

2.1.7. Size Exclusion Chromatography

1. Centrifugal concentrator, 10,000 Da mwco from Sartorius Stedim Biotech, Surrey, UK.
2. Superdex 75 10/300 GL column from GE Healthcare, Buckinghamshire, UK.
3. Labeling buffer: 20 mM Sodium Bicarbonate, 100 mM NaCl, and 1 mM EGTA at pH 7.0. Store at 4°C.

2.2. GST-Tagged Annexin V

2.2.1. Bacterial Growth and Induction of GST-Annexin V Expression

1. *E. coli* strain BL21 harboring a pGEX vector (GE Healthcare, Buckinghamshire, UK) containing the cDNA encoding annexin V cloned downstream and in frame with the GST gene.
2. LB Media: to 700 mL of dH_2O, add 10 g bacto-tryptone, 5 g yeast extract, and 10 g NaCl; adjust pH to 7.5 with NaOH and make up to 1 L with dH_2O. Autoclave in two 1 L conical flasks sealed with a bung and covered with foil.
3. Ampicillin: prepare at a concentration of 100 mg/mL in water. Store at –20°C (see Note 5).
4. IPTG: prepare a 1 M solution in water and sterilize using a 0.22 μm syringe filter. Store at –20°C.
5. GST-binding buffer: 1× PBS pH 7.4. Store at 4°C.

2.2.2. Harvesting and Lysis of Bacterial Cultures

1. GST-binding buffer: 1× PBS pH 7.4. Store at 4°C.
2. 0.22 μm syringe filter: Millipore, Watford, UK.

2.2.3. Binding of GST-Tagged ANXV to GSTrap FF Column

1. 5 mL GSTrap FF Column from GE Healthcare, Buckinghamshire, UK.
2. GST-binding buffer: 1× PBS pH 7.4. Store at 4°C.

2.2.4. On-Column Cleavage of GST-Tagged ANXV

1. Thrombin protease: prepared in 100 unit aliquots in GST-binding buffer. Store at –80°C.
2. GST-binding buffer: 1× PBS at pH 7.4. Store at 4°C.

2.2.5. Purification of Annexin V

1. GST-binding buffer: 1× PBS at pH 7.4. Store at 4°C.
2. GST-elution buffer: 50 mM Tris–HCl and 10 mM reduced glutathione at pH 8.0.

2.2.6. Analysis of Annexin V Purification

1. See Subheading 2.1.4.

2.2.7. Buffer Exchange of Annexin V	1. Labeling buffer: 20 mM Sodium Bicarbonate, 100 mM NaCl, and 1 mM EGTA, pH 7.0. Store at 4°C. 2. Hi Trap Desalting column from GE Healthcare, Buckinghamshire, UK.
2.3. Production of Labeled Annexin V *2.3.1. Annexin V Labeling Reaction*	1. Centrifugal concentrator, 10,000 Da mwco from Sartorius Stedim Biotech, Surrey, UK. 2. Bradford reagent from Sigma Aldrich, Dorset, UK. 3. Fluorescent dye: From Dyomics, Jena, Germany. Store at −20°C in the dark.
2.3.2. Removal of Unconjugated Dye	1. Disposable PD-10 Desalting Columns from GE Healthcare, Buckinghamshire, UK. 2. Labeling buffer: 20 mM Sodium Bicarbonate, 100 mM NaCl, and 1 mM EGTA at pH 7.0. Store at 4 °C.
2.4. Detection of Apoptotic Cells *2.4.1. In Vitro: FACS*	1. Fluorescent Annexin V – prepare a 100 μg/mL solution in 1× PBS at pH 7.4. Store in the dark at 4°C. 2. Propidium Iodide (PI) – prepare a 100 μg/mL solution in 1× PBS at pH 7.4 and filter sterilize using a 0.22 μm syringe filter. Store in the dark at 4°C. 3. 10× Binding buffer – 100 mM HEPES pH 7.4, 1.4 M NaCl, and 25 mM $CaCl_2$. Use at 1× by diluting 1 part of 10× binding buffer with nine parts of distilled water. 4. 1× PBS at pH 7.4. Store at 4°C.
2.4.2. In Vivo: DARC	1. Fluorescent Annexin V – 2.5 μg in 5 μL of 1× PBS. Store in the dark at 4°C.

3. Methods

3.1. Untagged Annexin V *3.1.1. Bacterial Growth and Induction of Untagged Annexin V Expression*	1. Inoculate 10 mL of sterile LB containing ampicillin (100 μg/mL) with a single colony of BL21 containing the annexin V expression vector. Incubate the starter culture overnight at 37°C and 180 rpm. 2. Add 5 mL of the starter culture to a 1 L flask containing 500 mL of LB/ampicillin (100 μg/mL) and incubate at 37°C and 180 rpm until an OD of 0.6 is reached at A_{600}. 3. To induce expression of annexin V, IPTG is added at a final concentration of 1 mM and the cultures are incubated for a further 2 h at 37°C and 180 rpm.

3.1.2. Harvesting and Lysis of Bacterial Cultures

1. Decant the cultures into centrifuge bottles and centrifuge for 20 min at $3,000 \times g$. Remove the supernatant and freeze the pellets at $-80°C$ until needed.
2. Resuspend the pellets in a total of 20 mL of prechilled lipid binding buffer and sonicate on ice for 45 s (see Note 6). Mix the bacterial suspension by inverting and repeat the sonication step twice more. Transfer the cell lysate to a centrifuge tube.
3. To separate the annexin rich insoluble fraction from the soluble fraction of the cell lysate the sample is centrifuged at $36,000 \times g$, $4°C$ for 25 min.
4. Remove the supernatant and retain for analysis and allow good drainage of the pellet to remove as much of the supernatant as possible (see Note 7).

3.1.3. Lipid Binding Enrichment of Annexin V

The following section of the purification method relies upon the ability of the annexin V to bind PS in a calcium-dependent manner. The endogenous PS of the bacterial lysate is utilized.

1. Fully resuspend the pellet in 20 mL of lipid binding buffer using a dounce homogeniser and transfer to a centrifuge tube (see Note 8).
2. Centrifuge the sample for 25 min at $36,000 \times g$, $4°C$. Remove the supernatant and retain for analysis.
3. Repeat the first two steps at least a further four times to remove any unbound proteins (see Note 9).
4. To elute the calcium-dependent lipid bound proteins from the pellet, resuspend it in 20 mL of lipid elution buffer (see Note 10) and centrifuge for 25 min at $36,000 \times g$, $4°C$.
5. The annexin will now be present in the supernatant (see Note 11). Retain both the supernatant and the pellet for analysis and further processing.

3.1.4. Analysis of Samples by SDS-PAGE

1. Add 10 µL of sample to 10 µL of 2× SDS-PAGE sample buffer and incubate at $95°C$ for 5 min. Repeat for all samples collected during the lipid binding enrichment step.
2. Briefly centrifuge the tubes to collect the sample in the bottom.
3. Load the 10 µL samples on to a 12% SDS-PAGE gel and run at 225 V for 45 min in 1× SDS-PAGE running buffer.
4. Remove the gel from the glass plates and carefully detach the stacking gel using a razor blade.
5. Place the resolving gel in coomassie blue and incubate on a rocker at room temperature for 30 min.
6. To destain the gel and allow visualization of the proteins, place the gel in water in a microwavable container and microwave on high power for 10 min. Ensure that the gel does not dry out (Fig. 1; see Note 12).

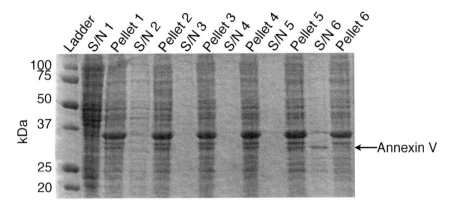

Fig. 1. SDS-PAGE analysis of annexin V enrichment by lipid binding. The annexin V rich insoluble fraction of the cell lysate was subjected to repeated washes with lipid binding buffer to remove any unbound protein. Calcium dependent-lipid binding proteins were eluted from the pellet using lipid elution buffer. The supernatant (S/N) and pellet from each wash step were analyzed on a 12% SDS-PAGE gel using coomassie blue to visualize the proteins. Annexin V (running at ~30 kDa) can be seen in pellets 1–5 and is then eluted into supernatant 6.

3.1.5. Desalting

1. Filter the remaining supernatant using a 0.22 μm syringe filter to remove any residual lipids before proceeding to the column purification steps (see Note 13).
2. To ensure the effective binding of proteins to the ion exchange column, removal of salt from the protein sample must be performed. The use of a desalting column is a quick and easy way to achieve this.
3. Concentrate the sample to 1.5 mL by centrifugation in a centrifugal concentrator at $3,000 \times g$, 4°C.
4. Manually equilibrate the HiTrap Desalting column with 20 mL of prechilled-ion exchange binding buffer, using a syringe (see Note 14).
5. Load the annexin V onto the column in exactly 1.5 mL (see Note 15). Discard the flow through.
6. Elute the annexin V from the column with 2 mL of prechilled ion exchange binding buffer. Collect the 2 mL-eluate.

3.1.6. Ion Exchange Chromatography

1. The remaining sections of the purification protocol assume the use of an FPLC to generate an accurate salt gradient and to monitor protein elution over the column runs.
2. Equilibrate the POROS HQ 50 μm column with 2 column volumes of ion exchange binding buffer at a flow rate of 3 mL/min.
3. Load the protein onto the column (see Note 16) and remove any unbound protein by washing the column with ion exchange binding buffer and monitor the elution of the unbound protein by absorbance at A_{280nm}. Stabilization of the

baseline indicates the removal of all unbound protein. Collect eluate in 1 mL fractions.

4. To elute the bound protein from the column gradually, a salt gradient is used. Elute the bound protein using a gradient of 0–100% elution buffer (0–1 M salt) over 200 column volumes (see Note 17). Collect eluate in 1 mL fractions.

5. Once the gradient has reached 100%, continue to run until all remaining protein has eluted from the column. Collect eluate in 1 mL fractions.

6. Analyze the fractions by SDS-PAGE, see Subheading 3.1.4 (Fig. 2).

3.1.7. Size Exclusion Chromatography

1. Concentrate the annexin V-containing fractions to 250–500 µL by centrifugation at $3,000 \times g$ and 4°C in a centrifugal concentrator.

2. Equilibrate the Superdex 75 column with 1.5 column volumes of labeling buffer, at a flow rate of 0.5 mL/min.

3. Load the protein on to the column and wash with 1.5 column volumes of labeling buffer, at a flow rate of 0.5 mL/min. Collect eluate in fractions of 0.5 mL.

4. Analyze the fractions by SDS-PAGE and Coomassie staining, see Subheading 3.1.4 (Fig. 3).

3.2. GST-Tagged Annexin V

3.2.1. Bacterial Growth and Induction of GST-Annexin V Expression

1. Inoculate 10 mL of sterile LB containing ampicillin (100 µg/mL) with a single colony of BL21 containing the GST-annexin V expression vector. Incubate the starter culture overnight at 37°C and 180 rpm.

2. Using 5 mL of the starter culture, inoculate two 1 L flasks each containing 500 mL of LB/ampicillin (100 µg/mL).

3. Incubate the cultures at 37°C and 180 rpm for 4–5 h.

4. To induce expression of annexin V, IPTG is added at a final concentration of 1 mM and the cultures are incubated overnight at 16°C and 180 rpm (see Note 18).

3.2.2. Harvesting and Lysis of Bacterial Cultures

1. Harvest and store the cultures according to Subheading 3.1.2.

2. Resuspend the pellet in 20 mL GST-binding buffer and lyse by sonication as described in Subheading 3.1.2.

3. Separate the soluble and insoluble fractions of the cell lysate by centrifugation at $36,000 \times g$ and 4°C for 25 min.

4. Pass the soluble fraction through a 0.22-µm syringe filter to further clarify the sample.

3.2.3. Binding of GST-Tagged ANXV to GSTrap FF Column

1. Manually equilibrate the GSTrap FF column with 20 mL pre-chilled GST-binding buffer using a syringe.

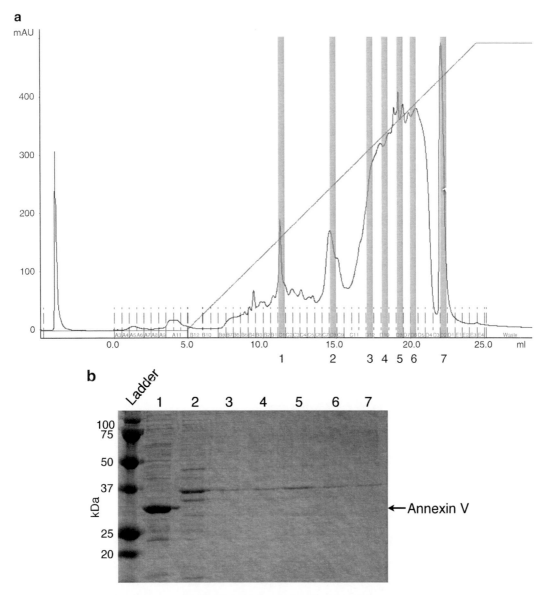

Fig. 2. Purification of annexin V using anion exchange chromatography and analysis by SDS-PAGE. (a) FPLC chromatogram showing the purification of annexin V using a POROS HQ 50 μm column. The y-axis shows absorbance at 280 nm while the x-axis shows the volume of buffer passed through the column in milliliters. The *trace represents* the elution of protein from the column measured at $A_{280\,nm}$. The *diagonal line* shows the progression of the elution gradient (0–1 M NaCl) over the run. The *broken vertical lines* indicate the regions of the run where fractions were collected and the *numbered gray boxes* show the fractions taken forward for analysis by SDS-PAGE. (b) SDS-PAGE analysis of the fractions indicated on the FPLC trace. The fractions were separated on a 12% SDS-PAGE gel, which was subsequently stained using coomassie blue. Fraction 1 was shown to contain annexin V.

2. Slowly apply the soluble fraction onto the column at approximately 1–5 mL/min to allow maximal binding of the GST-tagged annexin V to the resin. Collect the flow-through for analysis.

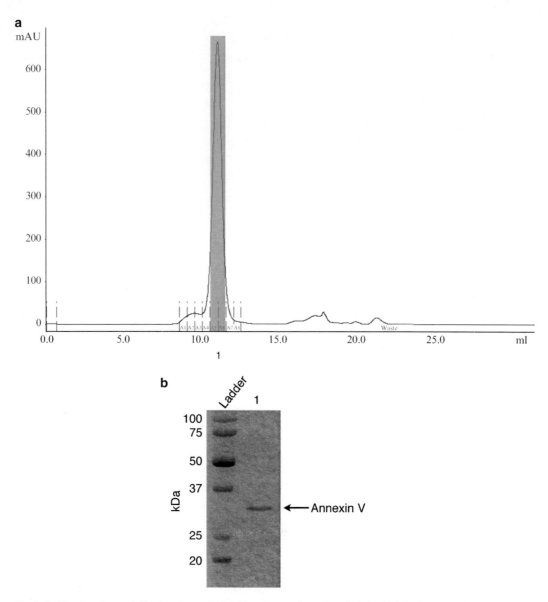

Fig. 3. Purification of annexin V using size exclusion chromatography and analysis by SDS-PAGE. (**a**) FPLC chromatogram of the size exclusion chromatography run using a Superdex 75 10/300 GL column. The *y*-axis shows the measurement of absorbance at 280 nm. The *x*-axis shows the volume of buffer passed through the column in milliliters. The *blue trace* represents the A_{280nm} and illustrates protein eluting from the column. The eluate collected as fractions is indicated by the *red lines* and the *gray box* shows those fractions analyzed by SDS-PAGE. (**b**) The fractions indicated by the *gray box* were pooled and analyzed on a 12% SDS-PAGE gel. Coomassie blue was used to visualize the protein.

3. Wash the column with 50 mL of prechilled GST-binding buffer to remove any unbound protein. Collect the flow-through fraction for analysis.

3.2.4. On-Column Cleavage of GST-Tagged ANXV

1. Make up 100 units of thrombin protease in 5 mL of prechilled GST-binding buffer (see Note 19).
2. Load the thrombin solution onto the GSTrap FF column using a syringe and discard the flow through.
3. Replace the column stoppers and incubate for 18–24 h at 22°C.

3.2.5. Purification of Annexin V

1. Elute the annexin V from the GSTrap FF column in 10 × 1 mL fractions of prechilled GST-binding buffer. Keep the fractions on ice.
2. Remove the GST and any remaining GST-tagged annexin V from the column by elution in 10 mL of prechilled GST-elution buffer. Retain the eluate for analysis.

3.2.6. Analysis of Annexin V Purification

1. Add 5 μL of sample to 5 μL of 2× SDS-PAGE sample buffer and incubate at 95°C for 5 min. Repeat for all samples collected during the purification process.
2. Analyze the sample according the method in Subheading 3.1.4 (see Note 20; Fig. 4).

3.2.7. Buffer Exchange of Annexin V

1. Manually equilibrate the HiTrap Desalting column with 20 mL of prechilled labeling buffer using a syringe.
2. Load the annexin V onto the column in exactly 1.5 mL (see Note 21). Discard the flow-through fraction.
3. Elute the annexin V from the column in 2 mL of prechilled-labeling buffer. Collect the 2 mL of eluate. The annexin V will now be in labeling buffer.

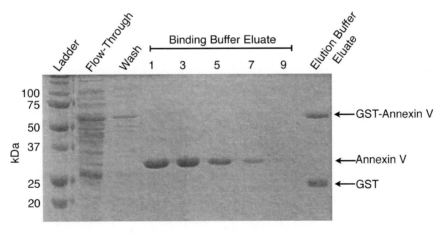

Fig. 4. SDS-PAGE analysis of annexin V purification via a GST-fusion protein. The soluble fraction of the cell lysate was applied to a GSTrap FF column and the flow through collected. Any unbound protein was eluted with binding buffer and retained. Bound GST-tagged annexin V was cleaved overnight with thrombin protease. Annexin V was eluted from the column in binding buffer while the GST-tag and any remaining GST-tagged annexin were eluted with elution buffer and retained. Fractions were analyzed on a 12% SDS-PAGE gel and visualized using coomassie blue. The flow through and wash samples show very little presence of GST-tagged annexin V (~62 kDa). Following cleavage, large amounts of annexin V can be seen in the binding buffer eluate, while the presence of GST-tagged annexin V in the elution buffer eluate indicate incomplete cleavage of the GST-tagged annexin V.

3.3. Production of Labeled Annexin V

There are many types of fluorescent dyes available with widely varying wavelengths and different conjugation chemistry (see Note 22). The dye described in this chapter was purchased from Dyomics and requires the use of the protocol below; however, if dye is purchased from an alternative source it is recommended that the manufacturer's instructions are followed.

3.3.1. Annexin V Labeling Reaction

1. Concentrate the annexin V to approximately 1–2 mg/mL by centrifugation at $3,000 \times g$ and 4°C in a centrifugal concentrator.
2. Resuspend 1 mg of the dye in 100 µL DMF and invert to mix.
3. Transfer the concentrated annexin into a foil wrapped tube. Add the dye to the annexin at a molar ratio of 3:1 and shake gently at room temperature for 2 h to allow conjugation.

3.3.2. Removal of Unconjugated Dye

1. Equilibrate the PD-10 column with 25 mL of labeling buffer, discarding the flow through.
2. Make the labeled annexin V up to 2.5 mL in labeling buffer.
3. Load the labeled annexin V into the PD-10 column, discarding the flow through.
4. Once 2.5 mL of solution has entered the column, completely add 3.5 mL of labeling buffer and collect the resulting 3.5 mL of column flow through. This should contain the labeled annexin, but no unconjugated dye.

3.4. Detection of Apoptotic Cells

3.4.1. In Vitro: FACS

1. Following the induction of apoptosis using your chosen method, wash the cells twice in ice cold 1× PBS.
2. Resuspend the cells in 100 µL of 1× binding buffer and count the number of cells using a hemocytometer, diluting accordingly.
3. Dilute cells to achieve a concentration of 1×10^6 cells/mL.
4. Transfer 100 µL of the cell suspension (1×10^5 cells) to a microfuge tube and add 5 µg of labeled-annexin V and 5 µL PI.
5. Mix the contents of the tube by gently vortexing and incubate for 15 min in the dark and at room temperature.
6. Add 400 µL of 1× binding buffer.
7. Analyze by flow cytometry within 1 h of preparation.

3.4.2. In Vivo: DARC

The following protocol describes the application of DARC in the rat – appropriate modifications would need to be made for other species, tissues, or tumors.

1. To generate a baseline image for comparison, image the retina using a confocal scanning laser ophthalmoscope, exciting at 495 nm and emitting at 520 nm (see Note 23) with imaging software to compensate for eye movements and improve the

signal-to-noise ratio Following the induction of apoptosis using your chosen method, inject 5 μL of fluorescently labeled-annexin V, into the vitreous of the eye using a 34-gauge needle. 5 μL of saline should be used as the control.

2. Allow 2–4 h for the protein to diffuse across the vitreous to the retina.

3. Reimage the retina using the settings mentioned previously (Fig. 5).

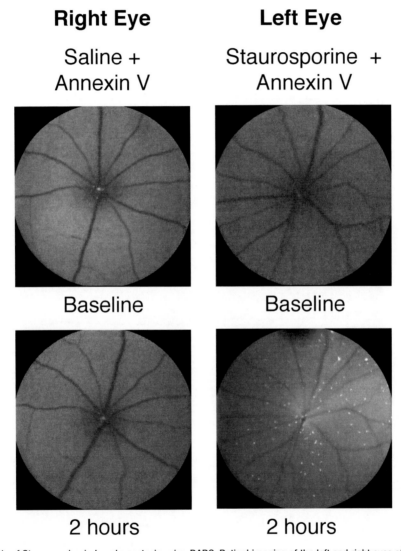

Fig. 5. Analysis of Staurosporine induced apoptosis using DARC. Retinal imaging of the left and right eyes of a Dark Agouti rat (200 g) was performed using a confocal scanning laser ophthalmoscope to obtain a baseline fluorescence (excitation – 495 nm; emission – 520 nm) reading. The left eye was injected with 2.5 μg of annexin V labeled with fluorescein isothiocyanate (annexin V-FITC) and 1 μg staurosporine (to induce apoptosis) in a total volume of 5 μL. The right eye was injected with 2.5 μg of annexin V-FITC made up to 5 μL in sterile saline. Retinal imaging was performed after 2 h. The *white spots*, indicating apoptosis of retinal ganglion cells, can be seen only in the eye treated with annexin V-FITC and staurosporine.

4. Notes

1. All buffers should be prechilled.
2. All fractions should be kept on ice when in use and stored at 4°C.
3. All fluorescent material should be kept in the dark.
4. Annexin V is known to aggregate at high concentrations; therefore it is not recommended to concentrate the protein any higher than 2 mg/mL. The protein is also prone to aggregation in the presence of high levels of calcium.
5. The use of ampicillin as an antibiotic for selection of transformed bacterial cells is only an example. Alternative expression vectors may require the use of a different antibiotic.
6. Do not allow the bacterial suspension to bubble during sonication as this will lead to inefficient cell lysis and protein denaturation.
7. Remove as much of the supernatant as possible as it contains the majority of the contaminating bacterial proteins.
8. Wash out the dounce homogeniser with a little excess buffer after each resuspension step to maximize yields.
9. More wash steps can be added to obtain a cleaner annexin V prep, but will result in low yields.
10. Less buffer can be used to resuspend the pellet in the elution step if a more concentrated annexin V sample is required.
11. Following the final centrifuge spin, the supernatant must be decanted from the pellet as soon as the run has ended to prevent the contamination of the sample with lipids. Failure to do this will result in difficulties when attempting to filter sterilize the sample in the next step.
12. Annexin V runs at approximately 30–32 kDa on an SDS-PAGE and not the expected 36 kDa.
13. Any residual lipids in the supernatant will clog the 0.22-μm filter during this clarification step. More than one filter is usually required.
14. Manual application of buffer and/or samples to columns with a syringe should be performed in a drip-wise manner.
15. The desalting can either be performed as a single step with the whole sample or the sample can be divided into 1.5 mL fractions and a number of desalting steps can be performed. If the latter option is chosen a superloop will be required to load the sample onto the POROS column. Alternatively, a further concentration step can be performed to reduce the sample volume.

16. The flow rate should be adjusted appropriately if a superloop is being used to load the sample, so as not to exceed the pressure limit.

17. Annexin V usually elutes from the column at 150 mM NaCl.

18. The culture incubation at 16°C helps to prevent the formation of inclusion bodies and results in a higher yield of soluble protein.

19. If complete cleavage of the GST tagged-annexin V protein is not seen, the amount of thrombin protease used in the reaction can be increased.

20. Molecular weights of proteins of interest: Annexin V – 36 kDa (but runs at approximately 30–32 kDa), GST-annexin V – 62 kDa, and GST – 26 kDa.

21. Depending on how much annexin V has been obtained from the purification step, buffer exchange can be performed as a single column run or using multiple runs.

22. Over-labeling of the annexin V molecule at the lysine residues using NHS-ester conjugation chemistry can interfere with PS binding.

23. For this particular method, annexin V labeled with a fluorophore, which excites and emits at the appropriate wavelengths is required.

References

1. Fadok, V. A., Voelker, D. R., Campbell, P. A., Cohen, J. J., Bratton, D. L. and Henson, P. M. (1992a) Exposure of phosphatidylserine on the surface of apoptotic lymphocytes triggers specific recognition and removal by macrophages *J Immunol* **148**, 2207–16.

2. Fadok, V. A., Savill, J. S., Haslett, C., Bratton, D. L., Doherty, D. E., Campbell, P. A. and Henson, P. M. (1992b) Different populations of macrophages use either the vitronectin receptor or the phosphatidylserine receptor to recognize and remove apoptotic cells *J Immunol* **149**, 4029–35.

3. Andree, H. A., Reutelingspergerm, C. P., Hauptmann, R., Hemker, H. C., Hermens, W. T. and Willems, G. M. (1990) Binding of vascular anticoagulant alpha (VAC alpha) to planar phospholipid bilayers *J Biol Chem* **265**, 4923–8.

4. Koopman, G., Reutelingsperger, C. P., Kuijten, G. A., Keehnen, R. M., Pals, S. T. and van Oers, M. H. Annexin V for flow cytometric detection of phosphatidylserine expression on B cells undergoing apoptosis *Blood* **84**, 1415–20.

5. Cordeiro, M. F., Guo, L., Luong, V., Harding, G., Wang, W., Jones, H. E., Moss, S. E., Sillito, A. M. and Fitzke, F. W. (2004) Real-time imaging of single nerve cell apoptosis in retinal neurodegeneration *Proc Natl Acad Sci USA* **101**, 13352–6.

6. Guo, L., Salt, T. E., Maass, A., Luong, V., Moss, S. E., Fitzke, F. W. and Cordeiro, M. F. (2006) Assessment of neuroprotective effects of glutamate modulation on glaucoma-related retinal ganglion cell apoptosis in vivo *Invest Ophthalmol Vis Sci.* **47**, 626–33.

7. Guo, L., Salt, T. E., Luong, V., Wood, N., Cheung, W., Maass, A., Ferrari, G., Russo-Marie, F., Sillito, A. M., Cheetham, M. E., Moss, S. E., Fitzke, F. W. and Cordeiro MF. (2007) Targeting amyloid-beta in glaucoma treatment *Proc Natl Acad Sci USA* **104**, 13444–9.

8. Hofstra, L., Liem, I. H., Dumont, E. A., Boersma, H. H., van Heerde, W. L., Doevendans, P. A., De Muinck, E., Wellens, H. J., Kemerink, G. J., Reutelingsperger, C. P. and Heidendal, G. A. (2000) Visualisation of cell death in vivo in patients with acute myocardial infarction *Lancet* **356**, 209–12.

9. Thimister, P. W., Hofstra, L., Liem, I. H., Boersma, H. H., Kemerink, G., Reutelingsperger, C. P. and Heidendal, G. A. (2003) In vivo detection of cell death in the area at risk in acute myocardial infarction *J Nucl Med* **44**, 391–6.

10. Belhocine, T., Steinmetz, N., Hustinx, R., Bartsch, P., Jerusalem, G., Seidel, L., Rigo, P. and Green, A. (2002) Increased uptake of the apoptosis-imaging agent (99 m) Tc recombinant human Annexin V in human tumors after one course of chemotherapy as a predictor of tumor response and patient prognosis *Clin Cancer Res* **8**, 2766–74.

11. Kartachova, M. S., Valdés Olmos, R. A., Haas, R. L., Hoebers, F. J., van Herk, M. and Verheij, M. (2008) 99mTc-HYNIC-rh-annexin-V scintigraphy: visual and quantitative evaluation of early treatment-induced apoptosis to predict treatment outcome *Nucl Med Commun* **29**, 39–44.

12. Kurihara, H., Yang, D. J., Cristofanilli, M., Erwin, W. D., Yu, D. F., Kohanim, S., Mendez, R. and Kim, E. E. (2008) Imaging and dosimetry of 99mTc EC annexin V: preliminary clinical study targeting apoptosis in breast tumors *Appl Radiat Isot* **66**, 1175–82.

Chapter 25

Measurement of DNA Damage in Individual Cells Using the Single Cell Gel Electrophoresis (Comet) Assay

Janet M. Hartley, Victoria J. Spanswick, and John A. Hartley

Abstract

The Single Cell Gel Electrophoresis (Comet) assay is a simple, versatile and sensitive method for measuring DNA damage in individual cells, allowing the determination of heterogeneity of response within a cell population. The basic alkaline technique described is for the determination of DNA strand break damage and its repair at a single cell level. Specific modifications to the method use a lower pH ('neutral' assay), or allow the measurement of DNA interstrand cross-links. It can be further adapted to, for example, study specific DNA repair mechanisms, be combined with fluorescent in situ hybridisation, or incorporate lesion specific enzymes.

Key words: Single cell gel electrophoresis (Comet) assay, Comet assay, DNA strand breaks, DNA interstrand cross-links, DNA repair

1. Introduction

The Single Cell Gel Electrophoresis (Comet) assay is a method for measuring single and double strand breaks in DNA. It was originally developed by Ostling and Johanson (1) as a method which allows visualisation of DNA damage in individual cells. In their assay, cells were lysed and electrophoresed at a pH of 10. A few years later Singh et al. (2) used a similar method under more alkaline conditions. Since then the method has become widely adopted for a range of applications from basic science through to clinical medicine (3–8). The principles of the assay have recently been reviewed in detail (9).

The basic alkaline technique described in this chapter is for the determination of DNA strand break damage and its repair at a single cell level. It is a simple, versatile, economic and sensitive method for measuring DNA damage and since analysis is made in individual cells, an important feature is that heterogeneity of response within a cell population can be determined. Another advantage is that relatively few cells are needed.

Specific modifications to the method use a lower pH ('neutral' assay), or allow the measurement of DNA interstrand cross-links such as those produced by a number of cancer chemotherapeutic agents (10). It can be further adapted to, for example, study specific DNA repair mechanisms (11), be combined with fluorescent in situ hybridisation (12, 13), or to incorporate lesion specific enzymes (14).

2. Materials

2.1. Cell Preparation and Drug Treatment

2.1.1. Suspension Cell Lines

1. Appropriate tissue culture medium.
2. L-glutamine (Autogen Bioclear, Calne, UK).
3. Foetal Calf Serum (FCS) (Autogen Bioclear).

2.1.2. Adherent Cell Lines

1. Appropriate tissue culture medium.
2. L-glutamine (Autogen Bioclear).
3. FCS (Autogen Bioclear).
4. Trypsin/ethylenediaminetetraacetic acid (EDTA) (1×) (Autogen Bioclear).

2.2. Single Cell Gel Electrophoresis (Comet) Assay

1. Single-frosted glass microscope slides and glass 24 × 40 mm coverslips.
2. Agarose, Type 1-A (Sigma, Poole, UK).
3. Agarose, Type VII: low gelling temperature (LGT) (Sigma).
4. Lysis buffer: 100 mM disodium EDTA, 2.5 M NaCl, 10 mM Tris–HCl; pH is adjusted to 10.5–11.0 with sodium hydroxide pellets. 1% triton X-100 to be added immediately before use. Store at 4°C.
5. Alkali buffer: 50 mM NaOH and 1 mM disodium EDTA at pH 12.5. Caution: Corrosive. Store at 4°C.
6. Neutralisation buffer: 0.5 M Tris–HCl at pH 7.5. Store at 4°C.
7. Phosphate buffered saline (PBS) at pH 7.4. Store at 4°C.
8. Flat bed electrophoresis. This should be of sufficient size to hold a large number of slides e.g. 30 × 25 cm gel tank from Flowgen Bioscience, Nottingham, UK, which holds up to 45 slides.

2.3. Staining and Visualisation

1. Propidium iodide (Sigma), 2.5 μg/mL. Make up fresh before use. Caution: Toxic and light sensitive.
2. Glass coverslips, 24 × 40 mm.

3. Double distilled water.

4. Epi-fluorescence microscope equipped with high pressure mercury light source using a 580 nm dichroic mirror, 535 nm excitation filter and 645 nm emission filter for propidium iodide staining (e.g. Olympus BX51 inverted microscope with Olympus U-RFL-T mercury lamp and Sony XCD-X710 digital camera).

5. Images are visualised, captured and analysed using a suitable image analysis system. Our laboratory uses Komet 5.5 analysis software from Andor Technology, formerly Kinetic Imaging (Belfast, UK) (see Note 1).

3. Methods

3.1. Cell Preparation and Drug Treatment

3.1.1. Suspension Cell Lines

1. Exponentially growing cells should be used at a density of 2.5–3.0×10^4 cells/mL in the appropriate medium containing 2 mM glutamine and 10% FCS.

2. Where appropriate, a minimum of 2 mL cells are treated with the DNA damaging agent and incubated for the appropriate time at 37°C in a humidified atmosphere with 5% carbon dioxide (see Note 2).

3. Pellet cells by centrifugation at $200 \times g$ for 5 min at room temperature.

4. Remove supernatant and re-suspend cells in 2 mL of fresh drug-free medium containing 2 mM glutamine and 10% FCS maintained at 4°C.

5. Alternatively, if required, re-suspend cells in the above medium at 37°C and incubate cells for the required post-treatment time, prior to step 4 (see Note 3).

6. Cells are now ready to be processed as described in Subheading 3.2, step 2.

7. Alternatively, cells can be frozen and stored at –80°C following treatment with the DNA damaging agent. This allows samples to be taken at different time points e.g. for repair studies (Step 5). The samples can then be processed in a single assay reducing the possibility of inter-assay variation. Following treatment, centrifuge the cells at $200 \times g$ for 5 min at 4°C. Discard the supernatant and re-suspend the pellet in 2 mL freezing mixture (FCS containing 10% dimethylsulphoxide) (Sigma). Aliquot into 2×1 mL freezing vials and freeze at –80°C (see Notes 4 and 5).

3.1.2. Adherent Cell Lines

1. Exponentially growing cells are treated with DNA damaging agent (where appropriate) and incubated for the required

time at 37°C in a humidified atmosphere with 5% carbon dioxide (see Note 2).

2. If a post-treatment incubation is required remove medium and replace with fresh medium at 37°C and incubate cells using the above conditions for the required time (see Note 3).

3. After the appropriate incubation, carefully remove the media and trypsinise cells with tryspin/EDTA solution until all cells have rounded up and detached (see Note 6).

4. Neutralise trypsinisation by the addition of fresh media containing 2 mM glutamine and 10% FCS.

5. Transfer cells to a universal tube, wash twice with media containing 2 mM glutamine and 10% FCS maintained at 4°C by centrifuging at $200 \times g$ for 5 min at 4°C.

6. Cells are now ready to be processed as described in Subheading 3.2, step 2.

7. Alternatively, cells can be frozen and stored at −80°C as described in Subheading 3.1.1, step 7 and the Single Cell Gel Electrophoresis (Comet) assay performed at a later date.

3.2. Alkaline Single Cell Gel Electrophoresis (Comet) Assay

Important: All stages of this assay should be carried out on ice under subdued lighting, solutions maintained at 4°C and incubations performed in the dark where indicated (see Note 7).

The method described uses slides prepared by the operator. Commercial kits are also available (see Note 8).

1. Pre-coat microscope slides with 1% type 1-A agarose in water by pipetting 1 mL of molten agarose onto the centre of the slide and place a coverslip on top. Allow to set and remove the coverslip. Slides are then allowed to dry overnight at room temperature. The slides must be dry before use (see Note 9).

2. Cells to be analysed are diluted to $2.5–3.0 \times 10^4$ cells/mL in the appropriate medium at 4°C as a single cell suspension. 1–2 mL of cells are required for each data point of the assay.

3. Take 0.5 mL of the cells and put in a 24-well plate on ice. Add 1 mL of molten 1% LGT agarose in water cooled to 40°C, mix, pipette 1 mL onto the centre of the slide on ice and place a coverslip on top (see Note 10). Once set, remove coverslip and place in a tray on ice. Duplicate slides should be prepared (see Note 11).

4. Add ice cold lysis buffer containing 1% triton X-100 ensuring that all slides are sufficiently covered.

5. Incubate on ice for 1 h in the dark.

6. Carefully remove lysis buffer ensuring that the gels are intact and remain on the slides (see Note 12).

7. Add ice cold double distilled water to completely cover the slides. Incubate on ice for 15 min in the dark. This should then be repeated a further three times.

8. Remove slides from tray and transfer carefully to an electrophoresis tank (see Note 13).

9. Cover slides with ice cold alkali buffer and incubate for 45 min in the dark (see Note 14).

10. Electrophorese for 25 min at 18 V (0.6 V/cm), 250 mA. This must be carried out in the dark (see Note 15).

11. Carefully remove slides from the buffer and place on a horizontal slide rack.

12. Flood each slide twice with 1 mL neutralisation buffer and incubate for 10 min.

13. Rinse slides twice with 1 mL PBS and incubate for 10 min.

14. Remove all excess liquid from slides and allow to dry overnight at room temperature.

3.3. Staining and Visualisation

1. Re-hydrate slides in double distilled water for 30 min.

2. Flood each slide twice with 1 mL 2.5 μg/mL propidium iodide solution and incubate for at least 30 min at room temperature in the dark (see Note 16).

3. Rinse slides twice with double distilled water for 10 min and once for 30 min.

4. Allow slides to dry at 40°C in the dark (see Note 17).

5. Once dry, place a few drops of distilled water onto the slide and cover with a coverslip (see Note 18).

6. Examine and record the data from individual cells at 20× magnification (Fig. 1) analysing a minimum of 25 images per duplicate slide (i.e. minimum 50 in total) (see Note 19).

7. Data can be expressed in a number of ways including:

    Tail length

    % DNA in the tail

    Tail moment (% DNA in tail × tail length)

    Olive tail moment (% DNA in tail × distance between the means of head and tail distributions) based on the definition of Olive et al. (15)

    Many of the commercial software packages will calculate and store these parameters (see Subheading 2.3, Step 5 and Note 1).

8. Remove the coverslip and store slides in a light-proof box at room temperature (see Note 20).

Undamaged cell

Cell irradiated with 12.5 Gy X-rays

Cell treated with a DNA cross-linking agent and irradiated with 12.5 Gy X-rays

Fig. 1. Typical comet images from human cells in the alkaline comet assay as described. The *upper panel* shows an undamaged cell with the majority of the DNA in the comet head. The *centre panel* shows a cell which has been irradiated with 12.5 Gy X-rays resulting in approximately 40% of the DNA in the tail. In the *lower panel* the cell was treated with a DNA cross-linking agent (melphalan) prior to irradiation with 12.5 Gy. The decrease in tail moment compared to irradiation alone is as a result of the interstrand cross-links retarding the migration of the fragmented DNA.

3.4. Modifications

1. 'Neutral' assay

    The assay is essentially the same as the alkaline assay, but cells are lysed at a lower pH and electrophoresed in either TBE or Tris acetate buffer pH 8–9 (16, 17).

2. Detection of DNA Interstrand Cross-links.

    (a) The comet assay is carried out as described in Subheading 3.2 but at step 2 cells are divided into two aliquots. One aliquot is irradiated with the required dose of X-rays to induce a fixed level of single strand breaks. The second aliquot is the unirradiated control for each sample (see Notes 21–23).

    (b) Following image analysis the % decrease in tail moment is calculated using the formula:

$$\%\text{Decrease in tail moment} = \left[1 - \left(\frac{\text{TMdi} - \text{TMcu}}{\text{TMci} - \text{TMcu}}\right)\right] \times 100$$

where TMdi, tail moment of post-dose irradiated sample; TMcu, tail moment of pre-dose unirradiated control; TMci, tail moment of pre-dose irradiated control.

(c) In samples treated under conditions that can also produce strand breaks e.g. combination of a cross-linking agent with an agent known to produce single strand breaks e.g. gemcitabine, cross-linking is expressed as percentage decrease in tail moment compared to irradiated controls calculated by the formula below. This formula is used to compensate for the additional single strand breaks induced by agents such as gemcitabine in addition to those produced by the irradiation step.

$$\% \text{ Decrease in tail moment} = \left[1 - \left(\frac{(TMdi - TMcu)}{(TMci - TMcu) + (TMdu - TMcu)}\right)\right] \times 100$$

where, TMdi is the tail moment of drug treated irradiated sample; TMcu, tail moment of untreated unirradiated control; TMci, tail moment of untreated irradiated control; and TMdu, tail moment of drug treated unirradiated sample.

(d) The percentage decrease in tail moment is proportional to the level of DNA cross-linking.

3.5. Applications

1. Repair assays

    (a) The time course of repair of strand breaks in DNA, or the formation and 'unhooking' of DNA interstrand cross-links, can be followed (see Notes 3, 4, and 24) (Fig. 2).

    (b) The excision repair ability of cells can be studied by the addition of viable cell extracts to the naked DNA on the slide following lysis (11).

2. Comet-FISH

    This technique combines the Comet assay with fluorescence in situ hybridisation (FISH) allowing detection of region specific DNA damage and repair in individual cells. The hybridisation step is added following the lysis and electrophoresis steps of the Comet assay so that specific sequences can be labelled and assigned to either the damaged or undamaged part of the DNA comet (12, 13).

3. Enzyme-Comet Assay

    To reveal damage such as oxidised bases or UV-induced dimers, single strand breaks can be created at these sites using lesion specific endonucleases or glycosylases, after lysis of the cells. For example, endonuclease III recognises oxidised pyrimidines, formamidopyrimidine glycosylase (FPG) recognises oxidised purines and T4 endonuclease V, UV induced

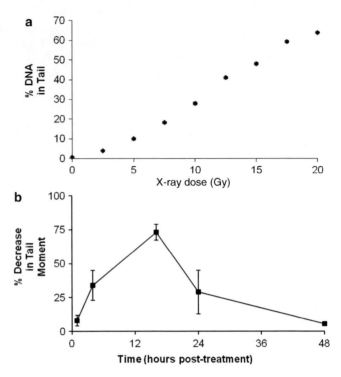

Fig. 2. (a) DNA strand breaks, expressed as % DNA in comet tail, following irradiation of human cells with increasing doses of X-ray up to 20 Gy. The results are the mean from 50 cells (25 cells from two duplicate slides) at each point. (b) The formation and repair ('unhooking') of melphalan-induced DNA interstrand cross-links in A549 human non-small cell lung cancer cells. Cells were treated with 50 μM melphalan for 1 h, then incubated in drug-free medium and samples taken at various time points. Results are expressed as percentage decrease in tail moment for 50 cells analysed (mean ± standard error). Peak of DNA interstrand crosslink formation (expressed as the % decrease in tail moment compared to the irradiated controls – see Subheading 3.4.2) observed 16 h following a 1-h treatment with melphalan. This is followed by significant repair of melphalan-induced DNA interstrand cross-links at 24 and 48 h post-treatment.

cyclobutane pyrimidine dimers. Following enzyme treatment on the slide the assay continues as usual with the electrophoresis, neutralisation and staining steps (14).

4. Notes

1. Other Comet analysis programmes are available incorporating all major measurement parameters such as percentage head/tail DNA, tail length and Olive tail moment, for example Comet Assay IV, Perceptive Instruments, Haverhill, UK and LAI Comet Analysis System (LACAS), Loats Associates Incorporated, Westminster, Maryland, USA. Automatic analysis programmes are available allowing unattended automatic comet acquisition and measurement of comet parameters, for

example Metafer CometScan, Metasystems, Altlussheim, Germany and AutoComet III, TriTek Corporation, Sumerduck, Virginia, USA.

2. If a DNA damaging agent is used, which needs to be reconstituted in solvents such as dimethylsulphoxide, the final concentration of solvent added to the cells should be no greater than 0.1%. This is to avoid any additional DNA damage or cytotoxicity.

3. The length of the post-treatment time is dependent on the DNA damaging agent used. A time-course experiment should be performed to ascertain this. For example, some agents produce DNA damage rapidly (X-rays) whereas others such as the cross-linking agent melphalan requires a post-incubation of 16 h to reach the peak of interstrand cross-link formation. For repair experiments the post-treatment time can be further extended beyond the peak of damage.

4. Samples should be re-suspended fully in the freezing mixture to prevent DNA damage or cytotoxicity. This should be carried out by gently resuspending the sample using a Pasteur pipette rather than vortexing.

5. Our laboratory has validated the effects of long-term storage at −80°C and samples can be stored up to 12 months without any detrimental effects to the integrity of the cells and DNA.

6. Trypsinise cells at 37°C as quickly as possible to avoid any additional DNA damage and to prevent any DNA repair. Alternatively a non-enzymatic preparation can be used such as cell disassociation solution (Sigma). It is imperative that a single cell suspension is achieved. If several cells migrate together through the gel, an overestimated comet tail moment can result.

7. It is imperative that the comet assay should be performed on ice to prevent any DNA repair, and in subdued lighting and incubations carried out in the dark where indicated to avoid light-induced DNA damage. All solutions should be ice cold and maintained at 4°C.

8. Reagent kits for the Single Cell Gel Electrophoresis (Comet) assay are available (CometAssay™, Trevigen Incorporated, Gaithersburg, Maryland, USA and CometAssay™, R & D Systems, Abingdon, UK). These kits contain precoated Comet slides, LGT agarose, lysis solution, EDTA and SYBR® Green I nucleic acid gel stain and provide enough reagents for 25 slides. The assay protocols for these kits have not however been optimised for individual requirements. For example, lysis, alkali denaturation and electrophoresis incubation times are not standardised and require significant optimisation by the user to achieve consistent results. Also such kits

are significantly more expensive when compared to purchasing and preparing reagents individually.

9. Precoated slides should be prepared and dried in advance. Slides can be stored dry at room temperature for up to 6 months in an airtight container.

10. When sample sizes are small, the sample and gel size may be reduced. Take 100 µL cell suspension at a final concentration of 2.5×10^4/mL and add to a 96-well plate. Add 200 µL of molten 1% LGT in water cooled to 40°C, mix and pipette 300 µL onto the centre of the slide. Place a 13-mm diameter circular coverslip on top. Once set, remove coverslip and place slide in a tray on ice. Continue assay from Subheading 3.2, step 4.

11. Molten 1% LGT agarose should be maintained at 40°C to aid uniform gel preparation. The thickness of the gel must be consistent between slides to ensure uniform DNA migration and reduce assay variability. All gels should have the dimensions of the coverslip. The gel should not flood the entire slide or the frosted section and should not contain air bubbles.

12. A number of protocols have stated that following lysis the slides can be kept overnight or even days in this solution prior to alkali DNA unwinding and electrophoresis. We do not advise that this should be carried out as the gels tend to break up and slip off the slide. It is therefore recommended that Subheading 3.2, steps 2–16 should be carried out in a single day.

13. Slides should be placed in a flat-bed electrophoresis tank lengthways with the frosted end towards the anode. It is essential that the tank is levelled and all slides face the same direction to ensure lack of variability between slides.

14. The volume of alkali buffer added to the electrophoresis tank should be consistent from one experiment to the next. It is advisable to measure the volume of buffer required ensuring that all slides are covered by at least 5 mm buffer.

15. These electrophoresis parameters are optimal for our equipment. The current can be adjusted to suit individual requirements.

16. The most commonly used fluorescent stains are propidium iodide and ethidium bromide. The highly sensitive fluorochrome SYBR® Green Inucleic acid gel stain has also been used successfully. It has the advantage of being far more sensitive that propidium iodide and produces no background fluorescence. However, it fades much more rapidly under intense UV light. Comet images can also be visualised using silver staining (CometAssay™ Silver Staining Kit, Trevigen Incoporated, Gaithersburg, Maryland, USA and CometAssay™ Silver kit,

R & D Systems Europe Limited, Abingdon, UK). This can be carried out after comets have been analysed using other staining methods and allows visualisation by standard light microscopy and provides a permanent staining for sample archiving.

17. Visualising the slides dry produces optimum results as all the cells are in the same plane giving clear cellular definition (18). This is favoured instead of the traditional wet slide method which can cause difficulties in focusing and quantitation.

18. Slides should be analysed as quickly as possible. If the coverslip is left on for a considerable length of time, it will become permanently stuck.

19. Each slide should ideally be scored blind to avoid any bias, taking care to ensure that comets are measured from the entire gel area and no part of the slide is analysed more than once.

20. Slides may be stored for at least 5 years and re-analysed at any time as described in Subheading 3.3, steps 5–9. Slides may also be re-stained as described in Subheading 3.3, steps 1–4 if the staining has faded during storage.

21. A standard curve for irradiation dose in non-drug treated cells should be performed to establish the optimum radiation dose for a given cell type. Ideally the dose should give a head to tail DNA ratio of approximately 1:1. Our laboratory finds 15 Gy X-rays optimum for most cell lines, lymphocytes and tumour samples (Fig. 2).

22. Our laboratory uses an AGO HS MP-1 X-ray machine with a Varian ND1-321 tube to produce X-rays at a dose rate of 2.5 Gy/min.

23. Each experiment should include an untreated un-irradiated control. In addition, an untreated irradiated control should also be included with every group of irradiated samples to allow for variation. It is also important to determine in the first instance if the drug under test will produce any detectable single strand breaks in addition to cross-links. This may be achieved by performing the comet assay on drug-treated cells but excluding the irradiation step. This also applies when samples are treated in combination with an agent known to cause single strand breaks.

24. When the time course of damage and repair of DNA strand breaks or formation and unhooking of interstrand cross-links is to be studied, it is advisable to freeze samples at each time point to allow samples to be analysed within the same assay. Appropriate controls to account for cell growth should be included for example at 12-h intervals.

References

1. Ostling, O., and Johanson, K.L. (1984) Microelectrophoretic study of radiation-induced DNA damages in individual mammalian cells. *Biochem. Biophys. Res. Commun.* **123**, 291–298.

2. Singh, N.P., McCoy, M.T., Tice, R.R. and Schneider, E.L. (1988) A simple technique for quantitation of low levels of DNA damage in individual cells. *Exp. Cell Res.* **175**, 184–191.

3. Olive, P.L., Banáth, J.P. (2006) The comet assay: a method to measure DNA damage in individual cells. *Nat. Protocols* **1**, 23–29.

4. Olive, P.L. (2009) Impact of the comet assay in radiobiology. *Mutat. Res.* **681**, 13–23.

5. Dhawan, A., Bajpayee, M. and Parmar, D. (2009) Comet assay: a reliable tool for the assessment of DNA damage in different models. *Cell Biol. Toxicol.* **25**, 5–32.

6. Jha, A.N. (2008) Ecotoxicological applications and significance of the comet assay. *Mutagenesis* **23**, 207–21.

7. Wasson, G.R., McKelvey-Martin, V.J. and Downes, C.S. (2009) The use of the comet assay in the study of human nutrition and cancer. *Mutat. Res.* **681**, 153–62.

8. McKenna DJ, Mckeown SR, McKelvey-Martin VJ (2009) Potential use of the comet assay in the clinical management of cancer. *Mutat. Res.* **681**, 183–90.

9. Collins, A.R., Oscoz, A.A., Brunborg, G., Gaivao, I., Giovannelli, L., Kruszewski, M., Smith, C.C. and Stetina, R. (2008) The comet assay: topical issues. *Mutagenesis* **23**, 143–151.

10. Spanswick, V.J., Hartley, J.M., and Hartley, J.A. (2010) Measurement of DNA interstrand crosslinking in individual cells using the single cell gel electrophoresis (comet) assay. *In:* Methods in Molecular Biology, Vol 613 Drug-DNA Interaction Protocols, 2nd Edn, Fox, K. (ed), Humana Press, Totowa, NJ, 267–282.

11. Gaivão, I., Piasek, A., Brevik, A., Shaposhnikov, S., Collins, A.R. (2009) Comet assay-based methods for measuring DNA repair *in vitro*; estimates of inter- and intra-individual variation. *Cell Biol. Toxicol.* **25**, 45–52.

12. Glei, M., Hovhannisyan, G., Pool-Zobel, B.L. (2009) Use of Comet-FISH in the study of DNA damage and repair: Review. *Mutat. Res.* **681**, 33–43.

13. Spivak, G., Cox, R.A. and Hanawalt, P.C. (2009) New applications of the Comet assay: Comet-FISH and transcription-coupled DNA repair. *Mutat. Res.* **681**, 44–50.

14. Collins, A.R. (2009) Investigating oxidative DNA damage and its repair using the comet assay. *Mutat. Res.* **681**, 24–32.

15. Olive, P.L., Banath, J.P., and Durand, R.E. (1990) Heterogeneity in radiation-induced DNA damage and repair in tumour and normal cells measured using the "comet" assay. *Radiat. Res.* **122**, 86–94.

16. Olive, P.L., Wlodek, D., Banáth, J.P. (1991) DNA double-strand breaks measured in individual cells subjected to gel electrophoresis. *Cancer Res.* **51**, 4671–4676.

17. Wojewodzka, M., Buraczewska, I., and Kruszewski, M. (2002) A modified neutral comet assay: elimination of lysis and high temperature and validation of the assay with anti-single-stranded DNA antibody. *Mutat. Res.* **518**, 9–20.

18. Klaude, M., Erikkson, S., Nygren, J., and Ahnstrom, G. (1996) The comet assay: mechanisms and technical considerations. *Mutat Res* **363**, 89–96.

Chapter 26

Molecular Breakpoint Analysis of Chromosome Translocations in Cancer Cell Lines by Long Distance Inverse-PCR

Björn Schneider, Hans G. Drexler, and Roderick A.F. MacLeod

Abstract

With conventional cytogenetic screening by fluorescence in situ hybridization (FISH) using genomic tilepath clones, identification of genes in oncogenic chromosome translocations is often laborious, notably if the region of interest is gene-dense. Conventional molecular methods for partner identification may also suffer severe limitations; for instance, genomic PCR screening requires prior knowledge of both sets of breakpoints, while rapid amplification of cDNA ends (RACE) is not only limited to translocations causing mRNA fusion, but also fails to provide potentially relevant breakpoint data. With Long Distance Inverse (LDI)-PCR, however, it is theoretically possible to identify unknown translocation partners and to map the breakpoints down to the base pair level. Implementing LDI-PCR only requires approximate sequence information on one partner, rendering it ideal for use in combination with frontline FISH analysis. The protocol described here has been tuned for use by those wishing to identify new cancer genes in tumor cell lines.

Key words: Breakpoint, BCL6, IGH, MLL, LDI-PCR, PAX5 translocation

1. Introduction

Chromosomal translocations are key events in tumorigenesis alongside copy number alterations, such as deletions and amplifications, respectively causing gene silencing and upregulation. Translocations have long been deemed diagnostically and prognostically significant in leukemia and lymphoma, and increasingly so in solid tumors as well (1). Consistent chromosomal rearrangements characterize many different types of cancer.

Hitherto, most cancer genes have been identified following analysis of recurrent chromosome translocations. The pathological significance and usefulness of such rearrangements depends on two key features. The first concerns whether rearrangements display distinct patterns of recurrence within specific tumors, e.g., t(8;14)(q24;q32) which is restricted to B-cell neoplasia; the second, how close together ("clustered") are the chromosomal breakpoints therein. The significance accorded to breakpoint data depends on their precision, from megabase and kilobase, down to single base pair levels, when ascertained by classical cytogenetics, fluorescence in situ hybridization (FISH), and sequence-based methods, respectively.

Chromosome translocations (and certain synonymous inversions or deletions which effect fusion/juxtaposition of genes lying on the same chromosome) fall into three broad categories. The first causes the physical fusion of the two mRNAs expressed by the genes involved, thus creating novel fusion proteins translated from exons emanating from both participant genes, e.g., BCR (at [chromosome]-22-[band]-q11] with ABL1 (at 9q34) fused by t(9;22)(q34;q11) in chronic myeloid leukemia (CML) and in some cases of acute lymphoblastic leukemias (ALL) (2, 3). The second category also fuses mRNA from genes at separate loci, but in this case, serves to deregulate a developmentally silenced partner by exchanging promoters with more active partners, e.g., BCL6 (at 3q27) which is activated by translocations with any one of many partners, chiefly in diffuse large B-cell lymphoma (DLBCL) (4). The third class of chromosome translocation again results in the activation of the normally silent partner, this time by juxtaposition with another constitutively active partner without mRNA fusion, e.g., the neighboring homeobox genes, TLX3 (at 5q35.1) and NKX2-5 (at 5q35.2). According to the breakpoint involved, either (but not both) genes may be activated in T-cell ALL by the recurrent t(5;14)(q35;q32.2) by which these are juxtaposed with regulatory regions from BCL11B (at 14q32.2) to stimulate transcription (5, 6).

A feature of cancer causing genes most frequently involved in translocations is "promiscuity" – the tendency of certain genes to engage with multiple partners. Among well-known promiscuous oncogenes are MLL with 64 known partners (7), BCL6 with 28 (4), RUNX1 with 39 (8), NUP98 with 29 (9), and the IgH-locus with 40 (10). Promiscuity reflects the dependence of tumors on the inappropriate expression of certain genes without overly caring how their deregulation is accomplished. However, the role of partner genes, previously neglected, has come under renewed scrutiny as, e.g., their involvement may reveal in which types of precancerous cells primary oncogenic rearrangements occur. In addition, biologically important genes, e.g., BCL11B, a key regulator of both differentiation and survival during thymocyte development, are often first rendered visible by their participation in cancer rearrangements (11).

It is useful to know the identities of both the partner genes and their precise breakpoints at the DNA base pair level, not only to characterize potential fusion genes/products, but also to ascertain whether additional non protein-coding genomic entities, until recently dismissed as "junk" DNA, such as chromosomal fragile sites (12), microRNA loci, putative (in silico) genes, unspliced "expressed sequence tags," or regulatory noncoding regions may be involved.

Breakpoint sequences themselves may provide clues to the biological mechanisms which generate chromosome rearrangements, including T- and B-cell receptor gene (VDJ) rearrangement, Alu-mediated recombination, nonhomologous end joining, etc. VDJ genes are flanked by recombination signal sequences composed of heptamers followed, in turn, by a spacer containing either 12 or 23 unconserved nucleotides, and a conserved nonamer. Spacers of 12 nucleotides undergo physiological recombination with those containing 23 in order to obey the so-called "12/23 rule." The presence of these or related sequences has been reported in connection with cancer translocations to reveal how physiologic processes may be abused to cause genomic rearrangements (13).

As an unexpected bonus, precise genomic fusion sequences may also be used to aid cell line authentication – an omnipresent problem confronting cell culturists, given that an unexpectedly (and unacceptably) high percentage of new cell lines has been misidentified, or cross-contaminated by older cell lines (14). While mRNA fusion sequences are constrained by splicing, their genomic equivalents allow sufficient variation to provide "fingerprints" unique to individual cell lines to serve as potential identifiers.

For those hunting new cancer genes, cancer gene promiscuity comes as a godsend as it enables oncogenic rearrangements likely to include cancer genes to be distinguished from random changes to facilitate the mining of cancer cell lines for novel cancer genes. Typically, FISH is initially used to confirm rearrangement of a contextually appropriate oncogene residing at the locus in question. Hence, a breakpoint at 9q34 might throw suspicion onto NOTCH1 in a T-cell neoplasia, ABL1 in myeloid neoplasia, and NUP214 in either entity. While such an approach is less highly developed among solid tumors, at this locus, TSC1 might be deemed a candidate in tuberous sclerosis cells. Even when the index breakpoint is precisely known, determination of its partner by FISH requires time-consuming and laborious procedures for those not afforded blanket tilepath-clone coverage with which to quarter the region of interest.

While in the light of the foregoing molecular approaches are clearly called for, "shotgun" PCR screening with hit-lists of known and potential partner genes is quite as laborious as FISH, and is liable to miss unknown translocation partners, or those with breakpoints lying outside their respective cluster regions. When there are reasonable grounds to suspect transcriptional fusion (as among partners of genes prone to this type of gene rearrangement, e.g., ABL1, ETV6, NUP98, etc.), an mRNA-based method, rapid amplification of cDNA ends (RACE), may be used to detect novel fusion partners (15). However, RACE provides no breakpoint information at the DNA level, and may, in addition, overlook certain splice variants. For these reasons, the technique of choice for identifying unknown partner genes and their breakpoints should both sidestep the need for prior knowledge of the partner gene, yet provide breakpoint data at the DNA base pair level.

Long Distance Inverse (LDI)-PCR is just such a technique. With the mining of oncogenic translocations as goal, LDI-PCR was developed from the earlier inverse-PCR (16) to allow the amplification of large DNA fragments comprised of known and unknown sequences (17). The template is a re-ligated circular restriction fragment. The primers are set in opposition within the known sequence. In the resultant amplicon, the unknown sequence is flanked on both sides by known sequences following re-ligation (Fig. 1). When a restriction fragment length polymorphism (RFLP) distinguishes the wild type and derivative alleles introduced by the genomic alteration, the two resulting amplicons should be separable by gel electrophoresis (Fig. 2), enabling their respective sequences to be compared. Sequencing with one of the PCR-primers directed towards the restriction site allows immediate identification of the partner gene. Sequencing in the other direction allows precise mapping of the breakpoint. Although easier to perform for genes with well-defined, short breakpoint cluster regions, LDI-PCR may be applied to any gene or region involved in a translocation and has, therefore, been applied to a wide variety of translocations involving numerous promiscuous oncogenes, including MLL (18), BCL6 (19), IGH (17, 20), PAX5 (21), etc.

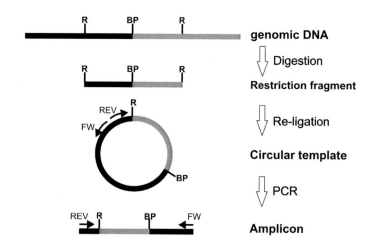

Fig. 1. Amplifying genomic fusions of unknown sequence. The schema summarizes how the genomic DNA is first restricted, then re-ligated to the circular template, and how the resulting amplicon should appear. Note the unknown region (*gray*) flanked by known sequences (*black*). *R* restriction site, *BP* breakpoint; *arrows*: forward (*FW*) and reverse (*REV*) primers.

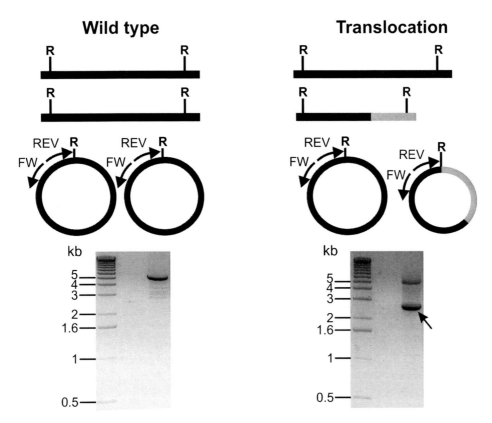

Fig. 2. How to interpret LDI-PCR gels. *Left figure* shows the wild type configuration where twin circular templates identical in size would yield a single band by agarose gel electrophoresis. Translocation bearing cells (*right figure*) yield both wild type and derivative templates, differing in size and detectible as two bands on the gel. The derivative band is indicated by an *arrow*. Known regions are outlined in *black* and unknown in *gray*; *R* Restriction site, *REV* reverse primer, *FW* forward primer. Below the gels, the identified fusion sequence is shown (*middle line*), aligned to chr. 3 (BCL6) and 7.

Limitations are set by the performance of the DNA polymerase since lengthier fragments may resist amplification, and by the placement of the RFLP, as fragments similar in size cannot be readily distinguished by gel electrophoresis.

In contrast to other PCR methods suitable for detection of unknown fusion sequences, such as panhandle PCR (22) or analogous techniques requiring adaptor ligations (reviewed in ref. 23), LDI-PCR is independent of any additional adaptors or anchors that have to be ligated to the restricted fragments, thus reducing the number of steps required, while remaining sufficiently flexible to allow a wide choice of restriction enzymes.

The following details the LDI-PCR protocol used at the DSMZ for identifying partners and breakpoints in cancer cell lines.

2. Materials

2.1. Design

1. Consult a genome browser to obtain sequence information on the index gene. The standard genome browsers listed below, to choose from, are largely esthetic. Our favorite is that hosted by UCSC the University of California at Santa Cruz (http://genome.cse.ucsc.edu/cgi-bin/hgGateway), but both ENSEMBL (http://www.ensembl.org/Homo_sapiens/Info/Index), and the NCBI Map Viewer (http://www.ncbi.nlm.nih.gov/projects/mapview/map_search.cgi?taxid=9606) have their adherents. Note that sequence data and coordinates can vary at each "release" (edition). To avoid future confusion as new releases come out, their edition numbers should be noted for the data set to which they refer.

2. If the partner locus is known approximately, e.g., from existing cytogenetic data, consult Mitelman Database of Chromosome Aberrations in Cancer (http://cgap.nci.nih.gov/Chromosomes/Mitelman) to verify whether likely partner gene candidates reside near the region of interest. For those wishing to mine translocations by seeking partners of promiscuous oncogenes in hematopoietic cell lines, see ref. 24 for a list of cell lines with recurrent translocations. For those wishing to mine all translocations in hematopoietic cell lines see refs. 25, 26 for karyotypic data, while for solid tumors the DSMZ website (www.dsmz.de) carries karyotypic details on their own holdings.

3. Restriction map generator hosted by BioEdit, Ibis Biosciences, Carlsbad, USA (http://www.mbio.ncsu.edu/BioEdit/bioedit.html), or Sequence Manipulation Suite (http://www.bioinformatics.org/sms2/rest_map.html).

4. Tool for calculating melting temperature (TM) and GC-content to facilitate calculation of melting temperatures (http://insilico.ehu.es/tm.php).

2.2. Template Preparation

1. High Pure PCR Template Preparation Kit (Roche, Mannheim, Germany).

2. 1 µg of high quality cell line DNA per digestion reaction. Store at 4–8°C. Do not freeze as repeated cycles of freezing and thawing may promote shearing of the DNA.

3. Restriction enzyme of choice (Fermentas, St. Leon-Rot, Germany) and appropriate 10× restriction buffer.

4. Nuclease-free sterilized MilliQ water.

5. QIAquick gel extraction kit (QIAGEN, Hilden, Germany).

6. T4 DNA ligase (5 U/µL) and appropriate ligation buffer (Fermentas).

2.3. PCR

1. Oligonucleotides (see Subheading 3.1, step 5 below regarding design parameters).
2. PCR kit: PCR Extender System (5Prime, Hamburg, Germany).
3. dNTP-mix containing all four nucleotides at 2.5 mM (Roche).

2.4. Gel Electrophoresis and Amplicon Purification

1. TAE buffer (50×): 242 g Tris (2 M) (Molekula, Gillingham, UK); 57.1 mL of glacial acetic acid (1 M) (Roth, Karlsruhe, Germany); 100 mL of 0.5 M EDTA solution at pH 8.0 (Roth); make up to 1 L in deionised water. Working solution is 1×. Store at room temperature.
2. Agarose, electrophoresis grade (Invitrogen, Karlsruhe, Germany).
3. Ethidium bromide, 5 mg/mL (Merck, Darmstadt, Germany). This is toxic and mutagenic!
4. Loading dye (6×): 25 mg bromphenol blue (Sigma, Steinheim, Germany), 25 mL xylencyanol (Merck), 5 mL glycerol (Sigma), 5 mL MilliQ water.
5. Size marker: 1 kb ladder (Invitrogen). Working solution, 0.1 µg/µL: 100 µL stock solution, 167 µL 6× loading dye, and 733 µL H_2O. Store at 4–8°C.
6. QIAquick Gel Extraction Kit (QIAGEN).

3. Methods

3.1. Design

1. Select sequence covering the genomic region of interest from the genome browser.
2. Paste selected sequence into the query box of the restriction map generator to highlight the restriction sites located within.
3. Choose restriction enzymes used for template preparation to obtain fragments in a size range of 2–5 kb (see Note 1).
4. Design the inverse primer pairs so that one primer is directed towards the restriction site and the other in the opposite direction (see Fig. 3). The gap between the primer tails should not be excessive as the intervening sequence lies outside the amplicon and, should the breakpoint lie within, cannot be detected unless there is another primer pair at the other end of the restriction fragment. If the fragment is sizeable (>5 Kbp, say), it might be advisable to design a primer set consisting of one forward and multiple reverse primers (see Fig. 3).

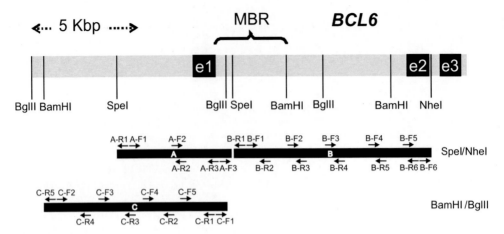

Fig. 3. How to design LDI-PCR reactions. Using BCL6 translocations as a model, the *upper figure* shows part of the BCL6 locus, extending 20 Kbp from the 5'-region to exon 3, encompassing the major breakpoint region (MBR), where most BCL6 breakpoints lie. Key restriction sites are indicated. *Lower figure* shows three restriction fragments (A–C) where *arrows* indicate the position and direction of the primers. Thus, a primer set for fragment A might consist of the primer pairs A-R1×A-F1, A-R1×A-F2, A-F3×A-R3, and A-F3×A-R2.

5. Oligonucleotides should be ca. 30 bp to balance uniqueness, heat stability, and hybridization efficiency. Check Tm and GC-content with an appropriate calculation tool (see Subheading 2.1, item 4). Tm should be around 65°C, and GC-content 40–60%.

3.2. Template Preparation

1. Prepare cell line DNA with the High Pure Template Preparation Kit (Roche) according to the manufacturer's protocol, and determine concentration and quality photometrically. Independent of concentration, the 260/280 value (measuring the ratio of DNA at 260 nm and protein/phenol/contaminants at 280 nm) should be between 1.8 and 2.0, as lower values indicate contamination. Check the integrity of 100 ng DNA by gel electrophoresis (see Subheading 3.4). Only one large band (~50 Kbp) should be visible. Laddering indicates DNA degradation (more often a problem with primary cell/tissue samples than cell lines) rendering the sample unsuitable for LDI-PCR.

2. For digestion, mix 1 μg DNA, 30–50 U of each restriction enzyme, 10 μL 10× buffer, and water up to a final volume of 100 μL (see Note 2).

3. While agitating, incubate the mixture for 3–4 h at the temperature indicated for the enzyme(s) chosen (usually 37°C). Where applicable, restriction enzyme(s) can be heat-inactivated thereafter.

4. Purify digested DNA fragments with gel extraction kit following the manufacturer's advices for purifying enzymatic reactions (see Note 3).

5. For re-ligation, mix purified restricted DNA with 8 μL 10× buffer, 1 μL T4 ligase (5 U/μL) in a total volume of 80 μL and leave overnight at 4–8°C. These conditions favor the desired self-ligation needed to form the circular template required for inverse PCR.

6. Terminate the ligation reaction by heat inactivation for 10 min at 65°C. Additional purification prior to PCR is not necessary. (See Fig. 1 for overview.)

3.3. PCR Reaction

1. Reaction mix: The following volumes are for a 50 μL reactions; for other volumes adjust the amounts of reagent and template accordingly.

    20 pmol primer at a final concentration of 400 nM

    10 μL dNTP mix (2.5 mM each nucleotide) for a final concentration of 500 μM

    5 μL of re-ligated, digested DNA (containing 62.5 ng) (see Note 4)

    5 μL 10× tuning buffer, 1 U DNA polymerase (0.2 μL) (see Note 5)

    Make up to 50 μL with MilliQ water.

2. Cycling conditions follow the manufacturer's protocol designed to yield amplicons of up to 20 kb.

In detail: Initial denaturation:	93°C, 3 min	
Cycles 1–10:	93°C, 15 s	Denaturation
	62°C, 30 s	Annealing (see Note 6)
	68°C, 18 min	Elongation
Cycles 11–35:	93°C, 15 s	Denaturation
	62°C, 30 s	Annealing (see Note 6)
	68°C, 18 min + 20 s increment each cycle	

3.4. Gel Electrophoresis, Amplicon Purification, and Analysis

1. For a 1.2% gel (ca. 11 × 14 cm), boil 1.2 g of agarose in 100 mL 1× TAE buffer until completely dissolved.

2. When the solution has cooled to a certain extent, add 6.5 μL ethidium bromide.

3. Pour gel carefully into the sleigh, add the comb, and then allow to solidify (ca. 45–60 min).

4. Add 8 μL of 6× loading dye to each PCR product.

5. Transfer the solid gel carefully into the electrophoresis chamber filled with 1× TAE buffer, extract the comb, and load the PCR reaction plus dye, and 8 μL size marker into the gel pockets (see Note 7).

6. Perform the electrophoresis run at 100–120 V for ca. 90 min (see Note 8).

7. After the run, visualize the gel bands under UV-light and identify discrepant bands (i.e., not corresponding to the calculated amplicons) as potential genomic fusion sequence products (Fig. 2). The wild type band serves as PCR control. For troubleshooting, see Note 9.

8. Carefully excise desired band(s) with a scalpel and purify with the gel extraction kit according to the manufacturer's protocol.

9. Subject purified amplicons to sequence analysis. Sequencing with the primer directed towards the restriction site reveals the translocation partner; while sequence read in the opposite direction shows the breakpoint. Depending on the fragment size and the read-length of the sequencer, additional primers may be necessary for breakpoint detection.

4. Notes

1. When performing double-digestions with restriction enzymes producing sticky ends, pairs must be checked for end-compatibility. These pairs should function in the same buffer and at the same temperature. This can be checked with the DoubleDigest™ tool on the Fermentas homepage (http://www.fermentas.com/doubledigest/index.html).

2. Mix water, then 10× buffer and restriction enzyme(s) by vortexing to dissolve the glycerol-embedded enzymes. As genomic DNA is readily sheared by vortexing, it should be added at the end and mixed manually by flicking tube ends.

3. Column-based purification of the restriction reaction is preferred to phenol/chloroform purification followed by sodium acetate/ethanol precipitation, as the latter is more laborious and time consuming, while residual phenol may disturb downstream enzymatic reactions.

4. According to the manufacturer's protocol, ~100–250 ng template DNA is recommended for targets up to 20 kb. We find that 62.5 ng are sufficient for LDI-PCR. In practice amplicons rarely exceed 10 kb.

5. Though the amount given is about half of that recommended by the manufacturer, PCR fidelity remains satisfactory. Reducing the amount of enzyme brings significant cost benefits.

6. Annealing temperature depends on oligonucleotide melting temperature, and should be 2–4°C below Tm.

7. If direct sequencing of the PCR product is intended, the entire 50 μL PCR reaction should be loaded onto the gel to provide

sufficient template for the sequencing reaction. If it is intended to clone the product into a vector and amplify it in transfected bacteria, a lesser amount (~10 µL, say) should be sufficient.

8. If the wild type and the derivative amplicon are both very large (> ca. 5 Kbp), additional time should be allowed for size separation to maximize separation.

9. Troubleshooting. Failure of LDI-PCR may have several causes.

If no PCR product is seen, first check PCR ingredients and conditions: primer sequences (avoidance of G or C repeats, repetitive sequences etc.), annealing temperature, or cycling conditions. Second, verify that the expected PCR product is not too long to be amplified successfully. Despite manufacturer's claims to the contrary, a realistic amplifiable limit is closer to 4–5 kb than 20 Kbp claimed. If wild type bands appear, to the exclusion of aberrant products, the false-negative result may be due either of the following: (a) the derivative fragment is too long and resists amplification; or, (b) the wild type and the derivative amplicon are similar in size and cannot be separated by gel electrophoreses. In both cases, other restriction enzymes or a double digest approach should be tried out to reduce the amplicon size and, thereby, increase the chances of differing fragment lengths. Finally, chromosome rearrangements in cancer do not always follow textbook balanced reciprocal exchange of DNA. Although the term "translocation" is commonly used, this assumes a mechanistic basis which is seldom, if ever, possible to verify experimentally. Extraneous sequences from apparently innocent chromosome regions are often detected at breakpoint junctions. Although these are often interpreted as artifacts, these may instead witness the cryptic involvement of other chromosomes. The importation of DNA repeats by such processes might undermine PCR amplification.

References

1. Mitelman, F., Johansson, B., and Mertens, F. (2007) The impact of translocations and gene fusions on cancer causation *Nat Rev Cancer* 7, 233–45.

2. Turhan, A.G. (2008) BCR (Breakpoint cluster region). *Atlas Genet Cytogenet Oncol Haematol.* URL: http://AtlasGeneticsOncology.org/Genes/BCR.html.

3. Turhan, A.G. (2008) ABL1 (v-abl Abelson murine leukemia viral oncogene homolog 1). *Atlas Genet Cytogenet Oncol Haematol.* URL: http://AtlasGeneticsOncology.org/Genes/ABL.html.

4. Knezevich, S. (2007) BCL6 (B-Cell Lymphoma 6). *Atlas Genet Cytogenet Oncol Hematol* URL: http://AtlasGeneticsOncology.org/Genes/BCL6ID20.html.

5. Bernard, O.A., Busson-LeConiat, M., Ballerini, P., Mauchauffé, M., Della Valle, V., Monni, R., et al. (2001) A new recurrent and specific cryptic translocation, t(5;14)(q35;q32), is associated with expression of the Hox11L2 gene in T acute lymphoblastic leukemia *Leukemia* 15, 1495–504.

6. Nagel, S., Scherr, M., Kel, A., Hornischer, K., Crawford, G.E., Kaufmann, M., et al. (2007)

Activation of TLX3 and NKX2-5 in t(5;14)(q35;q32) T-cell acute lymphoblastic leukemia by remote 3′-BCL11B enhancers and coregulation by PU.1 and HMGA1 *Cancer Res* **67**, 1461–71.

7. Meyer, C., Kowarz, E., Hofmann, J., Renneville, A., Zuna, J., Trka, J., et al. (2009) New insights to the MLL recombinome of acute leukemias *Leukemia* **23**, 1490–9.

8. Huret, J.L., and Senon, S. (2003) AML1 (acute myeloid leukemia 1); RUNX1 (runt-related transcription factor 1 (acute myeloid leukemia 1; aml1 oncogene)); CBFA2 (core binding factor A2) *Atlas Genet Cytogenet Oncol Haematol* URL: http://AtlasGeneticsOncology.org/Genes/AML1.html.

9. Kearney, L. (2002) NUP98 (nucleoporin 98 kDa) *Atlas Genet Cytogenet Oncol Haematol* URL: http://AtlasGeneticsOncology.org/Genes/NUP98.html.

10. Lefranc, M.P. (2003) IGH (Immunoglobulin Heavy) *Atlas Genet Cytogenet Oncol Haematol* URL: http://AtlasGeneticsOncology.org/Genes/IgHID40.html.

11. MacLeod, R.A.F., Nagel, S., Kaufmann, M., Janssen, J.W.G., and Drexler, H.G. (2003) Activation of HOX11L2 by juxtaposition with 3′-BCL11B in an acute lymphoblastic leukemia cell line (HPB-ALL) with t(5;14)(q35;q32.2) *Genes Chromosomes Cancer* **37**, 84–91.

12. Schneider, B., Nagel, S., Kaufmann, M., Winkelmann, S., Drexler, H.G., and MacLeod, R.A.F. (2008) T(3;7)(q27;q32) fuses BCL6 to a non-coding region at FRA7H near miR-29 *Leukemia* **22**, 1262–6.

13. Gu, Y., Cimino, G., Alder, H., Nakamura, T., Prasad, R., Canaani, O., et al. (1992) The (4;11)(q21;q23) chromosome translocations in acute leukemias involve the VDJ recombinase *Proc Natl Acad Sci USA* **89**, 10464–8.

14. MacLeod, R.A., Dirks, W.G., Matsuo, Y., Kaufmann, M., Milch, H., and Drexler, H.G. (1999) Widespread intraspecies cross-contamination of human tumor cell lines arising at source *Int J Cancer* **83**, 555–63.

15. Frohman, M.A., Dush, M.K., and Martin, G.R. (1988) Rapid production of full-length cDNAs from rare transcripts: amplification using a single gene-specific oligonucleotide primer *Proc Natl Acad Sci USA* **85**, 8998–9002.

16. Ochman, H., Gerber, A.S., and Hartl, D.L. (1988) Genetic applications of an inverse polymerase chain reaction *Genetics* **120**, 621–3.

17. Willis, T.G., Jadayel, D.M., Coignet, L.J., Abdul-Rauf, M., Treleaven, J.G., Catovsky, D., and Dyer, M.J. (1997) Rapid molecular cloning of rearrangements of the IGHJ locus using long-distance inverse polymerase chain reaction *Blood* **90**, 2456–64.

18. Meyer, C., Schneider, B., Reichel, M., Angermueller, S., Strehl, S., Schnittger, S., et al. (2005) Diagnostic tool for the identification of MLL rearrangements including unknown partner genes *Proc Natl Acad Sci USA* **102**, 449–54.

19. Akasaka, H., Akasaka, T., Kurata, M., Ueda, C., Shimizu, A., Uchiyama, T., and Hitoshi, O. (2000) Molecular anatomy of BCL6 translocations revealed by long-distance polymerase chain reaction-based assays *Cancer Res* **60**, 2335–41.

20. Sonoki, T., Willis, T.G., Oscier, D.G., Karran, E.L., Siebert, R., and Dyer, M.J. (2004) Rapid amplification of immunoglobulin heavy chain switch (IGHS) translocation breakpoints using long-distance inverse PCR *Leukemia* **18**, 2026–31.

21. An, Q., Wright, S.L., Konn, Z.J., Matheson, E., Minto, L., Moorman, A.V., et al. (2008) Variable breakpoints target PAX5 in patients with dicentric chromosomes: A model for the basis of unbalanced translocations in cancer *Proc Natl Acad Sci USA* **105**, 17050–4.

22. Megonigal, M.D., Rappaport, E.F., Wilson, R.B., Jones, D.H., Whitlock, J.A., et al. (2000) Panhandle PCR for cDNA: A rapid method for isolation of MLL fusion transcripts involving unknown partner genes *Proc Natl Acad Sci USA* **97**, 9597–602.

23. Tonooka, Y., and Jujishima, F. (2009) Comparison and critical evaluation of PCR-mediated methods to walk along the sequence of genomic DNA *Appl Microbiol Biotechnol* adv online pub, DOI 10.1007/s00253-009-2211-5.

24. MacLeod, R.A.F., Nagel, S., Scherr, M., Schneider, B., Dirks, W.G., Uphoff, C.C., et al. (2008) Human leukemia and lymphoma cell lines as models and resources *Curr Med Chem* **15**, 339–59.

25. Drexler, H.G. (2001) The Leukemia-Lymphoma Cell Line FactsBook, Academic Press, San Diego.

26. Drexler, H.G. (2009) Guide to Leukemia-Lymphoma Cell Lines, eBook (available from the author), DSMZ, Braunschweig, Germany.

Chapter 27

Cell Migration and Invasion Assays

Karwan A. Moutasim, Maria L. Nystrom, and Gareth J. Thomas

Abstract

A number of *in vitro* assays have been developed to study tumor cell motility. Historically, assays have been mainly monocellular, where carcinoma cells are studied in isolation. Scratch assays can be used to study the collective and directional movement of populations of cells, whereas two chamber assays lend themselves to the analysis of chemotactic/haptotactic migration and cell invasion. However, an inherent disadvantage of these assays is that they grossly oversimplify the complex process of invasion, lacking the tumor structural architecture and stromal components. Organotypic assays, where tumor cells are grown at an air/liquid interface on gels populated with stromal cells, are a more physiologically relevant method for studying 3-dimensional tumor invasion.

Key words: Invasion, Migration, Scratch assay, Transwell, Matrix, Organotypic culture

1. Introduction

Tumor cell migration and invasion *in vivo* is a complex process, and many *in vitro* assays have been developed in an attempt to recapitulate these processes. Historically, migration/invasion assays have been mainly monocellular, where carcinoma cells have been studied in isolation. Such single cell movement may be mesenchymal or amoeboid, and cells may switch rapidly between the different types of motility. Cells using mesenchymal motility are typically elongated and spindle-shaped, forming actin-rich filopodia or lamellipodia at the leading edge. This process is modulated by Rho GTPases, particularly Rac and cdc42, and involves integrins and proteolytic enzymes such as matrix metalloproteinases. Amoeboid invasion is characterized by cycles of expansion and contraction of the cell body, which allows the cell to squeeze through gaps in the extracellular matrix. Amoeboid invasion is promoted by the Rho/Rock signaling pathway and mediated by cortically located myosin and actin. Additionally, cells may move collectively, sprouting, branching, streaming, or moving as a sheet, and maintaining cell–cell contacts. Collective migration allows coordination of many cells and is a more complex process to study *in vitro*.

It is reported increasingly that tumor stroma has an invasion promoting effect. More physiologically relevant assays should, therefore, incorporate stromal components such as fibroblasts, endothelial cells, macrophages etc. and also reproduce 3-D characteristics of the relevant organ. This chapter aims to cover a range of different techniques that are used to assess tumor cell migration and invasion.

1.1. In Vitro Scratch Assay

This is a simple and inexpensive method to study cell migration *in vitro*. In this assay, a "wound" is created in a cell monolayer and the ability of cells to migrate, and thus, "close" the wound, is assessed by capturing images at different time points (Fig. 1).

The major advantages of this technique are its simplicity and relative low cost and also the ability to visualize cell movement in real time using time-lapse microscopy. Classically, it has been used to study the collective movement of populations of cells, for example, skin wound healing, where to some extent it mimics the migration of keratinocytes as an epithelial sheet. It is useful for studying cell–cell and cell–matrix interactions, and the role of gene overexpression or suppression can be investigated using standard transfection techniques, including microinjection. Analysis of directional migration can be carried out with fluorescently labeled cells using time-lapse microscopy and image analysis software. Cell signaling events can also be investigated by microscopic visualization of specific fluorescently labeled intracellular proteins.

The technique is not suitable for studying chemotaxis, but its relative simplicity and low cost, combined with the lack of need for specialist equipment, still make a popular method for studying cell movement.

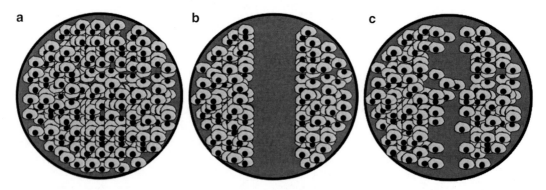

Fig. 1. Scratch assay, in which (**a**) represents a confluent culture, (**b**) the scratch on the surface, and (**c**) the healing scratch.

1.2. Modified Boyden Chamber Assay

In this assay, cells are added to an upper chamber from where they are allowed to invade through a cell permeable membrane toward an attractant placed in the lower chamber (1). Different types of stimuli can be investigated using this system; chemotaxis (movement toward a soluble gradient, e.g., growth factor in the lower chamber), haptotaxis [movement toward a gradient of substratum-bound attractant (e.g., matrix protein coated on the undersurface of the membrane)], or random migration (chemokinesis). After a defined period of time, the cells that have invaded through the membrane into the lower chamber may be counted, making this assay easily quantifiable (2).

Originally designed to study the migration of nonadherent inflammatory cells, the Boyden chamber has been modified in several ways to make it suitable to study invasion of adherent carcinoma cells (1). Firstly, the cell-permeable membrane was coated with a proteinaceous matrix or "basement membrane equivalent" (BME) to replicate the basement membrane through which carcinoma cells must invade *in vivo* (3). Secondly, since invaded carcinoma cells are adherent and remain attached to the undersurface of the cell permeable membrane, invasion cannot be quantified simply by counting the cells in the medium in the lower chamber using a spectrophotometer or a hemocytometer (1, 4). Instead, the cells fixed to the undersurface of the membrane must be either detached from the membrane with trypsin and counted or stained *in situ* in the membrane and counted using a microscope and eyepiece graticule (1, 5). Other methods to quantify this assay include fluorescence, radiolabeling, or colorimetric analysis using various cell dyes or cell viability markers (1, 6, 7). Modified Boyden chambers are commercially available and a widely used example is the Transwell® assay (Fig. 2) (8), which is available with filters of various pore sizes (reflecting the cell type under investigation) and benefits from being highly reproducible (8).

The methods described above examine cell migration. To investigate tumor cell invasion, a physical protein barrier is required. Transwell assays can also be used to study tumor cell invasion, by coating the upper surface of the cell-permeable membrane with an ECM protein gel (Fig. 2b). The protein composition of the gel may vary, but a substrate that has enjoyed widespread use is Matrigel®, isolated from the Englebreth-Holm-Swarm mouse sarcoma, a tumor rich in ECM proteins (9, 10). Matrigel® largely is composed of laminin, Type IV collagen, and heparan sulfate proteoglycans, which are the main constituents of basement membrane, which forms the initial barrier to carcinoma invasion (10). At high concentrations, Matrigel® will polymerise when warmed to 37°C such that it forms a barrier similar in composition to basement membrane. The concentration of Matrigel® should be sufficiently high to provide a physical barrier that will differentiate between invasive and noninvasive cells, but not so

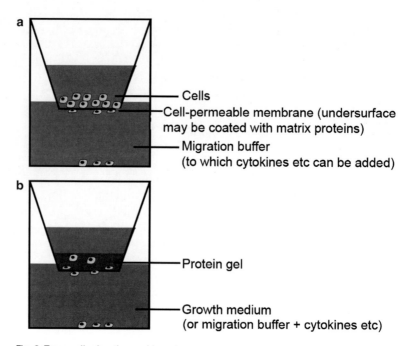

Fig. 2. Transwell migration and invasion assays.

high that it impedes penetration of even the most invasive cells (10). A dilution of 1 part Matrigel:2 parts base medium is usually sufficient (11).

1.3. Three-Dimensional Methods of Studying Invasion

The assays described previously in this chapter have "traded" physiological relevance for ease of repetition and their inherent disadvantage is that they grossly oversimplify the complex process of invasion. Transwell assays lack structural architecture and the cells are added as a single cell suspension. Another inherent disadvantage of these monocellular assays is the absence of stromal cells, which are of fundamental importance in carcinoma invasion. Although stromal paracrine influence can be studied in these simple monocellular assays, for example, by using fibroblast-conditioned medium (12, 13), it is recognized that tumors have a complex architecture with cells of different types in close juxtaposition and this direct physical contact between different cell types may be important. Multicellular invasion assays have been developed which yield information of greater clinical relevance to their monocellular counterparts.

1.4. Organotypic Cultures

Organotypic culture has been especially successful in the field of skin biology (11), but has also been developed to recapitulate other sites such as breast (14), lung (15), oral cavity (16) and pancreas (17). In this assay, epithelial cells are grown at an air/liquid interface on collagen matrices populated with fibroblasts (Fig. 3). Such models have allowed study of cell interactions, whether in the context of normal epithelial growth and differentiation, or between

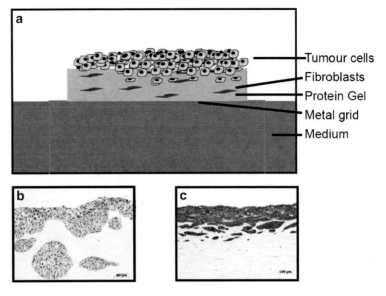

Fig. 3. Organotypic culture. (**a**) Schematic representation of an organotypic culture. (**b**) H+E-stained section of basal cell carcinoma organotypic culture. Note the surface epithelium and the invading islands of tumor. (**c**) A squamous carcinoma organotypic culture immunostained for cytokeratins to highlight the tumor cells.

tumor cells and fibroblasts (16, 18). Additionally, organotypic cultures can be transplanted into immunocompromised mice to study invasion *in vivo* (16).

2. Materials

2.1. In Vitro Scratch Assay

1. Tissue culture medium (e.g., DMEM) with supplements (fetal bovine serum, glutamine, antibiotics).
2. Trypsin/versene (EDTA).
3. Phosphate-buffered saline (PBS).
4. Tissue culture dishes (e.g. 60 mm).
5. Scalpel.
6. p200 pipette tips.
7. Phase-contrast microscope.

Additional materials:

1. Plasmid-encoding GFP or other markers.
2. CO_2-independent medium.
3. Stage incubator.
4. CO_2 supply.
5. Video camera.
6. Image analysis software.

2.2. Transwell® Migration Assay

1. Polycarbonate filters (Transwell®, 8-μm pore size, Becton Dickinson).
2. Attractant of choice [e.g., growth factor (chemotaxis), extracellular matrix protein (haptotaxis)].
3. PBS.
4. Migration buffer (tissue culture medium containing 0.1% BSA, adenine, and glutamine).
5. Tissue culture medium (e.g. DMEM).
6. Trypsin/versene.

2.3. Transwell® Invasion Assays

1. Polycarbonate filters (Transwell®, 8-μm pore size, Becton Dickinson).
2. Matrigel (BD Biosciences).
3. PBS.
4. Tissue culture medium (e.g. DMEM) with and without supplements.
5. Trypsin/versene.

2.4. Three-Dimensional Invasion Assays

1. Collagen type I (rat-tail, BD Biosciences).
2. Matrigel (BD Biosciences).
3. FBS.
4. 10× DMEM.
5. Fibroblast growth medium (e.g., DMEM supplemented with 10% FBS, glutamine with/without Pen/Strep).
6. 25% Glutaraldehyde.
7. Steel grids.
8. Nylon mesh (100-μm pore size).

3. Methods

3.1. In Vitro Scratch Assay

1. An appropriate tissue culture dish (e.g., 60 mm) is coated with the extracellular matrix substrate of choice (e.g. fibronectin), and an appropriate control (e.g., poly-D-lysine) and dishes are incubated for 2 h at 37°C.
2. Unbound is ECM is removed by washing with PBS.
3. Cells that have previously been grown to subconfluency are trypsinized and plated onto dish to create a confluent layer.
4. Plates are incubated for 6 h at 37°C to allow attachment and spreading over the substrate.
5. A "scratch" is created using a p200 pipette tip by scraping the monolayer in a neat, straight line. The plate is washed with

the growth medium to remove the debris, and fresh medium is added.

6. Reference points on the outer surface of the dish are placed using a scalpel, at the approximate areas of the scratch.

7. The dish is placed under a phase-contrast microscope (see Note 1) and the first image is acquired.

8. The dish is returned to the incubator at 37°C for 8–18 h. Examine the dish periodically and capture images at different time points to determine the rate of cell migration. It is critical to match the reference points each time the dish is placed at the microscope.

9. Further analysis and quantification is done using a computer software (e.g., Image Pro-Plus, Media Cybernetics) (19).

3.2. Transwell® Migration Assay (Haptotactic Migration)

1. 200 µL protein solution (e.g., fibronectin 10 µg/mL is added to the lower chamber and incubated for 1 h at 37°C).

2. The protein solution is removed and replaced with 200 µL migration buffer for 30 min at 37°C.

3. Cells are plated in the upper chamber of quadruplicate wells at a density of 5×10^4 in 50 µL of migration buffer for 8–24 h (depending on the cell line).

4. After 8–24 h, the cells in the lower chamber (including those attached to the undersurface of the membrane) are trypsinized (500 µL trypsin/versene is added to the lower well) and counted.

3.3. Transwell® Invasion Assays

1. 70 µL Matrigel (diluted 1:2 in serum-free growth medium) is added to the upper chamber and allowed to gel for 1 h at 37°C.

2. To act as a chemoattractant, 500 µL of complete (serum-containing) tissue culture medium is placed in the lower chamber.

3. Cells are plated in the upper chamber of quadruplicate wells at a density of 5×10^4 in 200 µL of serum-free medium and incubated at 37°C for 72 h.

4. After 72 h, the cells in the lower chamber (including those attached to the undersurface of the membrane) are trypsinized (500 µL trypsin/versene per well) and counted.

3.4. Three-Dimensional Invasion Assays

Day 1 – Gel Preparation

1. A fibroblast cell suspension is prepared at 500,000 cells/gel in FGM.

2. Gels are mixed on ice in the following ratios:

Collagen: matrigel (1:1)	7 volumes (3.5 collagen: 3.5 matrigel)
10× DMEM	1 volume
FBS	1 volume
Fibroblasts	1 volume

3. 1 mL of the above mix is aliquoted into each well of a 24-well plate and incubated at 37°C for 1 h.
4. After gel polymerisation, 1 mL of FGM is added per well and gels are left for 18 h at 37°C to equilibrate.

Day 2 – Plating out cells on top of gels

1. Medium is aspirated from the top of the gels.
2. Keratinocytes are plated out at 500,000 cells per gel in keratinocyte medium supplemented with 10% FCS and glutamine.
3. Gels are incubated overnight at 37°C.

Day 2 – Preparation of nylon sheets

1. Sterile nylon sheets (one sheet per each gel) are placed in a sterile dish.
2. Fibroblast-free gel solution is made in the following ratios:

Collagen	7 volumes
10× DMEM	1 volume
FCS	1 volume
10% DMEM	1 volume

3. The solution is mixed well, and if yellow, is neutralized with sterile NaOH 0.1 M until it turns pink.
4. 250 μL is added to each nylon sheet and incubated at 37°C for 15–30 min.
5. Fix in 10 mL 1% glutaraldehyde (in PBS). Incubate at 4°C for 1 h.
6. Gels are washed three times with PBS, once in FGM and then left in FGM at 4°C overnight.

Day 3 – Raising gels on to grids

1. A steel grid is placed into each well of a 6-well plate and a gel-coated nylon sheet is laid on top of it. Steel grids are made from 2.5 cm^2 of stainless steel mesh with the edges bent down to form 4- to 5-mm high legs.
2. Organotypic gels are removed from the 24-well plate using a sterile spatula and then placed on the collagen-coated nylon sheet that is resting on the steel grid (see Note 2).

3. The well with complete keratinocyte growth medium, such that it reaches the undersurface of the grid, thus allowing the epithelial layer to grow at an air–liquid interface, and incubated at 37°C (see Note 3).

4. Medium is replaced every 2 days.

5. Gels are harvested after 7–14 days of culture.

3.4.1. Processing of Gels

1. The organotypic culture is removed from the well *in toto* and cut in half with a sterile scalpel.

2. Both halves are fixed in a Universal tube containing 10% formol saline for 24 h.

3. Formol saline is replaced with 70% ethanol left overnight.

4. Gels can then be processed to paraffin blocks and sectioned for staining and/or immunochemistry.

3.4.2. Quantifying Invasion from Organotypic Cultures

Although organotypic cultures recapitulate the morphology of SCC *in vivo*, analysis generally has been subjective and restricted to description of the invasive appearance in stained sections. Some authors have attempted to quantify tumor cell invasive activity in three-dimensional assays. For example, confocal microscopy has been used to determine depth of invasion of fluorescently labeled cells into a gel (20). Consecutive images at increasing depths of focus (Z-sectioning) have been used to reconstruct three-dimensional images and the number of cells at each level determined by using computer software to count pixel number per slice.

Others have quantified invasion in three-dimensional cocultures by measuring the infiltration distance, i.e., the maximum distance between the epithelium-tumor frontier and the most distant tumor cells, in serial sections (15). However, this measurement does not take into account the pattern or amount of tumor invasion.

More recently, digital image analysis methods that assess tumor infiltration objectively have been developed (21). For example, Nystrom and colleagues analyzed the degree of invasion by generating an "Invasion Index" which was the product of the depth of invasion, and the number and area of invading tumor islands (21); such methods give an accurate representation of tumor invasiveness and are also applicable to analyzing tumor invasion *in vivo*.

4. Notes

1. Cells transfected with a marker plasmid (e.g., GFP) can be observed under a fluorescence microscope. If a time-lapse microscope is used, particular attention should be paid to the

growth medium if the stage incubator is equipped with a temperature control only, in which case cells should be kept in CO_2-independent (HEPES-buffered) medium (19).

2. Critical step: particular care should be taken when raising the gels onto the steel grids to avoid misorientation and destruction of the gel.

3. Note that it is this initial time point that is defined as day 1 of the organotypic culture.

References

1. Shaw, L. M. (2005) Tumour cell invasion assays. *Methods Mol. Biol.* **294**, 97–105.
2. Brown, N. S. & Bicknell, R. (2001) *Cell Migration and the Boyden Chamber*, Totowa, New Jersey, Humana Press Inc.
3. Iwamoto, Y. & Sugioka, Y. (1992) Use of a reconstituted basement membrane to study the invasiveness of tumour cells. *Adv. Exp. Med. Biol.* **324**, 141–149.
4. Grotendorst, G. R. (1987) Spectrophotometric assay for the quantitation of cell migration in the Boyden chamber chemotaxis assay. *Methods Enzymol.* **147**, 144–152.
5. Terranova, V. P., Hujanen, E. S., Loeb, D. M., Martin, G. R., Thornburg, L. & Glushko, V. (1986) Use of a reconstituted basement membrane to measure cell invasiveness and select for highly invasive tumour cells. *Proc. Natl. Acad. Sci. U.S.A.* **83**, 465–469.
6. Saito, K., Oku, T., Ata, N., Miyashiro, H., Hattori, M., & Saiki, I. (1997) A modified and convenient method for assessing tumour cell invasion and migration and its application to screening for inhibitors. *Biol. Pharm. Bull.* **20**, 345–348.
7. Albini, A. (1998) Tumour and endothelial cell invasion of basement membranes. The Matrigel chemoinvasion assay as a tool for dissecting molecular mechanisms. *Pathol. Oncol. Res.* **4**, 230–241.
8. Thomas, G. J., Lewis, M. P., Hart, I. R., Marshall, J. F., & Speight, P. M. (2001). Alphavbeta6 integrin promotes invasion of squamous carcinoma cells through upregulation of matrix metalloproteinase-9. *Int. J. Cancer* **92**, 641–650.
9. Kleinman, H. K., McGarvey, M. L., Liota, L. A., Robey, P. G., Martin, G. R. (1982). Isolation and characterization of type IV collagen, laminin and heparan sulfate proteoglycan from the EHS sarcoma. *Biochemistry*, **21**, 6188–6193.
10. Kleinman, H. K., Martin, G. R. (2005). Matrigel: basement membrane matrix with biological activity. *Semin. Cancer Biol.* **15**, 378–386.
11. Marsh D, Dickinson S, Neill GW, Marshall JF, IR Hart, Thomas GJ (2008). αvβ6 integrin promotes the invasion of morphoeic basal cell carcinoma through stromal modulation. *Cancer Res.* **68**, 3295–303.
12. De Wever, O., Nguyen, Q. D., Van Hoorde, L., Bracke, M., Bruyneel, E., Gespach, C. & Mareel, M. (2004). Tenascin-C and SF/HGF produced by myofibroblasts *in vitro* provide convergent pro-invasive signals to human colon cancer cells through RhoA and Rac. *FASEB J.* **18**, 1016–1018.
13. Lewis, M. P., Lygoe, K. A., Nystrom, M. L., Anderson, W. P., Speight, P. M., Marshall, J. F., & Thomas G. J. (2004) Tumour-derived TGF-beta1 modulates myofibroblast differentiation and promotes HGF/SF-dependent invasion of squamous carcinoma cells. *Br. J. Cancer*, **90**, 822–832.
14. Kim, J. B., Stein, R., & O'Hare, M. J. (2004) Three-dimensional in vitro tissue culture models of breast cancer – a review. *Breast Cancer Res. Treat.* **85**, 281–291.
15. Al-Batran, S. E., Astner, S. T., Supthut, M., Gamarra, F., Brueckner, K., Welsch, U., Knuechel, R. & Huber, R. M. (1999). Three-dimensional *in vitro* cocultivation of lung carcinoma cells with human bronchial organ culture as a model for bronchial carcinoma. *Am. J. Respir. Cell Mol. Biol.* **21**, 200–208.
16. Nystrom ML, McCullogh D, Weinreb P, Violette S, Speight PM, Marshall JF, Hart IR, Thomas GJ (2006). COX-2 inhibition suppresses αvβ6 integrin-dependent oral squamous carcinoma invasion. *Cancer Res.* **66**, 10833–42.
17. Froeling, F. E., Mirza, T. A., Feakins, R. M., Seedhar, A., Elia, G., Hart, I. R., Kocher, H. M. (2009) Organotypic culture model of pancreatic

cancer demonstrates that stromal cells modulate E-cadherin, beta-catenin and Ezrin expression in tumour cells. *Am. J. Pathol.* **175**, 636–648.
18. Boukamp, P., Breitkreutz, D., Stark, H. J. & Fusenig, N. E. (1990) Mesenchyme-mediated and endogenous regulation of growth and differentiation of human skin keratinocytes derived from different body sites. *Differentiation* **44**, 150–161.
19. Liang, C., Park A. Y., and Guan J. (2007) *In vitro* scratch assay: a convenient and inexpensive method for analysis of cell migration *in vitro*. *Nat. Prot.* **2**, 329–333.
20. Vial, E., Sahai, E., Marshall, C. J. (2003) ERK-MAPK signalling co-ordinately regulates activity of Rac1 and RhoA for tumour cell motility. *Cancer Cell* **4**, 67–79.
21. Nystrom, M. L., Thomas, G. J., Stone, M. L., Mackenzie, I. C., Marshall, J. F. (2005) Development of a quantitative method to analyse tumour cell invasion in organotypic culture. *J. Pathol.* **205**, 468–475.

Chapter 28

Angiogenesis Assays

V. Poulaki

Abstract

The angiogenic process is central in the pathogenesis of various diseases. The in vitro and in vivo monitoring of the neovascular process is essential for the development and evaluation of angiogenesis inhibitors or stimulators. Since no single method exists that can assess angiogenesis in a robust, reliable, and reproducible fashion, researchers often use a combination of assays to circumvent this problem. The experimental details of the most commonly in vitro, ex vivo, and in vivo assays are presented here.

Key words: Angiogenesis, Endothelial cells, Co-culture, Fibroblasts, Matrigel, Retinopathy, Chorioallantoic membrane

1. Introduction

Angiogenesis is a process which is essential for normal development and entails the creation of new blood vessels from an existing vasculature. Physiological angiogenesis is a highly organized concert of events that serve the ever changing tissue requirements as vascular initiation, formation, maturation, and remodeling or regression. Pathological angiogenesis is the hallmark of more than 50 diseases such as cancer, is not as meticulously orchestrated and although the first stages are the same, the vessels rarely mature or involute (1). The better understanding of the factors that are involved in the angiogenic process has a central role in the treatment of angiogenesis-dependent diseases. Therefore, the design of a "gold standard" angiogenesis study will allow the dissection of the mechanisms of these diseases and help toward the identification of inhibitors. This study should be rapid, sensitive, and specific, with capabilities of automated analysis and multiparameter assessments and should have a direct relevance to clinical findings. Despite the increasing variety of angiogenesis studies, the ideal method has yet to be developed, and therefore researchers often use a combination of techniques to circumvent that problem (2). In vitro studies are focusing more on the proliferation, differentiation, and migration of endothelial cells and their interaction with the surrounding supporting stromal cells and the extracellular matrix, while in vivo studies include more complex interactions of the endothelial cells with the surrounding tissue and blood.

2. Materials

2.1. Endothelial Cell Coculture with Fibroblasts

1. Human umbilical vein endothelial cells (HUVEC) (Invitrogen, Portland, OR) passage between 4 and 8.
2. Human dermal neonatal fibroblasts (Invitrogen) passage between 4 and 8.
3. DMEM supplemented with 5% fetal bovine serum (FBS, Hyclone), 2 mM glutamine, 1 mM sodium pyruvate, and 100 U/mL penicillin–streptomycin (Invitrogen).
4. Woven nylon mesh ring (Tetko, Inc, Monterey Park, CA).
5. 96-Well plates (BD Biosciences, San Jose, CA).
6. Type I rat tail collagen (BD Biosciences).
7. 10× M199 (Invitrogen).
8. Serum-free DMEM, FBS, M199, and PBS (Invitrogen).
9. 70% Alcohol.
10. Mouse anti-human PECAM antibody (R&D systems, Abingdon, UK), alkaline-phosphatase-conjugated anti-mouse antibody (R&D systems), 5-bromo-4-chloro-3-indolyl phosphate/nitro blue tetrazolium (BCIP/NBT) or Angio kit (TCS cell works Botolph Claydon, UK).
11. Microscope (Leica) with camera (Hamamatsu, Hamamatsu city Japan).

2.2. In Vitro Matrigel Invasion Study

1. Matrigel (Sigma Chemical, St Louis).
2. Pipettmen and prechilled sterile tips.
3. HUVEC (Invitrogen).
4. HUVEC medium (Invitrogen).
5. 24-Well transwell plates, 150 mm dishes (Invitrogen).
6. RPMI or DMEM medium (Invitrogen).
7. Table centrifuge for 50 mL conical tubes.
8. 50 mL conical tubes.
9. Hemocytometer.

10. Trypsin–EDTA solution with 0.5% trypsin, 0.53 mM EDTA (Sigma-Aldrich, St Louis, MO).
11. Trypsin inhibitor (Sigma-Aldrich).
12. Incubator with 95% air/5% CO_2.
13. 24-Well plates.
14. Diff-Quik fixative and solution II (Baxter Scientific Products, McGaw, IL).
15. Permount mounting medium (Fisher Scientific, Pittsburg, PA).

2.3. Chorioallantoic Membrane Assay

1. Fertilized chick eggs on cardboard egg crates (SPAFAS, Norwich, CT).
2. 3% CO_2 humidified incubator set at 37°C, thermometer.
3. Three Petri dishes with covers (Falcon).
4. 0.1% Methylcellulose liquid (Invitrogen).
5. Teflon pedestals (3–4 mm × 2 cm).
6. Petri dishes with covers (Invitrogen).
7. 1-cc Hamilton syringe with 26-gauge needle as applicator of sample or 10-μL pipette (Hamilton Nevada), tweezers.
8. Pure test growth factor and control protein (1–2 μg).
9. Dissecting microscope (Leica) with camera (Hamamatsu, Orca).

2.4. Rat Aortic Ring Assay

1. Male Fisher-344 rats, 8–12 weeks old.
2. Microdissection scissors (iris) and forceps.
3. 10× Eagles Medium (Invitrogen).
4. Serum-free culture medium, MCDB131 (Gibco), supplemented with glutamine (2 mM), penicillin (100 U/mL), streptomycin (100 μg/mL), and 25 mM $NaHCO_3$.
5. 1.5% Agarose (type VI-A) solution (Invitrogen).
6. Puncher with two concentric rings (diameter 10 and 17 mm).
7. Sterile aqueous 1 M NaOH.
8. Sterile bovine fibrinogen (Culture-tested Sigma).
9. Sterile bovine thrombin (Sigma).
10. Eagle's MEM (Gibco).
11. Fluorescein-labeled acetylated low-density lipoprotein (Dil-Ac-LDL, Biomedical Technologies).
12. Humidified incubator 5% CO_2/air.
13. Fluorescence microscope (Leica) with camera (Hamamatsu, Japan).
14. Digital camera.

2.5. Matrigel Plug Angiogenesis Assay

1. C57Bl/6N female mice.
2. Matrigel 10 mL at 15 mg/mL (stock), store at −20°C. Enough will be needed to inject 0.5–1 mL per mouse.
3. Prechilled 50-mL falcon tubes (Invitrogen).
4. Prechilled TB syringes with 25-gauge needles (Becton Dickinson).
5. 3-mL Syringes (Becton Dickinson).
6. Endothelial cell growth supplement (EGCS, Collaborative Biomedical BD, NJ). Store in aliquots at −20°C and prepare 1 mg/mL solution just before use in sterile distilled water.
7. DMEM prechilled (Invitrogen).
8. Scissors to dissect the skin, forceps.
9. Dispase (Sigma-Aldrich).
10. Specimen vials.

2.6. Mouse Corneal Angiogenesis Assay

1. CD-1 mice (Charles River Laboratories, Wilmington, MA).
2. Nembutal solution (Abbott Laboratories, North Chicago, IL).
3. Hamilton syringes 1 cc with 33-gauge needle.
4. Half an inch 30-gauge needle.
5. Dissecting microscope.
6. Growth factor and control protein.
7. Wescott scissors.
8. Forceps.
9. 10-mL Syringes and 16-gauge perfusion cannula, IV tubing.
10. Phosphate-buffered formalin (Fisher, Fairlawn, NJ).
11. FITC-coupled ConA lectin (Vector Laboratories, Burlingame, CA).
12. Microscope slides.
13. Fluorescent microscope (Leica) with camera (Hamamatsu, Japan).

2.7. Retinopathy of Prematurity Model

1. Newborn C57Bl/6 mice.
2. Ketamine (100 mg/mL) and xylazine (20 mg/mL).
3. Plexiglas chamber with oxymeter (GMH 3690 GL; Greisinger electronic GmBH, Regenstauf, Germany).
4. Wescott scissors.
5. Forceps.
6. 10-mL Syringes and 16-gauge perfusion cannula, IV tubing.
7. Phosphate-buffered formalin (Fisher, Fairlawn, NJ).

8. FITC-coupled ConA lectin (Vector Laboratories, Burlingame, CA).
9. Microscope slides.
10. Fluorescent microscope (Leica).

3. Methods

3.1. Coculture of Endothelial Cells with Fibroblasts (3, 4)

1. Preparation of media: Add 340 µL of type I rat tail collagen (BD Biosciences), 76 µL of 10× M199 medium, 136 µL of serum-free DMEM, 10 µL of FBS, and 340 µL of PBS and adjust the pH to 7.2 with NaOH. All the components should be kept sterile.

2. Culture of cells: Add 1×10^6 HUVEC and 5×10^5 HDFN (see Note 1) to the collagen mixture for a final collagen concentration of 1.25 mg/mL. Spot 30 µL of collagen/cell mixture on to a 5-mm woven nylon mesh ring (Tetko, Inc) to provide structural support (see Note 2). Allow the collagen to polymerize for 60 min at 37°C in a humidified 5% CO_2 incubator, after which transfer each ring to a 96-well plate and cover with the culture media (basic media supplemented with 1% FBS and 30 ng/mL VEGF-A 165). Incubate the cocultured cells for up to 14 days at 37°C in a 5% CO_2 in air humidified atmosphere and change the medium every 2–3 days.

3. Fixation and staining of cells: At the end of the coculture, fix the cells with 70% ice-cold ethanol. Wash the cells with PBS containing 2% BSA and incubate with a mouse monoclonal anti-human PECAM-1 antibody for 1 h at 37°C. After the end of the incubation period, wash the wells with PBS and incubate with an alkaline-phosphatase-conjugated anti-mouse antibody and stain using 5-bromo-4-chloro-3-indolyl phosphate/nitro blue tetrazolium (BCIP/NBT) as recommended by the manufacturer. You can alternatively use all of the above as a kit (Angiokit, see Subheading 2).

4. Image capture and tubule quantification: Photograph five randomly selected fields of view in each of the triplicate wells per condition (see Note 3). Capture the images using a digital camera attached to a microscope on low magnification (×40) and save them as TIFF files. Import the images into Scion Image (NIH) and convert them to binary format. Calculate the tubule area as the total number of pixels in thresholded images. Measure the area occupied by the endothelial cell aggregates in interconnecting tubules in the same way but with the images of tubules deleted from each image. Assess the tubule length by drawing a line along each tubule and measure the length of the line in pixels, and count

the branch point manually. Convert the measurements into micrometer after calibration of the microscope/camera set up by taking the image with a 50-μm graticule.

3.2. In Vitro Matrigel Invasion Study (5–9)

1. Preparation of the endothelial cells: Rinse a confluent 150-mm dish or equivalent HUVECs in 5 mL of RPMI or DMEM. Wash cells three times with RPMI and culture them in serum-reduced medium (1% FBS) overnight.
2. When the chamber is ready (see below), add 5 mL of trypsin EDTA solution and incubate the cells at 37°C until detached from the plate (approximately 5 min).
3. Neutralize the trypsin with equal amounts of trypsin inhibitor (Soybean). Place the cells in a 50-mL conical sterile tube and centrifuge for 5 min at $50 \times g$.
4. Remove the supernatant and wash them three times with RPMI with 1% FBS and reconstitute them in RPMI with 1% FBS at a final concentration of 10^6 cells/mL (see Note 4).
5. Preparation of the matrigel: Thaw matrigel at 4°C overnight and dilute it (5–1 mg/mL) in serum-free cold culture media (RPMI 1640 or DMEM). Under sterile conditions, add 100 μL of the diluted matrigel into the upper chamber of the 24-well transwell and incubate the transwell at 37°C for at least 4–5 h for gelling. Wash the gelled matrigel with warmed serum-free media.
6. Culture of cells in the invasion chamber: Add 100 μL of the cell suspension into the upper part of the matrigel-containing transwell and fill the lower chamber of the transwell with 600 μL of culture media with 5 μg/mL fibronectin (see Notes 5 and 6).
7. Incubate for 24–72 h at 37°C. Stop the assay by adding 200 μL of the Diff-Quik fixative per well and incubate for 30 min. Remove cells from the top of the filter using a Q tip twice. Aspirate the fixative and stain the cells for 2 min with Diff-Quik solution II that has been diluted 1/1 with water (double distilled). Dip the inserts 1 min for each solution (fix solution, solution I and II). Dip the insert in the water to wash out the dye. After the filter dries, remove from insert with a scalpel and mount the filter on the slide with cell side face down using permount. Count the result under microscope (20× objective) (see Notes 3 and 7–9).

3.3. Chorioallantoic Membrane Assay (9–11)

1. Preparation of fertilized eggs: Fill the lower pan of the incubator with double-distilled water to ensure a humidified environment. Incubate fertilized chick eggs in the incubator (37°C, 3% CO_2) for 3 days on egg crates. Gently roll them 3–4 times a day (see Note 10).

2. When the eggs crack, remove some of the eggshell with the tweezers and carefully pour the yolk into the Petri dishes without disturbing it (see Note 11).

3. Preparation of dry test pellets: Instill 10 µL of liquid methylcellulose and 1 µg of purified growth factor or purified control protein into the Teflon pedestals and dry for 1 h (see Note 12).

4. Insertion of dry pellets and photographing of the chorioallantoic membrane assay (CAM): Insert the dry pellets on the P6 developing chick embryo, photograph the CAM, and quantify the vessels using a computer software, comparing vessels for embryos treated with the control substance versus the growth factor.

5. Alternatively, the branch points can be quantified manually with the use of several blinded observers. A typical scale might be as follows: less than 75 branches "0," 75–150 "1," 150–300 "2," 300–600 "3," and more than 600 "4."

3.4. Rat Aortic Ring Assay (12–16)

1. Preparation of the Aorta: Euthanize the animal and open the chest with dissecting instruments (scissors, forceps). Isolate the aorta carefully avoiding trauma to its body.

2. Transfer the aorta to a culture dish containing ice-cold MCDB 131 culture medium or balanced salt solution. Flush the aortas with ice-cold medium with a 1-mL syringe fitted with a 23-gauge needle until the medium runs clear, and the aorta is free of clotted blood. Cut the proximal and distal 2 mm segments and remove the surrounding fibro-adipose tissue.

3. Cut the aorta into 1 mm ring sections (see Note 13) and rinse in 5–8 washes of ice-cold MCDB 131 culture medium. Remove carefully under the dissecting microscope the fibroadipose tissue and the collateral vessels with the microdissection scissors and forceps (see Note 14). Make sure that the rings do not dry out. Transfer thse rings to 10 mL of fresh ice-cold medium in a sterile culture tube and wash rings by replacing the medium five times. Touch the rings onto a clean Petri dish to remove excess medium.

4. Making the Agarose Culture Wells: Prepare a 1.5% w/v aqueous solution of agarose and autoclave to dissolve and sterilize. Pour 30 mL of agarose solution into each culture dish and let it solidify.

5. Make 6–7 concentric rings/dish with the puncher and use them as culture wells.

6. Remove the central portion of the rings with a bent spatula and transfer them into bacteriological Petri dishes (using these dishes improves the adherence of the agarose to them).

7. Preparation of the aortic ring cultures: Dissolve the fibrinogen in MEM and filter the solution through a 0.4-μm sterile filter to remove clumps, which will interfere with its polymerization. Mix 1 volume of the fibrinogen solution with 0.02 volumes of the thrombin (50 U/mL) and vortex. Promptly add 200 μL of the collagen solution to coat the bottom of each agarose well and allow it to gel in the incubator.

8. Add the rings that were prepared as previously and arrange them in the bottom of each agarose well horizontally. Fill each well with 200 μL of the fibrinogen solution to cover the rings. Add 30 mL of MCDB131 medium containing L-glutamine (1.5 mg/mL), penicillin (100 U/mL), streptomycin (100 μg/mL), amphotericin B (0.25 μg/mL), and 300 μg/mL ε-amino-*n*-caproic acid (to prevent spontaneous fibrinolysis) to each dish and incubate at 37°C in the incubator.

9. Culturing of the rings: Incubate the cultures at 37°C with 5% CO_2 and change the medium every 2 days. Fibroblasts grow out of the aortic explant within 2 days of culture, whereas endothelial cells migrate into the matrix as microvessels by day 4. The growth combination throughout the first week (number and length of the microvessels), whereas in the second week the vessels begin to regress. On day 12, carefully remove the aortic segments and incubate the remaining cells (microvessels) for 4 h at 37°C with a solution of Dil-Ac-LDL (10 μg/mL) in MEM medium followed by three washings of 1 h each in fresh medium. The uptake of the Dil-Ac-LDL is strong from the endothelial cells and minimal from the fibroblasts. Visualize stained endothelial cells in a fluorescent microscope (Leica) and photograph them (see Notes 15–17).

3.5. Matrigel Plug Angiogenesis Assay (17–20)

1. Preparation of the matrigel/growth factor injections: Thaw matrigel on ice overnight at 4°C and the next day mix on vortex. Be careful to not allow the matrigel to be warmed at room temperature (see Note 18). Pipette equal amounts of matrigel in separate 14 mL tubes, calculating that each mouse should require 0.5–1 mL and each test condition should need 3–4 mice.

2. Each test condition should have a positive control containing 150 ng/mL FGS and a negative control containing only matrigel. Add the desired concentration of the growth factor that will be tested (see Note 19), equalize the volumes with cold PBS and mix in cold room with careful inversion and avoid bubbles.

3. Animal injections and monitoring: Anesthetize the animals and load the prechilled tuberculin syringes, without the needle, with the mixture of matrigel and growth factor (or control). Inject in each animal 500 μL of the matrigel mixture

into the ventral area of each mouse (see Note 20). The typical maximal response is 1 week to 10 days later (the peak of the angiogenesis response is around the seventh day) after the injection. The plugs appear as bumps on the central side of the animal.

4. Removal of the plugs: Inject 0.2 mL of 25 mg/mL FITC-dextran (1×10^6 MW) in PBS (30 mg/mL) through the lateral tail vein and allowed to circulate for 30 min. Euthanize the mice with cervical dislocation or CO_2 and collect blood samples by cardiac puncture into heparinized tubes. Centrifuge immediately after the collection, collect the upper phase (plasma) and store at 4°C protected from light. Resect the matrigel plugs, place in tubes containing 1 mL of 1/10 dispase and incubate in a shaker in the dark, at 37°C overnight. The following day homogenize the plugs, centrifuge at $3,000 \times g$ for 10 min and measure the fluorescein emission with a fluorescence plate reader using a standard curve created by serial dilution of FITC-dextran. The angiogenic response can be expressed as a ratio of matrigel plug fluorescence and plasma fluorescence (see Note 21).

3.6. Mouse Corneal Angiogenesis Assay (21–22)

1. Preparation of mouse for injection: Anesthetize mice with 70–80 mg/kg nembutal sodium solution intraperitoneally. Place mouse under the microscope and use your fingers as an ophthalmic retractor to expose the cornea. Under direct observation, under the microscope make a nick in the epithelium and anterior stroma in the midperiphery with a half an inch 30-gauge needle (Becton Dickinson, Franklin Lakes, NJ) (see Note 22).

2. Inject 2 µL of the growth factor or control protein with a half an inch 33-gauge needle with a 30° bevel on a 10-µL gas tight Hamilton syringe (Hamilton, Reno, NV). Take pictures with the dissecting microscope and document vessel growth with the microscope on days 1, 3, 7, and 21. Quantify the vessel length with computer software (density slicing, see below, Improvision, Openlab).

3. Perfusion of the animals and enucleation of the eyes: Anesthetize mice deeply as above and open the chest cavity with the use of dissecting scissors. Introduce a 16-gauge perfusion cannula attached to a 10-mL syringe filled with PBS into the left ventricle and place a cut on the right atrium to allow drainage of the blood. Slowly perfuse PBS into the left ventricle following by 1% paraformaldehyde and FITC-coupled Con A lectin (20 µg/mL in PBS, pH 7.4, 5 mg/kg BW) at physiologic pressure.

4. Fixing and staining of the corneas: Enucleate the eyes, dissect the corneas, and mount them onto microscope slides.

Use the fluorescent microscope to obtain pictures of the fluorescein-labeled vessels. Outline the total surface of the retina using the outermost vessel of the arcade near the ora serrata. Set a threshold of fluorescence above which only the fluorescence in the vessels will be captured (density slicing) and a level of fluorescence above which the whole cornea will be captured. Calculate the ratio of the vascular area and the total corneal area for each eye. The blood vessel tufts can be counted manually by analyzing enlarged images of the cornea.

3.7. Retinopathy of Prematurity Model (23–25)

1. Hypoxia–hyperoxia induction: Keep newborns with their nursing mother either in room air or in the Plexiglas chamber from postnatal day P0–P7. Maintain the oxygen concentration of the chamber between 74.5 and 76%. Return the mice in room air for 5 days from P7 until P17 (see Notes 23–25).

2. Perfusion of the animals: Anesthetize the mice deeply by an intraperitoneal injection of 9 µL/g mouse BW of a solution containing 1 mL ketamine (100 mg/mL), 1 mL xylazine (20 mg/mL), and 5 mL sterile saline. Make a sternotomy with the dissecting scissors and penetrate the left ventricle with a 20-gauge 1½ cannula (BD). Perforate the right atrium to establish blood outflow. Perfuse the mice with 1.5 mL of NaCl 0.9% followed by 1 mL of paraformaldehyde 1%, 1.5 mL fluorescein concanavalin A 5% solution, and finally 1.5 mL PBS maintaining physiologic pressure.

3. Preparation of retinas: Enucleate the eyes and remove the cornea, the lens, and the vitreous. Fixate the eye cups in 1% PFA solution for 20 min. Make four radial cuts that will allow the spread of the tissue as a clover leaf. Separate the retina from the underlying choroid and sclera and mount on a microscope slide using fluorescence mounting medium.

4. View the retinas with a fluorescence microscope and photograph by using a digital camera. Capture the images on a computer and analyze, for example with Open Lab software (Improvision, Inc).

5. Convert the images to TIFF files (tagged information file format). Outline the total surface of the retina using the outermost vessel of the arcade near the ora serrata. Set a threshold of fluorescence above which only the fluorescence in the vessels will be captured (density slicing) and a level of fluorescence above which the whole retina will be captured. Calculate the ratio of the vascular area and the total retinal area for each eye. The blood vessel tufts can be counted manually by analyzing enlarged images of the retina.

4. Notes

1. HUVECs and fibroblasts should not be older than six passages because they are primary cultures and they tend to lose their differentiated characteristics, confluence should be less than 70% (otherwise the endothelial cells stop proliferating-contact inhibition). Carry the experiments soon after the plating of the cells. Because the endothelial cells are usually quiescent, it is often necessary to stimulate proliferation by serum starvation (1% FBS) and then reintroduce serum with the test substances (you have to be careful not to induce apoptosis).
2. If the coculture system is used to test an angiogenic or antiangiogenic substance, several concentrations should be tested at least in triplicate (three positive wells and three negative). When handling the transwell for the in vitro matrigel assay, avoid bubbles at both sides of the filter.
3. Although the dishes with the stained cells can be stored up to a week at 4°C, the pictures should be taken within few days because the matrigel tends to dry up and the morphology is not preserved.
4. The lowest amount of serum concentration with which endothelial cells from tubes in vitro in matrigel should be used.
5. To perform assays on smaller plates, the amount of matrigel and cells that are plated has to be reduced accordingly. For example, 48-well dishes the wells are coated with 200 µL of matrigel and 24,000 cells are plated in 150 µL of medium. If the cells aggregate during the invasion study, reduce the density of the cell suspension.
6. Fresh media with and without the test substances have to be replenished every 72 h and the plates have to be tested for tube formation every 24 h. Matrigel tends to form a gel quickly at room temperature, so the pipettes and the tips that will come in contact with it need to be prechilled to prevent it from solidifying. Do not use a concentration of matrigel bellow 1 mg/mL because it will not solidify.
7. Although the endothelial tubes can be seen even in unstained cultures, to determine the extent of tubule formation, it is best to fix the cells with 70% alcohol and stain with endothelial markers. Do not stain the matrigel with Diff-Quik solution II for longer than 2 min without previous dilution because it will stain excessively the matrigel and the identification of the tubular structures will not be easily observed.

8. Collagen-embedded cells were fixed in 4% formaldehyde overnight and incubated with 10 μg/mL tetramethylrhodamine isothiocyanate (TRITC)-labeled lectin (Ulex europaeus UEA-I Sigma) for 1 h. For prelabeling fibroblasts, cells were loaded with CellTracker Dye Green DMFDA (10 μm Molecular Probes) for 30 min in serum-free medium and used in the angiogenesis assays. Cells were mounted in Aqua Mount (Lerner Labs) and visualized on a Leica DM5000B microscope.

9. An alternative method of growing vessel quantification is the measurement of hemoglobin content with a commercial kit (Drabkin reagent 525 kit, Sigma, St Louis, MO).

10. Rotate the eggs periodically and do not let them sit for the whole 3 days. Use 3–4 dozen eggs on the crates because 10% of eggs do not develop embryologically.

11. Make sure that when you pour the eggs into the dishes the yolk remains intact, otherwise the embryo will not develop normally.

12. Always use a control protein because the inflammatory response from the procedure itself can mimic the angiogenesis process.

13. Embed the aortic rings on their sides and not on their ends, because this provides optimal visualization of the microvessels (if the rings are on their ends, the vessels appear to grow toward the observer). The optimal length of the rings should be 1 mm, shorter lengths are more difficult to handle and vascularize less, whereas longer lengths have lower yields per aorta.

14. To reduce the adventitial fibroblast content in the culture, the intimal layer can be teased out with fine forceps to about half thickness of the aorta. Rat endothelium is very well adhered to the intima and unlike larger species rinsing of the aortas and removal of clotted blood will not lead to stripping of the endothelium.

15. When the cultures are supplemented with growth factors, the vessel growth increases exponentially (50 ng/mL FGF or 0.5% serum). This makes manual quantification difficult. Computer-assisted quantification is preferable because it takes into account the length and width of each vessel.

16. It is necessary to keep the same microscopic setting throughout the entire experiment and the same illumination setting, otherwise biases are introduced into the computer-assisted quantification to decrease the background fluorescence Lab-Tek Permanox 8-well chamber slides (Nunc International) can alternatively be used.

17. The aortic rat study is most suitable to study the efficacy of angiogenic agents. This is true because spontaneous angiogenesis in the rat rings is minimal and significantly enhanced by angiogenic compounds. For the study of antiangiogenic agents, it is best to supplement with growth factors (50 ng/mL FGF or 0.5% serum).

18. *Notes for the matrigel in vivo assay* – To minimize bubbles from the vortexed Matrigel and minimize the injection problems, allow the tubes to stand on ice for 10 min.

19. When screening putative antiangiogenic compounds add 150 ng/mL bFGF instead of ECGS to all the tubes except the negative controls.

20. Allow the matrigel to gel for 10 min before injecting it to the animal and avoid puncturing the skin because the material will flow out. Always use a sterile technique to prepare the injections and perform them.

21. If there are bubbles, cracks, or artifacts in the picture, use the drawing tool to select a portion of the picture without them because they will be picked up by the density slice feature and skew the results. Draw the same size box on all sections of the experiment.

22. The mice need to be anesthetized deeply because they need to be completely immobilized to ensure a successful injection. Make the cut in the cornea in correspondence to the pupil and orient the injection toward the lower lid. It is sometimes useful to drain a small amount of aqueous fluid to reduce corneal tension. It is better to position the injection 1 mm from the limbus to allow for the diffusion of the test substances and the subsequent formation of a gradient for the endothelial cells of the limbal vessels.

23. *Notes for the hyperoxia-induced retinopathy model.* C57BL/6J mice should be preferably used because the pathology of other strains differs when this protocol is used. Some strains such as BALB/cByJ do not develop hyperoxia-induced vascular obliteration.

24. The chamber that will be used should allow the circulation of air throughout and should have an automated oxygen monitor that controls the oxygen influx in the chamber according to the levels of oxygen. The chamber should be air tight so that the desirable (75%) levels of oxygen are maintained but at the same time not to allow lethal concentrations of carbon dioxide to be reached.

25. Hyperoxia can cause problems to the mothers. Monitor the mothers closely during the last days of hyperoxia and the first days of normoxia and replace with surrogate mothers, if necessary.

References

1. Carmeliet, P. and Jain R.K. (2000) Angiogenesis in cancer and other diseases. *Nature* **407**, 249–257.
2. Carolyn A. Staton, Malcolm W. R. Reed and Nicola J. Brown. (2009) A critical analysis of current in vitro and in vivo angiogenesis assays. *Int. J. Exp. Path.* **90**, 195–221.
3. Bishop, E.T., Bell, G.T., Bloor, S. Broom, I.J., Hendry, N.F., and Wheatley, D.N. (1999) An in vitro model of angiogenesis : basic features. *Angiogenesis* **3**, 335–344.
4. Donovan, D., Brown, N.J., Bishop, E.T., Lewis C.E. (2001) Comparison of three in vitro human "angiogenesis" assays with capillaries formed in vivo, Angiogenesis **4**, 113–121.
5. Tolboom, T.C., Huizinga, T.W. In vitro matrigel fibroblast invasion study *Meth. Mol. Med.* **135**, 413–421.
6. Lochter, A., Srebrow, A., Sympson, C.J., Terracio, N., Werb, Z., and Bissell, M.J. (1997) Misregulation of stromelysin-1 expression in mouse mammary tumor cells accompanies acquisition of stromelysin-1-dependent invasive properties. *J. Biol. Chem.* **272**, 5007–5015.
7. Knutson, J.R., Iida, J., Fields, G.B., and McCarthy, J.B. (1996) CD44/chondroitin sulfate proteoglycan and alpha 2 beta 1 integrin mediate human melanoma cell migration on type IV collagen and invasion of basement membranes. *Mol. Biol.Cell.* **7**, 383–396.
8. Setsuko, K. (2000) Chambers In vitro invasion assays. *Meth. Mol. Med.* **39**, 179–185.
9. Storgard, C., Mikolon, D., and Stupack, D.G. (2004) Angiogenesis assays in the chick CAM. *Meth. Mol. Biol.* **294**, 123–136.
10. Mydlo, J. (2001) Angiogenesis assays. *Meth. Mol. Med.* **53**, 265–275.
11. Nguyen, M., Shing, Y., and Folkman, J. (1994) Quantitation of angiogenesis and antiangiogenesis in the chick embryo chorioallantoic membrane. *Microvasc. Res.* **47**, 31–38.
12. Cockerill, G. W., Gamble, J. R., and Vadas, M. A. (1995) Angiogenesis: models and modulators. *Int. Rev. Cytol.* **159**, 113–160.
13. Nicosia, R. F. and Ottinetti, A. (1990) Growth of microvessels in serum-free matrix culture of rat aorta: a quantitative assay of angiogenesis in vitro. *Lab.Invest.* **63**, 115–122.
14. O'Reilly, M. S., Holmgren, L., Shing, Y., Chen, C., Rosenthal, R.A., Moses, M., Lane, W.S., Cao, Y., Sage, E.H., and Folkman, J. (1994) Angiostatin: a novel angiogenesis inhibitor that mediates the suppression of metastases by a Lewis lung carcinoma. *Cell* **79**, 315–328.
15. Ferrara, N. and Alitalo, K. (1999) Clinical applications of angiogenic growth factors and their inhibitors. *Nat. Med.* **5**, 1359–1364.
16. Go, R.S. and Owen, W.G. (2000) Very low concentrations of rat plasma and rat serum stimulate angiogenesis in the rat aortic ring assay. *Fibrinolysis Proteol.* **19** (Suppl 1), 45.
17. Passaniti, A., Taylor, R. M., Pili, R., Guo, Y., Long, P. V., Haney, J. A., Pauly, R. R., Grant, D. G., and Martin, G. R. (1992) A simple, quantitative method for assessing angiogenesis and antiangiogenic agents using reconstituted basement membrane, heparin, and fibroblast growth factor. *Lab. Invest.* **67**, 519–528.
18. Grant, D. S., Kinsella, J. L., Fridman, R., Auerbach, R., Piasecki, B. A., Yamada, Y., Zain, M., and Kleinman, H. K. (1992) Interaction of endothelial cells with a laminin A chain peptide (SIKVAV) in vitro and induction of angiogenic behavior in vivo. *J. Cell Physiol.* **153**, 614–625.
19. Kibbey, M. C., Corcoran, M. L., Wahl, L. M., and Kleinman, H. K. (1994) Laminin SIKVAV peptide-induced angiogenesis in vivo is potentiated by neutrophils. *J. Cell Phys.* **160**, 185–193.
20. Malinda, K. (2008) In vivo matrigel migration and angiogenesis assay, *Meth. Mol. Biol.* **467**, 287–294.
21. Stechschulte, S.U., Joussen, A.M., von Recum, H.A., Poulaki, V., Moromizato, Y., Yuan, J., D'Amato, R.J., Kuo, C. and Adamis, A.P. (2001) Rapid ocular angiogenic control via naked DNA delivery to cornea. *Invest. Ophthalmol. Vis. Sci.* **42**, 1975–1979.
22. Joussen, A.M., Poulaki, V., Mitsiades, N., Stechschulte, S.U., Kirchhof, B., Dartt, D.A., Fong, G.H., Rudge, J., Wiegand, S.J., Yancopoulos, G.D. and Adamis, A.P. (2003) VEGF-dependent conjunctivalization of the corneal surface. *Invest. Ophthalmol. Vis. Sci.* **44**, 117–123.
23. Smith, L.E., Wesolowski, E., McLellan, A., Kostyk, S.K., D'Amato, R., Sullivan, R. and D'Amore, P.A. (1994) Oxygen-induced retinopathy in the mouse. *Invest. Ophthalmol. Vis. Sci.* **35**, 101–111.
24. Kociok, N., Krohne, T.U., Poulaki, V. and Joussen, A.M. (2007) Geldanamycin treatment reduces neovascularization in a mouse model of retinopathy of prematurity. *Graefes Arch. Clin. Exp. Ophthalmol.* **245**, 258–266.
25. Maier, A.K., Kociok, N., Zahn, G., Vossmeyer, D., Stragies, R., Muether, P.S. and Joussen, A.M. (2007) Modulation of hypoxia-induced neovascularization by JSM6427, an integrin alpha5beta1 inhibiting molecule. *Curr. Eye Res.* **32**, 801–812.

Chapter 29

Flow Cytometric DNA Analysis of Human Cancers and Cell Lines

Sarah A. Krueger and George D. Wilson

Abstract

Measurement of DNA content was one of the first applications to be developed in the use of flow cytometry and is still used routinely in many experimental and, to a lesser extent, clinical studies. The goal of this technique is to produce a high quality DNA profiles for accurate analysis of DNA content and cell cycle distribution. In this chapter, we describe three DNA measurement methods that satisfy this requirement in different situations. It is widely accepted that the Vindelov method produces the highest quality DNA profiles in nuclei from solid tumours or cell lines. However, in many situations, DNA content is combined with another marker, so we describe a method which produces high quality DNA profiles in intact cells. Third, because the Vindelov technique requires prompt processing of fresh tumours, so we also describe a technique that derives nuclei from ethanol fixed tumours providing the convenience of storage before processing.

Key words: Flow cytometry, Cell cycle analysis, DNA index, Propidium iodide

1. Introduction

Two of the most popular flow cytometric applications are the measurement of cellular DNA content and the analysis of the cell cycle. This has led to a diverse array of different protocols and methods of analysis that have been developed to accurately assess these parameters. However, the amount of information that can be extracted from a DNA distribution depends heavily on the technical quality of the preparation and the methods used for standardization and statistical analysis of the sample. In this chapter, we will describe a set of methods from the most rigorous to the most practical that can be used to satisfy different experimental needs in the setting of both cancer cell lines and solid tumors.

Measurement of the DNA content of individual cells provides information about ploidy and the distribution of cells across the cell cycle; both of which are relevant information in oncology (1). The fundamental tool of flow cytometric DNA analysis is the DNA histogram (Fig. 1). The aim of the methods described in this chapter is to provide reliable, robust, and reproducible methods to produce good quality DNA histograms. A good quality DNA histogram is characterized by a minimal amount of debris and a symmetrical G1 peak with a low coefficient of variation (CV). These attributes facilitate the analysis of the histogram to reliably measure the different components of information present in the profile such as the number of subpopulations with different DNA contents, including their DNA index and their proportion, and the percentages of cells in each phase of the cell cycle (G1, S and G2+M).

The DNA histogram for cell lines appears to be a relatively simple data set, which is usually characterized by two peaks separated by a trough. The first peak, which is generally larger, represents cells in G0/G1 and the second, which have twice the fluorescence intensity of the first, corresponds to cells with G2 or M DNA content. The cells between these peaks represent S-phase cells. In an ideal DNA histogram all G0/G1 cells and G2/M cells would reside in two single channels as they have the same DNA content. However, in practice, the data are distributed due to instrument-related, staining and biological factors, and this dispersion is assumed to be Gaussian. Unlike more complex tissues and tumors, it can be assumed that all cells in cell cultures have the same intermitotic cycle at the end of which they divide and produce two daughter cells. Thus all cells contribute to growth

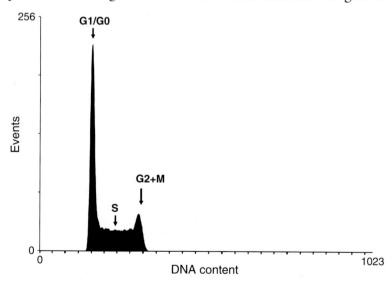

Fig. 1. A basic DNA histogram of propidium iodide staining of DNA content on the x-axis versus cell number on the y-axis. The major cell cycle populations are indicated.

and the cell population doubles in size every cell cycle time (T_c). A population of cells such as this will, unless synchronized, follow a smooth exponential growth curve with no periodicity; this is the characteristic of asynchronous growth. The equation to describe the growth of an asynchronous population is given by

$$N_t = N_0 \exp(bt)$$

where N_0 is the population size at some arbitrary time zero and b is the growth constant which is simply related to the cell cycle time in this situation,

$$b = \frac{\log_e 2}{T_c}$$

A feature of asynchronous cell populations that is rarely considered in experimental studies is that there is a distribution of cell ages is present. In a growing cell population, the age distribution cannot be rectangular as there must be more young cells in the population than old (2). If cell age is measured from the end of mitosis and, every cell produces two daughter cells, then the probability of finding a cell of zero age is twice the probability of finding a cell at age T_c (Fig. 2).

The age distribution is a crucial link between the time a cell spends in a particular state (or cell cycle phase) and the proportion of cells that will be found in that state. The S-phase area and shape of its upper surface from the DNA histogram will be identical to the upper S-phase boundary of the age distribution diagram

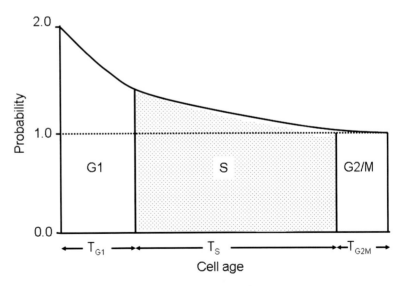

Fig. 2. A schematic representation of the age distribution of exponentially growing cells through the cell cycle. The distribution depicts that the probability of finding a cell at zero age (the beginning of the cycle) is twice that of finding a cell at age T_c (the end of the cycle).

(Fig. 2) if the rate of DNA synthesis is constant. This explains why rapidly dividing cell lines, such as V79 cells, have a skewed DNA profile as they have more cells in early S than in late S. Thus the shape of the DNA histogram reflects the age distribution and proliferation characteristics of the cell line.

The shape of the S-phase distribution is an important consideration for the analysis of cell-cycle phase distribution. The dispersion in the DNA histogram requires that statistical models need to be applied to deconvolute the overlap of the S-phase distribution into the G1 and G2 populations. In general, Gaussian peaks are fitted to the G1 and G2 populations but a variety of models including multiple Gaussians, rectilinear or polynomial have been applied to model different S-phase distributions (3).

The term ploidy is often used in flow cytometry to describe the quantity of DNA in a sample. However, ploidy is a cytogenetic term referring to the number of chromosomes which are not measured by simple DNA staining and flow cytometric analysis. The term "DNA index" is a more acceptable and realistic description of the information that is obtained from flow cytometric analysis of DNA. The DNA index is the ratio of the G1 peak position of a test population to the G1 peak position of a known diploid standard. The key aspects for rigorous DNA index analysis are the quality of sample preparation, stoichiometry of DNA dye binding and methods of standardization.

Three different methods for extracting high quality DNA profiles are described below for three specific situations (1) nuclei from cell lines – the Vindelov technique (4), (2) intact cells from cell lines and (3) nuclei from solid tumors. The production of good quality DNA profiles from cells extracted from solid tumors has always been problematical due to the need to use fresh tissue which is digested by enzymes such as trypsin, collagenase, and elastase. The method we will describe uses pepsin digestion of 70% ethanol-fixed tumours to produce nuclei (5).

2. Materials

2.1. Staining DNA in Nuclei from Cell Lines (Vindelov Technique)

1. Citrate buffer – 11.76 g Trisodium citrate·H_2O (40 mM) and 85.5 g sucrose (250 mM) dissolved in ~800 mL distilled water and 50 mL DMSO. Adjust pH to 7.6 and total volume to 1,000 mL with distilled water. Store at 4°C.

2. Stock solution – 1,000 mg trisodium citrate·$2H_2O$ (3.4 mM), 1,000 µL Nonidet P-40 (0.1% v/v) (Roche Applied Science), 522 mg spermine tetrahydrochloride (1.5 mM) (Sigma-Aldrich Co.), 61 mg Tris (hydroxymethyl)aminomethane (0.5 mM) (Sigma-Aldrich Co.) in distilled water to a final volume of 1,000 mL.

3. Solution A – 1,000 mL stock solution and 30 mg trypsin (Sigma-Aldrich Co.) adjusted to pH 7.6.

4. Solution B – 1,000 mL stock solution, 500 mg trypsin inhibitor (Sigma-Aldrich Co.), and 100 mg RNAse A (Sigma-Aldrich Co.) adjusted to pH 7.6.

5. Solution C – 1,000 mL stock solution, 416 mg propidium iodide (Fluka Chemicals), and 1,160 mg spermine tetrahydrochloride (Sigma-Aldrich Co.) adjusted to pH 7.6 (see Note 1).

The citrate buffer is stored at 4°C. The staining solutions are stored in aliquots of 5-mL in capped plastic tubes at –80°C. The tubes with solution C are wrapped in aluminium foil for light protection of the propidium iodide. Before use the solutions are thawed in a 37°C water bath but kept at room temperature. Solution C is kept in an ice-bath.

2.2. Staining DNA from Cultured Cell Lines

1. T25 flasks, 15-mL round-bottomed centrifuge tubes, and 5-mL FACS tubes.

2. Cell culture media: Choose the appropriate media for the cell line of interest.

3. Hank's Buffered Salt Solution (HBSS 10×, Gibco, Invitrogen, Carlsbad, CA, USA).

4. 0.05% Trypsin/0.02% EDTA. Weigh out 250 mg of trypsin and 100 mg of EDTA and dissolve in 500 mL HBSS. Filter and sterilize the solution and then aliquot and store at it –20°C until use.

5. Phosphate buffered saline (PBS 10×, Gibco, Invitrogen, Carlsbad, CA, USA).

6. 70% Ethanol. Measure out 350 mL of 200 proof pure ethanol and add 150 mL distilled water or PBS. Mix well and store tightly capped at –20°C.

7. Phosphate buffered saline (PBS 10×, Gibco, Invitrogen, Carlsbad, CA, USA).

8. Propidium iodide (stock solution, 400 µg/mL). Add 1.2 mg PI to 3 mL distilled water and vortex to mix. Store stock solution at 4° or aliquot and store at –20°C protected from light (see Note 1).

9. RNase A (stock solution, 1 mg/mL). Add 1 mg RNase A to 1 mL distilled water or PBS and vortex to mix. Aliquot and store at –20°C.

2.3. Staining DNA from Solid Tumours

1. Phosphate buffered saline (PBS 10×, Gibco, Invitrogen, Carlsbad, CA, USA).

2. 70% Ethanol (as above).

3. Pepsin solution: 0.4 mg/mL pepsin (Sigma-Aldrich Co.) in 0.1 M HCl. Dissolve the required amount of pepsin in

1–2 mL of PBS and then add the HCl. If the solution looks cloudy warm to 37°C.

4. 35 μm Nylon mesh.
5. Propidium iodide (stock solution, 400 μg/mL).

3. Methods

3.1. Staining DNA in Nuclei from Cell Lines (Vindelov Technique)

Staining is performed by stepwise addition of the staining solutions to 200 μL of sample cell suspension with the appropriate cell concentration (~10^6 cells total).

1. Add 1,800 μL of Solution A to the cells and invert the tube to mix contents gently. Incubate the tube at room temperature for 10 min with occasional inversion.
2. Add 1,500 μL of Solution B to the tube and invert the tube to mix contents gently. Incubate the tube at room temperature for 10 min with occasional inversion.
3. Add 1,500 μL of ice-cold Solution C to the tube and mix contents gently. Filter through 35 μM nylon mesh into tubes wrapped in aluminium foil for light protection of PI.
4. Keep on an ice-bath until analysis, preferably within 15 min to 3 h after adding Solution C.

3.2. Staining DNA from Cultured Cell Lines

1. Culture cells in T25 flasks until 60–70% confluency is reached (see Note 2).
2. Remove media and wash cells with 0.5 mL trypsin, swirling to coat flask surface evenly. Aspirate and add another 0.5 mL trypsin, placing the flask in an incubator for 2–3 min or until cells begin to detach from the flask surface with a gentle tilt or tap.
3. Add 3–4 mL of culture media to the flask to neutralize the trypsin and wash the cells from the flask surface. Transfer cell suspension to a 15-mL centrifuge tube and centrifuge at $200 \times g$ for 5 min at room temperature.
4. Aspirate media from tube being careful not to disturb the cell pellet and add 5 mL PBS (see Note 3). Mix cells gently to bring them back into suspension and then centrifuge at $200 \times g$ for 5 min. Repeat once.
5. After the second PBS wash, resuspend the cell pellet in 200 μL PBS and add 3–5 mL ice cold 70% ethanol dropwise while vortexing. Cap tightly and store samples at least overnight at 4°C or up to several weeks at −20°C (see Note 4).
6. When ready for staining, remove tubes from refrigerator or freezer and add 5 mL cold PBS to each.

7. Centrifuge at 700 ×g for 5 min at 4°C and aspirate off PBS/alcohol being careful not to disturb the cell pellet. Add 2–3 mL of cold PBS to resuspend the pellet and then centrifuge at 700 ×g for 5 min.

8. Aspirate the second PBS wash and add 200 µL PBS to resuspend the cell pellet. Transfer cells to a 5-mL FACS tube and pass through a 18 gauge needle 4–5 times to reduce clumping and produce a single-cell suspension (see Note 5).

9. Add 100 µL of the stock RNase A solution and 100 µL of the stock PI solution to the cell suspension and incubate protected from light for 30 min at room temperature.

3.3. Staining Nuclei from Solid Human Tumors

1. Mince fresh tumour specimens with scissors or leave as large pieces.

2. Add 10 mL of 70% ethanol and store at 4°C. Fix overnight or longer prior to staining.

3. Decant the ethanol and mince the tissue into small (1 mm) fragments.

4. Add 5–10 mL of the pepsin/HCl solution.

5. Incubate on a rotor or shaking water bath for 30–60 min depending on the tissue (see Note 6).

6. Further dissociate by pipetting.

7. Filter through the nylon mesh into a conical-bottomed 10-mL tube.

8. If required, count the suspension and remove the desired number of nuclei.

9. Centrifuge at 700 ×g for 5 min.

10. Resuspend the pellet in 1–2 mL of PBS containing 10 µg/mL propidium iodide (see Note 7).

3.4. The Use of DNA Standards

Determination of nuclear DNA content by flow cytometry requires comparison with a reference standard. The use of external standards such as lymphocytes or granulocytes is time-consuming and inaccurate. Chicken red blood cells (CRBC) have a DNA content of 35% of the human diploid value and have been widely used as internal standard. The ratio calculated on the basis of the peak channel numbers of the standard and the sample and used to indicate the DNA content is, however, very sensitive to changes in the zero level adjustment of the flow cytometer. If two internal standards are used the DNA ratio becomes independent of the zero level. Rainbow trout red blood cells (TRBC) have a DNA content of 80% of human diploid cells and their combination with CRBC provides the most consistent and robust method to standardize DNA index analysis (6). Ideally, the standards should be processed in the same way as the cells of interest.

The red cell concentrations should be adjusted by dilution with citrate buffer to $CRBC = 145 \times 10^4$/mL and $TRBC = 255 \times 10^4$/mL. Mixing these solutions in equal volumes gives a final concentration of 2×10^6/mL with a ratio of CRBC:TRBC of 4:7 which will produce equal size peaks in the histogram. If cell cycle analysis is the only parameter of interest, then there is no need to include standards.

3.5. Flow Cytometric Analysis

This protocol assumes that the user is familiar with the principles and practices of flow cytometry. This type of staining can be analyzed on any of the modern flow cytometers with the proviso that the machine is equipped with a pulse-processing facility to enable the discrimination of cell doublets. The procedure described below is specific to a Becton Dickinson FACS machine.

1. For the purposes of collecting data, all plots must be formatted for "Acquisition."
2. Create a two-parameter dot-plot of forward light scatter (FLS) vs. side scatter (SS).
3. Create a two-parameter dot-plot of FL2 (area) vs. FL2 (width) to monitor doublets (see Note 8).
4. Create a single-parameter FL2 (area) histogram with linear x-axis to illustrate relative DNA.
5. Select the signal threshold (the point at which a signal will be accepted as a positive event) to FL2 and then set an appropriate value to gate out debris.
6. Run the sample and adjust both FLS and SS photomultiplier tube (PMT) voltages so that the majority of dots are contained roughly within the center of the dot-plot.
7. Adjust the FL2 PMT voltage up or down until the peak appears in the graph. The convention for DNA index analysis is to adjust the G1 peak to channel 200 such that tetraploid populations of cells would be contained within the 1,024 channel scale. However, if the analysis is only concerned with cell cycle distribution of a cell line, it is better to adjust the G1 to channel 400 and provide more channels (9,400 versus 200) for subsequent computer analysis of the profile. The gain on FL-2 W may need to be increased to best discriminate the doublets (Fig. 3).
8. Collect at least 10,000 ungated events.

3.6. Data Analysis

1. The starting point for data analysis is to create the FL2-area versus FL2-width dot plot in the analysis portion of CellQuest (Becton Dickinson) or other proprietary (ModFit – Verity Software House Inc.) or free software (WinMDI 2.8) (see Note 9).

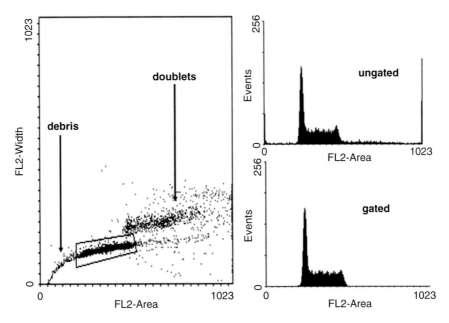

Fig. 3. The application of pulse processing to carry out doublet discrimination and allow the accurate analysis of the single cell population.

2. A region is created (Fig. 3) around the single cell population in the dot plot. A histogram of FL2-area is created and the region 1 gate applied to the data to exclude debris and doublets.

3. Simple analysis can be achieved by setting regions on the appropriate populations in the DNA histogram. However, accurate cell cycle and DNA index analysis should be performed using dedicated software supplied with the instrument being used. For the purposes of this protocol, analysis was carried out using the ModFit software (Verity Software, Topsham, ME, USA).

4. Open the ModFit program and select the appropriate **FILE**.

5. Choose the parameter for analysis; in this case select FL2A for relative DNA content.

6. Define "gate 1" by selecting FL2A (x) and FL2W (y). Drag each of the points of the gate (R1) to include the entire cell population of interest.

7. Choose a specific **MODEL** to analyze the data or use the suggested model according to specified parameters, such as whether samples were fresh or frozen or paraffin embedded; of diploid, aneuploid, or tetraploid DNA content; whether aggregates were present; or if there is a visible G2/M fraction. The model can also account for the presence of internal standards should they be included.

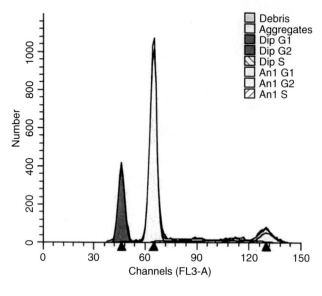

Fig. 4. The use of cell cycle fitting software (ModFit) to discriminate different populations based on DNA index and to analyze cell cycle distributions. An autoanalyze model has been used to determine the DNA index (DI: 1.40) and the percentage of cells in the diploid and aneuploid compartments of a head and neck cancer. The routine locates the main peaks (indicated by *triangles*) and fits an automated cell cycle analysis based on Gaussian distributions for the peak populations and will choose the best model to determine cell cycle phase analysis based on nonlinear least squares analysis. Many options are available in the software to choose specific ploidy models and S-phase distribution analyses that can be tailored to a specific histogram.

8. Check the position and **RANGE** of the markers that are automatically placed on the histogram and adjust their position if necessary (this may be the case, particularly if the S-phase fraction becomes relatively high).

9. Calculate the relative cell cycle distributions using the **FIT** option (Fig. 4).

10. Data can then be tabulated and exported to a suitable presentation package, for example, Excel.

4. Notes

1. Propidium iodide is a suspected carcinogen so contact with eyes, skin or mucous membranes should be avoided. Always wear proper protective clothing and gloves when preparing or handling this solution. PI should be handled cautiously at all times and stock solutions prepared in a fume hood.

2. This method is described using a T25 flask of adherent cells for simplicity. The size of flask and cell confluency may be

altered to suit each individual experiment or cell line, just scale up the amount of each reagent used based upon the numbers presented here. For nonadherent cells, skip to step 4 as they will not need to be trypsinized.

3. Aspiration (as opposed to pouring off) of the liquid is preferred so as to not accidentally dislodge the cell pellet. Aspiration also allows for better removal of the wash buffer and alcohol during the staining steps.

4. It is very important to resuspend the cell pellet before adding the ethanol. Also, adding the ethanol slowly while vortexing is necessary to ensure proper fixation.

5. Cells may be passed through a cell filter instead of using a needle, or, for particularly sticky cells, both may be utilized to ensure a single-cell suspension.

6. Different human or experimental tumours take different times to dissociate depending on their cellular make-up. The optimum time needs to be established for each tumour.

7. There is no requirement to add RNAse to the staining solution as the acid component of the dissociation solution hydrolyzes RNA in the nuclei.

8. Doublet discrimination or pulse processing utilizes the difference in the width component of the FL2 signal to separate doublet G1 cells from true G2+M cells; the doublets have a higher width signal. Some modern machines, such as the FACSAria, do not have the ability to discriminate doublets due to their beam profile and are not suited to routine DNA analysis.

9. In most benchtop flow cytometers, the propidium iodide signal can be collected into the FL2 (orange) or FL3 (red) detectors as the emission spectrum of the dye spans the collection range of both channels. Use of FL3 is recommended when propidium iodide staining is combined with a fluorochrome such as fluorescein isothiocyanate (FITC) to minimize any spectral overlap.

References

1. Ross, J. S. (1996) DNA ploidy and cell cycle analysis in cancer diagnosis and prognosis, *Oncology (Williston Park)* **10**, 867–882, 887; discussion 887–890.
2. Steel, G. G., and Bensted, J. P. (1965) In vitro studies of cell proliferation in tumours. I. Critical appraisal of methods and theoretical considerations, *Eur J Cancer* **1**, 275–279.
3. Gray, J. W., Dolbeare, F., Pallavicini, M. G., Beisker, W., and Waldman, F. (1986) Cell cycle analysis using flow cytometry, *Int J Radiat Biol Relat Stud Phys Chem Med* **49**, 237–255.
4. Vindelov, L., and Christensen, I. J. (1990) An integrated set of methods for routine flow cytometric DNA analysis, *Methods Cell Biol* **33**, 127–137.

5. Wilson, G. D. (2000) Analysis of DNA-Measurement of cell kinetics by the bromodeoxyuridine/anti-bromodeoxyuridine method., in *Flow Cytometry: A Practical Approach* (Ormerod, M. G., Ed.), pp 159–177, IRL Press, Oxford.

6. Vindelov, L. L., Christensen, I. J., and Nissen, N. I. (1983) Standardization of high-resolution flow cytometric DNA analysis by the simultaneous use of chicken and trout red blood cells as internal reference standards, *Cytometry* **3**, 328–331.

Chapter 30

Expression Analysis of Homeobox Genes in Leukemia/Lymphoma Cell Lines

Stefan Nagel and Hans G. Drexler

Abstract

Homeobox genes code for transcription factors which have a strong impact on cellular behavior, including differentiation, proliferation, and survival. Therefore, upon deregulation these genes may turn into oncogenes, contributing substantially to cancerogenesis. Among hematopoietic malignancies, including leukemias and lymphomas, several homeo-oncogenes have been described. Many of them have been identified in hematopoietic cell lines, which serve as useful tools for oncogene hunting and characterization. Here, we describe molecular methods for analysis and quantification of dysregulated homeobox gene expression in leukemia/lymphoma cell lines.

Key words: Homeobox, Leukemia, Lymphoma, Oncogene, Quantification

1. Introduction

Homeobox genes share a ~180 bp region, named homeobox, which codes for a DNA-binding and protein-interacting structure, termed homeodomain (1). Human genomes contain about 200 different homeobox genes (2). This gene family is evolutionary old and appeared along with the origin of multicellularity (3). According to their conserved homeobox sequences, homeobox genes were classified into classes and families. Antennapedia is a homeotic mutant in Drosophila and, according to the affected homeobox gene, the name of the most comprehensive class of homeobox genes which comprises several families, including NKL, HOX, and EHG (4).

Due to the high impact of homeobox genes in developmental and tissue-specific processes, mutations or dysregulations may result in severe congenital diseases or cancer. For example, members of the NKL-family of homeobox genes are involved in certain genetic diseases and a special form of leukemia, T-cell acute lymphoblastic leukemia (T-ALL): NKX2-5 mutations are responsible for heart malformations and the Wolf–Hirschhorn syndrome; TLX3/HOX11L2 mutations cause congenital central hypoventilation (5–7). In T-ALL both genes are ectopical activated via chromosomal aberrations and probably deregulate apoptosis and differentiation (8, 9). Other T-ALL associated homeobox genes include members of the clustered HOX genes, HOXA5, HOXA9, and HOXA10 (10). Expression analysis of Hodgkin lymphoma cell lines identified HOX-family member HOXB9 as the most prominent deregulated homeobox gene in this and related lymphoma entities (11). Finally, HLXB9 is a member of the EHG family of homeobox genes, mutated in the Currarino syndrome and additionally deregulated in particular types of leukemia and lymphoma (12–14).

Taken together, homeobox genes are important regulators for cell differentiation with major impact on normal and aberrant development. Expression analysis of this group of transcription factors is of interest for understanding differentiation processes and cancer, including leukemia and lymphoma (15). Cell lines represent an ideal tool for hunting and characterizing oncogenic homeobox genes. Subsequent knockdown and overexpression approaches in cell lines allow for the identification of downstream targets and cellular functions of particular homeobox genes (16, 17).

Here, we describe polymerase chain reaction (PCR)-based methods to analyze homeobox gene expressions in (leukemia/lymphoma) cell lines: qualitative analyses of particular genes by reverse transcription (RT)-PCR and quantitative analyses by real-time PCR and of several genes using degenerate oligonucleotides with subsequent subcloning.

2. Materials

2.1. RNA Extraction and cDNA Synthesis

1. PBS (8 g NaCl, 0.2 g KCl, 1.44 g Na_2HPO_4, 0.24 g KH_2PO_4, pH 7.2, per 1 L).
2. TRIzol reagent (Invitrogen, Karlsruhe, Germany).
3. Chloroform.
4. Isopropanol.
5. Sterile, deionized water.
6. Nanodrop 1000 Spectrophotometer (Peqlab, Erlangen, Germany).
7. Reverse transcriptase superscript II (Invitrogen).
8. Random hexa-oligonucleotides (Invitrogen).
9. dNTPs with 10 mM of each dNTP (Peqlab).

2.2. RT-PCR Analysis of Selected Homeobox Genes

1. TGradient PCR thermo-cycler (Biometra, Göttingen, Germany).
2. Taq-polymerase, 10× buffer and 5× Q-solution (Qiagen, Hilden, Germany).
3. 2.5 mM dNTPs (Takara, Potsdam, Germany).
4. Agarose.
5. TAE buffer (4.84 g Tris, 1.14 mL Na-acetate, 2 mL 0.5 M EDTA, pH 8.0).
6. 100 bp ladder (Invitrogen).
7. Ethidium bromide (stock with 20 mg/mL) or alternative dye.

2.3. Quantification of Homeobox Gene Expression by Subcloning

1. Teasy Vector System I, including vector and ligase (Promega, Mannheim, Germany).
2. Competent DH5alpha cells (Invitrogen).
3. LB–ampicillin agar plates (15 g agar, 10 g tryptone, 5 g yeast extract, 10 g NaCl/1 L, 80 µg/mL ampicillin).
4. Spin Miniprep Kit (Qiagen).
5. *Eco*RI with buffer (MBI Fermentas, St. Leon-Rot, Germany).

2.4. Quantification of Homeobox Gene Expression by Real-Time PCR

1. 7500 Real-Time PCR System (Applied Biosystems, Darmstadt, Germany).
2. MicroAmp 96-well reaction plates (Applied Biosystems).
3. Taqman universal PCR master mix (Applied Biosystems).
4. TBP endogenous control primers and probe (Applied Biosystems).

3. Methods

3.1. RNA Extraction and cDNA Synthesis

1. Cultivation of leukemia/lymphoma cell lines is performed as described previously (18). For RNA extraction 2×10^6 cells are washed in PBS and pelleted by centrifugation for 5 min at $485 \times g$. Discard the supernatant and lyse the cells subsequently in 1,000 µL TRIzol reagent and incubate at room temperature for 10 min (see Note 1).

2. After adding 500 µL chloroform, vortex the tube for 30 s and centrifuge subsequently at $15,000 \times g$ for 10 min at 4°C. Finally, transfer the upper phase into a new tube and precipitate the RNA with 900 µL isopropanol. After centrifugation at $15,000 \times g$ for 30 min at 4°C, dissolve the pellet in about 30 µL water and put the tube on ice. Alternatively, RNA preparation kits may be used instead of the TRIzol method (e.g., RNAeasy from Qiagen).

3. Quantify the RNA solution via spectrophotometer. Depending on the cell line, the yielded concentration of RNA is about 1 μg/μL. The RNA solution should be kept on ice and can be stored at −80°C for several years. However, frequent thawing and freezing should be avoided. The integrity of the RNA may be controlled via gel electrophoresis, loading 5 μg RNA per lane.

4. Use 5 μg RNA per reaction of cDNA synthesis (final volume of 20 μL), 45 ng random primer, 1.5 μL dNTPs, and 1 U superscript II reverse transcriptase. Stop the reaction after 1 h incubation at 37°C by chilling on ice. The cDNA is usable for several years if stored at −20°C.

3.2. RT-PCR Analysis of Selected Homeobox Genes

1. To design oligonucleotides for RT-PCR analysis the following genome browsers as sources for RNA/cDNA sequences are available (http://www.ncbi.nlm.nih.gov, http://genome.cse.ucsc.edu, http://www.ensembl.org). Although several online primer-design programs are recommendable (http://frodo.wi.mit.edu/primer3/, http://bioweb.uwlax.edu/GenWeb/Molecular/seq_anal/primer_design/primer_design.htm, http://eu.idtdna.com/scitools/scitools.aspx?adword=scitool_eh&gclid=cisyq_iq0pwcfruezgodnvudlq, http://ihg2.helmholtz-muenchen.de/ihg/ExonPrimer.html), the design of oligonucleotides without the help of algorithms is practicable. However, the following rules should be respected: the lengths of the oligonucleotides should be about 20 bp. Oligonucleotides should contain about 50% GC and their melting point about 55°C. Typically, their 3′-end should contain G or C. The length of the PCR product should be about 200–500 bp. The binding sites of the oligonucleotides should flank exon–intron borders to suppress the synthesis of PCR products obtained from genomic DNA which frequently contaminates the RNA preparation. Of note, take care of splice-variants which may create alternative product lengths or prevent product synthesis at all if the particular RNA of interest lacks a certain exon. Furthermore, if using oligo-dT primer for cDNA synthesis, the oligonucleotides should match the 3′-end of the RNA/cDNA, because of limited processing activity of the reverse transcriptase.

2. Sequences of oligonucleotides of selected homeobox genes, including TLX1/3, NKX2-5, MSX2, PITX2, GBX2, HLXB9, HOXB9, and HOXB13, are listed in Table 1. Additionally, use control oligonucleotides to check the cDNA quality. Applications of TEL or UBF as control genes are recommended (Table 1).

3. To find a positive control for the gene of interest it is helpful to analyze cells obtained from an adequate tissue (primary

Table 1
Oligonucleotides used for RT-PCR analysis

Gene	Acc No	Forward (5'–3')	Reverse (5'–3')	PCR product (bp)
UBF	X53390	CTGGAATGCATGAAGAACAACC	TCTTGGTTAGGTCCAGGTTGC	342
TEL	NM_001987	AGGCCAATTGACAGCAACAC	TGCACATTATCCACGGATGG	272
EN1	NM_001426	ACACGTTATTCGGATCGTCC	GTGGAGTGGTTGTACAGTCC	311
EN2	NM_001427	TCTACTGTACGCGCTACTCG	ATCTTGGCGCGCTTGTTCTG	247
GBX1	AC010973	GCCTCAGGATTCTGAACTTG	TTAGCTTCGGTTTAGGTCCC	246
GBX2	AF118452	AAGGCTTCCTGGCCAAAGAG	GGAACTCCTTCTCTAGCTCC	343
HLXB9	NM_005515	CTAAGATGCCCGACTTCAAC	TGCGTTTCCATTTCATCCGC	225
HOXB9	NM_024017	TGTCTAATCAAAGACCCGGC	ATACATCCCAGACAGCACTG	403
HOXB13	BC007092	CTGAGTTTGCCTTCTATCCG	GTACGGAATGCGTTTCTTGC	296
MSX2	XM_037646	AGATGGAGCGGGCGTGGATGC	ACTCTGCACGCTCTGCAATGG	194
NKX2-5	NM_004387	TCTATCCACGTGCCTACAGC	TGGACGTGAGTTTCAGCACG	264
PITX2	NM_153426	AAATGCTGTGTGGACCAAC	CATGCTCATGGACAGAGATAG	317
TLX1	XM046733	GCGTCAACAACCTCACTGGCC	GTGGAAGCGCTTCTCCAGCTC	170
TLX3	XM003705	GCGCATCGGCCACCCCTACCAGA	CCGCTCCGCCTCCCGCTCCTC	242
ANTP	n.a.	GCAGATATCGGARCTGGAGVRVSRVTT	CGAATTCRWNCKCCGGTTCTGRAACCA	130

Acc No accession number, *n.a.* not applicable

cells or cell lines). Moreover, expression profiling data may give a hint to find cells or cell lines which endogenously express the gene of interest.

4. The type of Taq-polymerase may be essential for a successful PCR reaction. Therefore, it is recommended to use the same type and supplier to get comparable results. Perform each reaction in a final volume of 25 µL, containing 1× PCR buffer, 1× Q-solution, 1 U Taq-polymerase, 50 pmol oligonucleotides of each, 2 µL cDNA (see above), and 2.5 mM dNTPs.

5. PCR thermo-cycler with a gradient function for temperature to optimize annealing temperatures may be helpful. The gradient should cover 47–62°C. However, an advisable RT-PCR protocol, lacking any hot-start step, is the following: 92°C for 0.5 min, 55°C for 1 min, 72°C for 2 min, 35 cycles.

6. The PCR products are subsequently analyzed by gel electrophoresis (1.5% agarose, 1× TAE buffer) in the presence of ethidium bromide, in comparison to a suitable length marker (100 bp ladder) (see Note 2).

3.3. Quantification of Homeobox Gene Expression by Subcloning

1. Degenerate oligonucleotides allow simultaneous amplification of several different homeobox genes. However, although the homeobox represents a conserved sequence, it is difficult to design functional oligonucleotides. Created degenerate oligonucleotides (14), named ANTP, which amplify HOX genes in addition to some nonclustered homeobox genes are listed in Table 1. The indicated regions bordering the homeobox represent the most conserved and hence promising region for the design (Table 2) (see Note 3).

2. Perform RT-PCR and gel electrophoresis as described above for gene-specific oligonucleotides. The expected PCR product is about 130 bp long. Optional, the PCR products may be precipitated with Na-acetate/ethanol and resuspended in 20 µL water.

3. For subcloning in T-tailed pGEMTeasy vector use the Teasy Vector System I: 8 µL PCR product, 1 µL vector, 1 µL ligase, 10 µL 2× ligation buffer. After incubation for about 3 h at room temperature, transform 10 µL of the ligation reaction into 100 µL of competent DH5alpha cells by heat shock at 42°C for 30 s in a water bath. Add 750 µL LB medium to the cells, shake the tubes at 37°C for 1 h, and put the cells on an LB–ampicillin plate overnight. The next day pick single clones to grow in 4 mL LB–ampicillin overnight.

4. Prepare plasmid DNA, using the Spin Miniprep Kit which yields 50 µL plasmid DNA solution. Restrict 3 µL of this solution with 10 U *Eco*RI for 2 h at 37°C and subsequently

Table 2
Design of degenerate oligonucleotides for amplification of homeobox genes

G		G GG GGGGGGG GG GGGGGGG			GG GG GGG G GGGGGGG		
A		AAA AAAAAA AAAAAAAAA A			AAAA AAA AAAA AAA AAAAAAAA		
T		TTTTTT TTT TT			TTT TTT TTT TT		
C		CC CCCCCCCCC CC CCCCC C			CC C CC CC C CC C CC		
ANTP-oligonucleotides		GCAGATATCGGA CTGGAG	TT		TGGTT CAGAACCGG G GAATTCG		
HOXA9	AAAAAGCGTGCCCCTATACAAAACACCAGACCCTGGAACTGGAGAAAGAGTT	TCTGTTCACATGTACCTCA	CCAGGGACCCGAGCTCAACCTCAACTGTCCTCACTCACCGAGAGGCAGGT	CAAGATCTGGTT	CCAGAACCGCAGGATGAAAATGAAG
HOXA11	AAAAAGCGCTGCCCCTATACAAGTACCAGATCCGAGAGCTGGAGAAAGAGTT	CTTCTTCAGCGTTCACATTCA	CATTGTCCGCATGCTCACATTAACAAAGAGAAGCGCTTCAACCTGATCGT	CAAGATCTGGTT	TCAGAACAGGAGAATGAAGGAAAAA
HOXB7	CGAGGCCGCCAGACCTACACCCGCTACCAGACCCTGGAGCTGGAGAAAGAATT	CACTACAATCGCTACCTGA	CCAGGGCCATGATCCGAGATCGCGCACACGCTCTGCCTCACGAAAGACAGAT	CAAGATTTGGTT	TCAGAACCGGCGCATGAAGTGGAAA
mHOXB7	CGAGGCCGCCAGACCTACACCCGCTACCAGACCCTGGAGCTGGAGAAGAATT	TCACTACAATCGCTACCTGA	CACACGCTCTCCCTCACCAAAGACAGATCGCGCACAGATCGCGCACGCTC	CAAGATCTGGTT	TCAGAACCGGCGCATGAAGTGGAAA
HOXC6	CGCGGCCGCCAGACCTACTCGCGGTACCAGACCCTGGAACTGGAGAAAGAATT	TCACTTCAATCGCTACCTGA	CCAGGGCCATGATCCGAGCCAGCCGCTTCAGACCGCTTGAATCTTAGTGA	CCAAGATCTGGTT	TCAGAACCGGCGAATGAAGTGGAAA
HOXC12	AAGAAGCGCAAGCCTATTCGAAGTTGCAACTGCAGAGCTGGAGGGCAGTT	TCATCAACCGAGAGTTCAT	CACACGCGCAGCAGGTCAAGATCTGGTTCAGAACCGGAGAATGAAAAGAAA		
HOXD1	TCCGCACGCACGAATTTCAGCACCAAGCAACTGACAGAACTGGAAAAAGAGTT	TCATTTCAATAAGTACTT	AACTGAGCCGGCCATCGTTGCACCTGAATGACACGCAAGTCAAATCTGGTT	CAGAATGAAACAGAAG	
HOXD13*	AAGAAGAGTGCCTTACACCAAACTGCAGCCCTAAAGAACTGGAGAACGAGTT	GCATTAACAAAATTCATTA	ACAAGGACAAGCGGCGGCGTATCTGGCTGCTACGAACCTATCTGAGAACAAGTGA	CCATTTGGTT	TCAGAACCGAAGAGTGAAGGACAAG
DLX1	AAACCCAGAGACGATTTATTCCAGTTTGCAGTTTGCAGGTTTGAACCGGAGGGTT	TCAGCAAACTCAGTACCTAG	CTCGCGGAGGGGCCGGACTGCCCGAGGACCTGCGCGCGAGACTGGGCCTG	CAAGATCTGGTT	CCAAAACAAGCGATCCAAGTTCAAG
DLX3	AAGCCGGTACAATCTACTCCAGCTACCAGCTGGCCGCCTACCAGGCCCCTGG	CGCCTCGGCCTCTGGGCC	TCAGCGGGCCCGTCGCCCGAGCGCCGCCCGCAGACACAGGTGAAAATCTGGTT	CCAGAACCGCCGTT	CCAAGTTCAAG
EN1	CGGCCGCGGACCGCGTTCACGGCCGAGCTGCAGAGCTCAAGGCGGAGTT	CCAGCCAGCAGTCGAAGCG	AGCGCCAAACGCCTACATCACGAGCAGGGGCCAGACCCTTGCCCAGGAACT	CAGCTTCAAGATCTGGTT	CCAGAACCAAGCGCCAGAGTCAAG
EVX1	CGTTACCGCACCGCCTTCACCGAGAGCAGATTGCCGGCTGGAAGAAATT	CTACCGGAGAACTACGTA	TTCAGGCCGCGGAGATGTGAGCTGGCGGCCGCCAAACTGCCGAAACCATCA	GGTGTGGTTCCAGAACCGG	CGCATGAAGGACAAG
HLXB9	CGGCCGGCCACCGCCCTTCACCAGCCAGCTGCTGGAGCTGGAGAAAGAGTTT	CAAGTTCAACAAGTACC	CTTCGAGCCGCCAAGCTTCACCCAGAGCCAGCGTTGAGTGCCACCTGCTCAT	GGTGAGATTTGGTT	CCAGAACCGGCGATGAAATGGAAA
mHLXB9	AGGCCTCGCACGGCCTTCACCAGCAGCAGCTGTTGGAGCTGGAACACCAGTT	CAAGCTCAACAAGCGAT	CCTTCGCCCCCAAGCTTTGAGGTGGCTACCTCGCTCACCGAGACTCAGGT	GAGATTTGGTT	CCAGAACCGCCGATGAAATGGAAA
NKX2-5	AAGCCGCGGTGCTCTTCTCGCAGGCGCAGTCTATGAGCGCAGTTCAAGCAGCA	GCGGTACTGTGCAGCGAC	CGCTGTCGGCCCCCAAGCGGTCTGAAAACTCACGGTCAGCCAGTTCAAGATCTGGTT	TCAGAACCGCGCTACAAGT	GCAAG
TLX3	AAGCCGCGCACGTCCTTTTTCCCGGGTGCAGATCTGCGAGCTGGAAAAAGCGCTT	CCATCGCCAGAGATGCCT	TCCCCCAGAAGTACCTGGGCCTTCGCGAGGGGGGCGGCTCCGAAGTCCCCAAGGTT	CAAGACCTGGTTCCAAAACGG	AGGACCAAGTGGCGG
MBIS2*	GGCATTTTCCCAAAGTAGCAACAAATATCATGAGAGCATGGCTTCTTCAGCATTCCACATCCACTCCACACATCCCAGTAAACACCAGACCCCAAGAA3CGAAGAAAACAGTAGCCAAGACGACGACAGCAGTTACAATTCTCCAGTAAACACTGGTTATTAATGCCAGAAGAAGAATACTA				

The table shows an alignment of homeobox sequences derived from clustered and nonclustered homeobox genes. Conserved base positions are shown in bold letters. The first rows show conserved parts which were summarized for each base (G, A, T, C) and used for oligonucleotide design. The sequences of the ANTP-oligonucleotides are shown below. Homeobox genes which are not predicted for amplification by ANTP-oligonucleotides are indicated by an asterix (*)

analyze the DNA fragments by gel electrophoresis. Analyze positive plasmid clones, containing a 130-bp insert, by DNA sequencing (MWG, Ebersberg, Germany).

5. Blast the sequence of the insert to the NCBI online genome browser (http://blast.ncbi.nlm.nih.gov/Blast.cgi), identifying the particular homeobox gene by sequence identity. The number of each identified gene correlates with its expression level in the analyzed cell line. The more clones analyzed, the higher are significance and spectrum of the identified homeobox genes.

In conclusion, this low-cost method allows simultaneous analysis of different homeobox genes. However, depending on the designed oligonucleotides, only selected genes/families are detectable.

3.4. Quantification of Homeobox Gene Expression by Real-Time PCR

1. Commercial primer sets are available for (nearly) all known homeobox genes, including PITX2 and HOXB9. Self-made primers including a labeled probe (NKX2-5) or excluding such a probe (PITX2, MSX2) have been designed (Table 1) (19, 20). The rules for oligonucleotide design resemble those described above, however, with the following differences: 60–70% GC-content, melting temperature about 61°C, PCR product length in a range of 100–300 bp, respecting exon–intron borders. The probe sequences should be designed with respect to higher melting temperature (68–70°C) and avoidance of G at the 5′-end. Of note, Taqman universal PCR master mix differs between oligonucleotide combinations with or without probes.

2. Perform each reaction in triplicate to obtain significant results. It is important to make a master mix for limitation of pipetting errors.

3. Use of the 7500 real-time PCR cycler from Applied Biosystems is advisable. Perform the program as recommended by the supplier. In conclusion, real-time PCR generates very exact data if performed appropriately.

3.5. Quantification of Homeobox Gene Expression by Expression Profiling

Expression profiling using DNA chips (e.g., Affymetrix) allows expression analysis of nearly all human genes. Naturally, this analysis includes (nearly) all homeobox genes (14). Therefore, this method permits the most comprehensive analysis of homeobox gene expression. However, expression profiling is just a screening method and the obtained data have to be validated by alternative methods, e.g., real-time PCR. The performance of this high-priced method requires much input of equipment, calculating capacity, and experience. The details of this method are described elsewhere.

4. Notes

1. At this stage, the lysate can be stored at −20°C, preserving the RNA stability for several years.
2. RT-PCR is a qualitative method. However, substantial differences in expression levels may be detectable in comparison to controls.
3. Most but not all clustered HOX genes are amplified by ANTP oligonucleotides, demonstrating differences in homeobox sequences.

References

1. Duboule, D. (1994) Guidebook to the homeobox genes. Oxford, United Kingdom: Oxford University Press.
2. Tupler, R., Perini, G. and Green M.R. (2001) Expressing the human genome. *Nature* **409**, 832–833.
3. Garcia-Fernàndez J. (2005) The genesis and evolution of homeobox gene clusters. *Nat. Rev. Genet.* **6**, 881–892.
4. Ryan, J.F., Burton, P.M., Mazza, M.E., Kwong, G.K., Mullikin, J.C. and Finnerty, J.R. (2006) The cnidarian-bilaterian ancestor possessed at least 56 homeoboxes: evidence from the starlet sea anemone, *Nematostella vectensis*. *Genome Biol.* **7**, R64.
5. Reamon-Buettner, S.M. and Borlak, J. (2004) Somatic NKX2-5 mutations as a novel mechanism of disease in complex congenital heart disease. *J. Med. Genet.* **41**, 684–690.
6. Nimura, K., Ura, K., Shiratori, H., Ikawa, M., Okabe, M., Schwartz, R.J., Kaneda, Y. (2009) A histone H3 lysine 36 trimethyltransferase links Nkx2-5 to Wolf-Hirschhorn syndrome. *Nature* **460**, 287–291.
7. Shirasawa, S., Arata, A., Onimaru, H., Roth, K.A., Brown, G.A., Horning, S., Arata, S., Okumura, K., Sasazuki, T. and Korsmeyer, S.J. (2000) Rnx deficiency results in congenital central hypoventilation. *Nat. Genet.* **24**, 287–290.
8. Nagel S, Kaufmann M, Drexler HG, MacLeod RA. The cardiac homeobox gene NKX2-5 is deregulated by juxtaposition with BCL11B in pediatric T-ALL cell lines via a novel t(5;14)(q35.1;q32.2). Cancer Res. 2003; 63:5329–5334.
9. MacLeod, R.A., Nagel, S., Kaufmann, M., Janssen, J.W. and Drexler, H.G. (2003) Activation of HOX11L2 by juxtaposition with 3′-BCL11B in an acute lymphoblastic leukemia cell line (HPB-ALL) with t(5;14)(q35;q32.2). *Genes Chromosomes Cancer* **37**, 84–91.
10. Soulier, J., Clappier, E., Cayuela, J.M., Regnault, A., García-Peydró, M., Dombret, H., Baruchel, A., Toribio, M.L. and Sigaux, F. (2005) HOXA genes are included in genetic and biologic networks defining human acute T-cell leukemia (T-ALL). *Blood* **106**, 274–286.
11. Nagel, S., Burek, C., Venturini, L., Scherr, M., Quentmeier, H., Meyer, C., Rosenwald, A., Drexler, H.G. and MacLeod, R.A. (2007) Comprehensive analysis of homeobox genes in Hodgkin lymphoma cell lines identifies dysregulated expression of HOXB9 mediated via ERK5 signaling and BMI1. *Blood* **109**, 3015–3023.
12. Lynch, S.A., Wang, Y., Strachan, T., Burn, J. and Lindsay, S. (2000) Autosomal dominant sacral agenesis: Currarino syndrome. *J. Med. Genet.* **37**, 561–566.
13. Nagel, S., Kaufmann, M., Scherr, M., Drexler, H.G. and MacLeod, R.A. (2005) Activation of HLXB9 by juxtaposition with MYB via formation of t(6;7)(q23;q36) in an AML-M4 cell line (GDM-1). *Genes Chromosomes Cancer* **42**, 170–178.
14. Nagel, S., Scherr, M., Quentmeier, H., Kaufmann, M., Zaborski, M., Drexler, H.G. and MacLeod, R.A. (2005) HLXB9 activates IL6 in Hodgkin lymphoma cell lines and is

regulated by PI3K signalling involving E2F3. *Leukemia* **19**, 841–846.
15. Argiropoulos, B. and Humphries, R.K. (2007) Hox genes in hematopoiesis and leukemogenesis. *Oncogene* **26**, 6766–6776.
16. Nagel, S., Meyer, C., Quentmeier, H., Kaufmann, M., Drexler, H.G. and MacLeod, R.A. (2008) MEF2C is activated by multiple mechanisms in a subset of T-acute lymphoblastic leukemia cell lines. *Leukemia* **22**, 600–607.
17. Nagel, S., Venturini, L., Przybylski, G.K., Grabarczyk, P., Schmidt, C.A., Meyer, C., Drexler, H.G., Macleod, R.A. and Scherr, M. (2009) Activation of miR-17-92 by NK-like homeodomain proteins suppresses apoptosis via reduction of E2F1 in T-cell acute lymphoblastic leukemia. *Leuk. Lymphoma* **50**, 101–108.
18. Drexler, H.G. (2005) Guide to leukemia-lymphoma cell lines. E-book on compact disk, Braunschweig.
19. Nagel, S., Scherr, M., Kel, A., Hornischer, K., Crawford, G.E., Kaufmann, M., Meyer, C., Drexler, H.G. and MacLeod, R.A. (2007) Activation of TLX3 and NKX2-5 in t(5;14)(q35;q32) T-cell acute lymphoblastic leukemia by remote 3'-BCL11B enhancers and coregulation by PU.1 and HMGA1. *Cancer Res.* **67**, 1461–1471.
20. Nagel, S., Venturini, L., Przybylski, G.K., Grabarczyk, P., Meyer, C., Kaufmann, M., Battmer, K., Schmidt, C.A., Drexler, H.G., Scherr, M. and Macleod, R.A. (2009) NK-like homeodomain proteins activate NOTCH3-signaling. *BMC Cancer* **9**, 371.

Chapter 31

Measuring Gene Expression from Cell Cultures by Quantitative Reverse-Transcriptase Polymerase Chain Reaction

Sharon Glaysher, Francis G. Gabriel, and Ian A. Cree

Abstract

Quantitative reverse transcriptase polymerase chain reaction (qRT-PCR) offers a robust method for the measurement of RNA levels for any gene within cells harvested at any point before or during cell culture. The key elements of RNA extraction followed by a two-step qRT-PCR method (reverse transcription and PCR) are described, followed by a brief section on analysis of the results. There are a number of excellent kits available commercially for much of this work, but it is essential to ensure that the quality and quantity of cDNA produced is adequate for the standard PCR or array to be used.

Key words: RNA, Cell culture, Gene expression, Quantitative RT-PCR, Taqman array

1. Introduction

The measurement of gene expression is now regarded as an essential adjunct to cell culture, and the boundaries between cell culture and molecular biology are no longer clearly defined. This chapter is a practical guide to the use of qRT-PCR in standard plates using cells obtained from cell culture. They are based on standard operating procedures and manufacturer's instructions, and the instructions given may need to be adjusted if different reagents or equipment are used. The stages of the procedure (Fig. 1) include RNA extraction from cell culture material, reverse transcription (RT), single gene PCR, and multiplex PCR using a Taqman Array™.

Cells in suspension are harvested by centrifugation, lysed with Macherey Nagel Lysis Buffer RA1 containing 2-mercaptoethanol (2ME), and stored at −80°C in this buffer until RNA extraction.

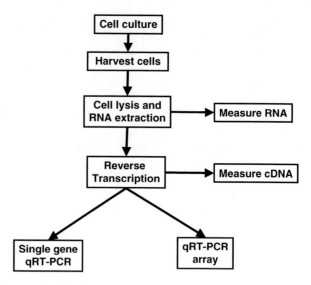

Fig. 1. Diagram of method, including cell culture, extraction, RT and PCR steps.

Total RNA is extracted from cells with a commercially available kit (Ambion Recoverall or similar) according to the manufacturer's instructions. The protocol includes a DNase digestion step to prevent carry-over of genomic DNA in further analysis. The nucleic acids are solubilised, while digested proteins are "salted out." Lysate buffer is added to the digest to increase the ionic/salt concentration and alkalinity ready for extraction by acidified silica/glass fiber method. The lysate buffer also aids the break up of cellular proteins and protects the RNA on storage.

This method follows the Ambion RecoverAll™ Total Nucleic Acid Isolation kit (Cat 1975) instruction manual (Version 0505) for full details with the following exceptions: (a) Breaking up sample with needle and syringe after addition of Ambion Isolation Additive (if necessary) and (b) freezing the sample after the addition of Isolation Additive (step 17).

Following this step, we use a NanoDrop scanning spectrophotometer to analyze 1 µL of the extracted RNA to assess the amount present and dilute this to 300 ng/mL in PCR mix for addition to either 96-well PCR plates for single gene estimations, or to a 384-well microfluidic Taqman Array card containing 48 reactions for each of the eight ports on the card (Fig. 2).

Two qRT-PCR methods are included in this chapter – the first allows the expression of a single gene to be measured (1), while the second can measure up to 384 genes in a microfluidic card (2). In both methods, reverse transcription is used to produce cDNA, which is then amplified exponentially in the presence of Taq Polymerase with specific primers via cycling through annealing, extension, and melting temperatures. TaqMan fluorescence labeled probes specific for the desired target are used to

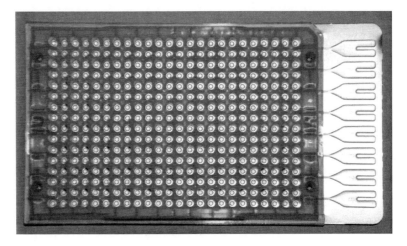

Fig. 2. A Taqman Array. This represents a 384-well microfluidic sealed array with eight ports, each of which serves 48 wells.

detect the presence of the amplified sequence. If the probe hydrolyses to the target, the fluorophore is cleaved by the Taq and is free to fluoresce when excited with light at particular wavelength. The emission is then measured at another wavelength. The amount of fluorescence is proportional to the amount of amplified nucleic acid target. Using housekeeping gene targets, the amplified nucleic acid target can be quantified relative to these reference genes.

2. Materials

2.1. Cell Lysis

1. 2-Mercaptoethanol (2ME).
2. Macherey Nagel RA1 Lysis Buffer.

2.2. RNA Extraction

1. 0.8 mm, 21 gauge, sterile needle.
2. 1-mL sterile polypropylene syringe.
3. Heating block(s) for 1.5-mL tubes (temperature range 95°C).
4. NanoDrop Spectrophotometer.
5. Molecular Biology Grade 100% Ethanol (BDH, 437433T).
6. Water, Nuclease free (Promega, P1193).
7. RecoverAll™ Total Nucleic Acid Isolation kit (Ambion, 1975).
8. Sterile fine tip polypropylene pasteurs.
9. 50 mL polypropylene tubes.
10. 1.5-mL polypropylene Eppendorf tubes.

11. 2-mL polypropylene conical tubes.
12. 0.6-mL polypropylene microfuge tubes.
13. 0.2-mL polypropylene PCR tubes.
14. Assorted Sterile filtered ART micropipette tips.

2.3. RNA Measurement

1. The NanoDrop® ND-1000 full-spectrum (220–750 nm) spectrophotometer or equivalent capable of measuring small samples (1 μL volume) with high accuracy and reproducibility.
2. Soft tissue wipes.

2.4. Reverse Transcription

1. Thermal Cycler.
2. Chilled cooling block.
3. NanoDrop Spectrophotometer.
4. 1.5-mL polypropylene Eppendorf tubes.
5. 0.6-mL polypropylene microfuge tubes.
6. 0.2-mL polypropylene PCR tubes.
7. Applied BioSystems High-Capacity cDNA Kit – 200 reactions (4368814) (for 200 reverse transcription reactions at 100 μL/reaction). Store at –15 to –25°C.
8. Promega nuclease-free water (P1193).

2.5. Single Gene PCR

1. 96-well PCR machine (e.g. Biorad i-cycler, Applied Biosystems 7500).
2. Taqman master mix.
3. Nuclease-free water.
4. cDNA samples from RT step.
5. DNAZap solutions 1 & 2.

2.6. Taqman Array PCR

1. Applied Biosystems 7900HT.
2. Taqman master mix.
3. Nuclease-free water.
4. cDNA samples from RT step.
5. DNAZap solutions 1 & 2.

2.7. General Equipment and Consumables

1. Benchtop refrigerated centrifuge.
2. Microfuge.
3. Vortexer.
4. Assorted pipettes covering ranges 0.5–1,000 μL – a set is required for each stage and should not be mixed.
5. Fridge (+4°C).

6. Freezers (−20°C and −80°C).
7. Assorted Sterile filtered ART micropipette tips.
8. Nitrile gloves (see Note 1).
9. Balance.

3. Methods

3.1. Cell Lysis

1. Add +3.5 mL of 2-mercaptoethanol (2ME) to 350 mL Macherey Nagel RA1 Lysis Buffer to make RA1-2ME lysis buffer.
2. Resuspend cell pellet in 0.7 mL RA1-2ME lysis buffer and store frozen, ideally at −80°C for extraction.

3.2. RNA Extraction

1. Pre-warm heating block to 95°C.
2. Remove lysate samples from −80°C storage (previously harvested and stored in RA1-2ME) and thaw sample lysate slowly on ice or +4°C fridge to room temperature (see Note 2).
3. The remaining steps in this protocol use the Ambion® RecoverALL™ Nucleic acid Isolation kit: please see the protocol enclosed with kit for full details (see Note 3), but the main points are listed here.
4. Label one Ambion filter spin column and collection tube per sample, and assemble.
5. Add 500 µL of 100% ice cold ethanol to the thawed lysate and mix by pipetting. The temperature of the ethanol aids in the precipitation of nucleic acids.
6. Using a fresh 1-mL ART tip per sample, pipette 700 µL of sample/ethanol mix onto the center of the cartridge and close the lid. Microfuge at $10,000 \times g$ (usually 10,000 rpm) for 60 s.
7. Collect the flow-through and store in a labeled tube for future DNA extraction (if required). Reinsert the filter cartridge into the same collection tube.
8. Repeat steps 4–6 at least twice more until all the lysate has been processed for each sample paired tube.
9. Add 700 µL of Wash 1 to Filter Cartridge and microfuge for 30 s at $10,000 \times g$.
10. Discard flow-through, and reinsert the Filter Cartridge in the same collection tube.
11. Add 500 µL of Wash 2/3 to the Filter Cartridge and microfuge for 30 s at $10,000 \times g$.

12. Discard flow-through, and reinsert the Filter Cartridge in the same collection tube.
13. Microfuge for 30 s at 10,000 ×g to remove residual fluid from the filter.
14. *For RNA isolation*, prepare the following DNase mix (a master mix may be prepared): 6 µL of 10× DNase Buffer, 4 µL of DNase, and 50 µL of nuclease-free water per sample.
15. Gently mix and add using an ART tip pipette 60 µL of the DNase mix to the center of each Filter Cartridge.
16. Cap tube and incubate at 30 min at room temperature (time critical).
17. During the incubation heat an aliquot of nuclease-free water (Promega) 150 µL per sample, to 95°C.
18. Following DNA digestion, add 700 µL of Wash 1 to the Filter Cartridge.
19. Incubate for 60 s at room temperature (time critical).
20. Microfuge for 30 s at 10,000 ×g.
21. Discard flow-through, and reinsert the Filter Cartridge in the same collection tube.
22. Add 500 µL of Wash 2/3 to the Filter Cartridge and microfuge for 30 s at 10,000 ×g.
23. Discard flow-through, and reinsert the Filter Cartridge in the same collection tube.
24. Repeat steps 21–22 for a second time with 500 µL of Wash 2/3.
25. Place the Filter Cartridge into a fresh Collection tube.
26. Apply 110 µL of heated (95°C) nuclease-free water to the centre of the filter (critical) and close cap. Seal the remaining aliquot of nuclease-free water and place back into the 95°C heating block.
27. Incubate at room temperature for 60 s (time critical).
28. Microfuge for 60 s at 13,000 ×g.
29. Reload the eluent onto the filter and repeat steps 26–27.
30. Elute volume collected will be approximately 100–110 µL for each nucleic acid.

3.2.1. RNA Measurement

1. After extracting and eluting purified nucleic acid, a small aliquot is taken for quantification and checking purity.
2. Switch on the NanoDrop 1000 and the assisting computer.
3. With the sampling arm in the down position, start the NanoDrop software by clicking on the icon on the computer screen.

4. From the drop-down menu choose the user account; all data taken will automatically be logged in the appropriate archive file; if not chosen, all the data is automatically saved in the folder "default" (C:\program files\nanodrop\nanodrop data\default).

5. Choose the "nucleic acid" module by clicking on the box with module name on the screen.

6. Before making any measurements, ensure that measurement pedestal surfaces are clean and with the sampling arm open, pipette a water sample onto the lower measurement pedestal. At this stage water cannot be replaced by any other solution or buffer.

7. When the water drop is placed on a pedestal drop down the sampling arm, click "OK." When the message "Initializing Spectrometer – please wait" disappears, the instrument is ready for use.

8. When the measurement is complete, open the sampling arm and wipe the sample from both the upper and lower pedestals using a soft laboratory wipe.

9. From the drop-down menu select the type of nucleic acid being measured – "DNA-50" for dsDNA, "RNA-40" for RNA, "ssDNA-33" for single-stranded DNA, or "Other" for other nucleic acids.

10. Dispense 1.3 µL of blank sample (the buffer, solvent, or carrier liquid used with your samples) onto the lower measurement pedestal and lower the sampling arm into the "down" position.

11. Click on the "Blank" (F3) button.

12. Wipe the blanking buffer from both pedestals using a laboratory soft tissue wipe.

13. Analyze an aliquot of water as though it were a sample. This is done using the "Measure" button (F1). The result should be a spectrum with a relatively flat baseline. Wipe the blank from both measurement pedestal surfaces and repeat the process until the spectrum is flat.

14. Mix well all the samples before measurements, as it ensures the small amount taken for measurements gives the right values.

15. Open the sampling arm; wipe off any fluid remaining on the pedestals.

16. Type in the sample ID into the highlighted window on the screen and using a fresh filtered tip fitted pipette, 1.3 µL of each undiluted sample onto the lower measurements pedestal.

17. Lower the sampling arm into the "down" position and click on the "measure" button.

18. When the measurement is complete repeat steps 15–17 for all samples, performing blank measurement every 10–20 samples.

19. During measurements, the full spectrum of absorbance (wave length 220–750 nm) is obtained, and a graph is displayed after the measurement is completed. In the panel on the right side of the screen quality and quantity values are displayed as well after each completed measurement, giving A_{260}, A_{280}, ratios A_{260}/A_{280} and A_{260}/A_{230} (defining the purity of the sample) and sample concentration in ng/µL based on absorbance at 260 nm and the selected analysis constant. Purity (A_{260}/A_{280} ratio based on using RNA in 10 mM Tris–Cl pH 7.5) should be 1.9–2.3. Quantity is based on a conversion factor where one unit of absorption, $A_{260} = 40$ µg/mL.

20. All measurement data taken are automatically saved in appropriate archive file, along with the respective absorbance spectrums.

21. After completing the measurements clean the upper and lower pedestals with a soft tissue moistened with water. Then using a dry soft tissue wipe upper arm and lower pedestal.

22. Click on exit button. The program will ask to view the measurement data; this can be either accepted or rejected. The measurement data can be viewed on a later date by opening "data viewer" from the main module screen.

23. Switch off the NanoDrop 1000 and the computer.

3.3. Reverse Transcription

1. Bring ABI High-Capacity cDNA Reverse Transcription Kits to room temperature.

2. Determine the concentration and volume of reagents required.

3. Save and print spectrophotometry results.

4. Immediately convert all the RNA (dilute where necessary in nuclease-free water to give a working concentration of 100–150 ng/µL) into cDNA using ABI High Capacity cDNA conversion kit. Do not attempt to store RNA.

5. A typical Reverse transcription mix is 75 µL RNA at 100 ng/µL to 75 µL cDNA 2× Master Mix and 15 µL RNA to 15 µL RT Negative 2× Master Mix (check for genomic contamination). Adjust mix according to RNA quantities.

6. Quantify cDNA by NanoDrop spectrophotometer.

7. If yields are low, perform 3 M sodium acetate and ethanol precipitation, reconstitute in a smaller volume of nuclease-free water and remeasure cDNA.

8. Store samples overnight at +4°C or −20°C for long-term storage (see Note 4).

9. Calculate enough for an extra sample to account for error.
10. Label a 1.5-mL Eppendorf tube or 0.6-mL PCR tube 2× RT Neg Master Mix (RTN_MM). This serves to check for genomic (DNA) contamination in the purified RNA sample.
11. Label a second 1.5-mL Eppendorf tube or 0.6-mL PCR tube as 2× cDNA Master Mix (cDNA_MM). This will be the reaction mixture for converting your sample RNA to a single strand of cDNA.
12. Using a fresh filtered pipette tip for each reagent, dispense the reagents in the order. Reagents can be dispensed in parallel for both RT_MM and cDNA_MM (see Note 5).
13. Seal tubes, mix gently, and pulse spin.
14. Using black marker for cDNA positive tubes and blue marker for RT negative tubes, label 0.2-mL Eppendorf tubes with samples name, and appropriate identifiers.
15. Using a fresh filter tip for each Master Mix, aliquot 15 µL of RT Negative mix into the appropriate labeled 0.2-mL PCR tube.
16. Using a fresh filter tip dispense 75 µL volumes of cDNA 2× Master mix for the corresponding 0.2 mL cDNA_MM sample tube.
17. Using a fresh pipette tip for each sample, dispense 15 µL of purified RNA sample (diluted if necessary in nuclease-free water) into the corresponding 0.2 mL RT N_MM sample tube. Gentle pipette up and down two times to mix.
18. Using a fresh pipette tip for each sample, dispense 75 µL of purified RNA sample (diluted if necessary in nuclease-free water) into corresponding 0.2 mL cDNA_MM sample tube.
19. Seal tubes and pulse microfuge at $11,000 \times g$ for 30 s to spin down the contents and to eliminate any air bubbles.
20. Place the tubes in a chilled cooling block until you are ready to load the thermal cycler.
21. Place 0.2-mL tubes in a thermal cycler, select and run appropriate program.
22. Cycling conditions for each reaction mix: step 1, temp 25°C for 10 min; step 2, 37°C for 120 min; step 3, 85°C for 5 s; step 4, 4°C for infinity.
23. Return ABI cDNA Reverse Transcription kit to −20°C freezer.
24. After 2 h 10 min, remove tubes and pulse microfuge at $11,000 \times g$ for 30 s.
25. Either store overnight at +4°C and measure cDNA the following morning using NanoDrop spectrophotometer or measure straight away. Determine the quantity required for qPCR.

26. Proceed to single gene qRT-PCR or Taqman assay (see Note 6), adjusting the input cDNA. Typically 5 μL of cDNA at 300 ng/mL in 4 μL of nuclease-free water (total 9 μL diluted cDNA) to 11 μL TaqMan mix with probe per well (see Note 7).

3.4. Single Gene PCR

1. Set up the PCR Machine (see Note 8).
2. Place the required number of Eppendorfs into a PCR rack and label the primers and master mix tubes "MM." NB. 1×0.2 mL Eppendorf/sample, 1×0.6 mL Eppendorf/primer. (MM), 2×0.6 mL Eppendorfs (primers gene 1 and primers gene 2).
3. Get 25 mM $MgCl_2$ from the Core Reagents kit and defrost.
4. Add the appropriate volume of water to all tubes, including primer master mix tubes, cDNA tubes, and cDNA negative tubes.
5. Add appropriate volume $MgCl_2$ to the primer MM tubes (see Note 9).
6. Add appropriate volume SYBR GREEN Master Mix (ABI or Sigma) to the primer MM tubes (See sheet).
7. Remove working concentrations (10 μM) primers from fridge and vortex.
8. Add appropriate volume of positive and negative primers to the primer MM tubes.
9. Vortex primer MM tubes well and pulse for a few seconds.
10. Pipette the primers MM (20 μL) into the appropriate well in the PCR plate, taking care to pipette onto one side of the wall of the well and remembering to pipette the primer MM into the cDNA negative wells.
11. Remove the cDNA samples from the −80°C freezer and place into the green PCR rack.
12. Turn the PCR plate around – you can then pipette the cDNA on to the other wall of the well.
13. Make the cDNA dilutions and pipette 2.5 μL cDNA into the 0.2 μL tube containing 10 μL water and mix.
14. Keeping the same pipette tip, alter the volume of the pipette to 5 μL and pipette the sample into each of the two appropriate PCR wells (see Note 10).
15. Put optical sealant onto the plate, ensuring that fingers do not touch the surface. Use the backing paper to seal the cover around the edges.
16. Centrifuge the plate briefly at $200 \times g$ for 2 min.
17. Place into PCR machine and continue run.
18. Enter the C_t value, μL of cDNA used/well, and μL of RNA used to make cDNA into the sighting shot database for analysis.

3.5. Taqman Array PCR

1. Each array consists of one loading port, and the card is loaded with six test samples, a quality control, and a negative control (see Note 11).
2. Switch on the 7900HT PCR machine 30 min before use to enable the laser to warm up.
3. Using filtered tips at all times, prepare the mastermix to be aliquoted into pre-labeled sterile DNase/RNase-free 600 µL Eppendorf tubes (see Note 12).
4. Prepare the loading volumes from the cDNA values obtained from the Nanodrop readings done previously.
5. Seal the Eppendorf tubes and pipette the cDNA into the smaller 200 µL Eppendorf tubes (which contain the nuclease-free water), seal the lids, and vortex to mix.
6. Pipette the contents of the small 200 µL Eppendorf tubes (water and cDNA) into the large Eppendorf tubes (mastermix). Vortex to mix the tubes.
7. Centrifuge (Sorvall using microfluidic card buckets) for 1 min at $300 \times g$ ensuring the centrifuge is correctly balanced.
8. Check the sample ports for air bubbles before repeating Subheading 3, step 7. If there are any air bubbles present, holding the card upright and gently tapping on the bench usually dislodges them (see Note 13).
9. Using the plate sealer, seal the microfluidic card by swiping the handle in one direction only away from you.
10. Cut off the loading port strip and discard (discarded port strips can potentially puncture a waste bag, so disposal in a sharps bin is preferable).
11. Open a new SDS file by clicking the SDS 2.2.2 icon on the desktop. Then select: Abs Quantification (drop-down menu).
12. Open relevant program from the drop-down menu.
13. Scan the barcode (make a note of it for future reference) and click "OK."
14. Select instrument tab and click "Connect."
15. The door of the 7900T will then open, place the plate (with barcode facing forwards) on the carrier, click "Close door" and then "Start."
16. Ensure the run has started successfully; the machine can then be left to complete the PCR. It takes approximately 2 h per card.
17. Once the run has completed, remove the card and discard in a clinical waste bag. Then switch off the PCR machine (see Note 14).

4. Notes

1. Health and Safety. Avoid acidification of additive reagent. Wash buffers contain guanidinium isothiocyanate (GTC). Dispose of sealed waste tubes via cytotoxic waste disposal route. Dispose of needles and syringes in sharps bin, and take care when handling needles. Wear nitrile gloves and a clean lab coat, use molecular biology grade nuclease-free reagents and gamma-irradiated polypropylene plastics, and dedicated pipettes with sterile aerosol resistant tips (ART). Avoid cross-contamination by using a new fresh Pasteur pipette and ART tip for each sample. Discard reagents via the appropriate waste disposal route.

2. Lysate samples can be stored for short-term at −20°C, for long-term storage before extraction −80°C is preferable.

3. When a new Ambion Recoverall kit is first opened, prepare the Wash 1 and Wash 2/3 buffers as per manual instructions before commencing the nucleic acid isolation. Add appropriate volume of ethanol (wash 1 – 42 mL; wash 2 – 48 mL) and indicate this, with date, on the reagent bottles. We use the RecoverAll™ Total Nucleic Acid Isolation kit with the following exceptions: pre-filtering the lysed sample before addition of ethanol, and eluting with nuclease-free water rather than elution buffer.

4. When storing molecular biology reagents and samples in freezers, beware of cycling of frost free freezers, which are not really suitable for laboratory use, as internal temperatures can be erratic.

5. No MultiScribe is added for the RT Negative Master Mix. The excluded volume is replaced with nuclease-free water.

6. If required, perform "sighting shot" PCR (pre-assay evaluation) of the sample for two appropriate house-keeping genes using SyBr Green with disassociation curves or a Taqman assay as suggested for single gene qRT-PCR. Also, run the RT Negative samples against the most highly expressed housekeeping gene to check for genomic contamination.

7. Choice of housekeeping gene depends on cell/tissue type – we usually use PBGD/HMBS as our preferred housekeeping gene.

8. cDNA can be used straight away in qPCR assays; stored overnight at +4°C and used the following day; or stored long-term at −20°C (not frost free). If using the NanoDrop spectrophotometer, a data text file can be opened with Excel and a workbook can be saved. Concentrations can then be copied into other worksheets.

9. For all PCR experiments we use a standardized Excel spreadsheet, which facilitates planning of the experiment. A template sheet is resaved with a new name for each experiment.

10. Remember to change the pipette volume back to 2.5 μL when you have pipetted the sample into the plate.

11. It may be useful to make a batch of positive quality control samples to run against your unknowns. This can be run alongside a non-template control containing no cDNA which has been adjusted to contain only mastermix and water. These can be run as positive and negative controls for each experiment.

12. To prevent cross contamination of mastermix solutions and cDNA samples, preparation of these solutions should ideally be done in separate locations (designated areas/labs for clean/reagent preparation and cDNA preparation), or in the very least, making sure you decontaminate your work area, change labcoat and gloves.

13. If you wish to prepare cards in advance, this can be done in the morning (usually a maximum of 4 per normal working day), but all plates must be run the same day. The cards are stored at this point (sealing them first to prevent cross contamination), clearly labeled, and placed with ports up in the fridge and will need to be re-spun prior to running the card. Remembering to prepare one card at a time in the hood.

14. It is very important to minimize the number of times the machine is switched on and off to preserve the life of the laser, so plan your runs efficiently.

Acknowledgments

This work was funded by CanTech Ltd.

References

1. Di Nicolantonio, F., Mercer, S. J., Knight, L. A., Gabriel, F. G., Whitehouse, P. A., Sharma, S., Fernando, A., Glaysher, S., Di Palma, S., Johnson, P., Somers, S. S., Toh, S., Higgins, B., Lamont, A., Gulliford, T., Hurren, J., Yiangou, C., and Cree, I. A. (2005) Cancer cell adaptation to chemotherapy, *BMC Cancer* **5**, 78.

2. Glaysher, S., Yiannakis, D., Gabriel, F. G., Johnson, P., Polak, M. E., Knight, L. A., Goldthorpe, Z., Peregrin, K., Gyi, M., Modi, P., Rahamim, J., Smith, M. E., Amer, K., Addis, B., Poole, M., Narayanan, A., Gulliford, T. J., Andreotti, P. E., and Cree, I. A. (2009) Resistance gene expression determines the in vitro chemosensitivity of non-small cell lung cancer (NSCLC), *BMC Cancer* **9**, 300.

Chapter 32

Proteomic Evaluation of Cancer Cells: Identification of Cell Surface Proteins

Samantha Larkin and Claire Aukim-Hastie

Abstract

The plasma membrane proteome can be defined as the entire complement of proteins present in the plasma membrane at a specific time. The process of carcinogenesis leads to changes in the array of proteins present in the plasma membrane proteome. Analysis of differential expression of such proteins in cancer is extremely important; due to their position on the cell surface they have a potential for use as diagnostic and/or prognostic markers and therapeutic targets. Biotin labelling followed by avidin chromatography can be used to obtain membrane protein enriched lysates from cell lines, which can then be resolved using SDS–PAGE, coomassie staining and mass spectrometry.

Key words: SDS–PAGE, Affinity chromatography, Biotin, Avidin, Plasma membrane proteome

1. Introduction

The plasma membrane proteome is frequently altered during carcinogenesis with changes often seen in the number or activity of growth factor receptors heightening growth factor responsiveness, in the number or type of cell adhesion molecules expressed, e.g. integrins, or in the number/activity of proteases present on the cell surface. Analysis of the differential expression of these proteins is vital as they may represent novel diagnostic/prognostic markers or even therapeutic targets. In the latter capacity, it may be possible to tailor treatments more specifically to the cancer cells and reduce the harmful side effects of anticancer treatments. The plasma membrane proteome, therefore, could not only be an indicator of the malignant status of the cell, but also a resource for targeted cancer therapy. Research has lead to the identification of a number of cancer associated plasma membrane proteins, which have been known to account for 70% of all known pharmaceutical drug targets, some of which are in clinical use (1).

The analysis of the plasma membrane proteome requires strategies optimised for the enrichment of rare, relatively insoluble proteins. Many strategies have been employed to analyse cell surface proteins from stable isotope labelling (2), to complex three-dimensional crystallisation analyses (3). However, vectorial labelling with less hazardous compounds, such as fluorophores and biotin, followed by affinity chromatography has been employed in the enrichment of this protein fraction (4).

A long chain sulphonamide ester of biotin, called Sulfo-NHS-LC-Biotin, has been successfully used to study spermatozoa cell surface proteins. In this study, 98 sperm cell surface proteins were identified, 22 of which were phosphorylated, including several novel protein identifications (5). Tagging with biotin derivatives, as opposed to His, FLAG or HA, is advantageous, as the addition of the sulpho group makes the biotin impermeable to the cell, thus limiting unwanted intracellular labelling. In one of the first studies to employ biotin labelling, it was found that this labelling technique minimised cytoplasmic contamination (6). The study of membrane proteins is commonly hampered by their insoluble nature; biotin labelling also increases the solubility of the compounds it is associated with, a quality which is a distinct advantage, as success in purification of membrane proteins depends on solubility (6).

The use of biotin tags such as Sulfo-NHS-LC-Biotin is however, of limited use in proteomics as the tag itself can lead to the generation of multiple isoforms due to the negative charge it carries. This can be avoided by the use of a reducible biotin derivative, such as Sulfo-NHS-SS-Biotin, which can be removed before protein separation. This reducible form of vectorial labelling has been successfully used to analyse cell surface protein composition, one of the first applications being the analysis of cell surface proteins on rat hepatocytes (6).

Biotin is a small coenzyme synthesised by bacteria, plants, and some fungi (7). Biotinylated proteins are rare in nature, it is estimated that there are only four mammalian proteins with this modification (8, 9). These are acetyl CoA carboxylase, pyruvate carboxylase, methylcrotonyl-CoA carboxylase and propionyl-CoA carboxylase (8). The rarity of endogenous biotinylated proteins indicates that biotin labelling provides a specific way of tagging surface proteins with little to no background from endogenously biotinylated molecules (10). Avidin has an extremely high affinity for biotin, therefore avidin–biotin affinity chromatography is an efficient, high-yield method of cell surface protein purification.

Pre-fractionation steps, such as affinity chromatography, are often used in proteomics to decrease the diversity and complexity of a protein mixture, giving a sample enriched in the number and concentration of a desired subset of proteins (11). Biotin–avidin affinity chromatography has been improved by the engineering of monomeric avidin columns with a decreased affinity for biotin, allowing competitive elution with free biotin. These columns however, are prone to degradation by protease and reducing agents. Therefore a streptavidin tetrameric column has been developed which has superior stability (10).

Here, we present an optimised protocol for cell surface protein analysis, which has been applied to the analysis of surface protein composition in the prostate isogenic cell lines, 1542 NPX (normal), and CP3TX (cancerous).

2. Materials

2.1. Biotin Labelling

1. 1× PBS (Invitrogen, Paisley, UK).
2. Biotin: 5 mg/mL EZ-Link Sulfo-NHS-LC-Biotin (Perbio, Tattenhall, UK) in 1× PBS or 5 mg/mL EZ-Link Sulfo-NHS-SS-Biotin (Perbio, Tattenhall, UK) in 1× PBS.
3. Blocking buffer: 50 mM NH_4Cl (Sigma, Poole, UK) in 1× PBS.
4. 1% Octylglucopyranoside (OGP): 1% OGP (Sigma, Poole, UK) in 1× PBS with one complete protease inhibitor tablet (Amersham [GE Healthcare], Little Chalfont, UK) per 50 mL.

2.2. Cell Lysis

1. 1× PBS (Invitrogen, Paisley, UK).
2. 1% Octylglucopyranoside (OGP): 1% OGP (Sigma, Poole, UK) in 1× PBS with one complete protease inhibitor tablet (Amersham [GE Healthcare], Little Chalfont, UK) per 50 mL.
3. Rubber cell scrapers (VWR, Poole, UK).

2.3. Affinity Chromatography

1. Binding buffer: 1% OGP lysis buffer.
2. Elution buffer: 100 mM DTT or 8 M Guanine HCl (Sigma, Poole, UK).
3. Regeneration buffer: 8 M Guanine HCl.
4. Wash buffer: 1% OGP, 1% NP40.

2.4. SDS–Polyacrylamide Gel Electrophoresis (SDS–PAGE)

1. 30% Bis-acrylamide (Bio-Rad, Hemel Hempstead, UK).
2. Tris–HCl (Sigma, Poole, UK): Prepare 1.5 M at pH 8.8 and 0.5 M at pH 6.8 solutions in advance.
3. SDS (Sigma, Poole, UK): Prepare 10% solution in water, in advance.
4. APS (Bio-Rad, Hemel Hempstead, UK): prepare 10% solution in water immediately before use.

5. Temed (Sigma, Poole, UK).
6. Sample Buffer: 62.5 mM Tris–HCl at pH 6.8 (Sigma, Poole, UK), 20% Glycerol (Sigma, Poole, UK) and 2% SDS (Sigma, Poole, UK) (see Note 1).
7. 1× Running Buffer at pH 8.3: 25 mM Tris, 192 mM Glycine (Sigma, Poole, UK) and 0.1% SDS.
8. Prestained molecular weight markers such as the Broad Range Rainbow Marker (Amersham [GE Healthcare], Little Chalfont, UK).
9. Coomassie staining solution: 0.2% R250 (Sigma, Poole, UK), 40% methanol and 10% Acetic acid (VWR, Poole, UK) made up in double distilled water (ddH$_2$O). Once prepared, filter and store at room temperature. This solution can be reused several times and stored for months, but may need re-filtering.
10. Coomassie de-staining solution: 10% MeOH and 5% acetic acid (VWR, Poole, UK) made up in ddH$_2$O. Once prepared, store at room temperature. This should not be reused.

3. Methods

3.1. Biotin Labelling

To selectively enrich the cell surface protein fraction of the sample, it is necessary to label the surface proteins so that they can be purified by affinity chromatography. The label chosen here is a biotin construct that binds to proteins at their lysine residues. Two different types of biotin tags have been trialled, a long chain Sulfo-NHS ester, EZ-Link Sulfo-NHS-LC-Biotin and a reducible form with a cleavable disulphide bond in the spacer arm, EZ-Link Sulfo-NHS-SS-Biotin (Perbio, Tattenhall, UK). The cell labelling protocol detailed below can be used for both forms, but the reducible form EZ-Link Sulfo-NHS-SS-Biotin (Perbio, Tattenhall, UK) gives a higher yield of membrane proteins (Fig. 1). This protocol was adapted from (4).

1. The required solutions, biotin, blocking buffer, 1% OGP and 1× PBS, should be chilled in advance.
2. The cells should then be washed three times in PBS to minimise contamination from media and cell debris during labelling.
3. During the final wash, the cells should be observed for any loss of adhesion and then the wash solution is removed.
4. Biotin solution (5 mL) should then be added and incubated for 5 min on ice; then excess biotin solution is removed.

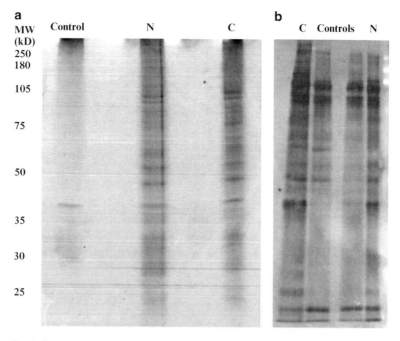

Fig. 1. Cell surface fractionated NP and CPT cell lysate labelled with (**a**) SS biotin and (**b**) LC biotin. The SS biotin gave a labelled protein yield visible on staining with coomassie blue, whereas the LC required avidin blotting to visualise labelled proteins.

5. The quench solution should be added and incubated for 10 min on ice. Cells should again be checked for signs of detachment.
6. Cells should again be washed three times in PBS and left in the final wash solution.
7. The final PBS wash can then be carefully removed to prevent dilution of the lysis buffer, and lysis of the cells can then be performed.

3.2. Cell Lysis

Although several lysis buffers can be used (see Note 2), depending on the different requirements of the experimental procedure, the basic protocol for lysis is identical. A control sample of cell lysate should be prepared following the same method, but omitting the biotin labelling. This protocol was adapted from (12).

1. The cells should be observed for their degree of confluence and providing they are at least 70% confluent, they are suitable for lysis and should be put on ice.
2. The old medium should be removed and the cells are washed three times with cold 1× PBS.
3. Following the final wash, all traces of PBS must be removed by aspiration.
4. Lysis buffer (1% OGP) can then be added. The amount of lysis buffer added is dependent on the area of the growing

surface; typically 500 μL should suffice for a T75 flask and three flasks should be used to give optimal protein concentration.

5. The lysis buffer should then be allowed to run over the growing surface several times within 1 min.

6. The cells can be removed from the growing surface into the buffer using a rubber scraper.

7. Subsequently, the lysis buffer containing the cells should be aliquotted into an Eppendorf and centrifuged for 1 min at $11{,}000 \times g$.

8. The supernatant, containing the membrane proteins, are then transferred to a clean Eppendorf, labelled and then stored frozen or used immediately.

9. The pellet, containing nuclear components and cell debris can also be retained for other analyses.

3.3. Affinity Chromatography

Several methods of affinity chromatography can be used and the method illustrated here was optimised for membrane protein analysis. The elution step varies depending on the type of biotin used. The two alternatives that have been mentioned before are described.

1. All solutions should be allowed to equilibrate to room temperature.

2. Meanwhile immobilised Neutravidin beads (Perbio, Tattenhall, UK) should be packed into three 1-mL columns as follows (12):

    (a) Secure bottom cap on the column tip and secure the column upright.

    (b) Add sufficient buffer to the column to fill up to the reservoir portion, tap to dislodge air bubbles.

    (c) Float porous disc of correct diameter on the buffer, then push to the bottom of the column using the open end of a serum separator.

    (d) Decant the column, return bottom cap.

    (e) Add a sufficient volume of gel slurry to obtain desired bed volume.

    (f) Allow gel to settle for 30 min.

    (g) Position a second porous disc on the settled gel bed using the serum separator, leaving a gap of 1–2 mm from top of the column bed.

    (h) Wash the inner top of the column with buffer; the column is now ready for use or storage.

3. Prepared columns are then clamped upright and attached to a peristaltic pump at a flow rate of 0.5 mL/min and equilibrated by the addition of 10 mL binding buffer.

4. The sample should then be applied to the column and the effluent collected.

5. The sample is left to incubate in the column for 30 min.

6. The effluent can then be re-applied and passed through the column.

7. This process is repeated to maximise column binding and the final effluent collected.

8. The column is then washed with 10 mL binding buffer of which two 5 mL aliquots of the effluent should be retained.

9. Elution of the sample from the column can be performed in one of the two ways described as follows:

    (a) For SS biotin labelled cells:

    (i) 1 mL of 100 mM DTT elution buffer should be added and left to incubate on the column for 15 min.

    (ii) This buffer is then drawn off, reserved and further 1 mL of elution buffer applied.

    (iii) This process should be repeated once more and finally 1 mL of regeneration buffer is added and the flow through is retained.

    (iv) The fractions should be pooled to give a final elution fraction of 4 mL.

    (v) The column is regenerated by the addition of 9 mL of regeneration buffer and then stored by adding 5 mL of storage buffer, at 4°C.

    (b) For the LC biotin-labelled samples:

    (i) The elution protocol is the same as above except for the use of 5 mL of 8 M Guanine HCl as the elution buffer.

The protein concentration of the elution fractions is generally quite dilute and, therefore, dialysis and concentration may be necessary (see Note 3).

3.4. SDS–PAGE

This protocol assumes the use of a 16 cm × 20 cm Bio-Rad (Hemel Hempstead, UK) system as this large format system is able to offer maximum resolution of the surface protein enriched samples. The protocol is, however, easily adaptable to other systems such as the 18 cm × 24 cm Hoefer system (Amersham [GE Healthcare], Little Chalfont, UK) by alteration of reagent volumes, but not proportions. This protocol was adapted from (13).

1. Prepare a 1.5-mm thick, 12.5% resolving gel:
   (a) Combine 12.5 mL of 30% Bis-acrylamide (Bio-Rad, Hemel Hempstead, UK), 7.5 mL of 1.5 M Tris–HCl at pH 8.8 (Sigma, Poole, UK), 0.3 mL of 10% SDS (Sigma, Poole, UK), 9.5 mL of ddH$_2$O and 0.15 mL of 10% APS (Bio-Rad, Hemel Hempstead, UK).
   (b) Mix well and then add 0.02 mL Temed (Sigma, Poole, UK).
   (c) Pipette the mixture between clamped plates of appropriate size, pre-cleaned with ddH$_2$O, set in the casting system.
   (d) Gel can then be overlaid with enough ddH$_2$O to cover the exposed top of the gel and to create a straight edge.
   (e) The gel should then be left to set for 1 h.
2. Prepare a 4.5% stacking gel:
   (a) Combine 1.35 mL of 30% Bis-acrylamide (Bio-Rad, Hemel Hempstead, UK), 2.25 mL 0.5 M Tris–HCl pH 6.8 (Sigma, Poole, UK), 0.09 mL of 10% SDS (Sigma, Poole, UK), 5.35 mL of ddH$_2$O and 0.038 mL of 10% APS (Bio-Rad, Hemel Hempstead, UK).
   (b) Mix well and then add 0.011 mL of Temed (Sigma, Poole, UK).
   (c) Remove water from the top of the resolving gel.
   (d) Pipette the mixture between clamped plates to the top of the resolving gel and insert a sample comb with an appropriate number of wells.
   (e) The gel should then be left to set for 30 min.
3. Accurate protein determination of samples will enable equal total protein loading for each sample. To prepare the samples, 3× sample buffer and the appropriate volume of sample should be combined in a ratio of 1:2.
4. The samples should then be mixed and heated to 95°C for 1 min.
5. Once the stacking gel has set, the sample comb can be removed.
6. The gel can then be attached to the cooling unit of the electrophoresis system and placed in the electrophoresis tank.
7. 1× Running buffer is then added such that both the exposed ends of the gel are covered and the top reservoir full.
8. The samples can then be applied to the gel, taking care not to cause any overspill to the neighbouring wells. Allow one well for a pre-stained molecular weight marker.

9. The electrophoresis conditions are dependent on the size of gel and the system used. Generally, the Bio-Rad gels are run for 1 h at 40 mA per gel, with cooling, and then overnight at 7 mA, or they can be run quickly at 50 mA for 4 h with cooling.
10. Once electrophoresis is complete (the dye front should have progressed to approximately 1 in. from the bottom of the glass plates), the gel should be removed from the electrophoresis apparatus and glass plates and the stacking gel are removed.
11. The gel is then put into a clean plastic tray, covered with 100 mL of filtered coomassie blue solution and left to incubate overnight at room temperature on a rocking platform at slow speed.
12. Remove the coomassie solution carefully and wash the gel in de-stain solution until the background is reasonably clear.
13. The gel can then be imaged, bands quantified, or bands cut out for mass spectroscopy (see Note 4). The latter requires all the preparation, running and staining of gels to be performed in a clean room environment to reduce the risk of contamination (see Note 5).

4. Notes

1. If using nonreducible biotin label, EZ-Link Sulfo-NHS-LC-Biotin, 5% β-Mercaptoethanol (Sigma, Poole, UK) can be added to sample buffer as a reductant. If using EZ-Link Sulfo-NHS-SS-Biotin, β- Mercaptoethanol must not be included in sample buffer.
2. Several lysis buffers can be used, but we tested 1% NP40 and 1% OGP. Both of these buffers provide a gentler method of membrane solubilisation than the more frequently used detergent SDS. Protein concentrations were higher in NP40 lysed cells at 2.0 mg/mL, compared to 0.8 mg/mL in OGP treated cells. However, avidin blotting showed that 1% OGP solubilised a greater number and range of biotinylated cell surface proteins. The NP40 lysis appeared to yield biotinylated proteins in the range of 75–160 kDa, whereas OGP lysis allowed resolution of biotinylated proteins from 15 to 160 kDa. The greater yield in range and quantity of biotinylated proteins using OGP lysis buffer is probably attributable to its ability to solubilise membrane proteins sensitive to hydrophobic environments.

3. To dialyse and concentrate the membrane proteins, we found 10,000 Da molecular weight dialysis cassettes (Perbio, Tattenhall, UK), followed by vacuum concentration overnight on a low heat setting, to be sufficient. The fractions can then be stored as a frozen pellet, or, if used immediately, reconstituted in ddH$_2$O. Prior to SDS–PAGE, samples should be quantified using an appropriate assay.

4. Sample bands cut from gels can be identified by Nano-HPLC-ESI-MS/MS, or similar, as a service from many proteomics laboratories.

5. Running and staining of SDS–PAGE gels for protein identification (via mass spectrometry) should be performed in a clean room environment. This location is necessary to minimise the protein contamination of bands cut from the gels, so that any samples being sent for mass spectroscopy analysis are free from contaminating protein as possible. Gel slices should be kept as small as possible whilst still encompassing the whole band of interest, transferred to a microcentrifuge tube and a little running buffer is added to ensure that the gel doesn't dry out in transit. It is often advisable to wrap the top of the tube with parafilm also.

References

1. Wu, C. C., and Yates, J. R., 3rd. (2003) The application of mass spectrometry to membrane proteomics, *Nat Biotechnol* 21, 262–267.
2. Lund,R. Leth-Larsen,R. Jensen, ON. Ditzel, HJ. (2009). Efficient isolation and quantitative proteomic analysis of cancer cell plasma membrane proteins for identification of metastasis-associated cell surface markers. *J Proteome Res.* 8(6):3078–90.
3. Cleverley, R. Saleem, M. Kean, J. Ford, R. Derrick, J. Prince, S. (2008) Selection of membrane protein targets for crystallization using PFO-PAGE electrophoresis. Mol.Mem.Biol. 25, 8, 625–630.
4. Shetty, J. Herr, J. (2009) Methods of Analysis of Sperm Antigens Related to Fertility. Immune Infertility. pp 13–31.
5. Naaby-Hansen, S., Flickinger, C. J., and Herr, J. C. (1997) Two-dimensional gel electrophoretic analysis of vectorially labeled surface proteins of human spermatozoa, *Biol Reprod* 56, 771–787.
6. Busch, G., Hoder, D., Reutter, W., and Tauber, R. (1989) Selective isolation of individual cell surface proteins from tissue culture cells by a cleavable biotin label, *Eur J Cell Biol* 50, 257–262.
7. Zempleni, J. Wijeratne, SS. Hassan, YI. (2009) Biotin. *Biofactors.* Jan-Feb, 35 (1) 36–46.
8. Chandler, C. S., and Ballard, F. J. (1988) Regulation of the breakdown rates of biotin-containing proteins in Swiss 3T3-L1 cells, *Biochem J* 251, 749–755.
9. Robinson, B. H., Oei, J., Saunders, M., and Gravel, R. (1983) [3H]biotin-labeled proteins in cultured human skin fibroblasts from patients with pyruvate carboxylase deficiency, *J Biol Chem* 258, 6660–6664.
10. Viard, M., Blumenthal, R., and Raviv, Y. (2002) Improved separation of integral membrane proteins by continuous elution electrophoresis with simultaneous detergent exchange: application to the purification of the fusion protein of the human immunodeficiency virus type 1, *Electrophoresis* 23, 1659–1666.
11. Naaby-Hansen, S., Waterfield, M. D., and Cramer, R. (2001) Proteomics--post-genomic cartography to understand gene function, *Trends Pharmacol Sci* 22, 376–384.

12. Nagano, K., Masters, J. R., Akpan, A., Yang, A., Corless, S., Wood, C., Hastie, C., Zvelebil, M., Cramer, R., and Naaby-Hansen, S. (2004) Differential protein synthesis and expression levels in normal and neoplastic human prostate cells and their regulation by type I and II interferons, *Oncogene* 23, 1693–1703.

13. Amersham. (2010) Amersham Gel Electrophoresis.

Chapter 33

Development of Rituximab-Resistant B-NHL Clones: An In Vitro Model for Studying Tumor Resistance to Monoclonal Antibody-Mediated Immunotherapy

Ali R. Jazirehi and Benjamin Bonavida

Abstract

Therapeutic strategies for cancer include chemotherapy, immunotherapy, and radiation. Such therapies result in significant short-term clinical responses; however, relapses and recurrences occur with no treatments. Targeted therapies using monoclonal antibodies have improved responses with minimal toxicities. For instance, Rituximab (chimeric anti-CD20 monoclonal antibody) was the first FDA-approved monoclonal antibody for the treatment of patients with non-Hodgkin's lymphoma (NHL). The clinical response was significantly improved when used in combination with chemotherapy. However, a subset of patients does not respond or becomes resistant to further treatment. Rituximab-resistant (RR) clones were used as a model to address the potential mechanisms of resistance. In this chapter, we discuss the underlying molecular mechanisms by which rituximab signals the cells and modifies several intracellular survival/antiapoptotic pathways, leading to its chemo/immunosensitizing activities. RR clones were developed to mimic in vivo resistance observed in patients. In comparison with the sensitive parental cells, the RR clones are refractory to rituximab-mediated cell signaling and chemosensitization. Noteworthy, interference with the hyperactivated survival/antiapoptotic pathways in the RR clones with various pharmacological inhibitors mimicked rituximab effects in the parental cells. The development of RR clones provides a paradigm for studying resistance by other anticancer monoclonal antibodies in various tumor models.

Key words: Rituximab-resistant clones, Bcl-2 family, Methods of resistance

Abbreviations

2MAM-A3	2-Methoxyantimycin-A3
ADCC	Antibody-dependent cell-mediated cytotoxicity
AP-1	Activator protein-1
ARL	Acquired immunodeficiency syndrome (AIDS)-related lymphoma
Bcl-2	B-cell lymphoma protein 2
Bcl-x_L	Bcl-2 related gene (long alternatively spliced variant of Bcl-x gene)
CDC	Complement-dependent cytotoxicity
DHMEQ	Dehydroxymethylepoxyquinomicin
DLBCL	Diffuse large B-cell lymphoma

ERK1/2 MAPK	Extracellular signal-regulated kinase1/2 mitogen activated protein kinase
FACS	Fluorescence-activated cell sorter
IKK	Inhibitor of kappa B (IkB) kinase complex
Mcl-1	Myeloid cell differentiation 1
RKIP	Raf-1 kinase inhibitor protein

1. Introduction

Non-Hodgkin's lymphomas (NHLs) consist of a group of lymphatic cancers of B-and T-cell origin, which are steadily increasing in prevalence worldwide. Although NHLs initially respond to a variety of therapeutic modalities, including combination chemotherapy, they exhibit a relapsing nature and are essentially considered incurable. The failure of standard therapies is due to the selective expansion and outgrowth of drug-resistant variants that also exhibit cross-resistance to other modalities, which highlights the urgent need for the design of new treatment regimens. In recent years, an alternative therapeutic approach to treat malignancies has been the usage of monoclonal antibodies (mAbs) targeted against specific surface markers, which are less systematically toxic and less myelosuppressive. About 80–85% of NHLs are of B-cell origin, and approximately 95% of these cells express surface CD20 (1). The B-cell-specific cell surface marker CD20 is an ideal target for immunotherapy of NHL, as it does not circulate in the plasma as a free protein, which could potentially block antibody (Ab) binding to the cells (2). Also, it is neither internalized upon Ab ligation (3) nor shed from the cell surface (4). Several Abs have been raised against CD20; however, in this chapter we limit our discussion to the role of rituximab, the first FDA-approved mAb for the treatment of NHL (5). While treatment with rituximab or rituximab in combination with CHOP resulted in significant clinical responses, however, a subset of patients initially does not respond or becomes refractory to further treatments. The mechanism of acquisition and/or developing resistance to rituximab is not known. In order to examine the underlying mechanism of resistance, we have adopted the approach of generating, in vitro, rituximab-resistant (RR) clones of B-NHL cell lines as a model that may mimic the development of resistance in patients. This chapter describes our findings comparing wild-type and resistant clones, regarding phenotypic and molecular signaling, sensitization to apoptosis by various chemotherapeutic drugs, and therapeutic interventions to reverse resistance. We also describe, briefly, the methodologies used to generate the rituximab-resistance clones. The findings that emanated from our studies with rituximab offer a model for other therapeutic antibodies currently being used against a variety of tumor cells (6–12).

2. Chimeric Mouse Anti-human CD20 mAb Rituximab (Rituxan, IDEC-C2B8)

Rituximab is a genetically engineered chimeric mouse anti-human CD20 mAb composed of murine light- and heavy-chain variable regions and human gamma 1 heavy-chain and kappa light-chain constant regions (IgG1κ). It binds with high affinity to CD20 expressing cells and is the first FDA-approved mAb for the treatment of NHL (5). Immunotherapy using Rituximab targeting CD20 has been successful in treating B-cell lymphoproliferative diseases as single-agent immunotherapy as well as in combination with chemotherapy (13). High response rates have been reported with rituximab in various NHL subtypes and chronic lymphocytic leukemia (CLL) (14). Various mechanisms have been postulated for the in vivo antilymphoma effects of rituximab including inhibition of cellular proliferation or triggering multiple cell-damaging mechanisms such as antibody-dependent cellular cytotoxicity (ADCC), complement-dependent cytotoxicity (CDC), and apoptosis induction (15). It also potentiates the cytotoxic effects of various chemotherapeutic drugs on drug-resistant NHL B-cells (8).

3. Molecular Mechanisms of Rituximab-Mediated Sensitization to Apoptosis by Chemotherapeutic Drugs

The mechanisms of in vitro and in vivo effectiveness of rituximab in eradicating NHL cells (ADCC, CDC, and apoptosis) have been described elsewhere (16). The in vivo efficacy of the combination of rituximab and drugs in the treatment of drug-resistant tumors suggests that rituximab can modify the drug-resistant phenotype by interfering with signaling pathways and augments drug-induced apoptosis. However, the molecular mechanism by which rituximab interferes with the cellular signaling pathways has been the subject of our research investigations (6–12). To address this issue, we employed two model systems consisting of AIDS-related lymphoma (ARL) and low-grade follicular (FL) lymphoma cell lines and analyzed the molecular effects of rituximab-mediated cell signaling.

3.1. Inhibition of STAT3, p38 MAPK Pathways, and Bcl-2 Downregulation by Rituximab in ARL

Alas et al. (6) originally described one potential signaling pathway negatively regulated by rituximab in 2F7 ARL cell line, which revealed a concentration- and time-dependent downregulation of IL-10 following rituximab treatment. IL-10 is an antiapoptotic protective factor in ARL cells in response to cytotoxic drugs, which utilizes the JAK/STAT pathway mainly through the activation of STAT3. Rituximab decreases the phosphorylation and DNA-binding activity of STAT3, which correlates with a decrease in Bcl-2 expression. Furthermore, it was shown that rituximab significantly inhibits the p38 mitogen-activated protein kinase

(MAPK) and NF-κB survival signaling pathways and in the selective inhibition of the antiapoptotic gene product Bcl-2. The specificity of rituximab-mediated effects was corroborated using various specific pharmacological inhibitors as well as IL-10 neutralizing Abs, all of which reduced the apoptosis threshold of drug-resistant 2F7 ARL cells. Inhibition of Bcl-2 expression by rituximab was largely responsible for sensitizing the cells to apoptosis by various chemotherapeutic drugs such as fludarabine, adriamycin, cisplatin, and etoposide (6, 7, 11). These results suggest that rituximab can trigger the ARL cells and through negative regulation of the IL-10/STAT3/p38MAPK autoregulatory loop decrease the expression of antiapoptotic Bcl-2, thus increasing the sensitivity of the cells to apoptosis induced by various drugs.

3.2. Inhibition of the NF-κB and the ERK1/2 Pathways and Bcl-x_L Downregulation by Rituximab in Low-Grade Follicular Lymphoma

The above studies show that rituximab adversely modulates signaling pathways in EBV$^+$ DLBCL (e.g., ARL cells). Using an EBV$^-$ low-grade FL in vitro model, we conducted additional studies to further determine the signaling pathways and to identify the apoptosis-related gene product(s) regulated by monomeric rituximab. The results of these studies delineated rituximab-mediated sensitization to drug-induced apoptosis of non-ARL Ramos, Daudi, and Raji cells in a synergistic fashion, via downregulation of Bcl-x_L. The FL B-NHL cells underwent apoptosis in response to low concentrations of various chemotherapeutic drugs (paclitaxel, VP-16, CDDP) through the type II mitochondrial apoptotic pathway (8).

We further delineated the signaling pathway(s) used by rituximab for selective inhibition of Bcl-x_L (in Bcl-2 deficient Ramos cells and Bcl-2 expressing Daudi cells) NHL B-cells. Computer analysis revealed NF-κB and AP-1-binding sites in the Bcl-x promoter region, and NF-κB and AP-1, in part, regulate Bcl-x_L gene expression. Thus, we evaluated the regulation of AP-1 by rituximab and its role in chemosensitization. We showed that the ERK1/2 pathway is constitutively hyperactivated in the FL cells and rituximab decreased the phosphorylation-dependent state of the components of the ERK1/2 signaling pathway concomitant with the upregulation of Raf-1 kinase inhibitor protein (RKIP) expression. Induction of RKIP by rituximab enhances its physical association with Raf-1, resulting in decreased activity of the ERK1/2 pathway, diminished AP-1-DNA binding capacity, and downregulation of Bcl-x_L expression and subsequent chemosensitization of the NHL B-cells. Similarly, rituximab-mediated RKIP induction augmented the physical association of RKIP with endogenous NIK, TAK1, and IKK, resulting in decreased activity of the NF-κB pathway and diminished NF-κB transcriptional activity resulting in downregulation of Bcl-x_L expression and subsequent chemosensitization of the NHL B-cell lines (10).

Pharmacological inhibition of the ERK1/2 pathway (e.g., GW5074 [Raf-1 inhibitor], PD098059, U0126 [MEK1/2 inhibitors]), NF-κB pathway (SN-50, DHMEQ, Bay11-9075 [inhibitors of NF-κB nuclear translocation]) or functional impairment of Bcl-x_L (e.g., 2MAM-A3) mimicked the antiproliferative and chemosensitizing effects of rituximab. These findings revealed the ability of rituximab to interrupt the constitutively active ERK1/2 and NF-κB pathways in FL NHL B-cell lines (9, 10). We further determined the direct role of NF-κB in drug resistance by functional block of the NF-κB pathway using cell lines stably transfected with IκB-α superrepressor; these cells were sensitive to drug-induced apoptosis in the absence of rituximab. The presence of two-tandem NF-κB binding sites in the upstream promoter region of the *Bcl-x* gene supported the role of NF-κB in the regulation of Bcl-x_L expression, which was diminished by the deletion of the NF-κbinding sites. We also confirmed the pivotal role of Bcl-x_L in chemoresistance by using Bcl-x_L-overexpressing cells, which exhibited higher drug resistance and were not sensitized by rituximab (10).

Our findings demonstrate that rituximab-mediated Bcl-2 and Bcl-x_L downregulation sensitizes the cells to drug-induced apoptosis. The protective role of Bcl-2 and Bcl-x_L against chemotherapy-triggered apoptosis was confirmed by using 2MAM-A3, which binds to Bcl-2/Bcl-x_L at the hydrophobic groove formed by the highly conserved BH1, BH2, and BH3 domains, thus impairing the antiapoptotic ability of Bcl-2 and Bcl-x_L. In its optimal orientation, 2MAM-A3 and a proapoptotic BH3 peptide occupy overlapping spatial coordinates within the hydrophobic groove and compete for binding (17). Treatment of the ARL and FL cells with 2MAM-A3 significantly enhanced the cytotoxic effects of the drugs and tumor cells underwent apoptosis in response to low and clinically achievable concentrations of various drugs. The above studies have identified several potential targets for therapeutic intervention, namely, the components of the ERK1/2, STAT3, p38 MAPK pathways, Bcl-2, Bcl-x_L, and RKIP, and might provide a rational molecular basis for the therapeutic use of rituximab and/or the inhibitors of these pathways in combination with chemotherapeutic compounds to increase treatment efficacy (9–11).

4. Development of Rituximab-Resistant B-NHL Clones

Despite its well-established clinical efficacy and the superior efficacy of rituximab + CHOP compared to CHOP alone, a subpopulation of patients does not respond to rituximab and/or acquires resistance upon long-term rituximab therapy, and the molecular mechanisms of such resistance are not fully understood (15, 16). Treatment of relapsed disease with another course of rituximab in NHL patients who initially responded to rituximab results in a second response in only 40% of patients (18). Moreover, NHL patients who received eight weekly infusions of rituximab showed only a slightly higher response rate (57%) than a comparable cohort of patients who received four weekly infusions (48%) (19). Hence, insufficient drug dosing alone may not account for de novo or acquired resistance to rituximab in many patients. Thus, it is imperative to delineate the underlying mechanisms of resistance to define novel therapeutic interventions. Owing to difficulties in obtaining patient-derived specimens and delineating the underlying mechanism(s) of acquired resistance, we established an in vitro model of RR NHL cells.

4.1. Methods

RR clones were derived from the rituximab-responsive parental NHL B-cell lines (ARL: 2F7, FL: Ramos and Daudi) by continuous growth of the parental cells [in RPMI1640 + 10% heat-inactivated fetal bovine serum (FBS)] in the presence of stepwise increasing concentrations of rituximab (5–20 μg/mL for 10 weeks) schematically shown in Fig. 1. At the termination of the 10-week period, the cells were counted (trypan blue dye exclusion assay)

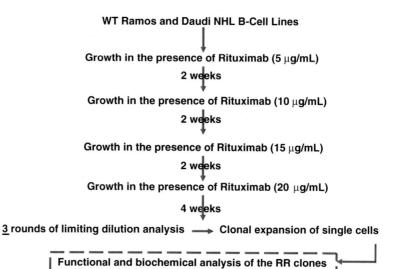

Fig. 1. *Establishment of rituximab-refractory (RR) B-NHL clones.* Rituximab-resistant (RR) clones were derived from the rituximab-sensitive parental NHL B-cell lines (ARL: 2F7, FL: Ramos and Daudi) by continuous growth of the parental cells [in RPMI1640 + 10% heat-inactivated fetal bovine serum (FBS)] in the presence of stepwise increasing concentrations of rituximab (5–20 μg/mL for 10 weeks). At the end of the 10-week period, the cells were counted by the trypan blue dye exclusion assay and were resuspended in fresh medium at a density of 10 cells/mL and 100 μL of the cell suspension was plated in flat-bottom 96-well plates (BD Biosciences) and incubated at 37°C incubator (5% CO_2) to initiate the growth of the cells from a single cell. Upon growth, the cells were again resuspended in medium and were plated in the wells of a flat-bottom 96-well tissue culture plate. A subsequent round of limiting dilution analysis (LDA) (a total of three consecutive rounds) ensured the homogeneity of the cell population. Thereafter, single cells were expanded and subjected to a battery of biochemical and functional assays to compare the parental and RR clones.

and were resuspended in fresh medium (10 cells/mL) and 100 µL of the cell suspension was plated in flat bottom 96-well plates (BD Biosciences) and incubated at 37°C incubator (5% CO_2). Two subsequent rounds of limiting dilution analysis (LDA) ensured the homogeneity of the cell population.

4.2. Molecular Analysis of the RR Clones

Using a battery of functional and biochemical assays [in vitro kinase assay, phospho-specific western blot, propidium iodide staining, FACS analysis, electrophoretic mobility shift assay (EMSA), XTT proliferation assay], representative clones were compared to their respective parental cells to examine alterations in rituximab-mediated effects. The following objectives were investigated: (1) phenotypic and functional properties of the RR clones (e.g., differences regarding CD20 surface expression, proliferation, CDC, cross-linked rituximab-mediated apoptosis), (2) chemosensitivity of the clones and chemosensitization by rituximab, (3) activation status of the survival signaling (STAT3, p38, ERK1/2, and NF-κB) pathways, (4) expression of the Bcl-2 family members (regulators of apoptosis sensitivity), and (5) effects of specific pharmacological inhibitors of the survival pathways on reversal of chemoresistance (10, 12).

4.2.1. Phenotypic and Functional Properties of the RR Clones (e.g., Differences Regarding CD20 Surface Expression, Proliferation, CDC, Cross-linked Rituximab-Mediated Apoptosis)

Initial analysis of a representative RR ARL and FL clone, namely, 2F7-RR1, Ramos-RR1, and Daudi-RR1, revealed reduced surface expression (~50%) of CD20 on the RR1 clones compared to their respective parental lines [as shown by FACS analysis of cells stained with 1 µg/mL anti-CD20 mAb 2B8 (IDEC Pharmaceuticals)]. As the cells were grown in rituximab-free medium for long periods of time prior to analysis, CD20 downregulation is not possibly due to masking of CD20. Similar observation was reported by others (20). Unraveling the significance of reduced surface CD20 expression needs further investigation.

Reducing the proliferation rate of NHL B cells is postulated as an in vivo antilymphoma mechanism of action of rituximab. In vitro, rituximab induces homotypic aggregation of the parental NHL B-cells and reduces their rate of proliferation (48–72%); however, the RR1 clones did not respond to growth inhibition by rituximab (even at fivefold higher concentration) possibly through a defective ceramide (CER)-acid sphingomyelinase (A-SMase) pathway (21). Parental cells underwent apoptosis in the presence of complement (as supplied by human AB serum), the extent of which was dramatically enhanced by rituximab (200 µg/mL) pretreatment; however, the clones were resistant to CDC and increasing the concentration of rituximab (up to 100 µg/mL) did not enhance their sensitivity. Membrane-bound complement regulators (CD46, CD55, and CD59) block complement-mediated cytolysis and are expressed by a variety of solid tumors. Of these, CD59 inhibits complement lysis by blocking the terminal phase of complement activation, which is reversible with an anti-CD59 neutralizing mAb (22). However, the therapeutic efficacy of rituximab is independent of the expression levels of the complement inhibitors (23).

Rituximab induces modest levels of apoptosis, the extent of which is significantly increased upon cross-linking rituximab. Cross-linking rituximab with a secondary goat anti-mouse antibody (0.5–1 μg/mL) induced significant levels of apoptosis in parental cells but not in the RR1 clones. Altogether, these data suggest that the development of RR1 clones is accompanied by phenotypic changes compared to parental cells; they express lower levels of surface CD20 and are no longer responsive to rituximab-mediated inhibition of proliferation, CDC, cross-linked rituximab-induced apoptosis.

4.2.2. Chemosensitivity of the RR1 B-NHL Clones and Chemosensitization by Rituximab

We have reported that rituximab sensitizes the NHL B-cells to chemotherapy-induced apoptosis (8). To investigate whether the RR clones can also be sensitized to chemotherapy, the RR1 clones were pretreated with the optimal concentration of rituximab (20 μg/mL for 24 h) followed by treatment with various drugs. The RR1 clones were not chemosensitized by rituximab even when higher concentrations of rituximab (up to 100 μg/mL) were used. Evaluating the sensitivity of the RR1 clones to various drugs revealed that RR1 exhibit higher drug resistance (1.41–5.1-fold) compared to their respective parental cells (24). A battery of anticancer drugs with distinct intracellular targets including DNA intercalating agents (e.g., CDDP), topoisomerase II inhibitor (e.g., VP-16), and microtubule targeting agents (e.g., paclitaxel and vincristine) were used, and the RR1 clones exhibited higher resistance to all of them, suggesting that the observed resistance might be due to the activation status of the clones on long-term rituximab exposure and/or aberrant cellular signaling, as deregulation of the signal transduction pathways such as the NF-κB, MAPKs, JAK/STAT, AKT/PI3 kinase or aberrant expression of the signaling molecules contribute to the acquired chemoresistance (25, 26).

4.2.3. Activation Status of the Cell Survival Signaling (STAT3, p38, ERK1/2, and NF-κB) Pathways

Rituximab redistributes CD20 to low-density, detergent-insoluble membrane lipid rafts, which serve as platforms for signaling cascades such as Src family kinases including lyn. We observed reduced levels of phospho-Lyn in parental 2F7 cell lines upon rituximab treatment, leading to inhibition of the p38 MAPK and NF-κB pathways and downstream transcription factors SP1 and STAT-3 leading to inhibition of Bcl-2 expression. In parental cells, tumor derived IL-10 acts as growth/antiapoptotic factor, activates STAT3 pathway and STAT3-dependent Bcl-2 expression. Interruption of this autoregulatory loop was largely responsible for increased sensitivity of tumor cells to drugs (11). These events were not observed in the 2F7-RR1 clone: we observed constitutive hyperactivation of the p38 MAPK, NF-κB pathways and increased STAT3 activity leading to overexpression of Bcl-2 and significantly higher resistance of the tumor cells to drug-induced apoptosis (12), and rituximab was incapable of reducing the activity of these pathways, suggesting that the RR1 clones have lost the ability to respond to rituximab-mediated effects. In parental FL cells (Ramos and Daudi), we identified two survival signaling pathways, namely, the ERK1/2 MAPK and the NF-κB, which are constitutively active, leading to the expression of Bcl-x_L. Rituximab (as discussed above) inhibited these pathways and various lines of evidence confirmed the protective role of Bcl-x_L in drug resistance of parental FL cell line (9, 10). Detailed analysis of the signaling pathways in the FL NHL B-clones (Ramos RR1 and Daudi-RR1) revealed that two major survival pathways (NF-κB and ERK1/2) are constitutively hyperactivated in the clones, leading to overexpression of Bcl-2, Bcl-x_L, and Mcl-1 antiapoptosis proteins, suggesting that the selective pressure applied by prolonged rituximab treatment has coselected for NHL B-cells that have constitutive hyperactivated signaling pathways and express higher levels of antiapoptotic proteins, which have lost the capacity to undergo apoptosis in response to various apoptotic stimuli.

4.2.4. Expression of the Bcl-2 Family Members (Regulators of Apoptosis Sensitivity)

In the parental 2F7, Ramos and Daudi, we observed high expression levels of antiapoptotic Bcl-2 and Bcl-x_L, respectively, which were reduced by rituximab pretreatment. Using various approaches, we identified that these antiapoptosis resistant factors were largely responsible for the drug-resistance phenotype of the parental cells and their functional block or reduced expression (by rituximab and inhibitors) sensitized the cells to drug-induced apoptosis (6–11). However, in the RR1 clones, we observed overexpression of the antiapoptotic Bcl-2 (in ARL RR1 clone) and Bcl-2 and Bcl-x_L (in FL RR1 clones) compared to their parental cells. Surprisingly, the expression levels of these antiapoptotic proteins were not modified upon rituximab treatment even at high concentrations. Overexpression of these resistant factors and the inability of rituximab to reduce their expression levels might account for the chemoresistant phenotype of the clones and not being sensitized to drug-induced apoptosis (12, 24), highlighting the need for alternative approaches to reduce the expression level of the antiapoptotic gene products and chemosensitize the drug- and rituximab-resistant NHL B-clones.

4.2.5. Effects of Specific Pharmacological Inhibitors of the Survival Pathways on Reversal of Chemoresistance

Failure of rituximab to reduce Bcl-2 and Bcl-x_L levels accounts for failure of rituximab to chemosensitize 2F7-RR1, Ramos-RR1, and Daudi-RR1 clones. Thus, we used specific pharmacological inhibitors of the hyperactivated signaling pathways. The proteasome inhibitor bortezomib is approved for the treatment of multiple myeloma and has significant single-agent activity against certain subtypes of NHL (27). The unique NF-κB inhibitor, DHMEQ, blocks the nuclear translocation, completely inhibits NF-κB DNA binding activity, inhibits the growth of human hormone-refractory prostate and bladder cancer cells, and at high concentrations induces apoptosis (28, 29). The ERK1/2 inhibitor, PD098059, exerts its effects by specifically binding to the inactive form of MEK1/2 and prevents its activation by Raf-1 (30). These inhibitors blocked the hyperactivated signaling pathways, reduced the Bcl-2 and Bcl-x_L levels, and sensitized the RR1 clones to apoptosis induced by low concentrations and clinically achievable doses of topoisomerase II inhibitor (VP-16), DNA intercalating agents (CDDP), and microtubule poisons (taxol, vincristine), suggesting that deregulated cell signaling leads to overexpression of antiapoptotic proteins in RR1 clones, which in turn results in cross-resistance to various drugs (12, 24). Further, the chemoprotective role of the Bcl-2 family members was confirmed by the use of 2MAM-A3, a specific inhibitor that binds to the hydrophobic groove formed by the highly conserved BH1, BH2, and BH3 domains, thus impairing the antiapoptotic function of Bcl-2, Bcl-x_L, and Mcl-1 (17), which proved necessary to reverse the drug-resistance phenotype.

5. Concluding Remarks

Based on the aforementioned experiments, rituximab can be used as a chemosensitizing as well as an immunosensitizing agent (not discussed here) to reduce the apoptosis threshold of NHL B cells of DLBCL and FL origin. This attribute might explain the superior efficacy of rituximab when used in combination with drugs as opposed to when it is used as a single agent. However, we also need to deal with the issue of acquired/de novo rituximab resistance. Our findings (Fig. 2) suggest that the development of resistance to rituximab may be due, in part, to inability of rituximab to signal the cells both at the cell membrane and intracellularly; thus, the resistant cells become unresponsive to rituximab-mediated cytostasis, CDC, apoptosis, or inhibition of antiapoptotic signaling pathways which lead to modification of the antiapoptotic-resistant factors. However, RR clones are still amenable to chemotherapy using specific molecular targeting of the components of deregulated pathways. These findings provide alternative approaches to address resistance and suggest that interference with specific pharmacologic inhibitors that can modulate survival pathways and downregulate antiapoptotic gene products is capable of reversing resistance when used in combination with subtoxic doses of drugs. These findings also suggest that the combination of the FDA-approved proteasome inhibitor

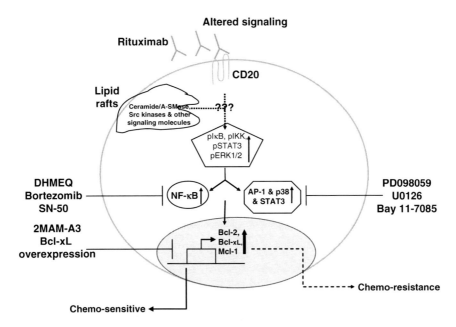

Fig. 2. *Proposed model of the development of RR in NHL B-cells.* Multiple mechanisms have been proposed for the acquired or inherent resistance to rituximab-mediated immunotherapy including (a) shedding of surface CD20 and the presence of circulating CD20 that could potentially bind prematurely to rituximab, (b) overexpression of complement inhibitors (blocking rituximab-mediated CDC), (c) transient or permanent loss of CD20, and (d) polymorphism in Fc receptor (blocks rituximab-mediated ADCC). We propose that altered cell signaling in the tumor cells can also contribute to rituximab resistance. In brief, in cells responding to rituximab-mediated effects inhibition of cellular survival pathways culminates in downregulation of antiapoptotic gene products and potentiation of chemotherapy-induced apoptosis. However, continuous long-term rituximab exposure leads to the development of NHL B-cells do not respond to rituximab's ability to regulate molecular switches leading to constitutive hyperactivation of the NF-κB, ERK1/2, p38, STAT3 survival pathways, resulting in overexpression of antiapoptotic gene products Bcl-2 and Bcl-x_L and increased apoptosis threshold. These cells exhibit significantly higher resistance to a battery of chemotherapeutic drugs and will no longer respond to rituximab-mediated chemosensitization even at high concentrations of rituximab. Targeted therapy using specific pharmacological inhibitors of deregulated cell survival pathways or overexpressed antiapoptotic gene products will reverse the resistance phenotype, and the clones can undergo apoptosis in response to low doses of drugs (see (16) for further information).

bortezomib (Velcade, PS-341) with low-dose chemotherapy may be applied in the treatment of B-NHL patients who are unresponsive to rituximab and/or to combination of rituximab and drugs.

References

1. Coiffier, B., Pfreundschuh, M., Stahel, R., Vose, J., and Zinzani, P.L. (2002) Aggressive lymphoma: improving treatment outcome with rituximab. *Anticancer Drugs* **2**, S43–50.
2. Andeson, K.C., Bates, M.P., Slaughenhoupt, B.L., Pinkus, G.S., Schlossman, S.F., and Nadler L.M. (1984) Expression of human B cell-associated antigens on leukemias and lymphomas: a model of human B cell differentiation. *Blood* **63**, 1424–33.
3. Press, O.W., Appelbaum, F., Ledbetter, J.A., et al. (1987) Monoclonal antibody 1F5 (anti-CD20) serotherapy of human B cell lymphomas. *Blood* **69**, 584–91.
4. Einfeld, D.A., Brown, J.P., Valentine, M.A., Clark, E.A., and Ledbetter, J.A. (1988) Molecular cloning of the human B cell CD20 receptor predicts a hydrophobic protein with multiple transmembrane domains. *EMBO J* **7**, 711–7.

5. Grillo-Lopez, A.J. (2002) Monoclonal antibody therapy for B-cell lymphoma. *Int J Hematol* **76**, 385–93.

6. Alas, S., Emmanouilides, C., and Bonavida, B. (2001) Inhibition of interleukin 10 by rituximab results in down-regulation of Bcl-2 and sensitization of B-cell non-Hodgkin's lymphoma to apoptosis. *Clin Cancer Res* **3**, 709–723.

7. Alas, S., and Bonavida B. (2003) Inhibition of constitutive STAT3 activity sensitizes resistant non-Hodgkin's lymphoma and multiple myeloma to chemotherapeutic drug-mediated apoptosis. *Clin Cancer Res* **1**, 316–326.

8. Jazirehi, A.R., Gan, X-H., De Vos, S., Emmanouilides, C., and Bonavida, B. (2003) Rituximab (anti-CD20) selectively modifies Bcl-$_{xL}$ and Apaf-1 expression and sensitizes human non-Hodgkin's lymphoma B cell lines to paclitaxel-induced apoptosis. *Mol Cancer Ther* **2**, 1183–93.

9. Jazirehi, A.R., Vega, M., Chatterjee, D., Goodglick, L., and Bonavida, B. (2004) Inhibition of the Raf-MEK1/2-ERK1/2 signaling pathway, Bcl-$_{xL}$ down-regulation, and chemo-sensitization of non-Hodgkin's lymphoma B-cells by Rituximab. *Cancer Res* **64**, 7117–26.

10. Jazirehi, A.R., Huerta, S., Cheng, G., and Bonavida, B. (2005) Rituximab (chimeric anti-CD20 mAb) inhibits the constitutive NF-κB signaling pathway in non-Hodgkin's lymphoma (NHL) B-cell lines: role in sensitization to chemotherapeutic drug-induced apoptosis. *Cancer Res* **65**, 264–76.

11. Vega, M.I., Huerta-Yepaz, S., Garban, H., Jazirehi, A.R., Emmanouilides, C., and Bonavida B. (2004) Rituximab inhibits p38 MAPK activity in 2F7 B NHL and decreases IL-10 transcription: pivotal role of p38 MAPK in drug resistance. *Oncogene* **23**, 3530–3540.

12. Vega, M.I., Martinez-Paniagua, M., Jazirehi, A.R., Huerta-Yepez, S., Umezawa, K., Martinez-Maza, O., and Bonavida, B. (2008) The NF-kappaB inhibitors (bortezomib and DHMEQ) sensitise rituximab-resistant AIDS-B-non-Hodgkin lymphoma to apoptosis by various chemotherapeutic drugs. *Leuk Lymphoma* **10**, 1982–94.

13. Czuczman, M.S., Fallon, A., Mohr, A., Stewart, C., Bernstein, Z.P., McCarthy, P., et al. (2002) Rituximab in combination with CHOP or fludarabine in low-grade lymphoma. *Semin Oncol* **1**, 36–40.

14. Coiffier, B., Haioun, C., Ketterer, N., Engert, A., Tilly, H., Ma D, et al. (1998) Rituximab (anti-CD20 monoclonal antibody) for the treatment of patients with relapsing or refractory aggressive lymphoma: a multicenter phase II study. *Blood* **92**, 1927–1932.

15. Maloney, D.G. (2005) Immunotherapy for non-Hodgkin's lymphoma: monoclonal antibodies and vaccines. *J Clin Oncol* **23**, 6421–8.

16. Jazirehi, A.R., and Bonavida, B. (2005) Molecular and cellular signal transduction pathways modulated by rituximab (rituxan, anti-CD20 mAb) in non-Hodgkin's lymphoma: mplications in chemo-sensitization and therapeutic. *Oncogene* **24**, 2121–43.

17. Tzung, S-P., Kim, C.M., Basanez, G., et al. (2001) Antimycin A mimics a cell death-inducing Bcl-2 homology domain 3. *Nat Cell Biol* **3**, 183–91.

18. Reff, M.E., Carner, K., Chambers, K.S., et al. (1994) Depletion of B cells in vivo by a chimeric mouse human monoclonal antibody to CD20. *Blood* **83**, 435–5.

19. McLaughlin, P., Grillo-Lopez, A.J., Link, B.K., Levy, R., Czuczman, M.S., Williams, M.E., et al. (1998) Rituximab chimeric anti-CD20 monoclonal antibody therapy for relapsed indolent lymphoma: half of patients respond to a four-dose treatment program. *J Clin Oncol* **16**, 2825–2833.

20. Pickartz, T., Ringel, F., Wedde, M., Renz, H., Klein, A., von, N.N., et al. (2001) Selection of B-cell chronic lymphocytic leukemia cell variants by therapy with anti-CD20 monoclonal antibody rituximab. *Exp Hematol* **12**: 1410–1416.

21. Bezombes, C., Grazide, S., Garret, C., et al. (2004) Rituximab antiproliferative effect in B-lymphoma cells is associated with acid-sphingomyelinase activation in raft microdomains. *Blood* **104**, 1166–73.

22. Treon, S.P., Mitsiades, C., Mitsiades, N., Young, G., Doss, D., Schlossman, R., et al. (2001) Tumor cell expression of CD59 is associated with resistance to CD20 serotherapy in patients with B-Cell malignancies. *J Immunother* **24**, 263–271.

23. Weng, W.K., and Levy, R. (2001) Expression of complement inhibitors CD46, CD55, and CD59 on tumor cells does not predict clinical outcome after rituximab treatment in follicular non-Hodgkin lymphoma. *Blood* **98**, 1352–1357.

24. Jazirehi, A.R., Vega, M.I., and Bonavida, B. (2007) Development of rituximab-resistant lymphoma clones with altered cell signaling and cross-resistance to chemotherapy. *Cancer Res* **67**, 1270–81.

25. Manshouri, T., Do, K.A., Wang, X., et al. (2003) Circulating CD20 is detectable in the plasma of patients with chronic lymphocytic leukemia and is of prognostic significance. *Blood* **101**, 2507–13.
26. Pommier, Y., Sordet, O., Antony, S., Hayward, R.L., and Kohn, K.W. (2004) Apoptosis defects and chemo-therapy resistance: molecular interaction maps and networks. *Oncogene* **23**, 2934–4.
27. O'Connor, O.A., Wright, J., Moskowitz, C., et al. (2005) Phase II clinical experience with the novel proteasome inhibitor bortezomib in patients with indolent non-Hodgkin's lymphoma and mantle cell lymphoma. *J Clin Oncol* **23**, 676–84.
28. Ariga, A., Namekawa, J., Matsumoto, N., Inoue, J., and Umezawa, K. (2002) Inhibition of tumor necrosis factor--induced nuclear translocation and activation of NF-B by dehydroxymethylepoxyquino- micin. *J Biol Chem* **277**, 24626–30.
29. Kikuchi, E., Horiguhi, Y., Nakashima, J., et al. (2003) Suppression of hormone-refractory prostate cancer by a novel nuclear factor B inhibitor in nude mice. *Cancer Res* **63**, 107–10.
30. Alessi, D.R., Cuenda, A., Cohen, P., Dudley, D.T., and Saltiel, A.R. (1995) PD098059 is a specific inhibitor of the activation of mitogen-activated protein kinase kinase in vitro and in vivo. *J Biol Chem* **270**, 27489–94.

Chapter 34

Analysis of Drug Interactions

Irene V. Bijnsdorp, Elisa Giovannetti, and Godefridus J. Peters

Abstract

Most of the current therapies against cancer, and also those against immune diseases or viral infections, consist of empirically designed combination strategies, combining a variety of therapeutic agents. Drug combinations are widely used because multiple drugs affect multiple targets and cell subpopulations. The primary aim is a mutual enhancement of the therapeutic effects, while other benefits may include decreased side effects and the delay or prevention of drug resistance. The large majority of combination regimens are being developed empirically and there are few experimental studies designed to explore thoroughly different drug combinations, using appropriate methods of analysis. However, the study of patterns of possible metabolic and biological interactions in preclinical models, as well as scheduling, should improve the development of most drug combinations. The definition of synergism is that the combination is more effective than each agent separately, e.g., one of the agents augments the actions of the second drug. The definition of antagonism is that the combination is less effective than the single agents, e.g. one of the agents counteracts the actions of the other. A combination can be studied by combining the two agents in various different ways, such as simultaneous or sequential combination schedules. It is essential to test the potency of a combination, before evaluation in the clinic, to prevent antagonistic actions. However, one should realize that an antagonistic action may be desired when toxicity is concerned, i.e. one drug decreases the side effects of another drug. Several attempts have been made to quantitatively measure the dose–effect relationship of each drug alone and its combinations and to determine whether a given combination would gain a synergistic effect. One of the most widely used ways to evaluate whether a combination is effective is the median-drug effect analysis method. Using this method, a combination index (CI) is calculated from drug cytotoxicity or growth inhibition curves. To calculate a CI, the computer software Calcusyn can be used, taking the entire shape of the growth inhibition curve into account for calculating whether a combination is synergistic, additive, or antagonistic. Here, we describe how combinations can be designed in vitro and how to analyze them using Calcusyn or Compusyn. Moreover, pitfalls, limitations, and advantages of using these combinations and Calcusyn/Compusyn are described.

Key words: Combination study, Synergism, Antagonism, Calcusyn, Drug cytotoxicity

1. Introduction

The objective of any drug combination is to achieve an improved therapeutic result. To optimally design new drug combinations, possible biological mechanisms of action serve as a basis for consideration (1). However, only under well-defined circumstances should the combination of selected drugs be followed from the hypothesis to laboratory experiments to clinical trials. In particular, preclinical studies should evaluate the possible molecular determinants of drug activity (i.e., targets and metabolism) and try to mimic clinical steps (i.e., with clinically achievable concentrations and scheduling). To study drug interactions, standard drug cytotoxicity assays can be performed, such as the sulphorhodamine B assay (2), the MTT assay (2), or the ATP assay (3) as described elsewhere. In these assays, various concentrations of the single agents and the combinations are tested, generating a growth inhibition curve. Several types of combinations can be studied (a) with a fixed constant ratio, (b) with a nonconstant ratio, (c) a simultaneous drug exposure, or (d) a sequential drug exposure. The tested dosing schedule and concentrations should depend on the biological mechanism of action. In addition, possible biological/metabolism interactions between drugs should be taken into account as well. Other endpoints that can be used are cell kill or modulation of a certain target. For some combinations, the aim is to achieve antagonism to decrease the side effects in vivo.

A combination is defined synergistic when the two drugs have a greater effect than the drugs separately (e.g., $1+1=>2$). A combination is antagonistic when the combination is less effective than one of the two agents (e.g., $1+1=\leq 1$). However, owing to the complexity of the whole-cell biological system, the evaluation of synergism and antagonism is not as straight forward as described by these calculations. Therefore, various mathematical methods and programs have been developed to calculate whether a combination is truly synergistic, including the isobologram (4), the fractional effect analysis (5), the response surface approach (6, 7), and the median-drug effect analysis (8, 9). All of these methods have their advantages and disadvantages.

1.1. Isobologram

An isobologram consists of classical plots of each drug alone (10), at a fixed concentration with a variable concentration of the other drug. From these classical curves, iso-effect values (e.g., IC_{50} values) are estimated. These values are used to generate a second curve, the isobologram (Fig. 1). When the combination is synergistic, the data points from the combination will be depicted at the left side of the curve, while the combination is antagonistic when these points are at the right side of the curve. This isobologram, however, does not allow any calculations

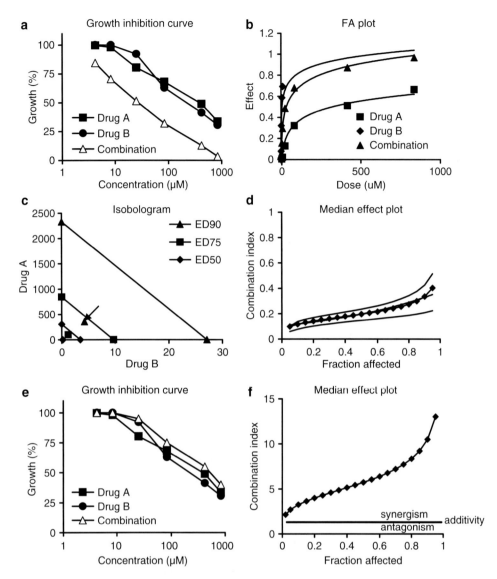

Fig. 1. Example of the growth inhibition curve and matching plots that are generated by Calcusyn. (a) Growth inhibition curve, showing the growth inhibition of drug A, drug B, and the combination of drug A with drug B at constant ratio. This combination is highly synergistic. (b) FA plot showing the dose vs. effect. The effect is synergistic when the effect is greater than the products of the effects of each individual agent, e.g., when $0.5 \times 0.5 < 0.25$. (c) Isobologram, showing synergism, because the observed datapoints are on the left side of the curve. (d) Median effect plot, calculated by Calcusyn, showing strong synergism. (e) Example of an antagonistic growth inhibition curve. (f) The corresponding median effect plot of the antagonistic growth inhibition curve.

about the extent of synergism and does not give confidence intervals. This isobologram can be generated by the computer program Calcusyn or Compusyn.

1.2. Fractional Effect Analysis

The fractional effect analysis (FA) method is one of the most straightforward methods to evaluate a combination (5). The effect is considered synergistic when the observed effect is greater

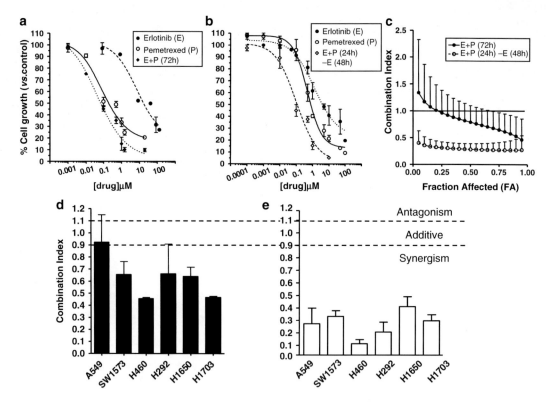

Fig. 2. Example of a presentation of drug interaction studies: cytotoxicity and pharmacological interaction of erlotinib and pemetrexed. (*Top*) (**a**) Representative curves of growth inhibitory effects of erlotinib, pemetrexed, simultaneous 72-h exposure and (**b**) simultaneous–sequential (24-h pemetrexed + erlotinib followed by 48-h erlotinib) combination, and (**c**) combination index (CI) fraction affected (FA) plot of erlotinib and pemetrexed combinations in SW1573 cells. (*Bottom*) Mean CI values of simultaneous (**d**) and simultaneous–sequential (**e**) erlotinib–pemetrexed combination in the panel of NSCLC cells. CI values at FA of 0.5, 0.75, and 0.9 were averaged for each experiment, and this value was used to calculate the mean between experiments. *Points and columns*, mean values obtained from three independent experiments; *bars*, SE. (13).

than the product of the effects of each individual agent, e.g. $0.5 \times 0.5 < 0.25$ (Fig. 2a, 3e). However, it assumes a linear concentration–effect relationship for each agent and the combination. Most of the concentration–effect curves have a sigmoidal shape. Therefore, this model has only a limited applicability. Moreover, confidence intervals are not given. The median-drug effect analysis uses the FA values to calculate synergism.

1.3. The Response Surface Approach

The response surface approach is based on the isobologram. It assumes that the data fit in a sigmoid concentration–effect relationship (7). Data are fitted into an equation using a computer modeling algorithm. Various parameters are evaluated (IC_{50}, sigmoidity, and confidence intervals), resulting in two- (contour plot) or three-dimensional (surface plot) graphical displays (7). These plots give interaction ranges for synergism and antagonism. The method is very complex and difficult to use reliably without

1.4. Median-Drug Effect Analysis (Calcusyn/Compusyn)

sufficient statistical knowledge. Because of this limitation, this method has not been widely used in the literature.

The median-drug effect analysis was originally described by Chou and Talalay (8), which is now the most widely used method in the literature for analyzing combinations. The program Calcusyn (now replaced by Compusyn) provides the software and is considered to be one of the simplest software programs for quantifying synergism or antagonism (9). It assumes that two or more drugs alone, as well as the combination, will result in sigmoid (and not linear) concentration–effect curves. The level of sigmoidity (m) and D_m (is the IC_{50}, which is 50% growth inhibition) can be estimated by first transforming the concentration–effect data to a logarithmic scale. Subsequently, a linear regression on the log-transformed data can be fitted to an equation in which the growth-inhibitory effect of each drug (FA), the sigmoidity of the curves, and the IC_{50} are included to calculate a combination index (CI). This CI is calculated for each FA. These CI values are indicated in the median-drug effect plot (Fig. 1). The program only allows entry of FA values in the range $0.01 < FA < 0.99$. From the median drug effect plots, the dose that reduced absorbance by 50% (D_x) and the slope (m) were calculated. The data were only applicable to this method of analysis when the linear correlation coefficient r of each obtained curve was >0.9. The program uses the formula $D_{1-FA} = D_x[FA/(1-FA)]^{1/m}$ to calculate the doses of the separate drugs and combination required to induce various levels of cytotoxicity. For each level of cytotoxicity a mutually nonexclusive combination index (CI) is calculated using the formula: $CI = [(D)_1/(D_{1-FA})_1] + [(D)_2/(D_{1-FA})_2] + [\alpha(D)_1(D)_2/(D_{1-FA})_1(D_{1-FA})_2]$. The parameters $(D)_1$ and $(D)_2$ represent the doses of the combination of drugs in a fixed ratio, whereas $(D_{1-FA})_1$ and $(D_{1-FA})_2$ are the doses of the individual drugs resulting in the effect $1-FA$ and $\alpha = 1$ for mutually nonexclusive drugs. Experimental conditions with an FA lower than 0.5 are generally considered as less clinically relevant, since it only represents a minor level of growth inhibition. Therefore, it is advisable to calculate the average of the CI values at FA values of 0.5, 0.75, and 0.9. These values can subsequently be evaluated for synergism, additivity, and antagonism (Table 1). It is, therefore, important to define whether the agents act mutually nonexclusive (the drugs have different mechanisms of action or are acting independently) or mutually exclusive (the drugs are operating on the same target, one agent may prevent the action of the other agent).

In this chapter, we describe with several examples and protocols, applications of this method for combination studies with a fixed constant ratio or a nonconstant ratio, with simultaneous drug exposure or sequential drug exposure schedules, and how to analyze these combinations using the program Calcusyn, the pitfalls, limitations, and the advantages of the models.

Table 1
Combination index values and their indication

CI		Synergism/antagonism
<0.1	+++++	Very strong synergism
0.1–0.3	++++	Strong synergism
0.3–0.7	+++	Synergism
0.7–0.85	++	Moderate synergism
0.85–0.9	+	Slight synergism
0.9–1.1		Nearly additive
1.1–1.2	–	Slight antagonism
1.2–1.45	– –	Moderate antagonism
1.45–3.3	– – –	Antagonism
3.3–10	– – – –	Strong antagonism
>10	– – – – –	Very strong antagonism

Simplified CI values and their indication

<0.8	Synergism
0.8–1.2	Additive
>1.2	Antagonism

2. Materials

2.1. Cytotoxicicity Assay

1. 96-Well plates (see Note 1).
2. Cell culture medium.
3. Drugs of interest (see Notes 2–5).
4. Plate reader with various filters for measuring the optical density (OD) or fluorescence intensity, depending on the type of cytotoxicity assay (see Note 5).

2.2. Computer Software Program Calcucyn

This program enables to evaluate drug combinations; it is now distributed by ComboSyn as Compusyn at WWW.ComboSyn.Com. Based on ISI Citation Index and PubMed 2009, the median-effect equation and plot (Chou), and the combination index equation and plot (Chou-Talalay) have been cited in over 3,970 scientific papers published in over 381 different biomedical journals. Applied to in vitro and in vivo data, single drug analysis or combinations up to seven drugs.

3. Methods

3.1. Combinations with a Fixed Constant Ratio

This type of combination is advised when the two agents have the same type of growth inhibition curves (Figs. 1 and 2), with concentrations over a comparable range around the IC_{50}.

1. Seed cells in the desired cell density in a 96-well plate (see Notes 2–4).
2. Determine a concentration range of the single agents to generate a sigmoid-shaped growth inhibition curve. It is preferable to have several data points above the IC_{50} and some points below the IC_{50}, to make the calculation more accurate (see Note 8).
3. It should be taken into account that the concentration range used in vitro is also achievable in vivo. Moreover, the concentrations should not result in a high rate of side effects in vivo.
4. Usually, combinations with a fixed ratio are based on the IC_{50} values of the single drugs (e.g., when the IC_{50} value of drug A is 1 nM and the IC_{50} value of drug B is 5 nM, the molar ratio will be 1:5). Use this ratio for each concentration tested (see Note 8). Dependent on the mechanism other types of ratios can be used, such as IC_{50}:IC_{10} based ratio.
5. In addition to the combination, always include the growth inhibition curves for each single agent.
6. Expose the cells to the drugs for the desired exposure time.
7. Determine the CI.

3.2. Combinations with a Non-fixed Ratio

This type of combination is used when one of the agents is much more active, or one of the agents has an action in a shorter time, while the second agent needs longer time to exert cytotoxicity, for example when one of the agents is given as a bolus injection, while the second by continuous infusion. Then the agents are administered in a non-fixed ratio. Use a fixed concentration for the drug that is administered as a continuous infusion and a non-fixed concentration for the drug that is given as a bolus.

1. Seed cells in the desired cell density in a 96-well plate (see Note 2–4).
2. Determine a concentration range of the single agents consisting of at least six different drug concentrations, generating a growth inhibition curve (Fig. 1) (see Note 7).
3. The concentration range for variable drug (e.g., drug B) should contain values that are nearby the IC_{50} concentration of drug B (when drug A is given at a fixed concentration, a concentration around the IC_{25} is desirable) (see Note 9).
4. Besides the combination, always include the growth inhibition curve for the single drug B and that of the fixed concentration of drug A.

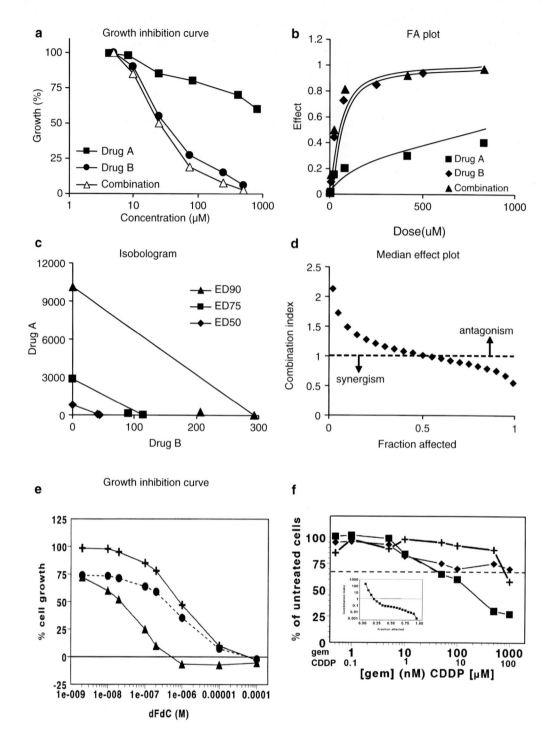

Fig. 3. Example of the plots that are generated by Calcusyn, when a combination is used of which one agent is cell cycle-dependent and needs longer time to exert its activity. (**a**) Growth inhibition curve. When the growth inhibition curves are like this example, it might be better to use a non-fixed concentration. (**b**) FA plot, belonging to the curves of **a**. (**c**) isobologram, belonging to the evaluation of **a** and **b**. (**d**) Median effect plot showing antagonism at the lower concentrations and synergism at the higher concentrations. (**e**) Example of a growth inhibition plot of a non-fixed combination. Concentration–effect curves of the gemcitabine–cisplatin (CDDP) combination. CDDP was used at causing a 25% growth inhibition (0.25 FA). The expected growth curve (—•—) was calculated by multiplying the effect of CDDP by that of gemcitabine (—+—) at each concentration.

5. Expose the cells to the drugs for the desired exposure time.
6. Determine the CI.

3.3. Combinations with a Sequential Schedule

This type of combination is used when one of the tested agents acts much more rapidly than the other. For example when drug A is cell cycle dependent, while drug B is not. Then give the drug A first (Fig. 2b and 3) (see Note 10).

1. Seed cells in the desired cell density in a 96-well plate (see Notes 2–6).
2. Determine a concentration range of the single agents consisting of at least six different drug concentrations, generating a growth inhibition curve (Fig. 1).
3. Determine the type of combination to examine, e.g., fixed ratio or non-fixed ratio.
4. The concentration range for the combination should contain values that are nearby the IC_{50} concentration.
5. Determine the exposure sequence (e.g., which drug first) and the time of preexposure of drug A, depending on the mechanism of action of the drug (see Note 11).
6. After the preincubation time, remove the medium from the plate and add the combination of drug A with B or only drug B. Alternatively, the combination is added to the medium (in a higher concentration because of the dilution). It is not required to wash your cells when the combination is added.
7. Besides the combination, always include the growth inhibition curve for the single agents under similar conditions and plate handling.
8. Make sure that all the controls are included for all the tested exposure times (e.g. when a 24 h preexposure is chosen after which the combination is added for 48 h, drug A as a single agent should be tested for 72 h and drug B as a single agent should be tested for 48 h).
9. Expose the cells to the drugs for the desired exposure time.
10. Determine the CI.

Fig. 3. (continued) Since the observed effect (—▲—) is below that of the expected line, synergy was concluded. Using the median drug effect analysis, the combination index was 0.4±0.02. Reproduced from Peters et al. (1), permission requested from the copyright holder. (f) Example of a non-fixed combination with a sequential drug exposure schedule (—■—). Evaluation of the drug interaction between gemcitabine (—♦—, exposure from 0-96 hr) and CDDP (—+—, from exposure 24-96 hr) when one of the drugs and/or the combination shows cell kill. For this purpose the Y-axis has been adapted, because a normal growth inhibition curve only depicts 0-100%, cell kill would be a negative value which can not be evaluated by the program. Therefore in this curve 0% represents background absorbance, and 100% is the same and represents control growth; the dotted horizontal line separates growth inhibition (higher values) from cell kill (lower values). Median drug effect analysis (insert with CI-FA plot) revealed a mean CI of 0.02. modified from Peters et al (2000), partially from data published in Padrón et al. (1999).

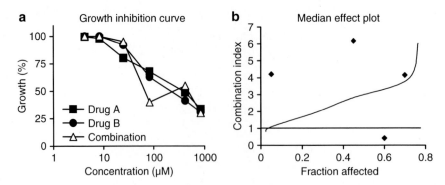

Fig. 4. Example of a poor growth inhibition curve and the corresponding median effect plot that is generated by Calcusyn. (**a**) Growth inhibition curve, where the combination is not representative for analysis, since the line is not smooth. When the growth inhibition curves are like this example, you can remove one data point (e.g., the point of 90 μM) or redo the experiment. (**b**) Median effect plot showing antagonism for most data points, and synergism at FA 0.6, because of the wrong data point.

3.4. Combination Analysis with Calcusyn

1. Calculate the percentage of growth from the growth inhibition graphs (Fig. 1) (see Note 12).
2. From the percentage growth, calculate the fraction affected (FA) by the following formula:

$$FA = 1 - \left(\frac{\% \text{ growth}}{100}\right).$$

3. In the computer program Calcusyn, the FA values need to be entered for each tested concentration of the single compounds and the combination(s) (Fig. 5).
4. When a fixed ratio is used, this can be filled in at the designated box (tick box "constant ratio"). When a non-fixed ratio is used, the ratios between the two drugs need to be entered for each concentration.
5. Each single agent and the combination need to be entered separately for each tested drug concentration. See Fig. 5 for an example to enter the FA in Calcusyn.
6. When the concentrations are entered, various graphs indicate the interaction of the combination. These graphs include the FA, isobologram, and combination index (Fig. 1). These are discussed in the next paragraphs.

3.4.1. FA curves

1. The FA curve is another way used by Calcusyn to depict a growth inhibition curve. The FA curve indicates the effect of each separate agent and the combination (Fig. 1b)
2. Do not include values that are too much at the end of the sigmoid shape of the curve. This is at the part of the graph where growth inhibition is hardly induced (100% growth) or at the maximum (0% growth).

3. Values higher than 100% or negative values cannot be used, e.g., when the growth is higher than 100% (FA = 0) or less than 0% (FA = 1) and cell kill (FA > 1) is induced. In this case, 0.01 or 0.99 can be entered, respectively.

3.4.2. Isobologram

1. This graph depicts the combinations of various doses; it can be used to illustrate additivity, synergism, or antagonism (Fig. 1c).
2. The ED90, ED75, and ED50 points in the graph indicate whether synergy or antagonism occurs. When the points are on the left side of the lines, the combination is synergistic. When the points are on the right side of the lines, the combination is antagonistic.

3.4.3. Median Effect Plot

1. This plot shows the combination index (Y-axis) to the FA values (X-axis) (Fig. 1d).
2. The standard deviation can be shown by inclusion of bars (Fig. 2c) or by an upper and lower line. When the standard deviation is very large, the entered points are possibly not in a smooth sigmoid shape, or the FA values are not entered correctly. If it concerns an outlier data point (Fig. 4), the outlier

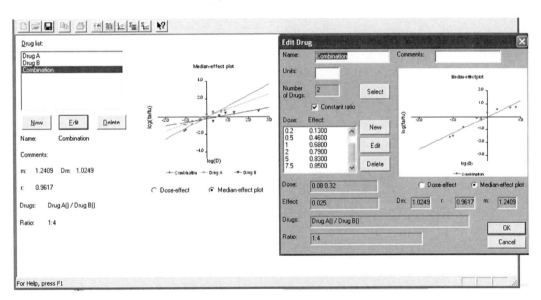

Fig. 5. Example of data points that have to be entered in the software program Calcusyn, showing a combination with a constant ratio of 1:4.

Procedure:
- Enter name of the experiment, the investigator and the date of the experiment.
- Add a new drug (drug A), enter the name of the drug and enter the concentrations and corresponding FA values.
- Add a new drug (drug B), enter the name of the drug and enter the concentrations and corresponding FA values.
- Add a new drug (the combination), enter the name of the combination.
- Select the drugs of the combination and tick the box constant ratio when a constant ratio combination experiment has been performed and enter the ratio.
- Enter the concentrations and the corresponding FA values for the combination.
- After completion, you can select the combination.
- The graphs and the data of all parameters are shown on separate pages

can be removed from the dataset or the experiment should be redone when it concerns more than one data point.

3. It is not preferable to delete any data points in the middle of the dose–effect curve. This can only be done when the data point is highly likely to be incorrect due to culturing or dilution errors (Fig. 4).

4. When the chosen data point resulted in none or very high levels of growth inhibition or even cell kill, these points can be unreliable due to the sensitivity of detection. Therefore, these data points can be deleted.

5. This multiple drug effect model does not allow the evaluation of combinations that lead to a complete growth inhibition or to cell kill. Such combinations can also be evaluated by a modification of growth inhibition plotting (12) (see Note 13; Fig. 3f).

6. In order to evaluate combinations indexes from different experiments, the CI values at FA 0.5, 0.75 and 0.9 of each experiment should be averaged for that specific experiment. This mean is subsequently averaged with the values of the other separate experiments, so that the various experiments can be presented as means ± SEM (Fig. 2d and e).

4. Notes

1. Other plates can also be used, such as a 24-well or a 6-well plate, depending on the type of assay. The 96-well plate is used to determine drug cytotoxicity using the SRB, MTT, or ATP assay. Other plates can be used for other types of assays, such as the 6-well plate for determining cell death with propidium iodine (PI)-stained cells.

2. Many compounds are light-sensitive. When adding such a drug, it is important to switch off the lights of the laminar flow.

3. Many drugs are instable, therefore avoid freeze-thaw cycles, but store aliquots.

4. Keep track of the expiration date of the drugs upon dissolving it, since some drugs cannot be stored a long time in the freezer.

5. For SRB and MTT, usually a filter of 540 or 492 nm is used. For fluorescence it depends on the excitation and emission of that specific dye.

6. To obtain reliable and reproducible results, it is important to start the experiment with cells that are growing within the log-phase. Seed cells in a density, high enough to end the experiment with subconfluent layers for the controls.

7. If the graphs do not have a comparable shape (for example see Fig. 3a), then use a nonconstant ratio concentration. Otherwise the calculation of synergy is inaccurate (Fig. 3).

8. The concentration range for the combination should contain values that are surrounding the IC_{50} concentrations of both drugs.

9. When a non-fixed ratio is used, use a fixed concentration of drug A that induces growth inhibition that is not too high, such as the IC_{25} concentration. When extremely high levels of growth inhibition are induced, interactions cannot accurately be determined. For drug B, make a concentration range which will result in a sigmoid-shaped curve.

10. When a drug needs for example 3 days to exert it cytotoxicity, do not use a 48 h drug exposure time, because the synergistic effect will be masked.

11. When a drug is cell cycle mediated (such as nucleoside analogs), a preincubation time of 24 h can be chosen, a time in which cells have completed at least one division cycle. For rapidly acting drugs, such as the TNF-related apoptosis inducing ligand (TRAIL), a shorter exposure time can be used, such as 24 h using the MTT assay; depending on the assay type used, these time points can vary and have to be determined in advance.

12. Note that negative growth values or values greater than 100% growth cannot be entered into the program Calcusyn for the calculation of the combination index.

13. Evaluation of a combination where cell kill occurs can be done by a different type of plotting. By this modified plotting, the whole of each concentration–effect curve (including the cell kill) is adjusted in a way that each data point will fall within the intervals of $0.01 < FA > 0.99$. This conversion can be done as follows: the entire growth inhibition curve is set to the percent of untreated cells at the end of the experiment, e.g., the untreated control is set to 100%. The background level of the readings of the spectrophotometer is set to 0%. This modification allows the inclusion of data points that otherwise would be excluded because they fall outside the range of effect that can be log-transformed. In this way, a combination that results in cell kill can be assessed, even though the single agents do not (12).

References

1. Peters, G.J., van der Wilt, C.L., van Moorsel, C.J., Kroep, J.R., Bergman, A.M. and Ackland, S.P. (2000) Basis for effective combination cancer chemotherapy with antimetabolites. *Pharmacol. Ther.* **87**, 227–53.

2. Keepers, Y.P., Pizao, P.E., Peters, G.J., van Ark-Otte, J., Winograd, B. and Pinedo, H.M. (1991) Comparison of the sulforhodamine B protein and tetrazolium (MTT) assays for in vitro chemosensitivity testing. *Eur. J. Cancer.* **27**, 897–900.

3. Cree, I.A. and Kurbacher, C.M. (1999) ATP-based tumor chemosensitivity testing: assisting new agent development. *Anticancer Drugs.* **10**, 431–5.

4. Berenbaum, M.C. (1989) What is synergy? (1989) *Pharmacol. Rev.* **41**, 93–141.

5. Webb, J.L. (1963) Effect of more than one inhibitor. In Enzymes and Metabolic Inhibitors. New York: Academic Press. pp. 66–79.

6. Greco, W.R., Bravo, G., and Parsons, J.C. (1995) The search for synergy: a critical review from a response surface perspective. *Pharmacol. Rev.* **47**, 331–385.

7. Kanzawa, F. and Saijo, N. (1997) In vitro interaction between gemcitabine and other anticancer drugs using a novel three-dimensional model. *Semin. Oncol.* **24**, :S7-8-S7-16.

8. Chou, T.C. and Talalay, P. (1984) Quantitative analysis of dose-effect relationships: the combined effects of multiple drugs or enzyme inhibitors. *Adv. Enzyme Regul.* **22**, 27–55.

9. Chou, T.C. (2010) Drug combination studies and their synergy quantification using the Chou-Talalay method. *Cancer Res.* **70**, 440–6.

10. Elion, G.B., Singer, S. and Hitchings G.H. (1954) Antagonists of nucleic acid derivatives: Part VIII. Synergism in combinations of biochemically related antimetabolites. *J. Biol. Chem.* **208**, 477–488.

11. Steel, G.G. and Peckham, M.J. (1979) Exploitable mechanisms in combined radiotherapy-chemotherapy: the concept of additivity. *Int. J. Radiat. Oncol. Biol. Phys.* **5**, 85–91.

12. Padrón, J.M., van Moorsel, C.J., Bergman, A.M., Smitskamp-Wilms, E., van der Wilt, C.L. and Peters, G.J. (1999) Selective cell kill of the combination of gemcitabine and cisplatin in multilayered postconfluent tumor cell cultures. *Anticancer Drugs* **10**, 445–52.

13. Giovannetti, E., Lemos, C., Tekle, C., Smid, K., Nannizzi, S., Rodriguez, J.A., Ricciardi, S., Danesi, R., Giaccone, G. and Peters, G.J. (2008) Molecular mechanisms underlying the synergistic interaction of erlotinib, an epidermal growth factor receptor tyrosine kinase inhibitor, with the multitargeted antifolate pemetrexed in non-small-cell lung cancer cells. *Mol. Pharmacol.* **73**, 1290–300.

Chapter 35

Transfection and DNA-Mediated Gene Transfer

Davide Zecchin and Federica Di Nicolantonio

Abstract

The advent of recent technologies such as gene expression microarrays and high-throughput sequencing methods has allowed for unveiling the molecular complexity of cancer. However, compared to the genomic discovery stage, the functional characterization of genes that have been found altered (by somatic mutations, rearrangements, or copy number variations) or differentially regulated at the expression level is still lagging behind. In the future, it is anticipated that efforts would be aimed at addressing the impact of such genes on several cancer traits, including tumor formation, dissemination, and response to therapies. These studies would likely have to rely on introducing the gene(s) of interest (in its wild-type or altered version) in cellular models. We describe here a number of techniques to introduce nucleic acids into eukaryotic cells, ranging from conventional plasmid transfection to lentiviral transduction and adeno-associated viral (AAV)-mediated DNA transfer.

Key words: Transfection, Plasmid, Lentiviral vectors, Targeted homologous recombination, Knock-in, Cancer mutations

1. Introduction

In the early 1980s, investigators had to face important questions related to the cause of human cancer. At that time it was clear that (1) some retrovirus-derived genes were characterized by transforming properties, (2) most naturally occurring human tumors, nevertheless, did not appear to be caused by viruses, and (3) the transforming genes of acute transforming retroviruses were homologous to DNA sequences present in normal uninfected cells (1).

These evidences raised the hypothesis that the cause of most human cancers can reside within the human genome and, in particular, in the alteration of genes normally present in the DNA of human cells. How to investigate, then, the transforming properties of sequences homologous to the retroviral oncogenes? The answer is this: by "introducing" such sequences within normal cells. Application of this strategy resulted in the discovery of one of the most important human oncogenes: ras (2).

Similar problems and questions, related to the molecular determinants causing human cancers, are still extant. In the last few years, in fact, high-throughput analysis of tumor genomes has led to the identification of a large number of cancer-associated alleles. Assessing their functional role in tumor progression by "introducing" them into appropriate cell models is the next challenge. The most popular methods used nowadays to deliver DNA sequences within target cells are briefly described in this chapter.

Two main approaches can be used to insert cancer-associated mutations in cell models. In the first approach (Fig. 1), the corresponding cDNAs of putative oncogenic alleles are introduced into cells by transfection or viral transduction. However, ectopic expression methods are often hampered by artificially high gene expression levels.

In the second approach (Fig. 1), mutations are introduced into the genomes of human normal or tumor cells using gene targeting methods. Although more laborious and time-consuming than overexpression-based methods, homologous recombination-mediated gene targeting in human somatic cells has proven valuable for a variety of purposes: to express oncogenic alleles from their endogenous promoters (3), to selectively delete the mutated allele of an oncogene (4), to knockout gene function by exon removal, and to delete both alleles of a tumor suppressor gene (5).

It should be noted that not all cell types are successfully amenable to manipulation by the techniques presented in this chapter. Therefore, the most suitable experimental approach has to be chosen in function of the target cells. Scientific literature and Web pages of companies selling transfection–transduction reagents are usually good sources of information for several cell types.

2. Materials

2.1. Transfection by Calcium Phosphate

1. Cell culture medium and supplements recommended for the target cells.
2. Incubator: 37°C, humidified atmosphere of 4–6% CO_2 in air.
3. Purified plasmid DNA.
4. 0.1× TE buffer: 1 mM Tris–HCl, 0.1 mM EDTA. Sterile. Store at +4°C.

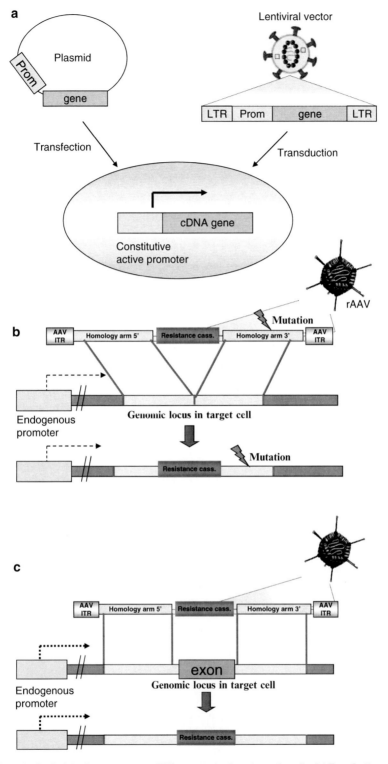

Fig. 1. Different strategies to introduce exogenous DNA sequences in eukaryotic cells. (**a**) Transfection of plasmids and transduction of lentiviral vectors are two strategies employed to introduce the cDNA of a transgene into a target cell. The expression of the cDNA is usually controlled by an exogenous constitutively active promoter collocated upstream of the gene. (**b**) rAAV vectors are able to introduce a specific DNA sequence in the genome of a cell by homologous recombination. This strategy can be employed to insert a genetic alteration into a target gene in a locus-specific manner. The resulting mutated gene will be expressed within the cell under the control of the endogenous promoter. (**c**) rAAV vectors can also be employed to delete a specific sequence from a target gene by homologous recombination. In the figure is represented the deletion of a codifying exon.

5. 2.5 M $CaCl_2$. Filtered using a 0.22-μm filter. Store at −20°C.
6. 2× HBS: 281 mM NaCl, 100 mM HEPES, 1.5 mM Na_2HPO_4, pH 7.12. Filtered using a 0.22-μm filter. Store at −20°C.

2.2. Transfection by Lipofectamine™ 2000

1. Steps 1–3 as in Subheading 2.1.
2. Opti-MEM® I Reduced Serum Media (Invitrogen Corporation, Grand Island, NY, USA).
3. Lipofectamine™ 2000 (Invitrogen).

2.3. Transfection by FuGENE® 6 Transfection Reagent

1. Steps 1–3 as in Subheading 2.1.
2. Dulbecco's Modified Eagle's Medium (DMEM) (Gibco/BRL, Bethesda, MD).
3. FuGENE® 6 transfection reagent (Roche Diagnostics Corp., Indianapolis, IN, USA).

2.4. Generation of Lentiviral Vectors

Important: The biosafety office at your institution must be notified prior to use of this system for permission and for further institution-specific instructions. BL2/(+) conditions should be used at all times when handling lentivirus.

1. Monolayer of 293T cells approximately at 30%–50% confluency.
2. Iscove's Modified Dulbecco's Medium (IMBM) (Invitrogen), containing 10% (v/v) fetal bovine serum (FBS) and antibiotics.
3. CMV-dR8.74 or pMDLg/pRRE + pRSV.REV purified packaging plasmids (Addgene Inc., Cambridge, MA 02139-1666, USA).
4. pMD_2VSV.G purified envelope plasmid (Addgene).
5. Suited transfer lentiviral purified vector.
6. 0.1× TE buffer: 1 mM Tris–HCl, 0.1 mM EDTA. Sterile. Store at +4°C.
7. 2.5 M $CaCl_2$. Filtered using a 0.22-μm filter. Store at −20°C.
8. 2× HBS: 281 mM NaCl, 10.0 mM HEPES, 1.5 mM Na_2HPO_4, pH 7.12. Filtered using a0.22-μm filter. Store at −20°C.

2.5. Concentration of Lentiviral Vectors

1. Ultracentrifuge Beckman Optima XL100K (Beckman Coulter Inc. Brea, CA, USA).
2. Ultracentrifuge tubes in polyallomer (Beckman Coulter).
3. Sterile Phosphate-Buffered Saline (PBS) with 1% (v/v) Bovine Serum Albumin (BSA).

2.6. Generation of rAAV

AAV vectors made in the absence of helper virus are considered BSL1, implying that no special precautions must be taken during handling. This system is now commercially available from Stratagene and has made constructing a rAAV more practicable in laboratories equipped for standard molecular biology and tissue culture procedures.

1. Monolayer of 293T cells approximately at 30%–50% confluency.
2. Iscove's Modified Dulbecco's Medium (IMBM) (Invitrogen), containing 10% (v/v) FBS and antibiotics.
3. Purified pHelper and pAAV RC plasmids (Stratagene, La Jolla, CA).
4. Recombinant pAAV vector harboring the homologous arms flanking a antibiotic cassette (pAAV-MCS backbone plasmid can be purchased from Stratagene).
5. 0.1× TE buffer: 1 mM Tris–HCl, 0.1 mM EDTA. Sterile. Store at +4°C.
6. 2.5 M $CaCl_2$. Filtered using a 0.22-μm filter. Store at −20°C.
7. 2× HBS: 281 mM NaCl, 100 mM HEPES, 1.5 mM Na_2HPO_4, pH 7.12. Filtered using a 0.22-μm filter. Store at −20°C.
8. Dry ice–ethanol bath and the 37°C water bath.

2.7. Infection of Target Cells by rAAV and Selection of Cells in Which the Construct is Integrated

1. Cell culture medium and supplements recommended for the target cells.
2. rAAV preparation from the previous step.
3. Sterile reagent reservoir (suitable for multichannel pipette).
4. 12 channel pipette, 200 μL.
5. 96-well tissue culture plates.
6. Antibiotic for selection (e.g., G418).

2.8. Screening of Site-Specific Recombinant

1. Lyse and Go reagent (Pierce, Rockford, IL, USA).
2. Platinum Taq, 50 mM $MgCl_2$, 10× PCR buffer, 100 mM dNTPs set (Invitrogen). Store at −20°C. Platinum Taq should be conserved within an enzyme cooler.
3. High-purity Dimethyl Sulfoxide (DMSO) solution (Sigma-Aldrich Co). Store at room temperature. Light sensitive.
4. Primers designed for detection of the locus-specific recombination of the construct.
5. 96-well PCR plates.
6. 12 channel pipettes, 200 and 20 μL.
7. PCR thermocycler suited for 96-well PCR plates.
8. Apparatus and reagents for DNA gel electrophoresis.

3. Methods

The process of introducing nucleic acids into eukaryotic cells by nonviral methods is defined as "transfection." This can be achieved by the following methods:

3.1. Nonviral Methods

3.1.1. Transfection by Calcium Phosphate Coprecipitation

Calcium phosphate coprecipitation became a popular transfection technique starting from the early 1970s (6). Calcium phosphate coprecipitation is widely used because the components are easily available and inexpensive, the protocol is easy-to-use, and many different types of cultured cells can be transfected. This technique is based on the combination between HEPES-buffered saline solution (HBS), containing phosphate ions, and a calcium chloride solution containing the DNA to be transfected. The mixture results in the formation of a calcium-phosphate fine precipitate binding the DNA on its surface. Target cells are able to take up some of the precipitate and, with it, also the DNA. This method is particularly suited for the generation of transient transfectants, but in some cases stable transfectants can also be produced.

All of the procedure should be performed under a laminar flow sterile hood.

1. One day before transfection, plate approximately $1–5 \times 10^6$ cells in a 10-cm dish so that cells will be 50–70% confluent at the time of transfection (see Note 1).
2. The next day, 2 h before the transfection, replace the medium using 9 mL of culture medium.
3. Prepare the calcium chloride–DNA solution as follows:
   (a) In a sterile 15-mL Falcon tube, pipette 450 μL of TE 0.1×/sterile water at a 2:1 ratio, containing 10 μg of DNA (see Note 2).
   (b) Add 50 μL of $CaCl_2$ 2.5 M to the solution from step 3a. Immediately vortex the tube and incubate the solution at room temperature for at least 5 min.
4. Add dropwise 500 μL of 2× HBS to the calcium chloride–DNA solution while the tube is vortexing. Immediately remove by pipette the new mixture (1 mL in total) and add it to the 9 mL of complete medium in the cell dish.
5. Gently swirl the dish.
6. Incubate cells at 37°C in a CO_2 incubator overnight.
7. Replace the medium with complete medium (see Note 3).
8. For stable cell lines: Passage cells at an appropriate dilution (e.g., 1:10) into fresh growth medium when they are very close to confluence. Add selective medium (if desired) the following day.

3.1.2. Transfection by Lipid-Based Methods

Lipid-based methods rely on the formation of complexes between the negatively charged backbone of nucleic acid molecules and the positively charged head groups of cationic lipids. These complexes, also called "liposomes," can then fuse with the cellular membrane, releasing the DNA within the target cell.

Lipid-based methods allow higher efficiency of transfection in several different cell types and higher reproducibility, and they can be employed to produce either stable transfectants or transient transfectants. Nevertheless, these approaches are much more expensive than calcium phosphate coprecipitation.

In first-generation commercial lipid methods, transfection should occur in the absence of serum. Next-generation products overcame these limitations, extending the range of target cells to those that are not able to tolerate serum-withdrawal conditions. These methods are currently among the most commonly employed ones for cancer cells, as these are usually resistant to the cytotoxicity induced by lipid compounds. The protocol for the transfection of cells using one of the most popular lipid-based reagents is reported below for a 24-well format:

3.1.2.1. First-Generation Products: Lipofectamine™ 2000

1. Adherent cells: One day before transfection, plate $0.5–2 \times 10^5$ cells in 500 µL of growth medium without antibiotics (see Note 4) so that the cells will be 80–90% confluent (see Note 5) at the time of transfection. In the case of cells in suspension, just prior to preparing complexes, plate $4–8 \times 10^5$ cells in 500 µL of growth medium without antibiotics.

2. For each transfection sample, prepare complexes as follows:

    (a) Preparation solution A: dilute 0.8 µg DNA in 50 µL of Opti-MEM® I Reduced Serum Medium without serum (or other medium without serum). Mix gently.

    (b) Preparation solution B: mix Lipofectamine™ 2000 gently before use and then dilute 2 µL of Lipofectamine™ in 50 µL of Opti-MEM® I Medium. Incubate for 5 min at room temperature. Proceed anyway to step c within 25 min.

    (c) After 5-min incubation, combine solution A with solution B (total volume = 100 µL). Mix gently and incubate for 20 min at room temperature (solution may appear cloudy).

3. Add the 100 µL of complexes to each well containing cells and medium. Mix gently by rocking the plate back and forth.

4. Incubate cells with DNA–Lipofectamine complexes at 37°C in a CO_2 incubator for 3–6 h (see Note 6).

5. Change medium using complete medium with the normal amount of serum and antibiotics.

6. *For stable cell lines.* Passage cells at a 1:10 (or higher dilution) into fresh growth medium 24 h after transfection. Add selective medium (if desired) the following day.

The latest generation lipidic compounds have overcome problems related to the cytotoxicity of the previous reagents. The use of these products has resulted in an increased efficiency of transfection, especially for cells requiring long-term incubations with the classical DNA–liposome complexes. On the other hand, such compounds are usually more expensive than first-generation reagents.

In the following lines, we report the protocol of transfection for one commercial reagent included in this latest list of transfection lipids.

3.1.2.2. Next-Generation Products: FuGENE® 6

1. One day before transfection, plate $1-3 \times 10^5$ cells in a 35-mm culture dish in 2 mL of medium (or a 6-well plate) so that cells will be 50–80% confluent at the time of transfection.
2. Prepare a sterile Eppendorf tube containing 97 μL of DMEM serum-free and antibiotics-free medium and then add 3 μL of FuGENE reagent to the medium. Vortex the solution for 1 s and incubate for 5 min at room temperature.
3. Add 2 μg of DNA solution to the prediluted FuGENE 6 from step 2. Gently tap the tube to mix the content (see Note 7).
4. Incubate 30 min at room temperature.
5. Add the transfection reagent–DNA complex to the cells in a dropwise manner. Swirl the wells or flasks to ensure distribution over the entire plate surface – there is no need to change the medium.
6. Extraction/selection after 48 h.

3.2. Transduction-Based Methods

Viruses have evolved specialized molecular mechanisms to efficiently transport their genomes into the cells they infect. Biologists adapted these mechanisms and developed viral vectors as a tool to stably introduce genetic material into the genome of target cells. Delivery of genes by a virus is termed transduction, and the infected cells are described as transduced.

3.2.1. Lentiviral Vectors

Lentiviruses are a subclass of retroviruses. They have recently been adapted as gene delivery vehicles (vectors), thanks to their ability to integrate into the genome of nondividing cells. The viral genome in the form of RNA is reverse-transcribed when the virus enters the cell to produce DNA, which is then inserted into the genome at a random position. Recombinant lentiviral particles harbor in their genome a transgenic cassette. After the infection of the target cells, the transgene is stably integrated into the genome and expressed under the control of an upstream promoter. Moreover, for safety reasons, virus particles are replication-deficient, so are designed to be unable to replicate in their host after they deliver their transgenic content.

3.2.2. Generation of Lentiviral Vector

Lentivector particles are generated by coexpressing the virion packaging elements and the vector genome in 293T human embryonic kidney cells, used as producer cells (7). In order to produce recombinant lentiviral particles, 293T cells are transiently cotransfected with:

- *Packaging plasmid(s)*. It encodes for elements required for vector packaging such as structural proteins and the enzymes that generate vector particles. A single plasmid (CMV-dR8.74) – second-generation packaging system – or two plasmids (pMDLg/pRRE and pRSV.REV) – third generation – can be used.
- *Transfer vector plasmid*. It is the only genetic material that will be transferred to the target cells. It typically comprises the transgene cassette flanked by *cis*-acting elements necessary for its encapsidation as single-strand RNA, reverse transcription and integration in the host cells.
- *Envelope plasmid pMD$_2$VSV.G*. It encodes for a G envelope protein derived from a heterologous virus, vesicular stomatitis virus (VSV), characterized by high stability and broad tropism for target cells.

This section reports the calcium phosphate-based protocol used to cotransfect 293T cells with three – second-generation packaging system – or four – third-generation plasmids. The entire procedure should be performed under a laminar flow sterile hood.

1. 24 h before transfection, seed 8×10^6 293T cells in a 15-cm dish in 22 mL of culture medium (see Note 8).
2. Change medium 2 h before transfection using IMDM medium complemented with heat-inactivated FBS 10%, glutamine, and antibiotics.
3. Prepare the DNA plasmid mix by adding:
   (a) *Packaging plasmid*. 16.25 μg of CMV-dR8.74 or 12.5 μg of pMDLg/pRRE + 6.25 μg of pRSV.REV (see Note 9).
   (b) *Envelope plasmid*. 9 μg of pMD$_2$VSV.G.
   (c) *Transfer vector plasmid*. 25 μg for transgene + promoter length 1,500 bp 32 μg for transgene + promoter length 3,000 bp.

   The plasmid solution is made up to a final volume of 1.125 mL with 0.1× TE/dH$_2$O 2:1. Finally, add 125 μL of CaCl$_2$ 2.5 M. Leave the mix at room temperature for 5 min.
4. Add dropwise 1.250 mL of 2× HBS to the calcium chloride–DNA solution while the tube is vortexing. Immediately, pipette the new mixture (2.5 mL in total) and add it to the cell dish.
5. Gently swirl the dish.
6. Incubate cells at 37°C in a CO$_2$ incubator for 12–14 h.

7. Replace the medium using complete medium for virus collection to begin.
8. Collect cell supernatant at 24 and 48 h after changing the medium.

3.2.3. Concentration of Lentiviral Vectors (Optional)

1. Collect the 293T cells supernatant, centrifuge at $200 \times g$ for 5 min at room temperature, and filter the supernatant through a 0.22-μm pore filter.
2. Concentrate the conditioned medium by ultracentrifugation at $50,000 \times g$ for 140 min at room temperature.
3. Discard the supernatant by decanting and resuspend the pellet in a small volume (200 μL or less) of PBS containing 1% (v/v) BSA. Pool in a small tube and rotate on a wheel for 1 h at room temperature.
4. Subdivide in small aliquots (20 μL), store at –80°C, and titer after freezing.

3.2.4. Recombinant Adeno-Associated Virus (rAAV)-Based Vectors

The methods so far presented are conceived to induce the (over)expression of a specific gene transfecting or transducing the target cells with the corresponding cDNA. In these systems, the expression is usually driven by an exogenous promoter placed upstream of the cDNA.

In contrast, recombinant Adeno Associated Viruses (rAAVs) are able to mediate homologous recombination between specific sequences (homology arms, around 1,000 bp each) of their genome and the corresponding sequences within the genome of the transduced cell. This technique has been employed to introduce single-nucleotide substitutions or deletions within a target gene, preserving the general structure of the gene and allowing the expression of the altered allele to be controlled by the endogenous promoter (3, 5). These viruses are able to induce homologous recombination at high frequencies also in human cells (homologous recombination occurs in around 1/100 of the cells in which the vector is stably integrated), making the method also useful for human somatic cells.

Homology arms in the genome of the rAAV usually flank an Antibiotic Resistance Cassette that allows the selection of the cells in which the construct have been stably integrated. The locus-specific recombination of the vector is then assayed by PCR-based methods. An overview of the method is given in Fig. 2.

3.2.5. Generation of rAAV

Adeno-associated virus-2 (AAV-2), a small single-stranded DNA virus of the parvovirus family, is naturally replication-deficient, and the productive infection requires provision of several factors in *trans*. The production of viral particles relies on the cotransfection of:

- Recombinant vector, flanked by 2 Inverted Terminal Repeats – ITR – containing all the *cis*-acting elements necessary for replication and packaging.

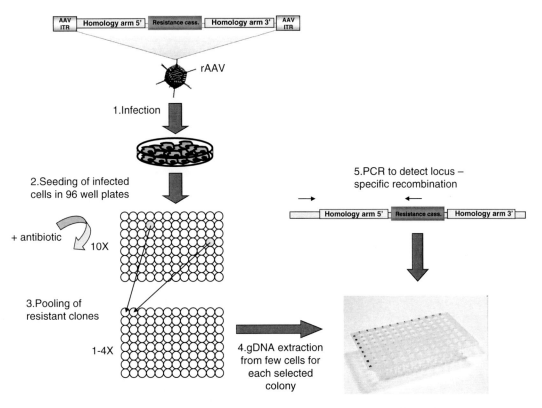

Fig. 2. Overview of the method used to transduce target cells by rAAV and to select recombinant clones.

- pHelper plasmid, carrying adenovirus-derived genes.
- pAAV-RC plasmid, carrying AAV-2 replication and capsid genes.

These three constructs, together, supply all of the factors required for recombinant AAV replication and packaging in the HEK293T cells.

The entire procedure should be performed under a laminar flow sterile hood.

1. 24 h before transfection, seed in two 10-cm dishes 3.5×10^6 293T cells for each.
2. Change medium 2 h before transfection using IMDM medium complemented with heat-inactivated FBS 10% (v/v), glutamine, and antibiotics.
3. Prepare the DNA plasmid mix by adding:
   (a) Recombinant Vector 20 µg
   (b) pHelper 20 µg
   (c) pAAV-RC 20 µg
4. The plasmid solution is made up to a final volume of 900 µL with $0.1 \times$ TE/dH$_2$O 2:1.
5. Finally, add 100 µL of 2.5 M CaCl$_2$.
6. Leave the mix at room temperature for 5 min.

7. Add dropwise 1 mL of 2× HBS to the calcium chloride–DNA solution while the tube is vortexing. Immediately, pipette the new mixture (2 mL in total) and add 1 mL of solution to each of the two dishes.
8. Gently swirl the dish.
9. Incubate cells in a 37°C incubator for 14–16 h.
10. Replace the medium using complete medium for virus collection to begin. Discard the medium as an infectious waste.
11. Return the plate to the 37°C incubator for an additional 48 h.
12. After the incubation, prepare a dry ice–ethanol bath and a 37°C water bath.
13. Collect cells from the plate by repeatedly pipetting vigorously 10 mL of IMBM growth medium on the monolayer while holding the plate inclined. Transfer the transfected cells plus IMBM growth medium to a 15-mL conical tube.
14. Subject the cell suspension to three rounds of freeze–thaw cycle by alternating the tubes between the dry ice–ethanol bath and the 37°C water bath, vortexing briefly after each thaw. Each freeze and each thaw will require approximately a 10-min incubation time.
15. Collect cellular debris by centrifugation at $10,000 \times g$ for 10 min at room temperature.
16. Aliquot the supernatant (primary virus stock, also called rAAV lysate) in sterilized cryovials (1 mL for each aliquot). Store at −80°C.

3.2.6. Infection of Target Cells

1. Day 1: Seed cells in a 10-cm dish to reach 60–80% confluence at day 2.
2. Day 2: Aspirate the media and add to the cells 20–200 μL of rAAV lysate (see Note 10) mixed with 5 mL of the appropriate growth media. The virus is allowed to infect cells at 37°C for 4 h.
3. After 4 h, add 5 mL of growth media to the cells.
4. Day 3: After 16-h incubation, replace rAAV containing media with fresh media.
5. Day 5: Detach infected cells with 1 mL Trypsin 2.5%, EDTA 1%.
6. Resuspend the detached cells in 10 mL serum-containing medium.
7. Transfer 5 mL of cell suspension to a sterilized reagent reservoir and add 95 mL of complete medium containing the appropriate antibiotic for selection (e.g., G418) (see Note 11).
8. Seed the infected cells in five 96-well plates (200 μL/well) using a 12-channel pipettor.
9. Repeat steps 7 and 8 for the remaining 5 mL of cell suspension.

3.2.7. Selection of Cells in Which the Construct is Integrated

1. Drug-resistant colonies have to be selected for 2–3 weeks: during the selection period, change the antibiotic-containing media after time intervals suited for the used antibiotic (e.g., around 2 weeks for G418).
2. To discard the medium, quickly invert the plate over the sink and shake off the medium. Distribute 200 µL/well of fresh medium containing antibiotics using a 12-channel pipettor.
3. At the end of the period of selection, pool together resistant colonies in new 96-well plates.
4. Detach each resistant clone from the original well using 35 µL of Trypsin–EDTA, and transfer 10–15 µL of detached cell solution to a new 96-well-plate containing complete medium with a selection antibiotic (see Note 12).

3.2.8. Screening of Site-Specific Recombinants

1. Detach each selected clone with 35 µL of Trypsin–EDTA (see Notes 13 and 14).
2. Once cells are detached, mix Trypsin–EDTA by pipetting up and down, and transfer 2 µL of detached cells-containing solution in 5 µL of Lyse and Go Lysis Buffer (for extraction of gDNA from single clones). The Lyse and Go buffer should be previously aliquoted in 96-well PCR plates, and the cell suspension is directly added to the buffer within the wells (see Note 15).
3. Add 200 µL of serum-containing medium to the culture wells still containing detached cells and place them back in the incubator at 37°C.
4. Perform the following thermal cycle program on cells + Lyse and Go solutions:

Step	Temperature (°C)	Time (s)
1	65	30
2	8	30
3	65	90
4	97	180
5	8	60
6	65	180
7	97	60
8	65	60
9	80	300

5. After thermal cycle program, add 25 µL of water to each lysate well.
6. Set up the PCR amplification on lysates using a primer that anneals outside the homology region and another one that

anneals within the Selection Cassette. The PCR amplification step should be performed in 96-well PCR plates. Perform the following PCR protocol for each reaction:

10× PCR buffer–MgCl$_2$ (Invitrogen)	1 μL
MgCl$_2$ 50 mM	0.3 μL
DMSO	0.6 μL
dNTPs 2.5 mM (each)	1 μL
Primer F	1 μL
Primer R	1 μL
Water	3 μL
Platinum Taq DNA Polymerase 5 U/μl (Invitrogen Cat.#11304-029)	0.1 μL
Sample (lysate)	2 μL

7. Carry out the following touchdown PCR program on previous reactions (see Note 16):

	Temperature (°C)	Time
1 step	94	120 s
3 cycles	94 64 70	15 s 30 s 90 s
3 cycles	94 61 70	15 s 30 s 90 s
3 cycles	94 58 70	15 s 30 s 90 s
35 cycles	94 57 70	15 s 30 s 90 s
Final	70 4	5 min Forever

8. Run 5 μL of PCR product by agarose gel electrophoresis. A band of the expected size will point out clones in which site-specific recombination may have occurred.

3.2.9. Confirmation and Growing of PCR-Positive Clones

1. Set up two further independent rounds of amplification on PCR-positive cells using the same protocol and the same primers used before.

2. Grow up confirmed clones in 10-cm dishes and freeze.
3. Confirm by sequencing of the genomic DNA locus the site-specific integration of the construct.

4. Notes

1. The number of cells to be plated the day before transfection should be adapted to the rate of growing of the specific cell type employed. It is important to reach a confluency of 50–70% at the time of transfection.

2. Concentration of the plasmid should be determined using 260 nm absorbance. High purity DNA (260/280 nm ratio = 1.8) increases the efficiency of the transfection.

3. A fine precipitate covering the cells should be visible the day after transfection.

4. The use of Amphotericin B (fungizone) in the culture medium should be avoided, as this can result in increased cytotoxicity in the presence of liposomal agents.

5. To obtain highest transfection efficiency and low cytotoxicity, optimize transfection conditions by varying cell density as well as DNA and Lipofectamine™ 2000 concentrations. Increase cell confluency at more than 90% and vary DNA (µg)– Lipofectamine™ 2000 (µL) ratio from 1:0.5 to 1:5.

6. Depending on the resistance to the cytotoxicity induced by the lipidic complexes, some cells can be also incubated overnight to increase the DNA uptake.

7. Optimize transfection conditions by varying DNA–FuGENE ratio from 2:3 to 1:6.

8. Use low-passage 293T cells (not more than P12-15). Do not ever let the cells grow to confluency.

9. Choose the packaging strategy (second or third generation) based on the lentiviral transfer vector used. Lentivectors containing a wt 5′LTR can be packaged only using the second-generation packaging system. If you wish to use the third-generation packaging system, you need to have a lentivector with a chimeric 5′LTR (e.g., CCL-, RRL-, etc.) in which the HIV promoter is replaced with CMV or RSV.

10. When a viral preparation is employed for the first time, the optimal volume to be used for infection should be optimized. Try different doses of the same preparation on separated cell dishes. The optimal dose is the one producing a range of 100–300 resistant clones after antibiotic selection, with an average of one colony for well.

11. Optimal concentration of selective antibiotic should be determined for the considered cell line before the infection. A standard killing curve should be performed, and the lowest effective concentration should be employed for selection.

12. Resistant colonies should be detached and pooled just when they reach 60–90% of confluence within the 96 well.

13. Processing of the cells in 96-well plates, as well as lysis and PCR steps, should be performed using suited 12-channel pipettors to expedite the procedure.

14. Different clones, growing in different wells, have to be maintained rigorously separated during the culture process. Also, lysates and PCR reactions accounting for different colonies have to be maintained separated. Avoid cross-contaminations between different wells during the culture and the screening procedures! Then, once pipettor tips touch cells or cell lysates, replace them with new tips before proceeding to the next wells.

15. Maintain a strict correspondence of wells positions between the 96-well cell culture plate and the corresponding 96-well PCR plate during lysis and PCR processes. This is important to go back to the positive cellular clone looking at the PCR results.

16. The PCR program reported here is suitable for products around 1,000 bp, or even longer. For smaller PCR products, the program should be adapted.

Acknowledgments

This work was supported by a grant from Regione Piemonte n.30258/DB2001 to Dr F. Di Nicolantonio.

References

1. Bishop, J. M. (1981) Enemies within: the genesis of retrovirus oncogenes. *Cell* **23**, 5–6.
2. Cooper, G. M. (1982) Cellular transforming genes. *Science* **217**, 801–6.
3. Di Nicolantonio F, Arena S, Gallicchio M, et al. (2008) Replacement of normal with mutant alleles in the genome of normal human cells unveils mutation-specific drug responses. *Proc Natl Acad Sci USA*.**105**, 20864–9.
4. Shirasawa S, et al. (1993) Altered growth of human colon cancer cell lines disrupted at activated Ki-ras. *Science* **260**, 85–8.
5. Kohli M, et al. (2004) Facile methods for generating human somatic cell gene knockouts using recombinant adeno-associated viruses. *Nucleic Acids Res.* **32**, e3.
6. Graham FL, van der Eb AJ. (1973) A new technique for the assay of infectivity of human adenovirus 5 DNA. *Virology* **52**, 456–67.
7. Dull T, et al. (1998) A third-generation lentivirus vector with a conditional packaging system. *J Virol.* **72**, 873–80.

Chapter 36

Drug Design and Testing: Profiling of Antiproliferative Agents for Cancer Therapy Using a Cell-Based Methyl-[3H]-Thymidine Incorporation Assay

Matthew Griffiths and Hardy Sundaram

Abstract

Drug design is an iterative process requiring cycles of compound synthesis and testing, with each successive synthesis phase yielding molecules predicted to have improved characteristics over the previous set of compounds. In the field of cancer drug discovery, a key early-stage element of the drug design and testing process usually involves the screening of compounds in cell-based in vitro assays. One of the most frequent parameters assessed in cancer drug discovery is the impact of a given molecule on the proliferation of a cancer cell. The methyl-[3H]-thymidine incorporation assay is a widely used, gold standard, method for measuring inhibition of cell proliferation and has been used successfully to screen and optimize potential new cancer drugs. The assay is based on measuring incorporation of methyl-[3H]-thymidine (the radiolabeled form of the DNA precursor thymidine) into the DNA of dividing cancer cells. The screen is used to generate concentration effect relationships for test compounds and for the derivation of IC50 values. IC50 value is defined as the concentration of a test compound required to achieve half maximal inhibition of methyl-[3H]-thymidine incorporation, a parameter that is indicative of antiproliferative potency. IC50 values derived from cell-based assays help drive the medicinal chemistry efforts toward improved drug design, and it is, therefore, critical that the screen provides consistent, robust data over the lifetime of the project – a requirement that necessitates good-quality cell culture practices. The methyl-[3H]-thymidine incorporation assay has been adapted to high-throughput format to facilitate screening of large numbers of compounds. The detailed description of this method, exemplified using the COLO-205 colorectal cancer cell line in a 96-well format, should give the reader a thorough account of how to conduct proliferation assays, as well as some notes and tips on how to ensure success and avoid potential pitfalls.

Key words: Structure–activity relationship, Proliferation assay, Methyl-[3H]-thymidine incorporation, Mitosis, COLO-205 cells

1. Introduction

Drug design and testing is a fundamental process in the discovery of small-molecule drugs. The process often utilizes high-throughput assays to support hit and lead optimization in medicinal chemistry projects. A typical drug design project may last up to several years and involves the use of an enzyme or receptor assay as an initial screen, followed by phenotypic assays to confirm activity of test compounds on the molecular target in an intact cellular system. It is critical that all screens are robust, generating consistent data over the lifetime of the project. The type of screens used to support early stage drug discovery projects are tailored to the disease area and molecular target under interrogation. This process generates structure–activity relationships (SAR), allowing the medicinal chemist to optimize molecules to address a specific molecular property such as cellular potency, efficacy, selectivity, solubility, etc.

One aim of cancer drug discovery is to assess the impact of a molecule on the proliferation of cancer cells. The methyl-[3H]-thymidine incorporation assay is used as an indicator of cellular proliferation and is particularly attractive for screening antimitotic agents for cancer, e.g., identifying inhibitors of the Aurora and Polo-Like kinases. Methyl-[3H]-thymidine is a radiolabelled DNA precursor incorporated into newly synthesized DNA during S-phase of the cell cycle (1, 2). The amount of incorporated methyl-[3H]-thymidine is related to the rate of proliferation; the higher the rate, the higher the signal detected using a scintillation counter. An inhibitor of proliferation (e.g., an antimitotic anticancer agent) will decrease the methyl-[3H]-thymidine signal compared to untreated cells. The methyl-[3H]-thymidine incorporation assay is a powerful and robust method for identifying antiproliferative agents and has been used successfully to screen and optimize potential new drugs (3–5). The success of cell-based screens such as the methyl-[3H]-thymidine incorporation assay is highly dependent on the quality of the cell culture methods employed. We describe in detail the methodologies required for performing a high-throughput (96-well) cell-based methyl-[3H]-thymidine screen using the colorectal cancer cell line COLO-205. This screen has been used to optimize Aurora kinase inhibitors, a novel class of antimitotic agents currently being evaluated in clinical trials (6). Following an iterative drug design and testing program, inhibitors of Aurora kinase were synthesized with improvements in cellular potency using the COLO-205 proliferation assay (7).

2. Materials

The methodologies described below have been optimized for the human COLO-205 cancer cell line. However, these procedures can also be applied to several adherent and nonadherent cell lines, following optimization for the particular cell type of interest (see Note 1).

2.1. Cell Lines

1. COLO-205 cells were obtained from the Health Protection Agency Culture Collections (HPACC), Salisbury, UK (formerly ECACC).
2. Alternative cell line depositories include the American Type Culture Collection (ATCC), Manassas, USA, and the German Collection of Microorganisms and Cell Cultures (DSMZ), Braunschweig, Germany.

2.2. Cell Culture Reagents

1. Complete cell culture medium for COLO-205 cells consisting of RPMI-1640 (SIGMA), 2 mM L-glutamine (SIGMA), Penicillin (100 U/mL)/streptomycin (100 μg/mL) (SIGMA), and 10% fetal bovine serum (SAFC) (see Notes 2 and 3).
2. Dulbecco's phosphate-buffered saline (PBS from SIGMA).
3. Trypsin (1×) (SIGMA).
4. T-175 cm^2 sterile polystyrene surface-treated culture flasks (Corning).
5. 96-Well sterile polystyrene surface-treated culture plates with lids (Corning).
6. 50 mL Sterile "Falcon" tubes.
7. Sterile pipettes – 2–, 5–, 10–, 25–, and 50 mL.
8. Sterile pipette tips.
9. 12-Channel multichannel pipettor, e.g., Finnpipette.
10. Virkon antiviral and antibacterial disinfectant (Invitrogen).
11. Automated liquid aspirator/dispenser, e.g., Pipetboy (Invitrogen).

2.3. Cell Counting

1. Beckman Coulter Vi-CELL (Beckman, High Wycombe, UK).
2. Disposable counting pots (Beckman, High Wycombe, UK).

2.4. Compound Preparation for Screening

1. Reference compounds (see Note 4):
   Doxorubicin hydrochloride (SIGMA).
   cis-Diammineplatinum(II) dichloride (SIGMA) – Dissolved in PBS.
   Etoposide (SIGMA).
2. DMSO (SIGMA) (see Notes 5 and 6).
3. 96-Well sterile polypropylene plates with lids (Corning).

2.5. Methyl-[3H]-Thymidine Assay

1. Dedicated tray for radioactive use (with disposable cover).
2. Methyl-[3H]-thymidine, 370 GBq/mM, 1 mCi/mL (Perkin Elmer, MA, USA) (see Note 7).
3. Complete cell media (see Subheading 2.2).
4. Eppendorf Multipette.

5. Sterile Eppendorf Biopur Combitips (50 µL – Eppendorf, Cambridge, UK).
6. Trypsin (10×) (SIGMA).
7. GF/P30 Filtermat (Perkin Elmer, MA, USA).
8. TomTec Cell harvester (TomTec, CT, USA) (see Note 8).
9. MeltiLex solid scintillant (Perkin Elmer, MA, USA).
10. Wallac MeltiLex Heat Sealer (Perkin Elmer, MA, USA).
11. Plastic sample bag (Perkin Elmer, MA, USA).
12. Bag HeatSealer (Perkin Elmer, MA, USA).
13. MicroBeta Filtermat cassette (Perkin Elmer, MA, USA).
14. MicroBeta liquid scintillation counter (Perkin Elmer, MA, USA).

3. Methods

Several distinct steps are required to screen for potential antimitotic agents using the methyl-[3H]-thymidine incorporation assay. This section describes detailed protocols for each successive stage of the procedure, including methods for (1) culturing and plating COLO-205 cells, (2) assay optimization and validation, and (3) screening compounds.

3.1. Protocols for Growth and Passage of COLO-205 Cells in Cell Culture

Good cell culture practices are key to establishing a stable assay (see Notes 9 and 10).

1. Culture cells in T-175 cm^2 flasks containing 40 mL of media (RPMI-1640, 10% FCS, penicillin/streptomycin) in log phase for at least two doubling times (see Note 11). Check that the cells are healthy and contamination free.

2. Remove cell culture medium to a 50 mL Falcon and pellet the cells by centrifugation at $200 \times g$ for 5 min. Discard the supernatant and resuspend the cell pellet in 10 mL of fresh media.

3. Add 10 mL of PBS to remaining cells in the T-175 cm^2 flask; swirl the contents and discard PBS.

4. Add 5 mL of trypsin to the flask. Thoroughly distribute the trypsin across all cells by gently rocking or swirling the flask. Briefly, return flask to incubator to allow the cells to detach. Lightly tap the flask on each side to disperse the cells and check that the cells have detached from the flask. Add 15 mL of fresh media to inactivate the trypsin and resuspend the cells by repeatedly pipetting up and down with a 5- or 10-mL pipette. Check that a single-cell suspension has been achieved using a microscope (see Note 12).

5. Pool the cells from both the suspension and adherent fractions and count cell number and measure percentage viability. Cells can be subcultured at an appropriate ratio, depending on when the cells will be required. Typically, COLO-205 cells are split twice a week at ratios ranging from 1:8 to 1:12.

3.2. Protocols for Counting Cell Number and Measuring Cell Viability

Counting and viability measurement of COLO-205 cells can be done using an automated Vi-CELL counter. The Vi-CELL utilizes algorithms to identify and count images of live and dead cells, which have been exposed to the cell-impermeable dye, trypan blue. In cases where an automated method of cell counting is not available, a standard hemocytometer-based manual method can be used.

1. Ensure that cells are in the form of a single-cell suspension. Low levels of clumping can be removed by repeated drawing of the cells through a 5– or 10-mL pipette.
2. Measure out approximately 400 µL of cell suspension into a single counting tube. Place tube in the Vi-CELL carousel.
3. Register the sample for counting on the Vi-CELL by selecting the correct program, carousel position, and dilution range. The settings for each cell line should be carefully chosen to reflect differences between different cell lines, e.g., size range (upper and lower) and density (see Note 13).
4. Count the sample. Monitor whether the machine is correctly discriminating between live and dead cells and whether the sample is a single-cell suspension. Repeat the process with an alternative dilution if the resulting cell number is not within the recommended range. Replicate samples are not required as both the percentage viability and the total cell count are derived from an average of 50 independent counts.

3.3. Protocol for Optimizing COLO-205 Cell Seeding Densities to Ensure Log-Phase Growth

1. Prepare a single-cell suspension and measure cell counts/viability as discussed in Subheading 3.2.
2. Dilute cells to 160,000 cells/mL in complete media. Add 200 µL of cells to the top row of a 96-well plate. Aliquot 100 µL of complete media into all other wells. A small number of media-only control wells are required on each plate to act as a blank.
3. Repeatedly, dilute the cell preparation 1 part in 2 down the plate using a 12-well channel pipette, i.e., 100 µL cells added to 100 µL media in the row below. Then, add 50 µL of complete media to all wells. Cover the plate (see Note 14).
4. Incubate the plate overnight at 37°C, 5% CO_2 (see Note 15).
5. Add 50 µL of fresh media to the plate wells to achieve a final volume of 200 µL and incubate for 96 h at 37°C, 5% CO_2.

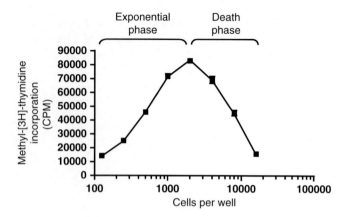

Fig. 1. Example log growth curve for COLO-205 cells. Two distinct phases of cell growth can be observed: exponential growth at low cell densities (<2,000 cells per well) and poor growth at higher seeding densities. Optimal incorporation of methyl-[3H]-thymidine is observed at 1,000 cells per well.

6. Measure methyl-[3H]-thymidine incorporation as described in Subheading 3.6.
7. Plot log cell number against methyl-[3H]-thymidine incorporation (CPM) to identify the concentration of cells at which log growth is achieved (Fig. 1) (see Notes 16 and 17).

Optimal incorporation of methyl-[3H]-thymidine is observed at 1,000 cells per well.

3.4. Protocol for Plating Cells for Screening (Day 1)

1. Prepare a single-cell suspension and measure cell counts/viability as discussed in Subheadings 3.1 and 3.2.
2. Dilute cells to 6,667 cells/mL in complete media. Add 150 µL of cells to columns 1–11 of a sterile 96-well tissue culture treated plate. This will result in 1,000 cells per well (the optimal cell number as determined in Subheading 3.3). The last column (12) is left blank as a control (Fig. 2) (see Notes 18 and 19).
3. Incubate the plate overnight at 37°C, 5% CO_2.

3.5. Protocol for Preparing Test and Control Compounds (Day 2)

This protocol is designed to produce an eight-point, concentration–effect curve. To achieve complete dose–response curves, compounds are typically screened from a final concentration of 20 µM and diluted 1 part in 5 across eight concentrations to give a final lowest concentration of 0.256 nM.

Dependent on the nature of the target or the compound, the serial dilutions or starting concentrations may need to be varied. Each compound is screened in duplicate. A typical plate layout is shown in Fig. 2 (see Note 20).

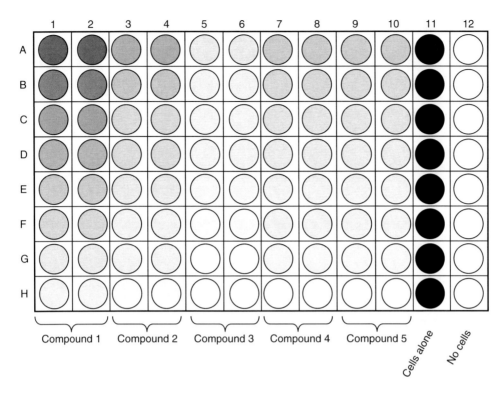

Fig. 2. Generic plate plan for a typical screen. Compounds are arranged in duplicate, with the highest concentrations (typically 20 µM) in row A and the lowest concentration in row H.

1. Dilute compound stock in media. For a 20 µM starting concentration, dilute 8 µL of 10 mM stock compound in DMSO in 1 mL of warmed fresh complete media (1 part in 125 dilution). This produces an 80 µM stock in media.
2. Dispense 200 µL of top concentration (80 µM) compound into the top two wells of a 96-well polypropylene plate (see Note 21).
3. Add 160 µL of complete media to all other wells.
4. Remove 40 µL from the top well with a 12-well pipette and add it to 160 µL in the well below. Repeatedly dilute the compound preparation 1:5 down the plate.
5. Dispense 50 µL of sample from the compound dilution plate into the cell plate. This further dilutes the compound 1 part in 4 to provide a top final concentration of 20 µM in the assay.
6. Incubate for 96 h at 37°C, 5% CO_2.

3.6. Generic Protocol: Methyl-[3H]-Thymidine Assay of COLO-205 Proliferation (Day 6)

1. Dilute methyl-[3H]-thymidine to 7.4 GBq/mM in complete media.
2. Remove cell plates from incubator and dispense 25 µL of diluted methyl-[3H]-thymidine into each well using a

Multipette stepper dispenser. This will achieve approximately 0.5 µCi per well. Replace lid (see Note 22).

3. Incubate for 4–8 h, 37°C, 5% CO_2 in an incubator designated for radioisotope work.
4. Harvest the cells onto a GF/P30 Filtermat using the cell harvester (see Note 23). The filters are prewashed once with deionized water and then washed three times further with the same, following capture of the cell particles on the filtermat.
5. Dry the filtermat at 50°C, 1 h.
6. Melt a MeltiLex solid scintillant strip onto the filtermat using the MeltiLex Heat Sealer.
7. Seal the filtermat into a sample bag with a plastic heat sealer. Cut off excess plastic.
8. Place the sealed bag containing the filtermat into a cassette and read the counts per minute (CPM) in a scintillation counter.

3.7. Data Analysis

3.7.1. Assay Robustness

Assay robustness should be assessed using a number of methods prior to establishing a routine screen. Typical parameters include (1) determining percent coefficient of variance (%CV) for reference compound IC50 values to ensure reproducibility across multiple screens (described in more detail in Subsection 3.7.3) and (2) assessing signal–background ratio and Z-prime (Z'), which provides an indication of overall assay performance. Once the screen is validated and runs routinely, these parameters are also used as quality-control measures (see Note 24). Total uptake (or incorporation) of methyl-[3H]-thymidine is typically in the range of 30,000–50,000 CPM, while the background levels (no cells) are less than 20 CPM. This particular assay generates a signal–background ratio of approximately 2,500:1, which is calculated by dividing the total radioactivity bound to the filter in the presence of COLO-205 cells by the total radioactivity bound to the filter in the absence of cells. However, absolute CPM values will vary depending on the methyl-[3H]-thymidine incubation time (the longer the samples are left, the more radioactivity is incorporated). Thus, although the signal–background ratio historically has been used as an indicator of assay performance, this value is potentially subject to day-to-day variation (see Note 25). The Z-prime (Z') calculation has been widely adopted in recent years to assess assay performance (8). The Z' is a dimensionless number between 0 and 1 and is defined from the mean and standard deviations of both the positive and negative controls. In the specific case of the methyl-[3H]-thymidine proliferation assay, the Z' value is calculated as follows:

$$Z' = 1 - \frac{(3 \times SD_{CPM-TU} + 3 \times SD_{CPM-FU})}{(Mean_{CPM-TU} - Mean_{CPM-FU})}, \quad (1)$$

where SD = standard deviation, CPM–TU = CPM values for total methyl-[3H]-thymidine uptake in the absence of test compound, CPM–FU = CPM values for methyl-[3H]-thymidine binding to filters. Values between 0.5 and 1 can be interpreted as good for use as an assay. Assays with Z′ values between 0.3 and 0.5 are marginal and require further work to minimize variability (8). Typically, the Z′ value for a COLO-205 proliferation screen in our hands is 0.7–0.9.

3.7.2. Generation of IC50 Values

The main objective of the studies outlined above is to generate IC50 values for test compounds. These values are used as an indicator of antiproliferative potency, an important parameter for generating SAR. Data are expressed as % specific thymidine incorporation for each inhibitor concentration as follows:

$$\%\text{Incorporation} = 100 \times \frac{\text{CPM (inhibitor concentration)} - \text{CPM (non-specific binding to filter)}}{\text{CPM (total binding to filter)} - \text{CPM (non-specific binding to filter)}}. \quad (2)$$

Conversion of the raw data to % values corrects for interplate variation, e.g., when comparing raw data for the same compounds across different plates and, furthermore, provides a more uniform way of presenting data during the lifetime of a drug design and testing project. The % values are then plotted against test compound concentration to yield concentration–effect curves. IC50 values are determined using a software package such as GraphPad Prism by applying the four-parameter logistic equation to generate sigmoidal curves (Fig. 3).

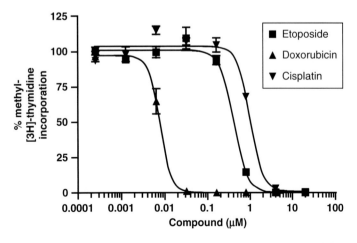

Fig. 3. Sigmoidal concentration–effect curve for control compounds tested on COLO-205 cells. See Table 1 for IC50 values.

Table 1
IC50 values generated for control compounds using the methyl-[3H]-thymidine proliferation assay

	IC50 value (µM)	Standard deviation (SD)	Coefficient of variance (% CV)
Etoposide	0.467 ($n=4$)	0.071	15
Cisplatin	1.050 ($n=4$)	0.078	7
Doxorubicin	0.008 ($n=3$)	<0.001	2

n represents the number of independent experiments for each control

3.7.3. Effect of Reference Compounds and Establishing SAR to Support Drug Design Efforts

The potency of a selection of reference compounds (Subheading 2.4) to inhibit COLO-205 cell proliferation is shown in Table 1. For routine screening, at least one reference compound should be included in each 96-well plate to ensure consistency and reproducibility. The percent coefficient of variance (%CV, defined as SD/mean × 100) for reference compounds should be as low as possible and the assays in which the reference compound deviates from the typical IC50 range should be rejected (see Note 24). Monitoring of reference compound IC50 values over different experiments for several months will help identify potential drift in assay performance – an effect that, if not identified early, will be damaging to the drug design efforts, since the resulting SAR may be misleading.

Data analysis and the determination of IC50 values from a large number of compounds constitute a time-consuming process (e.g., usually greater than 30 compounds are tested in a typical screen). Consequently, it is helpful to automate the process by utilizing databases capable of integrating the large datasets that can be produced from a full drug-discovery program, e.g., Genedata Screener (Genedata AG), ActivityBase (IDBS), or MDL Assay Explorer (Elsevier-MDL).

4. Notes

1. Care should be taken when selecting an individual cell line for a specific screen. The choice should be based on numerous factors such as relevance to disease area, target expression levels, availability of complementary assays used to support the project (e.g., viability assays), cell doubling time, and relevance of the assay for assessing potential patient populations (responder profiling). These factors can highlight a number of different cell lines that may require further testing to decide which

particular cell line is selected. As cell lines have different growth attributes, they all require individual characterization of factors that affect the assay such as doubling time, plating density, intrapassage variability, etc.

2. Different batches of serum can alter the growth characteristics of cells. Test the growth of cells in different batches of serum (some vendors will provide serum batches for evaluation prior to purchase) and ensure that sufficient stocks of an individual batch are secured purely for screening.

3. Regularly check for the presence of mycoplasma contamination. There are numerous tests that offer an accurate and rapid detection of mycoplasma in cultured cell media.

4. Choose readily available reference compounds that give robust, easily interpretable curves and obtain them in sufficient quantities to complete the screening campaign. Controls must always be run to assess both the intra- and interplate variability.

5. DMSO is used as a universal solvent to dissolve both reference and test compounds. It is easiest to prepare a stock dilution plate of test compounds in media, although in circumstances where compounds are poorly soluble in media an intermediate dilution step in DMSO may be required. DMSO stocks are stored at −80°C at 10 mM. Minimize the number of freeze–thaw cycles for each stock by preparing all stocks as single-use aliquots.

6. DMSO can affect cell growth from levels as low as 0.2%. Test the DMSO tolerance of each cell line and include a DMSO dilution curve in each screen.

7. Methyl-[3H]-thymidine emits low-energy beta particles as it decays with a physical half-life of 12.3 years. No shielding is required other than standard personal protective equipment (PPE) as the maximum range in air is 6 mm. However, since methyl-[3H]-thymidine is a DNA precursor and may cause damage if ingested, it is important that the local radioactivity rules are followed at all times.

8. Alternative 96-well harvesters include instruments from Brandell and Packard. The TomTec system described in this chapter has been used in our laboratory for several years and has the advantage that waste radioactivity can be flushed directly into the sink, without any manual intervention.

9. Safety considerations for the growth and culture of cell lines: Suitable PPE should be used at all times. All cell-culture work should be performed within a safety cabinet appropriate to the cell line classification, e.g., Class II. It is important to maintain a sterile working environment through aseptic techniques: single-use plastic pipettes, decontamination of work

surfaces and equipment with alcohol (e.g., 70% ethanol). All biological waste should be disposed of in accordance with local rules, e.g., autoclaving of solids and treating liquid waste with bleach.

10. Cell-culture procedures for efficient storing of cancer cell lines are described elsewhere. The procedures described in Subheading 3.1 assume that cells have been thawed and transferred to cell culture for a few initial passages. Check cell-line growth regularly and discard cell batches if the behavior deviates from normal. COLO-205 cells are "semi-adherent" and have a doubling time of approximately 25 h. The passaging procedures for COLO-205 cells are tailored to maintain the mixed adherent/suspension phenotype. The IncuCyte microscope (Essen Instruments) can be used to noninvasively monitor growth kinetics of cells growing on any clear plasticware, e.g., T-175 cm^2 flasks and is a useful tool for maintaining cells.

11. Prolonged passaging can alter the growth characteristics of COLO-205 cells. It is a good practice to define acceptable limits for culturing cancer cells; in the case of COLO-205s, these cells are grown for 15 cell passages after thawing. In our hands, cells grown within this passage range have behaved consistently in the methyl-[3H]-thymidine incorporation assay. A further tip to maintain consistency of data is to minimize the number of different batches of COLO-205 cells used. This can be achieved by freezing a large number (e.g., >30) of cell-line stocks under liquid nitrogen to last the duration of the screening campaign. Large volumes of viable cells can easily be prepared and frozen into multiple aliquots. Check that these cells recover rapidly after thawing and are viable.

12. Minimize the amount of time that the cells are in contact with trypsin. Some cell lines may require trypsin–EDTA for rapid detachment, while others will detach in EDTA alone.

13. Automated cell counting machines, e.g., Beckman Coulter Vi-CELL (High Wycombe, UK), offer a significant advantage over manual counting with a hemocytometer. In particular, the counting criteria are defined and used throughout a screening campaign by many different operators, thus removing the subjective element involved in manual cell counting. The COLO-205 counting parameters are as follows: size range 5–50 µm, cell brightness 85, cell sharpness 100, viable spot brightness 75%, and viable cell spot area 5%. The optimum counting range for the Vi-CELL is between 0.2 and 5.0×10^6 cells/mL.

14. This produces a 1:2 cell titration from 16,000 cells per well down to 125 cells per well. In most cases, this is sufficient to

identify the optimal cell seeding concentration for the proliferation assay.

15. The overnight incubation allows the cells to adhere to the surface of the 96-well plate and to resume normal growth and proliferation.

16. Choosing a time point for an end-point assay requires a number of scientific and practical factors to be taken into consideration, for example, the length of time it takes for inhibition of the molecular target to impact the cellular readout, signal to noise for the readout, and the number of screens to be run each week. To observe inhibition of proliferation, measurements should be taken at least after one doubling time: In the case of COLO-205 cells, the time course studies showed that time points greater than 48 h were suitable. We decided to use a 96 h incubation time as we were combining our assay with other screens that required a 96 h incubation period. However, both 48 and 72 h have produced good, quality results.

17. Using this procedure, the optimal cell density in our hands for COLO-205 cells is 1,000 cells per well for a 96-h incubation assay. Given that cell growth conditions can vary from lab to lab, it is advisable that this parameter is determined independently for each laboratory. For adapting the methyl-[3H]-thymidine uptake assay for other cell lines, cell titrations should be repeated for each individual cell type, as the growth and proliferation characteristics of individual cells can vary dramatically.

18. Pipette cells carefully to avoid bubbles forming in the well. This does not affect cell proliferation but can impact on the addition of compound.

19. Plating large numbers of cells can be rapidly achieved with an automated liquid dispenser, e.g., Multidrop (Thermo Labsystems). The protocols described in Subheadings 3.4 and 3.5 are suitable for transferring to liquid handling robots, e.g., Beckman, Caliper, PAA, and Tecan systems.

20. Five compounds in duplicate can be independently screened per plate across ten columns of a 96-well plate. The remaining wells are used as plate controls, both positive controls (cells alone with no compounds) and negative (no cells).

21. Check that the compound does not precipitate out of the solution after initial dilution into media and after long-term exposure on cells. Compound precipitation can be easily observed with a bright-field microscope at 20× or 40× magnification, often as either a high density of punctate speckles or alternatively as larger classical crystals. The formation of precipitate will affect the available concentration of free compound and should be noted. Prewarming all media to 37°C

before diluting compounds can reduce the issues with solubility. If this cannot be overcome, then an intermediate dilution curve can be performed in DMSO (or other solvent) prior to dilution in media.

22. Place a small radioactive spill tray covered with a disposable insert in the class II hood. Swab both the tray and surrounding hood to check for potential contamination after each procedure. Local radioactivity rules must be followed at all times.

23. COLO-205 cells are a semiadherent cell line and therefore do not require a pretreatment step with trypsin to dislodge the cells from the 96-well plate. If other adherent cells are used in this protocol, then typically the cells will require trypsinization prior to cell harvesting. Remove media and add 50 µL of trypsin 10× (SIGMA) to each well. Incubate at 37°C for 10 min. Then, harvest cells onto a filtermat.

24. Reference compound IC50 values are generated during the assay development phase (i.e., prior to routine screening) and compared to values reported in the literature (if available). Criteria used for defining acceptable IC50 values for reference compounds can vary depending on the nature of the assay and the rules or preferences employed by the investigator. In order to ensure reproducibility, it is common practice in our laboratory for an additional investigator to independently perform the screen to assess overall consistency in the IC50 values generated. In the case of the methyl-[3H]-thymidine assay, a threefold variation in IC50 value for reference and test compounds is within the typically observed range. Variations of threefold and above should be cause for concern and further investigation.

25. Fluctuations in CPM values from screen to screen or unexpectedly low CPM values may be indicative of problems such as low initial seeding density, culture contamination, or poor viability of cells.

References

1. Friedman, H.M., and Glaubiger, D.L. (1982) Assessment of *in vitro* drug sensitivity of human tumor cells using thymidine incorporation in a modified human tumor stem cell. *Cancer Res.* **42**, 4683–4689.
2. Naito, K., Skog, S., Tribukait. B., Andersson. L., Hisazumi. H. (1987) Cell cycle related [3H]-thymidine uptake and its significance for the incorporation into DNA. *Cell Tissue Kinet.* **20**, 447–457.
3. Harrington, E.A., Bebbington, D., Moore, J., Rasmussen, R.K., Ajose-Abeogun, A., Nakayama, T., Graham, J.A., Demur, C., Hercend, T., Diu-Hercend, A., Su, M., Golec, J.M., Miller, K.M. (2004). VX-680, a potent and selective small-molecule inhibitor of the Aurora kinases suppresses tumour growth in vivo. *Nature Med.* **10**, 262–267.
4. Hennequin, L.F., Thomas, A.P., Johnstone, C., Stokes, E.S.E., Ple, P.A., Lohmann, J.M., Ogilve, D.J., Dukes, M., Wedge, S.R., Curwen, J.O., Kendrew, J., Lambert-van der Bempt, C. (1999). Design and structure-activity relationship of a new class of potent VEGF receptor tyrosine kinase inhibitors. *J. Med. Chem.* **42**, 5369–5389.

5. Ogita, H., Isobe, Y., Takaku, H., Sekine, R., Goto, Y., Misawa, S., Hayashi, H. (2001). Synthesis and structure-activity relationship of diarylamide derivatives as selective inhibitors of the proliferation of human coronary artery smooth muscle cells. *Bioorganic & Med. Chem. Lett.* **11**, 549–551.

6. Boss DS, Beijnen JH and Schellens JH (2009) Clinical experience with aurora kinase inhibitors: a review. *Oncologist* **14**, 780–793.

7. Bebbington, D., Binch, H., Charrier, J.D., Everitt, S., Fraysse, D., Golec, J., Kay, D., Knegtel, R., Mak, C., Mazzei, F., Miller, A., Mortimore, M., O'Donnell, M., Patel, S., Pierard, F., Pinder, J., Pollard, J., Ramaya, S., Robinson, D., Rutherford, A., Studley, J., Westcott, J. (2009) The discovery of the potent aurora inhibitor MK-0457 (VX-680) *Bioorg Med Chem Lett.* **19**, 3586–3592.

8. Zhang, J., Chung, D.Y., Oldenburg, K.R (1999). A simple statistical parameter for use in evaluation and validation of high throughput screening assays. *J. Biomol. Screen.* **4**, 67–73.

Chapter 37

Feeder Layers: Co-culture with Nonneoplastic Cells

Celine Pourreyron, Karin J. Purdie, Stephen A. Watt, and Andrew P. South

Abstract

Maintenance of a mitotically inactive feeder layer which is able to provide extracellular matrix and growth factors can be critical in establishing and maintaining primary tumor cells. How feeder cells are handled and processed is crucial for providing trouble-free support for primary tumor cells and spontaneously immortalized lines.

Key words: 3T3 cells, Feeder cells, Feeder layer, Co-culture

1. Introduction

Feeder layers have long been used for the effective cultivation of primary cells such as epidermal keratinocytes (1) and more recently to support embryonic stem cells (2). Although serum-free, feeder layer free, cell systems have been developed, many laboratories still choose to use feeder cells to support growth and long-term maintenance of primary cells (3). We have used 3T3 feeder layers to effectively support the initial growth of cutaneous squamous cell carcinoma cells which eventually become feeder-independent after a number of passages. The feeder layer also inhibits the proliferation of isolated fibroblasts which have the potential to contaminate and outgrow primary keratinocyte cultures. The method described here is based on the pioneering work of Rheinwald and Green (1). Originally, a clone of Swiss 3T3 cells, J2, was selected for its ability to provide optimal support of keratinocyte growth when used as a feeder layer. However, J2 3T3 cells are hard to come by and NIH 3T3 or Balb/C 3T3 lines work adequately (4). Essentially 3T3 cells are mitotically inactivated, either chemically or by irradiation (see Note 1). The cells survive for a short period of time during which they are able to provide extracellular matrix and growth factor support for the mitotically active primary cells.

2. Materials

1. 3T3 medium: Dulbecco's Modified Eagle's Medium (DMEM) with l-glutamine, 4,500 mg/L d-glucose, 110 mg/L sodium pyruvate (GIBCO/Invitrogen Ltd, Paisley, UK. Cat. No. 41966–029) supplemented with 10% newborn calf serum (NCS, GIBCO/Invitrogen Ltd, Paisley, UK).
2. 1× Trypsin/EDTA solution (GIBCO/Invitrogen, Paisley, UK).
3. Mytomycin C (light sensitive) (2 g, Sigma-Aldrich, Poole, UK).
4. NALGENE Cryo 1°C Freezing Container, "Mr. Frosty" (Thermo Fisher Scientific, Loughborough, UK).

3. Methods

Cultivation of 3T3 Cells

1. On receipt of 3T3 cells they should be rapidly expanded and aliquots frozen down for future use. Cells should be frozen at 10^6 or 2×10^6 cells/mL in 90% NCS/10% dimethylsulfoxide as a cryo-preservant. Freezing is best achieved at a rate of −1°C/min at −80°C with the aid of a cryo freezing container before transfer to liquid nitrogen. 3T3 cells should be maintained in culture for 8–12 weeks and should not be continuously cultured for longer period as they have a tendency to lose contact inhibition and transform or senesce and no longer provide adequate primary cell support.
2. 3T3 cells are grown in 3T3 medium at 37°C in a 5% CO_2 incubator. Cells are disaggregated by incubation with trypsin/EDTA solution at 37°C for 5 min or until the cells are visibly rounded and can be easily detached from the culture dish. The cells are routinely split 1:5 two to three times a week.

Preparation of Feeder Layer: Mitomycin C

1. Adherent cells are incubated with 7 µg/mL mitomycin C for 3–4 h at 37°C in a 5% CO_2 incubator. Cells are then washed three times, harvested with a centrifuge at $500 \times g$ and plated at a density of 10^6 cells/100-mm flask.

Preparation of Feeder Layer: Irradiation

1. Cultures are detached and collected by incubation with trypsin/EDTA solution at 37°C for 5 min in a 5% CO_2 incubator and harvested in a centrifuge at $500 \times g$.
2. Detached cells are irradiated with 6,000 rad (cobalt^{60}) and plated at a density of 10^6 cells/100-mm flask (see Note 2).

Removing Feeder Layers

1. 2–3 days postplating 3T3 feeders will lose their ability to support primary cells (see Notes 3 and 4). Spent feeders can be removed by incubation with EDTA solution at 37°C in a 5% CO_2 incubator.
2. Freshly seeded feeders may take longer period to remove using EDTA or may need a weak trypsin/EDTA solution for 2–3 min. Care is needed not to remove primary cells if trypsin is used.

4. Notes

1. It is important to mitotically inactivate 3T3 cells efficiently. Mitotic 3T3 cells can quickly outgrow primary cultures and can be near impossible to remove without single cell cloning. Once 3T3 cells transform they become resistant to mitomycin C treatment and irradiation at the stated doses. It is important to grow 3T3 cells in 3T3 medium as the presence of FBS and growth factors will encourage transformation. Transformation can be observed after the loss of contact inhibition, the formation of foci, high proliferation, and smaller morphology. Figure 1 depicts foci forming 3T3 cells after continuous growth in keratinocyte medium.
2. Feeders can be stored in suspension at +4°C for up to 48 h after preparation but the length of time they can support primary cultures will be reduced.
3. The number of feeders needed to support primary keratinocyte growth is dependent on keratinocyte density. 3T3 cells are seeded at a maximum density of 10^6 cells/100-mm flask but as the keratinocytes proliferate, this number can be reduced. When keratinocyte cultures are over 70% confluent, feeders can be removed. The keratinocytes will then reach 100% confluence in 24–48 h.
4. Determining whether neoplastic cells are feeder-independent is a process of trial and error and is very much cell line-dependent. In many cases, feeder cells may continue to stimulate cell proliferation even when a line has lost its absolute requirement

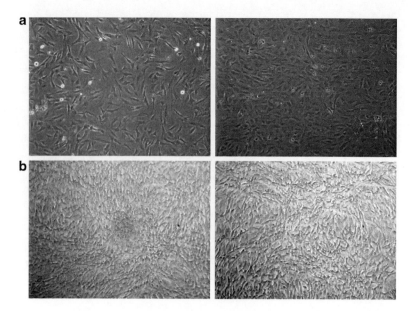

Fig. 1. *3T3 feeders can transform in culture over time.* Examples of mitomycin C responsive 3T3 cells (*top panel*) compared with mitomycin unresponsive 3T3 cells (*bottom panel*). Unresponsive 3T3 cells lose contact inhibition and increase proliferation.

for their presence. As a rule, once the tumor cells are growing well, cultures are split into flasks with and without feeders. Feeder-dependent cultures are apparent after 24 h culture on plastic alone.

Acknowledgments

CP, SAW, and APS are funded by the Dystrophic Epidermolysis Bullosa Research Association. KJP is funded by Cancer Research UK.

References

1. Rheinwald, J. G., and Green, H. (1975) Serial cultivation of strains of human epidermal keratinocytes: the formation of keratinizing colonies from single cells. *Cell* **6**, 331–43.
2. Thomson, J. A., Itskovitz-Eldor, J., Shapiro, S. S., Waknitz, M. A., Swiergiel, J. J., Marshall, V. S., and Jones, J. M. (1998) Embryonic stem cell lines derived from human blastocysts. *Science* **282**, 1145–7.
3. Cobleigh, M. A., Kennedy, J. L., Wong, A. C., Hill, J. H., Lindholm, K. M., Tiesenga, J. E., Kiang, R., Applebaum, E. L., and McGuire, W. P. (1987) Primary culture of squamous head and neck cancer with and without 3T3 fibroblasts and effect of clinical tumor characteristics on growth in vitro. *Cancer* **59**, 1732–8.
4. Rheinwald, J. G. (1989) Human epidermal keratinocyte cell culture and xenograft systems: applications in the detection of potential chemical carcinogens and the study of epidermal transformation. *Prog Clin Biol Res* **298**, 113–25.

Chapter 38

Xenotransplantation of Breast Cancers

Massimiliano Cariati, Rebecca Marlow, and Gabriela Dontu

Abstract

Three experimental systems based on mouse models are currently used to study breast cancer: transgenic mice, carcinogen-induced models, and xenografts of breast cancers. Each of these models has advantages and limitations. This chapter focuses on xenotransplantation of breast cancers and reviews the techniques used so far in establishing this model, the advantages and limitations compared to other experimental systems, and finally, the technical questions that remain to be answered.

Key words: Breast cancer, Xenografts, Xenotransplantation, Animal models

1. Introduction

Animal models have contributed significantly to our understanding of breast development and breast cancer biology. The uniqueness of animal models lies in the ability to reproduce to some extent the systemic and local environment of the developing mammary gland or tumour. The presence of the endocrine, immunologic, and stromal context provides the ideal setting for the study of developmental stages of mammary gland formation, especially thanks to a unique feature of mammary gland growth; postnatal development (1, 2). Animal models have also been instrumental in developing new therapeutic strategies, particularly in understanding host-related pharmacologic/pharmacokinetic effects (1).

The gold standard for the in vivo study of mammary development is represented by the "cleared mammary fat pad" transplantation, a technique routinely used since the pioneering work of Deome and colleagues (3). In the fourth abdominal mammary fat pad of the 3-week-old mouse, the mammary epithelium is still

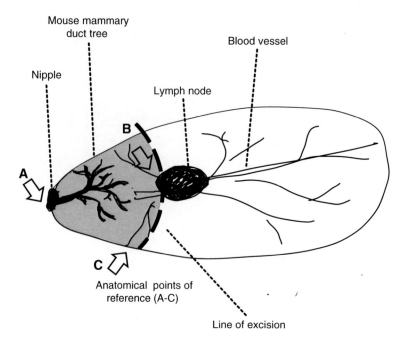

Fig. 1. Clearing of the mouse mammary fat pad. In pubescent (3-week-old) female mice, the developing duct tree does not extend past the lymph node into the fourth pair of mammary fat pads. Removal of the triangular area (*shadowed*), delineated by the nipple, the junction of lymph node and blood vessel, and the blood vessel and the edge of the fat pad, towards the ventral bridge between the two 4th pads (indicated witharrows A–C). Removal of this area is done by careful excision. The remaining fat pad tissue represents the "cleared" mammary fat pad into which cells can be transplanted without interference from the host's mammary gland.

concentrated in the nipple area and it has not grown yet beyond the mammary lymph node (Fig. 1). This allows the fat pad to be "cleared" by surgically removing the area from the nipple to the lymph node, therefore leaving the fat pad free of endogenous epithelium. Transplantation of cells in limiting dilutions in the cleared fat pad has become a key assay for measuring in vivo cell proliferation potential in mammary stem cell biology (4). This model was instrumental in validating the phenotype associated with mammary stem cells of mouse and human origin (5–10).

Three experimental systems based on mouse models are currently used to study breast cancer: transgenic mice, carcinogen-induced models, and xenografts of breast cancers. Each of these models has advantages and limitations. This chapter focuses on xenotransplantation of breast cancers and reviews the techniques used so far in establishing this model, the advantages and limitations compared to other experimental systems, and finally, the technical questions that remain to be answered.

The transplantation of human established cell lines and tumour samples into immunodeficient mice (xenografting) has the advantage of using material originating from clinical samples and has become gold standard for the study of tumour progression and the development of therapeutic strategies (1).

In order to achieve successful xenograft survival, the mouse immune system has to be suppressed to prevent a reaction against allogenic antigens. To achieve this immunosuppression, mutations at approximately 30 loci have been shown to reduce immune function in mice (11). The immunodeficient mouse strains most commonly used to generate hosts for xenografting are nude (nu), beige (bg), severe combined immunodeficiency (scid), and X-linked immunodeficiency (xid) (Table 1). More recently NOD/scid (non-obese diabetic scid) and NOG mice (NOD/Shi-*scid*/IL-2Rγ^{null}) mice were increasingly used in xenotransplantation studies. The success rate of xeno-engraftment is considerably increased in these mice. In NOG mice, all tumour subfractions appear to have similar tumorigenic potential, at least for certain cancers, particularly melanoma (12).

Human tumour xenografts can be generated through several different approaches. The most commonly and successfully explored utilizes the injection of established human breast cancer cell lines, re-suspended in suitable media, metatopically (subcutaneous) or orthotopically (in the mammary fat pad). The mammary fat pad provides an orthotopic site that is readily accessible. Orthotopically implanted tumours seem to display a more biologically relevant phenotype and a better model for drug testing. (13, 14). During the late 1970s, studies using direct tumour implantation into

Table 1
Strains of mice used xenotransplantation of breast cancers

Strain	Inbred/outbred	Mutation	Deficit	Reference
Balb/c Nude	Inbred	Nu	T-cell	(25)
Nu/nu	Outbred	Nu	T-cell	(26)
NIH-III	Outbred	Nu/bg/xid	T-cell, NK, B-cell immaturity	(27)
SCID	Inbred	SCID	B-cell, T-cell	(28)
SCID beige	Inbred	SCID/bg	B-cell, T-cell, NK	(27)
NOD/SCID	Inbred	NOD/SCID	B-cell, T-cell, less "leaky" mutants	(29)
NOG	Inbred	NOD/Shi-*scid*/IL-2Rγ^{null}	T-cell, B-cell, NK, macrophage and dendritic cell dysfunction, no leakiness	(30)

nude mice resulted in xenografts that were able to grow and preserve the characteristics of the original tumour such as histological traits and enzymatic electrophoretic patterns (15). Recent studies demonstrate that global expression profiles and global genomic aberrations are also conserved (16, 17).

Since the preliminary work by Outzen and colleagues, who were the first to achieve survival of primary breast cancer xenografts in nude mice (18), much progress has been made towards increasing reproducibility and success rates of engraftment (17, 19–22).

The cleared/humanized mammary fat pad model has proven to be the most successful in producing consistent rates of tumour engraftment: ~20% for xenografts that can be passaged more than five times and ~40% for xenografts with limited proliferation in the fat pad. The advantage of this technique over other approaches is the applicability to studies on normal human mammary epithelium. In light of these considerations, we choose this model to describe this model in this chapter.

2. Materials

Tissue procurement and handling:

1. Tissue transport medium (DMEM/F12 + 5% FBS + 5 µg/mL insulin + 0.5 µg/mL hydrocortisone and penicillin/streptomycin).
2. Sterile containers of appropriate volume – 50 mL tubes.
3. Ice container.

Tumour dissociation:

1. 60 mm Tissue culture dish; 250 mL Erlenmeyer flask (tissue dissociation flask, Stemcell Technologies); Sterile scalpel blades (number 10 and number 15); and 100 µm cell strainers.
2. EpiCult-B medium, Collagenase/Hyaluronidase 10× solution (Stemcell Technologies).
3. Matrigel (BD Biosciences), phosphate-buffered solution (PBS).
4. Freezing medium (RPMI 1640, 15% foetal bovine serum, and 5% DMSO).

Mammary fat pad clearance and humanization:

1. Dissecting scissors, scalpels, fine forceps, trocars, and sterile swabs (Fine Science Tools or Stoelting).
2. 0.5 mg 90-day release oestrogen pellets (Innovative Research of America).

3. Cork board and surgical tape.
4. Hypnorm, Midazolam, and Vetergesic.
5. Surgical adhesive, surgical clips, and clip applicator.
6. Hamilton syringes and 1 and 0.3 mL insulin syringes.
7. Human mammary fibroblasts – untreated and irradiated.
8. Ear tags (National Brand and Tag Company).

3. Methods

Tissue procurement and handling:

1. Tumour tissue is procured following the Ethical Committee or Institution Review Board regulations for research in human subjects.
2. Initial dissection of specimens must be carried out using sterile instruments in a containment cabinet with laminar air-flow. The tumour fragment procured should ideally be representative of the entire tumour. The sample is transferred to a 50 mL falcon tube in 5 mL transport medium, stored on ice, and transported as fast as possible to the laboratory to be processed.

Tumour dissociation:

1. Under a containment cabinet, the tumour tissue sample is transferred to a sterile 60 mm tissue culture dish containing 3 mL Epicult-B Medium. Using two number 10 or 15 sterile blades, tumour sample is minced to as fine a grain as possible. This will speed up the enzymatic phase of tissue digestion, therefore increasing cell viability at the end of the procedure.
2. The tumour mince is transferred to a 250 mL tissue dissociation flask. The remaining tissue in the culture dish is rinsed with 2 mL Epicult-B Medium and added to the tissue dissociation flask. The final volume of the tumour mince suspension should be 9 mL. One millilitre of 10× Collagenase/Hyaluronidase mixture is then added to the tissue dissociation flask.
3. The flask is sealed with aluminium foil and parafilm and transferred to a rotary shaker/incubator (previously set at 37°C and 80 rpm) and incubated for 15–60 min until no visible lumps are present. The tumour digestion mixture is transferred to a 50 mL falcon tube and triturated using a blunt 18-gauge needle and a 10 mL syringe, until the mixture flows freely. The enzymatic solution is then inactivated by adding 0.5 mL of foetal bovine serum.

4. The digested mixture is sieved through a 100-μm cell strainer sieve into a fresh 50 mL falcon tube. The strained cell suspension is then centrifuged at $300 \times g$ for 5 min. The supernatant is discarded and the cell pellet washed with pre-warmed PBS (37°C).

5. If at the end of step 4 a large quantity of tissue has not filtered through the cell strainer, the mixture is transferred back to the tissue dissociation flask and steps 2–4 are repeated until most of the tissue has been successfully dissociated.

6. Cells are counted and cell viability assessed using a haemocytometer and vital stain, e.g. Trypan Blue.

7. The cell pellet is re-suspended, together with a mixture of 2.5×10^5 unirradiated and 2.5×10^5 irradiated (four grey) human mammary fibroblasts (see Note 1), in a 1:1 solution of PBS and Matrigel™. Depending on the number of cells to be implanted, concentration of cells is adjusted so the total volume to be injected is 50–100 μL. Typically, for establishing a new xenograft from a clinical sample, at least 5×10^5 tumour cells or 5×10^4 normal mammary epithelial cells are needed. It is important to change the medium immediately after irradiation of fibroblasts and perform the irradiation 2–3 h before they are used for humanization.

Mammary fat pad clearing and humanization of the mouse mammary fat pad:

1. 21-Day-old immunodeficient female mice are anaesthetized by intra-peritoneal injection of a solution of Hypnorm (Fentanyl-Fluansione – 0.4 mL/kg) and Diazepam or Midazolam (5 mL/kg). Typically, a 15 g female mouse will require 10 μL/g of a solution made of three parts: water for injection, one part Hypnorm, and one part Midazolam.

2. The abdominal region and hindquarters of the mouse are shaved using dedicated hair clippers, the mouse is laid ventral side up and paws are taped onto cork board, and the abdominal skin is prepared using a non-alcoholic disinfectant, e.g., Tricine or Betadine.

3. The abdominal wall is picked up on the mid-line using a pair of fine rat-tooth forceps and a small incision is performed, using fine dissecting scissors, through the skin and subcutaneous tissue. This incision is then extended on the mid-line towards the thorax and on both sides towards the hindquarters, so as to create a reverse-Y incision.

4. Using blunt dissection or a sterile swab, the skin and subcutaneous layer are separated from the abdominal wall bilaterally, therefore exposing the fourth mammary fat pads and rudimental glands. The Y-shaped convergence of mammary

fat pad vessels and the mammary lymph node are easily visualized at this point (Fig. 1).

5. The part of the fat pad that contains mammary ducts is situated medially to the lymph node. This part is excised carefully without removing lymph node, in order to avoid lymphoedema (Fig. 1). Also, resecting the femoral vein should be avoided in order to prevent unnecessary significant bleeding. This dissection is then repeated on the contra lateral side.

6. The remaining fat pad is lifted using a pair of smooth forceps, and a suspension of 2.5×10^5 irradiated (4 Gy) and 2.5×10^5 non-irradiated human fibroblasts (total volume 50–100 µL) is injected (with care to avoid rupturing the connective tissue surrounding the fat pad) using a Hamilton or Insulin syringe depending on the volume.

7. The incision is closed using three surgical clips (one for each arm of the incision) and surgical adhesive. A 0.5 mg 90-day release oestrogen pellet (Innovative Research of America) is inserted subcutaneously in the back of each animal using a trocar. Analgesia is administered by subcutaneous injection of 100 µL of a 4.2 µg/mL solution of Buprenorphine. Mice are transferred to a warm post-operative chamber to prevent hypothermia and allowed to recover (see Notes 2–4).

3.1. Breast Cancer Xenografting

1. A single cell suspension of primary breast cancer cells obtained as described in the "Tumour dissociation" protocol is re-suspended in a 1:1 mixture of PBS and phenol-red-free Matrigel, aiming to achieve a concentration of 10^7 cells/mL. The total volume should not exceed 100 µL. This is mixed with 100,000 irradiated and non-irradiated human fibroblasts suspended in 1:1 PBS:Matrigel at approximately 10^6 cells/mL.

2. The above cell suspension must be kept on ice at all times, to prevent Matrigel jellification.

3. Minimum 14 days should be allowed after mammary fat pad clearance and humanization. For xenotransplantation, female mice are anaesthetized via an IP injection of Hypnorm/Midazolam as described in the previous protocol.

4. Following induction of anaesthesia, a reverse Y-shaped incision is made ventrally through skin and subcutaneous tissue in order to expose the humanized fat pads.

5. Humanized fat pads are picked up with a pair of smooth forceps and 50–100 µL of the previously prepared cell suspension is injected into each fat pad, with care to avoid rupturing the surrounding connective tissue.

6. The incision is closed using three surgical clips (one for each arm of the incision) and surgical adhesive. Analgesia is

administered by subcutaneous injection of 100 μL of a 4.2 μg/mL solution of Buprenorphine. Mice are ear tagged to allow identification and transferred to a warm post-operative chamber to prevent hypothermia and allowed to recover.

3.2. Re-passaging of Established Xenografts

1. Maintenance of an in vivo breast cancer xenograft model involves re-implantation of the tumour into further generations of animals, once a tumour has reached its maximum allowed size. Once a tumour has reached 10–15 mm in its greatest diameter, mice are humanely killed using an appropriate method.

2. Using a ventral incision identical to the one described previously, tumours are dissected and excised whole. Part of the specimen is fixed and stored according to the specific research aims of the study (i.e. in 4% formalin for future histopathological analysis, snap-frozen by dipping in liquid nitrogen, or in RNAlater, etc.). The rest of the specimen is transferred to tissue transport medium and stored on ice for transfer to the laboratory.

3. Tumour samples are dissociated following the same steps described in the dissociation of primary tumour specimens (steps 1–6). Typically, xenograft samples will be easier to dissociate than primary tumours and hence will require shorter enzymatic dissociation times (20–30 min).

4. Following digestion, xenografts are immediately re-established by injecting the above cell suspension in humanized mammary fat pads using the procedure and method described in Subheading 3.1.

5. Typically, take rate and growth rate of re-passaged xenografts improve when compared with those of the first-generation tumours (see Notes 5 and 6).

4. Notes

1. Typically, immortalized mammary fibroblasts are used for humanization of the fat pad, in order to facilitate standardization of experiments. Such lines immortalized by stable expression of telomerase and labelled by GFP expression have been used in several studies (22–24). Kuperwasser and colleagues also used an HGF overexpressing line of fibroblasts, which appears to stimulate the growth of epithelial cells (22). Irradiation of fibroblasts is similarly presumed to stimulate the growth of the epithelium through secretion of growth factors post-irradiation. One precaution regarding this

protocol is related to transformation of the fibroblasts lines, which occurs after multiple passaging in vitro. This will lead to in vivo growth of sarcomas, which will frequently overgrow breast tumours. Fibroblasts should be maintained in high-density cultures and at passage numbers below 15.

2. Xenografting of human mammary epithelium, either normal or malignant, can be performed 2–4 weeks after clearing and humanization. Whereas humanization is necessary to support the growth of the normal mammary epithelium, engraftment of tumours does not always depend on it. Successful tumour xenografting was obtained without humanization and in metatopic implantation (16, 17). No study has been published so far, directly comparing the engraftment success of the same breast tumours, in fat pads with and without humanization. Comparing success rates reported by different groups, using different strains of mice, different breast tumours, and procurement of samples (biopsy cores, post-surgery fragments, before after therapy, and primary or metastatic lesions) and different technical protocols (with and without oestrogen pellet implantation, with and without clearing and/or humanization, single cell suspensions versus tumour mince or fragments, and orthotopic versus metatopic), the only factor that increases the success of engraftment is the strain of mice. NOG mice appear to develop tumours in 40% of cases, whereas all the other strains develop tumours in 10–20% of cases.

3. The proliferation of normal mammary epithelium is dependent on humanization of the fat pad and the implantation of oestrogen pellets. Several investigators have reported, however, that oestrogen pellets may lead to slower growth of both ER-positive and ER-negative tumours. Taking into consideration that the majority of ER-positive tumours will originate from post-menopausal women, women under preventive hormonal therapy or women who had neo-adjuvant hormonal therapy, the use of oestrogen pellet may not model the hormonal milieu in human patients. Moreover, the levels of circulating oestrogen in mice are not monitored and may be higher than that in pre-menopausal women. It is possible that oestrogen levels reach those levels used therapeutically in breast cancers (a therapeutic intervention used before the introduction of tamoxifen). We, and others, also observed the loss of ER-positivity in xenografted tumours, after the first implantation or after several passages.

4. The use of slow release oestrogen and progesterone pellets or pellets with more than 0.5 mg 90-day release may lead to genito-urinary side effects, sometimes severe enough to require euthanasia.

5. Advantages of the orthotropic xenograft model:
   (a) The xenograft model is the experimental system that closely recapitulates the in vivo complexity and natural environment of human normal or malignant mammary epithelium.
   (b) For studies regarding therapeutics, pharmacokinetic/pharmacodynamic considerations can be included in the study.
   (c) Cancer cells do not have an evolution in vitro, which can introduce artificial constraints and changes. At low passage, cancer cells will have a short history out of the natural human environment.

6. Limitations of the orthotropic xenograft model:
   (a) Establishing this model is technically challenging, costly, and time and effort consuming, particularly due to the low rate of success.
   (b) Using this model relies on the availability of clinical samples.
   (c) The histoclinical heterogeneity of breast cancers may not be reflected in the tumours, due to bias in success of engraftment in favour of more aggressive cancers.

Acknowledgment

R.M. is supported by the Breakthrough Breast Cancer Research Unit, Guy's Hospital, King's Health Partners, AHSC, London, UK.

References

1. Clarke, R., and Johnson, M.D. (2000) Chapter 22: Animal models, in *Diseases of the Breast* (Harris, J.R., Lippman, M.E., Morrow, M., Hellman, S., Ed.) 2nd ed., pp 319–333, J. B. Lippincott Co., Philadelphia.
2. Smalley, M., and Ashworth, A. (2003) Stem cells and breast cancer: a field in transit, *Nat Rev Cancer 3*, 832–844.
3. Deome, K.B., Faulkin, L.J., Jr., Bern, H.A., and Blair, P.B. (1959) Development of mammary tumors from hyperplastic alveolar nodules transplanted into gland-free mammary fat pads of female C3H mice, *Cancer Res 19*, 515–520.
4. Smith, G.H., Strickland, P., and Daniel, C.W. (2002) Putative epithelial stem cell loss corresponds with mammary growth senescence, *Cell Tissue Res 310*, 313–320.
5. Stingl, J., Eirew, P., Ricketson, I., Shackleton, M., Vaillant, F., Choi, D., Li, H.I., and Eaves, C.J. (2006) Purification and unique properties of mammary epithelial stem cells, *Nature 439*, 993–997.
6. Shackleton, M., Vaillant, F., Simpson, K.J., Stingl, J., Smyth, G.K., Asselin-Labat, M.L., Wu, L., Lindeman, G.J., and Visvader, J.E. (2006) Generation of a functional mammary gland from a single stem cell, *Nature 439*, 84–88.
7. Sleeman, K.E., Kendrick, H., Ashworth, A., Isacke, C.M., and Smalley, M.J. (2006)

CD24 staining of mouse mammary gland cells defines luminal epithelial, myoepithelial/basal and non-epithelial cells, *Breast Cancer Res 8*, R7.

8. Ginestier, C., Hur, M.H., Charafe-Jauffret, E., Monville, F., Dutcher, J., Brown, M., Jacquemier, J., Viens, P., Kleer, C.G., Liu, S., Schott, A., Hayes, D., Birnbaum, D., Wicha, M.S., and Dontu, G. (2007) ALDH1 is a marker of normal and malignant human mammary stem cells and a predictor of poor clinical outcome, *Cell Stem Cell 1*, 555–567.

9. Cicalese, A., Bonizzi, G., Pasi, C.E., Faretta, M., Ronzoni, S., Giulini, B., Brisken, C., Minucci, S., Di Fiore, P.P., and Pelicci, P.G. (2009) The tumor suppressor p53 regulates polarity of self-renewing divisions in mammary stem cells, *Cell 138*, 1083–1095.

10. Pece, S., Tosoni, D., Confalonieri, S., Mazzarol, G., Vecchi, M., Ronzoni, S., Bernard, L., Viale, G., Pelicci, P.G., and Di Fiore, P.P. (2010) Biological and molecular heterogeneity of breast cancers correlates with their cancer stem cell content, *Cell 140*, 62–73.

11. Shultz, L.D., Schweitzer, P.A., Hall, E.J., Sundberg, J.P., Taylor, S., and Walzer, P.D. (1989) *Pneumocystis carinii* pneumonia in scid/scid mice, *Curr Top Microbiol Immunol 152*, 243–249.

12. Quintana, E., Shackleton, M., Sabel, M.S., Fullen, D.R., Johnson, T.M., and Morrison, S.J. (2008) Efficient tumour formation by single human melanoma cells, *Nature 456*, 593–598.

13. Meyvisch, C. (1983) Influence of implantation site on formation of metastases, *Cancer Metastasis Rev 2*, 295–306.

14. Volpe, J.P., and Milas, L. (1990) Influence of tumor transplantation methods on tumor growth rate and metastatic potential of solitary tumors derived from metastases, *Clin Exp Metastasis 8*, 381–389.

15. Grant, A.G., Duke, D., and Hermon-Taylor, J. (1979) Establishment and characterization of primary human pancreatic carcinoma in continuous cell culture and in nude mice, *Br J Cancer 39*, 143–151.

16. Bergamaschi, A., Hjortland, G.O., Triulzi, T., Sorlie, T., Johnsen, H., Ree, A.H., Russnes, H.G., Tronnes, S., Maelandsmo, G.M., Fodstad, O., Borresen-Dale, A.L., and Engebraaten, O. (2009) Molecular profiling and characterization of luminal-like and basal-like in vivo breast cancer xenograft models, *Mol Oncol 3*, 469–482.

17. Marangoni, E., Vincent-Salomon, A., Auger, N., Degeorges, A., Assayag, F., de Cremoux, P., de Plater, L., Guyader, C., De Pinieux, G., Judde, J.G., Rebucci, M., Tran-Perennou, C., Sastre-Garau, X., Sigal-Zafrani, B., Delattre, O., Dieras, V., and Poupon, M.F. (2007) A new model of patient tumor-derived breast cancer xenografts for preclinical assays, *Clin Cancer Res 13*, 3989–3998.

18. Outzen, H.C., and Custer, R.P. (1975) Growth of human normal and neoplastic mammary tissues in the cleared mammary fat pad of the nude mouse, *J Natl Cancer Inst 55*, 1461–1466.

19. Li, Z., Huang, X., Li, J., Ke, Y., Yang, L., Wang, Y., Yao, L., and Lu, Y. (2002) Human breast carcinoma xenografts in nude mice, *Chin Med J (Engl) 115*, 222–226.

20. Sakakibara, T., Xu, Y., Bumpers, H.L., Chen, F.A., Bankert, R.B., Arredondo, M.A., Edge, S.B., and Repasky, E.A. (1996) Growth and metastasis of surgical specimens of human breast carcinomas in SCID mice, *Cancer J Sci Am 2*, 291–300.

21. Visonneau, S., Cesano, A., Torosian, M.H., Miller, E.J., and Santoli, D. (1998) Growth characteristics and metastatic properties of human breast cancer xenografts in immunodeficient mice, *Am J Pathol 152*, 1299–1311.

22. Kuperwasser, C., Chavarria, T., Wu, M., Magrane, G., Gray, J.W., Carey, L., Richardson, A., and Weinberg, R.A. (2004) Reconstruction of functionally normal and malignant human breast tissues in mice, *Proc Natl Acad Sci USA 101*, 4966–4971.

23. Lim, E., Vaillant, F., Wu, D., Forrest, N.C., Pal, B., Hart, A.H., Asselin-Labat, M.L., Gyorki, D.E., Ward, T., Partanen, A., Feleppa, F., Huschtscha, L.I., Thorne, H.J., Fox, S.B., Yan, M., French, J.D., Brown, M.A., Smyth, G.K., Visvader, J.E., and Lindeman, G.J. (2009) Aberrant luminal progenitors as the candidate target population for basal tumor development in BRCA1 mutation carriers, *Nat Med 15*, 907–913.

24. Stingl, J., and Caldas, C. (2007) Molecular heterogeneity of breast carcinomas and the cancer stem cell hypothesis, *Nat Rev Cancer 7*, 791–799.

25. Rygaard, J., and Povlsen, C.O. (1969) Heterotransplantation of a human malignant tumour to "Nude" mice, *Acta Pathol Microbiol Scand 77*, 758–760.

26. Rygaard, J., and Povlsen, C.O. (1974) Effects of homozygosity of the nude (NU) gene in three inbred strains of mice. A detailed study of mice of three genetic backgrounds (BALB-c, C3H, C57-BL-6) with congenital absence of the thymus (nude mice) at a stage in the gene transfer, *Acta Pathol Microbiol Scand A 82*, 48–70.

27. Clarke, R. (1996) Human breast cancer cell line xenografts as models of breast cancer. The immunobiologies of recipient mice and the characteristics of several tumorigenic cell lines, *Breast Cancer Res Treat 39*, 69–86.

28. Bosma, G.C., Custer, R.P., and Bosma, M.J. (1983) A severe combined immunodeficiency mutation in the mouse, *Nature 301*, 527–530.

29. Prochazka, M., Gaskins, H.R., Shultz, L.D., and Leiter, E.H. (1992) The nonobese diabetic scid mouse: model for spontaneous thymomagenesis associated with immunodeficiency, *Proc Natl Acad Sci USA 89*, 3290–3294.

30. Ito, A., Ishida, T., Yano, H., Inagaki, A., Suzuki, S., Sato, F., Takino, H., Mori, F., Ri, M., Kusumoto, S., Komatsu, H., Iida, S., Inagaki, H., Ueda, R. (2008) Defucosylated anti-CCR4 monoclonal antibody excercises potent ADCC-mediated anti-tumor effect in the novel tumor-bearing humanized NOD/Shi-scid IL2R gamma (null) mouse model, *Cancer Immunol Immunother 58*, 1195–206.

31. Meyer, M.J., Fleming, J.M., Lin, A.F., Hussnain, S.A., Ginsberg, E., Vonderhaar, B.K. (2010) CD44posCD49fhiCD133/2hi defines xenograft-initiating cells in oestrogen receptor negative breast cancer, *Cancer Res 70*, 4624–33.

Appendix A

Formulations of Commonly Used Cell Culture Media

Medium (component)	BME (mg/L)	DMEM (mg/L)	Iscove's DMEM (mg/L)	Ham's F12 (mg/L)	RPMI 1640 (mg/L)	Medium 199 (mg/L)
Calcium chloride	200	200	165.3	33.3	–	139.6
Calcium nitrate·4H$_2$O	–	–	–	–	100	–
Cupric sulfate·5H$_2$O	–	–	–	0.0025	–	–
Ferrous sulfate·7H$_2$O	–	–	–	0.834	–	–
Ferric nitrate·9H$_2$O	–	0.1	–	–	–	0.72
Magnesium sulfate (anhydrous)	97.67	97.67	97.67	57.6	48.84	97.67
Potassium chloride	400	400	330	224	400	400
Potassium phosphate monobasic	–	–	–	–	–	60
Potassium nitrate	–	–	0.076	–	2,000	–
Sodium acetate (anhydrous)	–	–	–	–	–	50
Sodium bicarbonate[a]	2,200	3,700	3,024	1,176	6,000	2,200
Sodium chloride	6,800	6,400	4,505	7,599	800	8,000
Sodium phosphate monobasic (anhydrous)	122	109	109	142.04	–	47.88
Sodium selenite	–	–	0.017	–	–	–
Zinc sulfate·7H$_2$O	–	–	–	0.863	–	–
L-Alanine	–	–	25	9	–	25

(continued)

Medium (component)	BME (mg/L)	DMEM (mg/L)	Iscove's DMEM (mg/L)	Ham's F12 (mg/L)	RPMI 1640 (mg/L)	Medium 199 (mg/L)
L-Alanyl-L-Glutamine[a]	–	–	–	–	434.4	70
L-Arginine·HCl	21	84	84	211	200	–
L-Asparagine·H$_2$O	–	–	28.4	15.01	50	–
L-Aspartic acid	–	–	30	13.3	20	30
L-Cysteine·HCl·H$_2$O	0	0	0	0	0	0.11
L-Cystine·2HCl	15.65	30	91.24	35	65.2	26
L-Glutamic acid	0	0	75	14.7	20	66.8
L-Glutamine[a]	292	584	584	146	300	100
Glycine	0	0	30	7.51	10	50
L-Histidine (free base)	8	105	42	20.96	15	21.88
Hydroxy-L-proline	–	–	–	–	20	10
L-Isoleucine	26	105	105	3.94	50	20
L-Leucine	26	1,460	105	13.1	50	60
L-Lysine·HCl	36.47	–	146	36.5	40	70
L-Methionine	7.5	66	30	4.48	15	15
L-Phenylalanine	16.5	42	66	4.96	15	25
L-Proline	–	–	40	34.5	20	40
L-Serine	–	95	42	10.5	30	25
L-Threonine	24	16	95	11.9	20	30
L-Tryptophan	4	103.79	16	2.04	5	10
L-Tyrosine·2Na·2H$_2$O	25.95	94	103.79	7.78	28.83	57.66
L-Valine	23.5	–	94	11.7	20	25
Ascorbic acid·Na	0	0	0	0	0	0.0566
D-Biotin	1	4	0.013	0.0073	0.2	0.01
Calciferol	0	0	0	0	0	0.1
Choline chloride	1	4	4	13.96	3	0.5
Folic acid	1	7.2	4	1.32	1	0.01
Menadione (sodium bisulfite)	0	0	0	0	0	0.016
myo-Inositol	2	4	7.2	18	35	0.05
Niacinamide	1	4	4	0.037	1	0.025
Nicotinic acid	0	0	0	0	0	0.025
p-Aminobenzoic acid	0	0	0	0	0	0.05
D-Pantothenic acid (hemicalcium)	1	–	4	0.48	1	0.01

(continued)

Medium (component)	BME (mg/L)	DMEM (mg/L)	Iscove's DMEM (mg/L)	Ham's F12 (mg/L)	RPMI 1640 (mg/L)	Medium 199 (mg/L)
Pyridoxal·HCl	1	4	4	–	0.25	0.025
Pyridoxine·HCl	–	0.4	–	0.062	1	0.025
Retinol acetate	0	0	0	0	0	0.14
Riboflavin	0.1	4	0.4	0.038	0.2	0.01
DL-Tocopherol phosphate·Na	0	0	0	0	0	0.01
Thiamine·HCl	1	–	4	0.34	1	0.01
Vitamin B12	–	4,500	0.013	1.36	0.005	–
	–	–	–	–	–	–
Adenine sulfate	–	–	–	–	–	10
Adenosine triphosphate·2Na	–	–	–	–	–	1
Adenosine monophosphate·Na	–	–	–	–	–	0.2385
Cholesterol	–	–	–	–	–	0.2
Deoxyribose	–	–	–	–	–	0.5
D-Glucose	1,000	15.9	4,500	1,802	2,000	1,000
HEPES	–	–	5,958	–	–	–
Hypoxanthine	–	–	–	4.08	–	0.3
Glutathione (reduced)	–	–	–	–	1	0.05
Guanine·HCl	0	0	0	0	0	0.3
Linoleic acid	–	–	–	0.084	–	–
Phenol red·Na[a]	11	110	16	1.3	5.3	21.3
Putrescine·HCl	–	–	–	0.161	–	–
Pyruvic acid·Na	–	–	110	110	–	–
Polyoxyethylenesorbitan monooleate (TWEEN 80)	–	–	–	–	–	20
Ribose	–	–	–	–	–	0.5
Thioctic acid	–	–	–	0.21	–	–
Thymidine	–	–	–	0.73	–	–
Thymine	–	–	–	–	–	0.3
Uracil	–	–	–	–	–	0.3
Xanthine·Na	–	–	–	–	–	0.344

[a] May be left out if desired

Appendix B

Human Cancer Cell Lines Available from the ATCC and DSMZ

& ECACC/Sigma.

The list below is not exhaustive, but gives a brief guide to some of those lines available from the ATCC and DSMZ. The ATCC (http://www.atcc.org) holds over 950 cancer cell lines (all species) while the DSMZ (http://www.dsmz.de) currently lists 657 human cancer cell lines. Most variants are not included.

Bladder cancer	5637	Hs 172.T	JMSU-1	SW-1710
	647-V	Hs 195.T	KU-19-19	T-24
	BC-3C	Hs 228.T	RT-112	TCC-SUP
	BFTC-905	HT-1197	RT-4	UM-UC-3
	CAL-29	HT-1376	SCaBER	VM-CUB1
	ECV-304	J82	SW 780	
Breast cancer	BT-20	HCC1395	Hs 319.T	Hs 748.T
	BT-474	HCC1419	Hs 329.T	Hs 841.T
	BT-483	HCC1500	Hs 343.T	Hs 849.T
	BT-549	HCC1569	Hs 344.T	Hs 851.T
	CAL-120	HCC1599	Hs 350.T	Hs 861.T
	CAL-148	HCC1937	Hs 362.T	Hs 905.T
	CAL-51	HCC1954	Hs 371.T	HT 762.T
	CAL-85-1	HCC202	Hs 479.T	JIMT-1
	COLO-824	HCC2157	Hs 540.T	MCF-7
	DU4475	HCC2218	Hs 566(B).T	MDA-MB-453
	DU-4475	HCC38	Hs 574.T	MFM-223
	EFM-19	HCC70	Hs 578T	MT-3
	EFM-192A	HDQ-P1	Hs 588.T	NCI-H548
	EFM-192B	Hs 190.T	Hs 605.T	UACC-812
	EFM-192C	Hs 255.T	Hs 606	UACC-893
	EVSA-T	Hs 274.T	Hs 739.T	
	HCC1143	Hs 280.T	Hs 741.T	
	HCC1187	Hs 281.T	Hs 742.T	
Cervical cancer	C-4 I	4510	BT-B	HeLa S3
	C-4 II	C-33 A	HeLa	H1HeLa
	DoTc2	SiHa	HeLa 229	SISO

(continued)

Colorectal cancer	SW837 C2BBe1 Caco-2 CL-11 CL-14 CL-34 CL-40 COLO-206F COLO-320DM COLO-320HSR COLO-678	DLD-1 HCT-116 HCT-15 HCT-8 (HRT-18) Hs 200.T Hs 207.T Hs 219.T Hs 257.T Hs 587.Int Hs 675.T LOVO	Hs 722.T HT-29 LS 174T LS 180 LS1034 LS123 LS411N LS513 NCI-H498 NCI-H508 NCI-H716	NCI-H747 SNU-C2A SNU-C2B SW1116 SW1417 SW1463 SW403 SW48 SW480 SW948 WiDr
Endometrial cancer	KLE HEC-1-A HEC-1-B	RL95-2 AN3-CA EN	EFE-184 MFE-280 MFE-296	MFE-319
Gastric	23132/87 AGS	Hs 740.T MKN-45	RF-1 SNU-1	
Glioma	42-MG-BA 8-MG-BA A172 CCF-STTG1 DBTRG-05MG DK-MG	GAMG GMS-10 GOS-3 H4 Hs 683 LN-18	LN-229 LN-405 M059J M059K SNB-19 SW 1088	SW 1783 T98G U-118-MG U-138-MG U-87-MG
Kidney	786-O 769-P A-498	A704 ACHN	Caki-2 CAL-54	SW 156 SW 839
Leukemia/ lymphoma	380 697 ALL-SIL AML-193 AP-1060 ARH-77 BC-1 BC-2 BC-3 BCP-1 BD-215 BE-13 BL-2 BL-41 BL-70 BLUE-1 BONNA-12 BV-173 CA-46 CCRF-CEM CCRF-HSB-2 CCRF-SB CESS CI-1 CMK	HD-MY-Z HEL HH HL-60 HNT-34 HPB-ALL Hs 313.T Hs 445 Hs 491.T Hs 505.T Hs 518.T Hs 602 Hs 604.T Hs 611.T Hs 616.T Hs 751.T Hs 777.T HSB-2 HS-Sultan HT HT 1417 HuT 102 HuT 78 IM-9 J.CaM1.6	LAMA-84 LAMA-87 LOUCY M-07e MC-116 ME-1 MEC-1 MEC-2 MEG-01 MHH-CALL-2 MHH-CALL-3 MHH-CALL-4 MHH-PREB-1 ML-2 MN-60 Mo-B MOLM-13 MOLM-16 MOLM-20 MOLM-6 MOLT-13 MOLT-14 MOLT-16 MOLT-17 MOLT-3	REH RI-1 RO ROS-50 RPMI 6666 RPMI-8402 RS4;11 SC-1 SD-1 SEM SET-2 SH-2 SHI-1 SIG-M5 SKM-1 SK-MM-2 SKW-3 SPI-801 SPI-802 SR-786 SU-DHL-1 SU-DHL-10 SU-DHL-16 SU-DHL-4 SU-DHL-5

(continued)

	CML-T1	J.RT3-T3.5	MOLT-4	SU-DHL-6
	CRO-AP2	J45.01	MONO-MAC-1	SU-DHL-8
	CRO-AP3	JEKO-1	MONO-MAC-6	SUP-B15
	CRO-AP5	JIYOYE	MOTN-1	SUP-HD1
	CRO-AP6	JJN-3	MUTZ-2	SUP-M2
	CTV-1	JK-1	MUTZ-3	SUP-T1
	DAUDI	JURKAT	MUTZ-5	SUP-T11
	DB	JURL-MK1	MV4-11	TALL-1
	DEL	JURL-MK2	NALM-1	TANOUE
	DERL-2	JVM-13	NALM-19	TE 161.T
	DERL-7	JVM-2	NALM-6	TE 175.T
	DG-75	JVM-3	NAMALWA	TF-1
	DND-39	K-562	NB-4	THP-1
	DND-41	KARPAS-1106P	NC-37	TMM
	DOGKIT	KARPAS-231	NK-92	Toledo
	DOGUM	KARPAS-299	NOMO-1	TOM-1
	DOHH-2	KARPAS-422	NTERA-2	TUR
	EB-1	KARPAS-45	NU-DHL-1	U-2932
	EB-3	KARPAS-620	NU-DUL-1	U-2940
	EHEB	KASUMI-1	OCI-AML2	U-2973
	EM-2	KASUMI-2	OCI-AML3	U-698-M
	EM-3	KCL-22	OCI-AML5	U-937
	EOL-1	KE-37	OCI-LY-19	U-H01
	F-36P	KG-1	OCI-M1	ULA
	FKH-1	KM-H2	OCI-M2	UT-7
	GA-10	KMOE-2	P12-ICHIKAWA	VAL
	GDM-1	KOPN-8	PEER	WILL-1
	GF-D8	KU-812	PF-382	WILL-2
	GRANTA-519	KYO-1	PL-21	WSU-DLCL2
	GUMBUS	L-1236	PLB-985	WSU-FSCCL
	H9	L-363	RAJI	WSU-NHL
	HAL-01	L-428	RAMOS	YT
	HC-1	L-540	RCH-ACV	
	HDLM-2	L-591	RC-K8	
	RL	L-82	REC-1	
Liver	HEP-3B	SNU-398	SNU-182	SNU-387
	Hep G2	SNU-449	SNU-475	SNU-423
	SK-HEP-1			
Lung	A-427	HCC-44	NCI-H2342	NCI-H23
	A-549	HCC-78	NCI-H2347	NCI-H2444
	BEN	HCC-827	NCI-H1703	NCI-H292
	CAL-12T	Hs 229.T	NCI-H1734	NCI-H358
	COLO-699	Hs 573.T	NCI-H1793	NCI-H510A
	CPC-N	Hs 618.T	NCI-H1836	NCI-H520
	DMS 114	LCLC-103H	NCI-H1838	NCI-H522
	DMS 53	LCLC-97TM1	NCI-H1963	NCI-H596
	DMS 79	LOU-NH91	NCI-H1975	NCI-H810
	DV-90	LXF-289	NCI-H2066	NCI-H82
	EPLC-272H	NCI-H1373	NCI-H2073	NCI-N417
	H-1184	NCI-H1395	NCI-H2085	SCLC-21H
	H-1339	NCI-H1417	NCI-H2126	SCLC-22H
	H-1963	NCI-H1435	NCI-H2135	SHP-77

(continued)

	H-209	NCI-H1563	NCI-H2170	SK-LU-1
	H-2171	NCI-H1581	NCI-H2172	SK-MES-1
	HCC-15	NCI-H1651	NCI-H2227	SW 1271
	HCC-33	NCI-H1672	NCI-H2228	SW 1573
	HCC-366	NCI-H1688	NCI-H2286	SW 900
Melanoma	A101D	Hs 600.T	Hs 940.T	SK-MEL-1
	A-375	Hs 688(A).T	HT-144	SK-MEL-24
	C32	Hs 839.T	IGR-1	SK-MEL-28
	COLO 829	Hs 852.T	IGR-37	SK-MEL-3
	COLO-679	Hs 906(A).T	IGR-39	SK-MEL-30
	COLO-783	Hs 906(B).T	IPC-298	SK-MEL-31
	COLO-800	Hs 908.Sk	Malme-3M	SK-MEL-5
	COLO-818	Hs 934.T	MEL-HO	WM-115
	COLO-849	Hs 935.T	MEL-JUSO	
	G-361	Hs 936.T	RPMI-7951	
	Hs 432.T	Hs 939.T	RVH-421	
Myeloma	AMO-1	MOLP-2	MOLP-8	RPMI 8226
	EJM	KMS-12-PE	OPM-2	U266
	KMS-12-BM	LP-1	NCI-H929	
Ovary	Caov-3	ES-2	EFO-27	Hs 38.T
	COLO-704	EFO-21	FU-OV-1	Hs 571.T
Pancreas	BxPC-3	HUP-T3	Panc 02.13	PA-TU-8988S
	CAPAN-1	HUP-T4	Panc 03.27	PA-TU-8988T
	CAPAN-2	KCI-MOH1	Panc 08.13	PL45
	DAN-G	MPanc-96	Panc 10.05	YAPC
	HPAC	PaCa-2	PANC-1	
	HPAF-II	Panc 02.03	PA-TU-8902	
Prostate	22Rv1	DU-145	LNCAP	PC-3
Sarcoma	A-204	Hs 704.T	Hs 889.T	SAOS-2
	A-673	Hs 706.T	Hs 890.T	Saos-2
	C-433	Hs 707(A).T	Hs 926.T	SJSA-1
	CADO-ES1	Hs 729	Hs 93.T	SK-ES-1
	CAL-72	Hs 729.T	Hs 94.T	SK-LMS-1
	CAL-78	Hs 735.T	Hs 941.T	SK-UT-1
	ESS-1	Hs 737.T	HT 728.T	SK-UT-1B
	G-292	Hs 755(B).T	HT-1080	SW 684
	G-402	Hs 778(A).T	KHOS/NP	SW 872
	HOS	Hs 778(B).T	KHOS-240S	SW 982
	Hs 127.	Hs 781.T	KHOS-321H	T1-73
	Hs 132.T	Hs 792(B).T	MES-SA	TC-71
	Hs 14.T	Hs 805.T	MES-SA/Dx5	TE 125.T
	Hs 15.T	Hs 811.T	MES-SA/MX2	TE 149.T
	Hs 184.T	Hs 814.T	MG-63	TE 159.T
	Hs 188.T	Hs 821.T	MHH-ES-1	TE 381.T
	Hs 295.T	Hs 822.T	MNNG/HOS	TE 417.T
	Hs 3.T	Hs 846.T	Murphy	TE 418.T
	Hs 357.T	Hs 856.T	R-970-5	TE 441.
	Hs 387.T	Hs 863.T	RD	TE 617.T
	Hs 39.T	Hs 864.T	RD-ES	TE-671

(continued)

Hs 5.T	Hs 866.T	RH-1	TO 203.T
Hs 51.T	Hs 870.T	RH-18	U-2 OS
Hs 57.T	Hs 871.T	RH-30	U-2197
Hs 63.T	Hs 88.T	RH-41	VA-ES-BJ
Hs 701.T	Hs 883.T	S-117	

Index

A

Acetylation 6
Acholeplasma 38, 94, 99, 100
Actin 333, 334
Adenosine triphosphate (ATP) 221, 223, 224, 229, 230, 248, 249, 251–253, 255, 260–262, 277, 278, 422, 432, 485
Adhesion molecules
    E-cadherin ....
    E-selectin 217
    ICAM–1 9
    integrins 395
    PECAM 346
    VCAM 217
    VE-cadherin 217
Aerosol 22, 80, 82, 244, 254, 392
Affinity chromatography 396–398, 400–401
Agarose 52, 96, 99, 126, 226, 310, 312, 317, 318, 327, 329, 347, 351, 352, 373, 376
Agarose gel electrophoresis 54, 325, 327, 448
Alamar blue 221–222
Alamethicin 107
Albumin 17
Aldehydes 25
Aliquoted 48, 204, 205, 211, 340, 391, 447
Alu 67, 323
Amelogenin 40, 46, 47, 49, 53
Aminopeptidase ....
Amoeboid 333
Amphiregulin 4
Amphotericin 137, 170, 177, 210, 352, 449
Amplicon 39, 40, 46, 95, 100, 324, 327, 329–331
Anaesthesia/anaesthetized 476, 477
Angiogenesis 2, 8, 209, 210, 345–357
Annexin V 226, 293–307
Antibiotic(s)
    Amphotericin B (fungizone) 137, 170, 177, 210, 352, 449
    ampicillin 295–297, 300, 306, 373, 376
    antimycoplasma 107, 196
    ciprofloxacin 83, 107, 109, 110, 113
    enrofloxacin 107, 109, 110, 113
    fluoroqinolones 107, 109, 112, 113
    gentamycin/gentamicin 136
    macrolide(s) 107
    metronidazole 137, 170, 177
    minocycline 111
    penicillin 25, 100, 106, 112, 136, 137, 142, 152, 153, 168, 169, 192, 196, 202, 210, 346, 347, 352, 453, 454, 474
    streptomycin 25, 100, 106, 112, 137, 192, 196
    tetracyclines 107
Antibody-dependent cellular cytotoxicity (ADCC) 9, 409, 417
Antineoplastic 58
Apoptosis
    annexin V 294
    apoptotic bodies 294
    bcl-2 .... 409–411, 415
    bcl-x 410, 411, 415
    caspase activity 225, 294
    DNA laddering 54, 226, 227, 294
    fas 6
    fas-L 6
    mcl–1 .... 415, 416
    TUNEL 227, 294
Arginine 106, 483
Ascites 163–166, 194, 251, 253
Ascorbate ....
Aseptic technique 30, 32, 79, 80, 83–85, 113, 461
ATP assay 223, 224, 229, 230, 260, 278, 422, 432
ATP binding site ....
Audit 116, 117, 119–123
Authentication 35–41, 45–47, 187, 193, 323
Autocrine 4, 132
Autofluorescence 71, 75
Autophagy ....
Avidin 396, 399, 403
    streptavidin 397

B

Bacterial artificial chromosome (BACs) ... 60, 69, 70, 75, 76
Barcoded 28, 33, 391
Basic fibroblast growth factor (bFGF) 8, 210, 357

Bax...3, 6
Benchtop...22, 136, 137, 239, 369, 384
Bevacizumab...210, 260, 275
BH3 peptide...411
Bioluminescence...221, 249
Biotherapeutics...87
Biotin...70, 396–401, 403
Bisphosphonates...248
Blood...8, 15, 37, 81, 88, 162, 171, 185, 193–195, 209, 210, 260, 264, 275, 345, 346, 351, 353, 354, 356, 365, 472
Bone marrow...59, 69, 185, 191, 193–195, 231, 260
Bortezomib...415, 417
Bovine serum albumin (BSA)...142, 202, 211, 338, 349, 438, 444
Boyden chamber...335–336
Breast cancer...39, 231, 267, 269, 270, 274, 471–480, 486
Bromophenol...95, 295
Buprenorphine...477, 478

C

Calcium phosphate...436, 438, 440, 441, 443
Capillary electrophoresis (CE)...47–49, 62
Carcinogenesis/carcinogen...3, 7, 288, 368, 386, 395, 472
Caspase...223, 226–227, 289, 293, 294
Catenin...3, 6
CD20...408, 409, 413–414, 417
Cell culture media
    CMRL 1066...16, 168
    Dulbecco's minimum essential medium (DMEM)...16
    Dulbecco's phosphate buffered saline...142, 168, 169, 453
    Eagle's basal medium (BME)...16, 483–485
    Eagle's minimum essential medium (MEM)...16, 195, 347
    Epicult B Medium...474, 475
    Ham's F10...14, 16, 195
    Ham's F12...14, 16, 152, 195, 483–485
    Hank's Balanced Salt Solution (HBSS)...29, 31, 168, 363
    Iscove's...16, 184, 195, 483–485
    Leibovitz L-15...168
    McCoy's...16, 168, 169, 176, 184, 195
    MCDB 131...351, 352
    Medium 199...16, 483–485
    RPMI 1640...107, 142, 148, 168, 169, 176, 184, 186, 195, 253, 350, 453, 454, 483–485
Cell cycle...6, 58, 63, 220, 229, 289, 359–362, 366–368, 428, 429, 433, 452
Cellometer...137, 139
Centrifugation...21, 22, 27–29, 50, 71, 73, 74, 109, 111, 112, 135, 155, 170–172, 174, 185, 188, 215, 290, 299, 300, 304, 311, 373, 381, 446, 454

Centromeres...67
Ceramide...413
Cerebrospinal fluid...194
Cetuximab...277
Chaotropic...50
Chelation...28
Chemical diversity...219
Chemiluminescence...
Chemokinesis...335
Chemoresistance...164, 249, 411, 413–416
Chemosensitivity...136, 163, 164, 167, 176, 219–233, 247–255, 413, 414
Chemotaxis...334, 335, 338
Chemotherapy...125, 163, 164, 210, 248, 261, 263, 265, 267, 269, 271–273, 276, 408, 409, 411, 414, 416, 417
Chloroform...95, 101, 153, 155, 330, 372, 373
Cholera toxin...142, 152, 153
Chorioallantoic membrane assay (CAM)...347, 350–351
Choriomeningitis...81
Chromatids...64, 73
Chromium release assay...270
Chromosome...46, 49, 57–60, 62–67, 69, 70, 72–76, 147, 158, 321–331, 362
    chromosome fragility...58, 69
Clastogenesis...58
Clonal/clones...15, 60, 69, 70, 75, 76, 107, 126, 128–132, 164, 197, 213, 216, 228, 323, 331, 376, 378, 407–417, 445, 447–450, 467
Clonogenic assays...167, 176, 220, 224, 228, 229, 261, 262
Cluster of differentiation (CD) antigens...205
Cobalt^{60}...469
Co-culture...8, 467–470
Coefficient of variance (%CV)...458, 460
Colcemid...60, 63–65, 73
Collagen...9, 18, 335, 336, 338, 340, 346, 349, 352, 356
Collagenase. *See* Matrix metalloproteinase
Colorectal cancer (CRC)...2, 4, 6, 135, 136, 164, 177, 210, 213, 452, 487
Combinatorial chemistry...219
Comet assay...307–319
Commensal...80
Complement...9, 17, 221, 413, 414, 417
Concentration-effect...424, 425, 428, 433, 456, 459
Confluency...31, 36, 127–129, 131, 132, 147, 156, 213, 215, 355, 364, 368, 399, 438–440, 446, 449, 450, 469
Confocal microscopy...227, 341
Conjugation...202, 304, 307
Contact inhibition...36, 41, 132, 243, 355, 468–470
Contamination
    cross-contamination...39–41, 46, 47, 54, 57, 59, 77, 88, 115, 192, 253–255, 323, 392, 393, 450
    intra-species...39, 46

microbial 24, 36, 37, 41, 79, 80, 85, 169, 175
Coomassie staining 300, 398
Cornea
    angiogenesis 348, 353–354, 357
    limbus 357
Counterstain 67, 72, 213, 215
Coverslip 16, 62, 71, 72, 75, 198, 310, 312, 313, 318, 319
Coxiella burnetti 81
Cryopreservation 110, 113, 131, 174, 185, 188
Cryostorage
    cryopreservation 87, 174
    cryovial(s) 87
Cyanol 95
Cyclin 5
Cytocentrifuge 197, 264
Cytochrome c 39, 228
Cytogenetics 46, 57–77, 158, 182, 210, 322, 326, 362
Cytokines
    basic fibroblast growth factor (bFGF) 8
    ECGS 357
    endothelin 1 143
    erythropoietin (EPO) 184
    granulocyte colony-stimulating factor (G-CSF) 184
    granulocyte-macrophage colony-stimulating factor (GM-CSF) 184
    hepatocyte growth factor (HGF) 478
    interferons 270
    interleukin 2 (IL-2) 184, 271
    interleukin 3 (IL-3) 184, 196
    interleukin 6 (IL-6) 184
    interleukin 10 (IL-10) 409, 410, 414
    stem cell factor (SCF) 184
    thrombopoietin (TPO) 184
    TNF related apoptosis inducing ligand (TRAIL) 6, 433
    transforming growth factor beta (TGF-β) 9
    tumour necrosis factor (TNF) 433
    vascular endothelial growth factor (VEGF) 8
Cytopathic 37, 38, 83, 87
Cytostatic 220, 222–225, 228, 229, 244, 248
Cytotoxic
    batzelline ....
    camptothecin 289
    carboplatin 267, 272, 273
    cisplatin (CDDP) 267–270, 272, 273
    cyclophosphamide 267, 268, 272, 273
    cytarabine 267, 268
    docetaxel (Taxotere) 126
    doxorubicin (Adriamycin) 267
    etoposide (VP16) 410
    5-fluorouracil 267
    gemcitabine 248, 428, 429
    irinotecan ....
    mitomycin C 145
    mitoxantrone 248
    oxaliplatin 268
    paclitaxel (Taxol) 261, 272, 273
    podophyllotoxin ....
    salinomycin ....
    taxanes 261, 277
    topotecan 164
    treosulfan 248
    vinca alkaloids (vincristine, vindesine, vinblastine, vinorelbine) 267, 268, 271
Cytotoxicity 219–233, 237, 259–278, 285–290, 317, 422, 424–427, 432, 433, 441, 442, 449

D

Deacetyl-N-methylcolchicine 60
Debride 135
Denaturation 69, 71, 75, 198, 306, 317, 329
Dendritic 9, 201, 203, 204, 207, 473
Density centrifugation 21, 135
Deoxynucleotides
    deoxyadenosine (dATP) 95
    deoxycytidine (dCTP) 95
    deoxyguanosine (dGTP) 95
    deoxythymidine (dTTP) 95
Deproteinize 71
Desalting 295, 297, 299, 303, 306
Dewar 27, 29, 30, 32, 33, 167
Dexamethasone 267, 268
Dextran 62, 353
Diaphorase 220, 221
Diazepam 476
Differentiation 6, 106, 157, 162, 182, 198, 205, 322, 336, 346, 372, 408
Digoxigenin 70, 71
Dimethylsulphoxide (DMSO) 143–145, 148, 149, 154, 169, 174, 185, 187, 198, 239, 264, 286, 287, 311, 317, 362, 439, 448, 453, 457, 461, 464, 474
Dinucleotide 52
DiSC assay 259–278
Disinfectants
    alcohol 25
    betadine 476
    formalin 25
    glutaraldehyde 25, 338, 340
    hypochlorites 25, 26
    phenolic 25
    tricine 476
    virkon 137, 453
Dissociation 21, 28, 29, 170, 369, 474–478
DNA 6, 7, 19, 37, 40, 45–54, 58–60, 62, 67, 69–72, 75, 77, 82, 94–102, 106–108, 153, 155, 157, 158, 193, 220, 226–229, 232, 266, 277, 286, 293, 294, 309–319, 322–326, 328–331,

359–369, 371, 374, 376, 378, 382, 385–387, 389, 409, 414, 416, 435–450, 452, 461
DNA repair
    ATM 7
    cross-links 310
Drosophila 371
Drug combination
    antagonism 244
    combination index (CI) 425, 426, 430, 431, 433
    fractional effect analysis 422–424
    isobologram 422–424, 428, 430, 431
    median drug effect analysis 422, 424–426, 429
    response surface analysis 422, 424–425
    sequential 422, 425, 429
    simultaneous 422, 425
    synergism 244, 422–426, 428, 430, 431
Dysregulation 6, 58, 371

E

E-cadherin ....
E. Coli 295, 296
EcoRI 373, 376
Efficiency 21, 94, 184, 228, 328, 441, 442, 449
Elastase 362
Electrochemiluminescence ....
Electrophoretic mobility shift assay (EMSA) 413
Eluate 97, 299, 300, 302, 303
Elution buffer 51, 295, 296, 298–300, 303, 392, 397, 401
Embryologically 9, 356
Endometrial adenocarcinoma 268
Endometrium 6, 162
Endonucleases 100, 226, 227, 293, 315
Endothelial cells (EC) 8, 13, 209–217, 275, 334, 346, 348–350, 352, 355, 357
Enrichment 201, 203–205, 211, 213–216, 295, 298, 299, 396
Epidermal growth factor receptor (EGFR) 3, 4, 20, 276
Epidermis 146, 149
Epigenetic 2
Epiregulin 4
Epithelial to mesenchymal transition (EMT) ....
Epithelium 161, 162, 230, 337, 341, 353, 471, 472, 474, 478–480
Eppendorf 50, 97, 139, 286, 287, 383, 384, 389–391, 400, 442, 453, 454
Erlotinib 275, 276, 424
Ethidium 99, 318, 327, 329, 373, 376
Ethylenediaminetetraacetic acid (EDTA) 28, 48, 61, 65, 95, 96, 126, 144, 146–148, 152–155, 157, 169, 172, 196, 202, 211, 310, 312, 317, 327, 337, 347, 350, 363, 373, 436, 438, 439, 446, 447, 462, 468, 469
Eubacteria 106
European Centre for the Validation of Alternative Methods (ECVAM) 116, 219

Evolution 2, 480
Exon 322, 328, 374, 378, 436, 437
Explant 15, 142, 148, 151, 154, 352
Exponential growth 361, 456
Extracellular matrix
    fibronectin 228, 338
    heparan 335
    laminin 335
    type IV collagen 335

F

Fas 6
FasL 9
Fast protein liquid chromatography (FPLC) 294, 299, 301, 302
Feeder layer 130, 151, 156, 183, 197, 467–470
Fentanyl 476
Fetal (foetal) bovine serum (FBS) 108, 112, 152–154, 168, 169, 172, 174, 184, 186–191, 195, 199, 202, 204, 205, 212, 337, 338, 340, 346, 349, 350, 355, 412, 438, 439, 443, 445, 453, 469, 474
Fibrinogen 347, 352
Fibrinolysis 352
Fibroblast
    immortalized 143, 152, 173, 183, 478
    neonatal 346
    3T3 .... 36, 41, 222
Fibronectin 17, 18, 228, 338, 339, 350
Ficoll-Hypaque (Lymphoprep) 169, 171, 184–186, 195, 202
Filopodia 333
Fixative
    Diff-Quik 347, 350
    ethanol 369
    formalin 25, 478
    glutaraldehyde 340
    paraformaldehyde 353, 354
Flavonoids
    quercetagetin ....
    staurosprine 305
Flow cytometry
    DNA histogram 360, 361, 366, 367
    fluorescence activated cell sorter (FACS) 286, 287, 366
    forward scatter 290, 366
    side scatter 366
    Vindelov method 362–364
Fluorescein diacetate assay 224, 260
Fluorescence in situ hybridization (FISH) 60–76, 175, 315, 322, 323
Fluorescent cytoprint assay (FCA) 167, 176
Formamide 52, 62, 70, 71
Formazan 220, 224, 237, 239–241
Fosmid 60

Freezers
    −20 C 17, 23, 28, 48, 61–64, 66, 71, 73, 95–97, 101,
        108, 140, 167–169, 185, 191, 196, 211, 212,
        239, 240, 244, 249, 295–297, 348, 363, 364,
        374, 379, 385, 388, 389, 392, 438, 439
    −80 C 23, 30, 64, 65, 70, 74, 108, 142, 143, 145, 154,
        167, 174, 188, 198, 296, 298, 311, 312, 317,
        363, 374, 381, 385, 390, 392, 444, 446, 461, 468
    frost free 392
Fungating 136
Fusion genes 323
    BCR-ABL 58

G

G-banding 59–61, 63–67, 69, 72–74, 76, 77
Gefitinib 248, 271, 275, 276
Gelatine 211, 213, 215
Geldanamycin ....
Gel electrophoresis 323, 324, 325, 327–331
    single cell 309–319
Gene expression 20, 38, 58, 221, 232, 233, 372, 373,
        376–378, 381–393, 410, 436
Genetic 1–3, 20, 85, 88, 106, 155, 157, 182, 190,
        371, 437, 442, 443
Genotoxic 83
Germinative 7
Giemsa 61, 63, 66, 67, 73, 171, 197, 264
Gleevec 58
Glucose 6-phosphate dehydrogenase (G6PD) 221, 229
Glutathione S-transferase (GST) 294, 296–297,
        300–303, 307
Glyceraldehyde–3-phosphate dehydrogenase
        (GAPDH) 221, 229
Good laboratory practice 23, 115–123, 192
Green fluorescent protein (GFP) 20, 337, 341, 478
Growth
    conditions 36, 37, 41, 132, 463
    curve 36–37, 240–243, 263, 361, 422–424,
        427–430, 433, 456
    inhibition 106, 220, 222, 238, 241, 242, 413,
        422, 423, 425, 427–430, 432, 433
Guanidinium 50
Guineapigs 81
Gyrase 106, 108

H

HaeIII 101
Haemocytometer 23, 139, 203–206, 476
Hantaan 81
Haptotaxis 335, 338
Hayflick phenomenon 7
Heat shock protein 6
HeLa 14, 39, 46, 50, 51,
        57, 486
Hematoxylin-eosin (HE) 171, 264

Heparanase ....
Heparin 4, 170, 195
Hepatocyte 230, 231, 396
HEPES 16, 168, 253, 297, 342, 438–440, 485
Her2 .... 269, 270
Heterozygosity 53
High efficiency particulate air (HEPA) 21, 22, 86
High throughput screening (HTS) 219, 220, 230,
        232, 290
Histone 6, 226–227
Hoechst stain 37, 38, 169
Homeobox genes 322, 371–379
Homologous recombination 436, 437, 444
Homologs 74, 75
*Hpa*II 100
hTERT 15, 19, 41
Humanization 474, 476–479
Human umbilical vein endothelial
        cells (HUVEC) 8, 214, 346, 349, 350, 355
Humoral 9
Hyaluronidase 474, 475
Hydrocortisone 17, 152, 153, 474
Hydroxylase 49
Hyperoxia 354, 357
Hyphae 24
Hypotonics 61, 63–66, 69, 73, 74
Hypoxia 354

I

IC50 .... 127, 128, 131, 458–460, 464
Image analysis 59, 61, 67, 69, 72, 311, 314, 334, 337, 341
Imatinib 58, 248, 275
Immortalization 15, 173, 183, 184
Immunity
    antibody-dependent cellular cytotoxicity (ADCC) 9
    complement-dependent cytotoxicity (CDC) ... 409, 417
    immunodeficiency 194, 473
    immunoediting 9
    immunosuppression 9, 273, 473
    immunotherapy 202, 407–417
Immunomagnetic beads 243
Implant 88, 473, 476, 479
Insulin 17, 148, 152, 153, 168, 169, 183, 474, 475, 477
Integrin 333, 395
Intracellularly 82, 97, 416
Intron 49
Invasion 2, 8–9, 126, 333–342, 346–347, 350, 355
Inverted terminal repeats (ITR) 444
Iodixanol 203
Ion exchange 295, 296, 299–300
Irradiation 18, 228, 314–316, 319, 468,
        469, 476, 478
Isobologram. *See* Drug combination
Isoenzymatic 46
Isoenzymes 46

Isogenic 397
Isopropanol 29, 32, 48, 51, 95, 97, 198, 239–241, 372, 373
Isotope labelling 396

K

Karyotyping (karyology, karyotype, karyotyped, karyotypes) 39, 58–60, 69, 74–77, 82, 190
Keratinocyte 142, 143, 151–158, 231, 334, 340, 341, 467, 469
Ketamine 348, 354
Kinase
    Aurora 452
    cSrc 414
    p38 alpha ....
    Polo-Like 452
    serine-threonine kinase ....
    tyrosine kinase 4, 58
Kinesins ....
Krebs cycle 260

L

Lactate dehydrogenase (LDH) release assay 220, 221, 223, 224, 229
Lamellipodia 333
Laminar flow hood 22, 23, 30, 137, 146, 170
Lectin
    concanavalin A (ConA) 348, 349
    Ulex europaeus UEA-I 356
Leukaemia 7, 19
Leukapheresis 185
L-Glutamine 152, 168, 169, 196, 202, 310, 352, 453, 468, 484
Lidocaine 265
Limiting dilution analysis (LDA) 412, 413
Liposomes
    FuGENE® 6 442
    lipofectamine 441–442
Luciferase 20, 221, 248, 249, 251–253, 255
Luciferin-luciferase assay 249, 251–253, 255
Luminometer...249, 251, 252, 255
Lung cancer 4, 20, 125–132, 267, 268, 271, 275, 316
Lymphatic 8, 262, 408
Lymphoblastoid 74, 182, 183, 191, 197
Lymphocytes 9, 58, 190, 191, 201, 203–205, 207, 262, 271, 319, 365
Lymphoedema 477
Lymphoma
    Burkitt's lymphoma 181
    diffuse large B cell 69, 322, 407
    T cell lymphoma................................77, 322, 372, 408
Lymphoprep 139, 202, 203
Lymphotrophic... 81
Lysis
    Macherey Nagel Lysis Buffer RA1 381

TRIzol reagent 373

M

Macrophages 183, 184, 203, 207, 271, 272, 334, 473
Magnetic beads........................ 21, 136, 202, 204–206, 243
Mammary fat pad 471–474, 476–478
Mass spectrometry.. 404
Mast cells..143, 201
Mastermix 51, 391, 393
Matrigel.................... 18, 126, 130, 131, 335, 336, 338–340, 346–348, 350, 352–353, 355, 357, 474, 476, 477
Matrix metalloproteinases (MMPs) 9, 333
Melanoma 7, 39, 141–149, 248, 473, 489
Membrane integrity................. 220, 223, 225, 226, 260, 286
Mendelian 46
Mercaptoethanol 196, 295, 381, 383, 385, 403
Mesenchymal 58, 333
Metabolomics 35
Metalloproteinases (MMPs)...................................... 9, 333
Metatopic 479
Methanol...61, 143, 197, 239, 295, 398
Methylation 3, 6
Methyl-[3H]-thymidine 451–464
Microarray 39, 220, 229, 232
Microcentrifuge 48, 50, 146, 167, 404
Microfluidic............220, 230–232, 286, 290, 382, 383, 391
Microinjection 334
Microplate(s) 18, 22, 52, 186, 199, 224, 239, 243, 248, 250–251, 255
Microsatellites 3, 46
Microtubules 414, 416
Midazolam 475–477
Migration 8, 286, 287, 289, 290, 314, 318, 333–342, 346
Minisatellite(s) 46, 47
Mitochondrion/mitochondrial.......... 38, 223–225, 227, 230, 232, 237, 243, 260, 277, 293, 410
Mollicutes...82, 93
Monochromosomal 67
Monoclonal antibodies 225–227, 408
Monocyte 183, 203–205, 207
Monolayers.............. 29, 38, 196, 224, 231, 278, 286, 334, 338, 438, 439, 446
Motility 8, 333
Mountant 72
Mr Frosty 28, 29, 152, 468
MTT assay 176, 223, 237–244, 260, 261, 277, 422, 433
Multiallelic 46
Multiple myeloma 164, 270, 416
Mutation 2, 4–7, 20, 38, 58, 183, 276, 371, 372, 436, 473
Myc 4
Mycobacteria 88
Mycoplasma
    mycoplasmal 113, 191
    mycoplasmas...............93, 94, 96, 97, 99, 100, 105–114, 174, 175

Myelosuppression 273
Myosin 334

N

Naevus/nevus
    benign 141
    dysplastic 141
Natural products 219
Necrosis 225, 227
Nembutal 348, 353
Neoadjuvant chemotherapy 163, 267
Neoplastic 4, 8, 13, 15, 58, 70, 136, 183, 186, 190, 193, 248, 262, 264, 467–470
Neutral red 222
Nitrogen
    dewar 27, 29, 30, 32, 33, 167
    liquid 19–21, 23, 27, 29–33, 87, 88, 108, 110, 112, 113, 143–145, 154, 167, 174, 182, 187, 188, 191, 462, 468, 478
    storage 20, 23, 27, 29, 31, 32, 87, 88, 113, 167, 174, 188
Nitrophenol 126, 127
NK cells 9, 201
Nonaplex 51–52
Normoxia 357
Notch 323
Noxa 6
Nucleoside 433
Nude mice 474
Nylon mesh 338, 346, 349, 364, 365

O

Oestrogen 474, 477, 479
Oligonucleotide 94, 98, 101, 232, 327, 328, 330, 372, 374–379
Oncogene
    APC 3, 6
    Bcl–2 ....
    Bcl-x ....
    beta-catenin 3, 6
    c-myc 4
    ERK 4
    hMDM2 5
    mitogen-activated protein kinase (MAPK) 4
    oncogenomic 58
    p15 5
    p16 5
    p21 6
    p53 .... 3, 5–7
    Raf 4
    Ras 4, 5, 436
Oncogenomics 58
Optiprep™ 202–205, 207
Organ culture 14
Orthotopic 473, 479

Ovarian carcinoma 39, 161, 164–166, 169–174, 176
Overexpress 20, 334, 372, 411, 415–417, 436, 478

P

Pancreatic cancer 274
Panning 202
Papilloma 5
Paracentesis 164, 169, 170
Paracrine 4, 336
Paraptosis 223
Passaging 38, 83, 88, 112, 114, 190, 191, 462, 478, 479
Pathway 4–7, 232, 333, 409–411, 413–417
Pazopanib 260
Penetrans 82
Pepsin 62, 70, 75, 362, 363, 365
Permeability 38, 210, 223–225, 227, 285
Peroxidase 49, 227
Pharmacodynamic 480
Pharmacokinetic 253, 262, 271, 480
Phase-contrast 23, 61, 62, 64, 66, 147, 167, 172, 262, 337, 339
Phenol red 16, 18, 24, 29, 140, 142, 168, 171, 176, 220, 254, 485
Phorbol myristate acetate (PMA) 142
Phosphatase 127, 346, 349
Phosphate buffered saline (PBS) 48, 50, 54, 61, 62, 65, 67, 75, 95, 97, 111, 126, 127, 129, 131, 132, 142, 153–155, 168, 169, 202, 211, 213, 215, 239, 240, 244, 286, 287, 296, 297, 304, 310, 313, 337, 338, 340, 346, 349, 352–354, 363–365, 372, 373, 397–399, 438, 444, 453, 454, 474, 476, 477
Phosphatidylserine 226, 294
Phototoxicity 222
Physalin B ....
Plasmid 337, 341, 376, 378, 436–439, 443, 445, 449
Plasticware 21, 23, 462
Pleural effusion 164–166, 193, 194
Ploidy
    diploid 362, 368
    haploid ....
    tetraploid 366, 367
Pluripotent 38
Polyacrylamide 52, 397–398
Poly-D-lysine 338
Polymerase chain reaction (PCR)
    long distance inverse (LDI)-PCR 321–331
    multiplex 39, 40, 47, 48, 51, 381
    primer 38, 46–48, 50, 99, 102, 324, 331, 390, 448
    quantitative 97, 372, 381–393
    reverse transcriptase PCR (RT-PCR) 381–393
    single gene 381, 382, 384, 390, 392
    Taqman 381, 382, 384, 390–392
Polyoma 81

Polypropylene 17, 28, 153, 155, 167, 171, 248–250, 252–254, 383, 384, 392, 453, 457
Polystyrene 17, 28, 145, 149, 251, 286, 453
Posthybridization.. 70
Potency...............................225, 228, 229, 231–233, 452, 459, 460
Precipitate................................155, 373, 376, 440, 449, 463
Primary cell culture.................3, 20–21, 25, 85, 86, 94, 136, 140, 248, 249, 278, 289
Progesterone 479
Programmed cell death (apoptosis) 1, 4–7, 58, 113, 224–229, 232, 260, 290, 293, 294, 304, 305, 355, 372, 408–411, 413–417, 433
Proliferation................. 4, 5, 38, 58, 106, 182, 183, 185, 190, 194, 196, 210, 220, 222, 228, 229, 233, 243, 264, 267, 268, 286, 287, 289, 290, 346, 355, 362, 409, 413–414, 452, 457–460, 463, 467, 469, 470, 472, 474, 479
Prometaphase 66
Propidium iodine (PI) 286, 287, 289, 290, 297, 304, 363–365, 368, 432
Prostaglandins 17
Prostate cancer... 3
Proteasome 415, 416
Protein A..205
Proteinase 50
Protein-protein interactions 75
Proteolytic 28, 333
Proteome/proteomics 35, 395, 396, 404
Pyruvate 16, 152, 220, 346, 396, 468

Q

Quality control (assurance)
    external ....
    internal ....
Quinacrine..

R

Radioactivity/radiation 262, 265, 319, 458, 461, 464
Radiosensitivity 224
Reamplifications.. 50, 98
Reanneal/reannealing 69
Recirculating.. 21
Red blood cells
    chicken 365
    trout...365
Renaturation... 69
Reovirus..81
Resazurin..221, 278
Restriction enzymes 100, 101, 325–328, 330, 331
Restriction fragment length polymorphism (RFLP)...... 46, 324, 325

Retina 304, 305, 354
Retinoblastoma... 2, 5
Rho GTP-ases
    cdc42 333
    Rac...333
Ribonucleic acid (RNA)
    extraction 372–374, 381, 383–388
    reverse transcription....................381, 388, 443
    RNase 369
Rituximab 277, 407–417
RNAlater..478

S

Sarcoma.................................161, 268, 335, 479, 489
Scintillant 454, 458
Scintillation counter 452, 454, 458
Scratch assay..334, 337–339
SDS-PAGE...295, 298–303, 306, 397–398, 401–404
Secretase ....
Selection.............2, 21, 26, 63, 86, 113, 126–132, 136–138, 184, 192, 210, 213, 215–217, 225, 306, 439, 442, 444, 446–450, 460
Selenite 17, 483
Sempervirine ....
Senescence..7, 14, 15
Sequence
    blast 378
    combination.. 378, 429
    DNA...................7, 46, 62, 67, 69, 72, 75, 95, 100–102, 315, 324, 374, 435–437
Serum
    AB serum 202, 413
    serum-free medium 65, 130, 248, 339, 356
*Sfu*I 101
Short tandem repeat (STR).........................39, 40, 45–54
Sieve 476
Single nucleotide polymorphism 40
Soft Agar method..197
Soybean trypsin inhibitor 143, 148
Spectrophotometric 60, 72, 222
Spermatozoa..396
S-phase 360–362, 368, 452
Spheroids...254
Spiroketals ....
Splice-variants 374
Squamous cell carcinoma.........................151–158, 467
18S RNA..38
16S rRna..38
Stem cells
    cancer... 3, 7, 164
    embryonic...79, 467
Sterilisation (sterilisable and autoclaved).............61, 81, 82, 86, 95, 106, 211

Streptomyces albus ....
Stroma 161, 162, 170, 176, 334, 353
Structure-activity relationships (SAR) 264, 452, 459, 460
Subclones 57, 58, 77, 126, 189, 194, 197
Subcutaneous 271–273, 473, 476–478
Sulphorhodamine B assay 422
Sulphydryl ....
Surfactin 107
SyBr Green 52, 390, 392
Syringe
    Hamilton 347, 348, 353, 475, 477
    insulin 475, 477
Sytox Green 286–290

T

Tamoxifen 231, 271, 479
Taq polymerase 51, 373, 376, 382
Telomerase 7, 14, 478
Telomere/telomeric 7, 69
Tetrameric 46, 397
Tetranucleotide 52
Tetraploidization 58
Tetrazolium 220, 237, 346, 349
Thermal cycler 48, 51, 95, 97, 98, 102, 384, 389
Three-dimensional cell culture 230, 231
Thymidine 61, 74, 95, 229, 267, 289, 451–464, 485
Thyronine 152
Tiamulin 108, 111
Tilepath 60, 70, 323
Time-lapse fluorescence imaging 287–289
Tissue inhibitors of metalloproteinases (TIMPs) 9
T4 ligase 329
Topoisomerase 414, 416
Toxicology 116, 182, 286
Transcription factor
    AP–1 4
    E2F 5
    Homeobox genes 372
    hypoxia-inducible factor 1 (HIF–1) ....
    Jak 4
    Kruppel-like factor 5 ....
    NF-kB 414
    Stat 4, 414
Transcriptome ....
Transduction 106, 414, 436, 437, 442–449
Transfection 15, 20, 126, 334, 435–450
Transferrin 17, 30, 88, 152, 153, 197, 212, 463
Transgenic mice 472
Translocation 41, 58–60, 67, 69, 76, 183, 321–331, 411, 416
Transplantation 471, 472

Transwell 127, 130, 335, 336, 338, 346, 350, 355
Trastuzumab (Herceptin) 277
Triiodothyronine 17
Trinucleotide 46
Triphosphate 95, 223, 485
Tris 48, 61, 96, 153, 295, 296, 310, 314, 327, 362, 373, 388, 397, 398, 402, 436, 438, 439
Tris buffered saline (TBS) ....
Trypan 54, 109, 111, 112, 139, 169, 184, 187–189, 195, 221–222, 226, 262–264, 289, 412, 455, 476
Trypsin
    TrypLE 28
    trypsinization 67, 146, 464

U

Ubiquitin-proteosome pathway ....
Ultracentrifuge 438
Uridine 61, 74

V

Variable number of tandem repeats (VNTRs) 46
Vascular mimicry 8
Vector 295–297, 300, 306, 331, 348, 349, 373, 376, 396, 437–439, 442–445, 449
Verapamil 265
Vessel co-option 8
Viability 20, 29, 54, 106, 109, 111, 112, 114, 123, 131, 132, 144, 149, 153, 171, 186, 187, 189, 195, 196, 207, 216, 220–224, 228–231, 249, 260, 274, 286–290, 335, 455, 456, 460, 464, 475, 476
Vi-CELL 453, 455, 462
Virus
    adeno-associated virus (AAV) 438, 444, 445
    bovine polyoma virus 81
    bovine viral diarrhoea virus (BVDV) 24, 81
    cytomegalovirus (CMV) 438, 443, 449
    epstein barr virus (EBV) 182, 183, 190, 191, 194, 410
    haemorrhagic fever virus 81
    hantaan virus 81
    hepatitis viruses 81, 194
    human herpesvirus–8 (HHV–8) 194
    human immunodeficiency virus (HIV) 25, 194, 449
    human papilloma virus (HPV) 5
    human T-cell leukemia virus (HTLV–1) 190, 194
    lentivirus 438, 442
    reovirus 3 81
    respiratory syncytial virus (RSV) 449
    retrovirus 82, 435, 442
    T-lymphotrophic viruses 81
    yellow fever virus 81

Vortexing 50, 97, 304, 317, 330, 364, 369, 440, 443, 446

W

Waterbaths 80, 84, 145, 184, 188, 195
Wnt pathway ....
Wortmannin ....
Wound 334

X

Xenografts 157, 472–474, 476–480
XTT proliferation assay 413
Xylene 95

Z

Zebrafish ....
Zoonotic 81
Z-prime (Z′) calculation 458

Printed by Books on Demand, Germany